P9-CJR-328

DATE DUE

DEMCO 38-296

OXFORD STUDIES IN NUCLEAR PHYSICS

GENERAL EDITOR
P. E. HODGSON

OXFORD STUDIES IN NUCLEAR PHYSICS

General Editor: P. E. Hodgson

1 J. McL. Emmerson: *Symmetry principles in particle physics* (1972)

2 J. M. Irvine: *Heavy nuclei, superheavy nuclei, and neutron stars* (1975)

3 I. S. Towner: *A shell-model description of light nuclei* (1977)

4 P. E. Hodgson: *Nuclear heavy-ion reactions* (1978)

5 R. D. Lawson: *Theory of the nuclear shell model* (1980)

6 W. E. Frahn: *Diffractive processes in nuclear physics* (1985)

7 S. S. M. Wong: *Nuclear statistical spectroscopy* (1986)

8 N. Ullah: *Matrix ensembles in the many-nucleon problem* (1987)

9 A. N. Antonov, P. E. Hodgson, and I. Zh. Petkov: *Nucleon momentum and density distributions in nuclei* (1988)

10 D. Bonatsos: *Interacting boson models of nuclear structure* (1988)

11 H. Ejiri and M. J. A. de Voigt: *Gamma-ray and electron spectroscopy in nuclear physics* (1989)

12 B. Castel and I. S. Towner: *Modern theories of nuclear moments* (1990)

13 R. F. Casten: *Nuclear structure* (1990)

14 I. Zh. Petkov and M. V. Stoitsov: *Nuclear density function theory* (1991)

15 E. Gadioli and P. E. Hodgson: *Pre-equilibrium nuclear reactions* (1992)

16 J. D. Walecka: *Theoretical nuclear and subnuclear physics* (1995)

17 D. N. Poenaru and W. Greiner (eds.): *Handbook of nuclear properties* (1995)

18 P. Fröbrich and R. Lipperheide: *Theory of nuclear reactions* (1996)

Handbook of Nuclear Properties

Edited by

DORIN POENARU

Institute of Theoretical Physics
J. W. Goethe University, Frankfurt am Main
and
Institute of Atomic Physics, Bucharest

and

WALTER GREINER

Institute of Theoretical Physics
J. W. Goethe University, Frankfurt am Main

CLARENDON PRESS • OXFORD
1996

Riverside Community College
Library
4800 Magnolia Avenue
Riverside, California 92506

QC 783 .H36 1996

Handbook of nuclear
properties

Walton Street, Oxford OX2 6DP
d New York
nd Bangkok Bombay
vn Dar es Salaam Delhi
Kong Istanbul Karachi
Kuala Lumpur Madras Madrid Melbourne
Mexico City Nairobi Paris Singapore
Taipei Tokyo Toronto
and associated companies in
Berlin Ibadan

Oxford is a trade mark of Oxford University Press

Published in the United States
by Oxford University Press Inc., New York

© The contributors listed on pp. xi–xii, 1996

All rights reserved. No part of this publication may be reproduced, stored in a retrieval system, or transmitted, in any form or by any means, without the prior permission in writing of Oxford University Press. Within the UK, exceptions are allowed in respect of any fair dealing for the purpose of research or private study, or criticism or review, as permitted under the Copyright, Designs and Patents Act, 1988, or in the case of reprographic reproduction in accordance with the terms of licences issued by the Copyright Licensing Agency. Enquiries concerning reproduction outside those terms and in other countries should be sent to the Rights Department, Oxford University Press, at the address above.

This book is sold subject to the condition that it shall not, by way of trade or otherwise, be lent, re-sold, hired out, or otherwise circulated without the publisher's prior consent in any form of binding or cover other than that in which it is published and without a similar condition including this condition being imposed on the subsequent purchaser.

A catalogue record for this book is available from the British Library

Library of Congress Cataloging in Publication Data
Handbook of nuclear properties/edited by Dorin N. Poenaru and Walter Greiner. – 1st ed.
(Oxford studies in nuclear physics; 17)
Includes bibliographical references.
1. Nuclear physics–Handbooks, manuals, etc. 2. Nuclear physics–Tables. I. Poenaru, Dorin, N. II. Greiner, Walter, 1935–. III. Series.
QC783.H36 1995 539.7'232–dc20 95-32248
ISBN 0 19 851779 3 (Hbk)

Typeset by the authors

Printed in Great Britain by
Biddles Ltd
Guildford & King's Lynn

PREFACE

Nuclear physics is a basic science in a continuous expansion. At present an increasing interest toward higher nuclear temperatures or densities, higher spin (super- and hyperdeformations, transitions from an ordered nuclear phase to a chaotic one), production and spectroscopy of proton- and neutron-rich nuclei far from stability, is clearly manifested. Radioactive beams from isotope separators and recoil spectrometers, and large multidetector arrays, are among the most important technical achievements making possible this development. By using a Penning trap mass spectrometer (see the corresponding chapter in the present handbook), an extremely high mass resolution has been obtained. The major heavy ion facilities in the world are now producing radioactive nuclear beams and are using more than half of the total machine time for the investigation of radioactive nuclei.

The limits of nuclear stability are determined either by hadron-decay modes, or (for heavy nuclei) by α-decay or spontaneous fission. The most important limits are the proton and neutron drip lines, where the proton or the neutron, respectively, has roughly zero binding energy. The accessibility of nuclides very close to the drip lines has increased considerably. Few proton emitters have been experimentally observed until now, but many β-delayed proton emitters.

Close to the neutron drip line, nuclei (like ^{11}Li) might exhibit a halo structure or neutron skin. In these nuclei mass and charge radii may differ by large amounts, and the molecular (or cluster) structure plays an important role. The density distributions show an extended tail at low density. On the neutron-rich side, it was observed that pairing is weakened for large values of $N - Z$.

The existence of very heavy nuclei is prevented by the spontaneous fission process. When the number of protons becomes very high, the Coulomb repulsion is no longer counterbalanced by the strong interaction, no matter how many neutrons are present, and the fission barrier height may approach zero. Between Th and the heavy Fm isotopes, the spontaneous fission half-life decreases by 30 orders of magnitude. Even at a finite positive (but small) barrier height the fission half-life could be very low in comparison with the partial half-life of any other disintegration mode. At GSI Darmstadt, it was shown that the shell stabilizing property, characteristic of the superheavy nuclei, is already present in the heaviest elements known to date: 107, 108, 109 up to 111. Owing to shell effects the fission barriers of these nuclei are so high that they decay mainly by α emission instead of spontaneous fission. According to recently performed calculations the superheavy nuclei do not form an island of stability surrounded by an instability sea. The shell effects of the predicted neutron deformed magic number $N = 162$ act to stabilize the neighbours, leading to a peninsula of superheavy nuclei.

Nuclear physics is also tightly linked to astrophysics, cosmology, and elementary particle physics, helping to unify our knowledge about the universe. The main site of nucleosynthesis for most of the elements is considered to be the stellar environment; a second one is the Big Bang (primordial nucleosynthesis). In both cases, the reactions of beta unstable nuclei play an important role. Nuclear reactions involving short-lived nuclei in explosive processes in the universe could be simulated in the laboratory.

The overlap between nuclear physics and elementary particle physics has become increasingly important. We admit that nucleons are made from quarks. There is a need to find some specific signatures of the quark structure in nuclear phenomena.

Various applications in industry, geophysics, nuclear medicine, the investigation of the environment, etc., have been developed.

Consequently, the literature on nuclear properties (ground or excited states) is enormous and is expanding at a rapid rate. The need for an updated review is a direct result of this large accumulation of new theoretical and experimental findings, followed or determined by technological developments and applications. The purpose of this handbook is to provide a reference for researchers, practitioners, and beginners interested in nuclear properties such as masses, deformations, spins and parities, decay modes, half-lives, etc. The trend toward intermediate and high energy nuclear physics is illustrated by including the decay modes of elementary particles.

The chapter on atomic masses presents the experimental aspects of this nuclear property. After a few basic definitions and a discussion concerning the influence of the nuclear structure (magic numbers, deformations), the most important experimental methods (double-focusing spectroscopes, radio-frequency spectrometers, ISOL facilities, mass separators, Penning trap spectrometers, etc.), allowing the determination of the masses of stable and radioactive nuclides, are briefly reviewed. The radioactivity (β, γ, α, proton, neutron, and β-delayed particle emission) and nuclear reactions are also useful for mass measurements. Finally, the extrapolation methods employed to estimate the masses when the experimental data are not available, are discussed.

A description of an intuitive way of understanding the empirical trends of masses, based on a microscopic theory, is given in the chapter on the shell model interpretation of nuclear masses. Empirical trends stressing the importance of magic numbers, odd–even effects, slope discontinuities of mass surfaces, and separation energies are observed in many systematics of spherical and deformed nuclei. The evidence for magic and submagic numbers, and the mutual support and weakening of magicities, are discussed. Then the shell model approach and its related applications (ground state and binding energy, seniority scheme, configuration interaction, deformation, the mass equation), as well as the interpretation of empirical trends, are presented in detail.

Penning traps can be used as modern mass spectrometers of high resolving power, accuracy, and sensitivity. After the basic principles (eigen frequencies of ions in the trap and the cyclotron resonance, ion injection and cooling), the authors of the chapter on this very successful kind of mass spectrometer describe the particular instrument ISOLTRAP, which is installed at the on-line mass separator ISOLDE at CERN, Geneva. The last section is devoted to the calibration, accuracy, efficiency, and applicability of the spectrometer.

The chapter on nuclear deformations gives a summary of basic theoretical concepts and experimental techniques. The parametrization of nuclear shapes (multipole expansion and others) introduces the main deformation coordinates for axially symmetric or triaxial systems. In the next two sections, the methods of extracting the deformations from the nuclear moments and the techniques used to measure the deformations (Coulomb excitation, strongly interacting probes, electron scattering, lifetime measurements, hyperfine studies, nuclear mean square radii) are discussed. Then the systematics of quadrupole, octupole, and higher order deformations are presented. The various examples of calculations (equilibrium deformations, variation with particle number, quadrupole, hexadecapole, and higher order moments, as well as octupole, dipole, and superdeformations) are compared with the data in a section which follows that on basic microscopic theories (mean-field, shell model, algebraic models, and cluster models).

We start the chapter on nuclear stability against various decay modes, by hadron and cluster emission, with a definition of metastable states, compared with stable or unstable ones. Besides the popular α-decay and spontaneous fission (from the ground state or from a shape isomeric state), the proton (p), ^{12}C, ^{14}C, ^{20}O, ^{23}F, $^{24-26}$Ne, ^{28,30}Mg, ^{32,34}Si, and other decay modes are introduced. Cold and bimodal fission, and particle accompanied fission are mentioned, among the new aspects of disintegration phenomena. There are two-step processes, *β-delayed particle emissions*, in which, after the population of an excited state by the β-decay of a precursor, a neutral (n, 2n, 3n, 4n) or a charged particle (p, 2p, 3p, d, t, α, light fission fragment, etc.) is rapidly emitted, the half-life being determined by β-decay. The half-life can be calculated within a dynamical model, developed by extending either the fission theory or the traditional approach of α-decay. Before any other model was published, we estimated, within our analytical superasymmetric fission model, the half-lives for more than 150 cluster decay modes, including all cases experimentally confirmed during 1984–1994. The measurements on cluster radioactivities are in good agreement with our predictions. The decay dynamics is studied in the one-, two-, and three-dimensional space of nuclear deformations. We describe the essential ideas of fragmentation theory, numerical and analytical superasymmetric fission models, universal curves, and semiempirical formulae for α-decay. The latest experimental results are compared with our predictions. The cluster decay of odd-mass nuclei (including the fine structure), and a new island of emitters around ^{100}Sn, are the last sections of this chapter.

In the latter part of the handbook a series of important tables are given. The first one contains fundamental constants (a main list, detailed recommended values, few maintained non-SI units) and energy conversion factors. The expanded matrix of variances, covariances, and correlation coefficients is necessary because the variables in a least-squares adjustment are statistically correlated.

In the latest version of the summary tables of particle properties, the Particle Data Group presents the decay modes of gauge and Higgs bosons, leptons, quarks, mesons, baryons, and searches for free quarks, monopoles, supersymmetries, compositeness, etc.

The tables of α-particle emitters (half-lives, branching ratios, energy, and emission probability) select some isotopes of Sm, Gd, Dy, Bi, Po, At, Rn, Fr, Ra, Ac, Th, Pa, U, Np, Pu, Am, Cm, Cf, and Es.

There are similar tables of selected, high accuracy, recommended data on γ-ray and X-ray standards used for calibrating detectors.

In the comprehensive table of all known nuclides (mainly ground states but also some isomeric states), the following properties are covered: spin and parity, mass excess, half-life (or abundance for stable nuclei), and the main decay modes with the corresponding branching ratios.

Top specialists directly involved in the development of new theoretical and experimental methods, and in the systematics of extremely rich data, accepted our invitation to contribute their expertise as authors of various chapters and tables. We very much appreciate their essential contributions and the effort they have made to produce updated reviews, fitting both the scientific purpose and the formal aspects useful for the unity of the final presentation.

We hope that the chapters on various modern theoretical and experimental aspects of nuclear masses, deformations, new decay modes, as well as the tables concerning fundamental constants, energy conversion factors, elementary particle decay modes, selected α-emitters, X-ray and γ-ray standards for detector calibration, and a table of nuclides will be of real help to our readers. Any comments and suggestions for future improvements would be appreciated.

The Summary Tables of Particle Properties is reproduced (with permission) from *Phys. Rev. D*. We would like to acknowledge the valuable help and professional cooperation of the staff at Oxford University Press. Our research work has been supported by the Bundesministerium für Forschung und Technologie, the Deutsche Forschungsgemeinschaft, the Gesellschaft für Schwerionenforschung Darmstadt, the Stabsabteilung Internationale Beziehungen der Kernforschungszentrum Karlsruhe, and the Institute of Atomic Physics, Bucharest. Without their help and steady support, our long-standing collaboration would not have been possible. One of us (DNP) received a donation of computer equipment from the Soros Foundation for an Open Society.

Dorin N. Poenaru and Walter Greiner

CONTENTS

CONTRIBUTORS

Georg Bollen, Ph.D., Researcher, CERN, PPE Division, CH-1211 Geneva 23, Switzerland, on leave from Gesellschaft für Schwerionenforschung (GSI), D-64291 Darmstadt, Germany

E. Richard Cohen, Ph.D., Distinguished Fellow, Rockwell International Science Center, Thousand Oaks, California 91360, USA

Walter Greiner, Ph.D., Dr.h.c.mult., Professor, Director of the Institut für Theoretische Physik der Johann Wolfgang Goethe Universität, Postfach 111932, D-60054 Frankfurt am Main, Germany

H.-J. Kluge, Ph.D., Gesellschaft für Schwerionenforschung (GSI), D-64291 Darmstadt and Physikalisches Institut, Universität Heidelberg, D-61290 Heidelberg, Germany

Alex Lorenz, M.S., Consultant, IAEA Vienna (Retired), 4891 Old Post Circle, Boulder, Colorado 80301, USA

Witold Nazarewicz, Ph.D., Professor, Institute of Theoretical Physics, University of Warsaw, PL-00681 Warsaw, Poland, and Joint Institute for Heavy Ion Research, Bldg. 6008, MS 6374, PO Box 2008, Oak Ridge, Tennessee 37831, USA

Alan L. Nichols, Ph.D., Department Head, AEA Technology, Harwell, Didcot, Oxon OX11 0RA, UK

Particle Data Group, Contact author: Ms Betty Armstrong, MS 50-308, Physics Division, Lawrence Berkeley Laboratory, 1 Cyclotron Road, Berkeley, California 94720, USA

Dorin N. Poenaru, Ph.D., Professor, Institute of Atomic Physics, P. O. Box MG-6, RO-76900 Bucharest, Romania, and Institut für Theoretische Physik der Johann Wolfgang Goethe Universität, Postfach 111932, D-60054 Frankfurt am Main, Germany

Ingemar Ragnarsson, Ph.D., Docent, Department of Mathematical Physics, Lund Institute of Technology, PO Box 118, S-22100 Lund, Sweden

Barry N. Taylor, Ph.D., Manager, Fundamental Constants Data Center, National Institute of Standards and Technology, Gaithersburg, Maryland 20899, USA

Jagdish K. Tuli, Ph.D., Physicist, National Nuclear Data Center, Bldg. 197D, Brookhaven National Laboratory, Upton, Long Island, New York 11973, USA

Aaldert H. Wapstra, Ph.D., Professor, National Institute of Nuclear and High Energy Physics, PO Box 41882, 1009 DB Amsterdam, The Netherlands

Nissan Zeldes, Ph.D., Professor, The Racah Institute of Physics, The Hebrew University of Jerusalem, Jerusalem 91904, Israel

1

ATOMIC MASSES

A. H. Wapstra

Table of contents

I. INTRODUCTION

Since the discovery that elements occur in various isotopes, the masses M of their atoms are known to be nearly integers A (the *mass number*) if expressed in suitable units. Since 1960, the unit u (or m_u) used is 1/12th of the mass of a ^{12}C atom in its nuclear and atomic ground states.

Mass values as given in tables (e.g. ref. 1) are those of atoms in such ground states (but for some exceptions where the nucleus is in an isomeric state). The reason for giving atomic masses rather than nuclear ones is that until very recently, measurements yielded more directly, and accurately, data on atomic masses than on nuclear ones, as will become clear from the discussions below.

The chemical properties of an element are determined by the number Z of electrons around it in a neutral atom, called the charge number of the nucleus. In this chapter we will use a description in which the nucleus contains Z protons and $N = A - Z$ neutrons (in reality, the situation is somewhat more complex). The difference $I = N - Z$ is called the *neutron excess*. Half this number is the third component $T_3 = I/2$ of the *isospin* T, a quantity of interest in classifying nuclear levels. Substances of which the atoms have the same Z, N, A (the other properties being different) are respectively called *isotopes*, *isotones*, *isobars*; if all three are the same, we talk of *nuclides*.

The difference $M - Am_u$ is called the *mass excess*. The quantity

$$B = NM_n + ZM_H - M(N, Z) \tag{1}$$

is referred to as the *nuclear binding energy* (M_n and M_H being the mass of a proton and of an H-atom respectively) though it contains less than a 0.1% part of the electronic binding energies of the atomic electrons.

Quantities concerning nuclear energetics are normally expressed in units of electron volts, eV. The relation between mass and energy units is 1 u = 931493.86 ± 0.06 keV*. (The asterisk indicates that the unit used is the one defined by a choice of the Josephson constant.)

II. MASSES AND NUCLEAR STRUCTURE

Several types of mass formula have been developed to describe the atomic masses as a function of Z and N; the reader is referred to two collections[2,3] of mass values as calculated from such formulas, with descriptions of the physics involved. The following may serve as a short introduction.

In a first approximation, the nucleus is found to behave as a spherical droplet of an incompressible liquid. As can then be expected, the volume is proportional to the number A of nucleons:

$$r = r_0 A^{1/3} \tag{2}$$

(Experimentally, the dependence on N is generally found only about half as fast, that on Z evidently somewhat faster.) The binding energy is then proportional to the volume, except for a defect proportional to the surface since the nuclear droplets are small. A Fermi gas calculation shows that a maximum in the binding energy at constant A must occur for $Z = N$, causing a "symmetry" term which, in approximation, is proportional to $(Z - N)^2/A$. The resulting tendency towards $Z = N$ is counteracted by the Coulomb term, given below as that of a homogeneously charged sphere. Finally, neutrons and protons both tend to occur in pairs, in which all quantum numbers except the intrinsic spin are the same, due to a "pairing energy". Thus in total, as first adopted by Bethe and Backer and Weizsäcker[4], the binding energy can be described as

$$B = a_v A - a_s A^{2/3} - a_I \frac{(Z - N)^2}{A} - a_c \frac{Z^2}{A^{1/3}} + \delta \tag{3}$$

$\delta = +(-)a_p/A^{1/2}$ for Z and N even (odd), 0 otherwise. Not unreasonable values for the constants are $a_v = 14.1$ MeV, $a_s = 13.0$ MeV, $a_c = 0.595$ MeV, $a_I = 76$ MeV, $a_p = 12$ MeV.

Values B are described by this formula with a precision of about 1%. This, however, is not sufficient. More important for nuclear physics purposes than the masses themselves are the differences in atomic masses (see below), and for these the precision is far from sufficient.

Several factors affect the validity of the formula. In the first place, the possible orbits of protons and neutrons in a nucleus are found to be arranged in shells. Near closures of major shells, at the "magic numbers" 2, 8, 20, 28, 50, 82 and 126, a very noticeable extra stability is found (the influence of minor shells can scarcely be observed except in a few special cases).

The other strong influence is that of nuclear deformation. As explained by the Nilsson version of the nuclear shell model, nuclei not close to magic numbers can gain energy by going over from a spherical shape to a quadrupole deformed one. This occurs, for example, rather suddenly by passing the neutron number $N = 88$, and gives rise there to a sudden increase in binding energy.

The course of the minimum mass excesses for odd values of A, see Fig. 1, shows the influences of the surface effect at low A, of the Coulomb effect at high A, and of shell and deformation effects in several places.

III. MASS SPECTROSCOPY OF STABLE NUCLIDES

In conventional (double-focusing) mass spectroscopes, electric and magnetic fields are combined in such a way that ions with the same mass to charge ratio, leaving an ion source with a certain distribution of angles and velocities, are focused on a line[5]. The position of this line, and a knowledge of the field

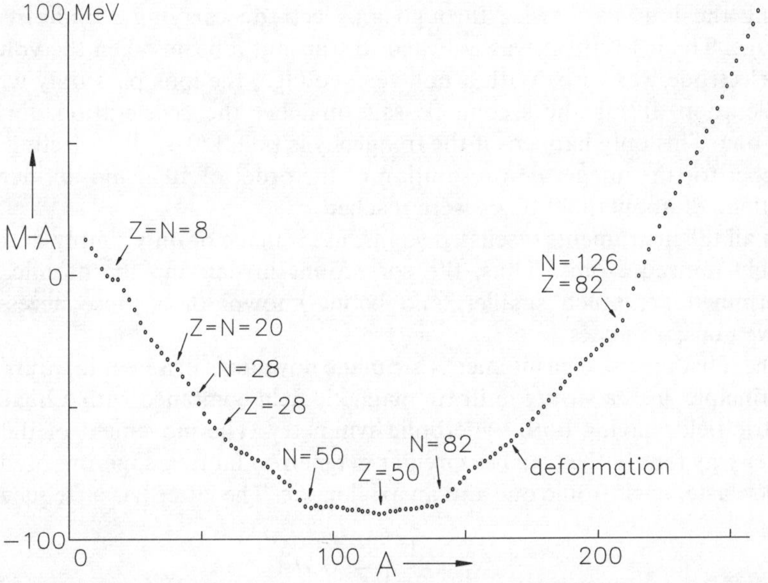

Fig. 1. Mass defects as a function of A. The values shown have been calculated from the same data as described for Fig. 2. Places where masses are influenced by shell closures and, in one place, deformation, are indicated.

strengths, then determine the mass if the charge state is known. Since, however, the field strengths cannot be measured with the required accuracy, one always compares the results of two kinds of ions with the same mass number to charge ratio (mass doublets). One uses, for instance, the (relativistically correct) property that when all magnetic fields are kept constant and all voltages changed proportionally, the foci for two different ion beams coincide if the relation of the voltages to the masses is $\Delta M = M\Delta V/V$. The ions then travel very nearly along the same path in the magnetic field of the spectrometer, which helps to avoid systematic errors. From the measured voltage ratio then, essentially, a mass ratio follows, but since one can always choose a comparison mass with a precision some four orders of magnitude higher than that in the voltage difference, the result is always expressed as a mass difference.

In radiofrequency mass spectrometers, ions are made to circulate in a homogeneous magnetic field, and their cyclotron frequency

$$\nu_c = \frac{B}{2\pi c} Q/M' \tag{4}$$

is studied. Here, M' is the relativistic mass

$$M' = M/\sqrt{1 - v^2/c^2} \tag{5}$$

A first type of such an instrument[6] used a conventional magnet and ion energies of a few times 10 keV. The cyclotron frequency was determined by letting the ions pass twice through an electrode carrying a radiofrequency voltage. The instrument was adjusted to transmit a beam when the voltage on the electrode was zero. With a non-zero voltage, the ions pass only when the acceleration during the second passage matches the deceleration during the first one. This only happens if the frequency is $(n + 1/2)\nu_c$. By selecting a large number for the integer n, a resolution of the order of 10^{-6} and an increase in accuracy of about 2000 times were reached.

In all the instruments discussed so far, use is made of only simply or, at most doubly ionized atoms. Thus, the corrections in deriving the atomic masses determined are much smaller, and better known, than those necessary to derive nuclear masses.

The most precise measurements are made nowadays in a Penning trap[7,8] that, in principle, uses a strong uniform magnetic field combined with a quadrupole electric field, having both hyperbolic symmetry. The movement of the rather low energy ions in the trap is a circular magnetron motion superimposed with a rather faster cyclotronic one and an axial mode. The latter has a frequency

$$\nu_z = \frac{1}{2\pi}\left(\frac{V}{d} Q/M\right)^{1/2} \tag{6}$$

V being a voltage and d characterizing the trap size. The magnetron and (disturbed) cyclotron frequencies are then

$$\nu_m = [\nu_c + (\nu_c^2 - 2\nu_z^2)^{1/2}]/2 \quad (7)$$

$$\nu_{c'} = [\nu_c - (\nu_c^2 - 2\nu_z^2)^{1/2}]/2 \quad (6)$$

At resonance with the (disturbed) cyclotron frequency, the axial movement – which can be detected via the image current in an end cap of the electrodes, even for single ions in the trap – is changed. In this way, measurements can be made with a resolution of the order of 10^{-8} and a precision a hundred times higher. Again, for precision measurements one compares the frequencies, and therefore the masses, of two kinds of ions. The resulting mass ratios can again best be expressed as mass differences.

Such Penning trap measurements have been made with completely stripped nuclei. The above remark, that measurements more directly yield values for atomic rather than nuclear masses, is then no longer valid.

IV. MASS SPECTROSCOPY OF UNSTABLE NUCLIDES

This branch of mass spectroscopy requires the technique of measuring on line with accelerators producing the desired nuclides. The first application[9] used a rather conventional (high transmission) mass spectrometer. The ions were obtained from a target in the proton beam of the CERN synchrocyclotron constructed in such a way that formed alkali atoms quickly diffused out, ionized by surface ionization. In later measurements[10], the alkali ions were first selected by the mass separator ISOLDE. This technique did not allow measurement of mass doublets, but (essentially) linear combinations of masses of three isotopes.[11]

Groups in Los Alamos[12] and Caen[13] have succeeded in making measurements directly on ions emerging from targets bombarded with high energy protons, or heavy ions, respectively. The technique used is a combination of magnetic deflection and time of flight measurements. The resolution obtained so far is of the order of one in a few thousand. The precision is a few hundred keV, poorer than with the more conventional techniques but the latter cannot be used for the very exotic nuclei measured this way.

The interpretation of both types of measurements just described is complicated by the fact that isomers could not be separated.

In principle, mass separators can also be used to determine atomic masses. The technique used is not much different from that with conventional mass spectrometers; only here "wide" doublets are used. (And, unnecessarily, the results are given as mass ratios rather than mass differences.) Recently[14] several such measurements have been reported, with a precision of about 100 keV. The detection is made by measuring the radiation emitted by the observed nuclei. Thus, separate measurements can be made on isomers.

A very important development is the application of ion trap cyclotron resonance techniques to unstable nuclei[15]. A special technique is then used to detect the resonance. The resolution has been found sufficient to separate not too narrow isomer pairs. The precision obtained is of the order of 10 keV.

V. NUCLEAR STABILITY; RADIOACTIVITY

For a treatment of the different kinds of radioactivity, the reader is referred to a subsequent chapter. For the present purpose, the following remarks may suffice.

In β-decay, a neutron in the nucleus changes into a proton on emission of an electron and an anti-neutrino, which carry away an energy

$$Q^- = M(A,Z) - M(A,Z+1) \tag{9}$$

Or, reversely, a proton captures an electron from an electron shell with electron binding energy B_e on emission of a neutrino with energy $E_\nu = Q^+ - B_e$ in which

$$Q^+ = M(A,Z) - M(A,Z-1) \tag{10}$$

Alternatively, if Q^+ is larger than twice the mass of an electron, a positron and a neutrino are emitted with a total energy $Q^+ - 2m_ec^2$. Half-lives for these processes decrease strongly with increasing Q. It is clear that the most β-stable nuclides occur at a minimum of the mass at constant A. Applying this to the Bethe–Weizsäcker formula, eq. (3), the line of (maximum) β-stability is found to occur for

$$\frac{I}{A} = \frac{1}{a_t A^{-2/3} + 1} \tag{11}$$

where $a_t = (M_n - M_H + a_I)/a_c$, approximately. This line, using a constant $a_t = 139$, is compared in Fig. 2 with values derived by fitting a parabola to the three lowest mass values at odd values of A (but where a magic number is crossed, two values may be presented, each derived from the magic isobar with two values on the same side of the magic number).

Nuclei can become unstable for emission of nucleons, or even nuclei. Most important for our purpose is α-radioactivity, the emission of ^{4}He nuclei. Owing to the strong binding of two protons and two neutrons in ^{4}He, the energy

$$Q_\alpha(Z,N) = M(Z,N) - M(Z-2,N-2) - M(^4He) \tag{12}$$

may become large and the decay relatively fast.

In this case, it is clear that the masses involved are more nearly atomic than nuclear ones. The total electron binding energy in the final nuclide is smaller than that in the parent atom; thus, the α particle must yield energy to the electron system. In fact, most often it yields just a little more than the difference in the binding energy and leaves the final atom in an excited or even ionized state. This fact has to be taken into account in precision measurements.

For neutron-rich nuclei, the neutron separation (or binding) energy

$$S_n(A,Z) = M(A-1,Z) + M_n - M(A,Z) \tag{13}$$

becomes negative at the *neutron drip line*. Beyond it, the decay is instantaneous. But in the analogous case, the proton decay of nuclides beyond the proton drip line is hindered by the Coulomb barrier between the nucleus and the outer world. The half-life for this decay may be measurable though it decreases rapidly with increasing $-S_p$.

Owing to the influence of pairing energies, it is possible that even-N nuclides are stable for the emission of one neutron, but not for that of a neutron pair; again the decay is then practically instantaneous. The two-proton activity is of course possible.

Both proton and α-decay are most probable for proton-rich nuclides, as shown in Fig. 2. In addition, the occurrence of proton shell closures, and for α-decay also neutron ones, has a pronounced influence.
Measured proton- and α-particle energies E_p and E_α are of course smaller than the decay energies, due to the nuclear recoil

$$Q_\alpha = \frac{E_\alpha M(A,Z)}{[M(A,Z) - M_\alpha]} \frac{1}{\sqrt{1 - 2M(A,Z)E_\alpha/[M(A,Z) - M_\alpha]^2}} \tag{14}$$

(except for a small correction for electron binding energies), and similarly for protons.

Fairly recently[16] nuclides have been found to decay by the emission of heavier nuclei, e.g. ${}^{223}_{88}\mathrm{Ra} \rightarrow {}^{209}_{82}\mathrm{Pb} + {}^{14}_{6}\mathrm{C}$. Other than in the cases above, the study of such events did not really contribute to our knowledge of atomic masses.

In all these cases, the decay need not necessarily yield nuclei in their ground states. Measurement of γ-rays following, for example, observed α-decays may then give the total decay energies. Also, decays may be observed starting from

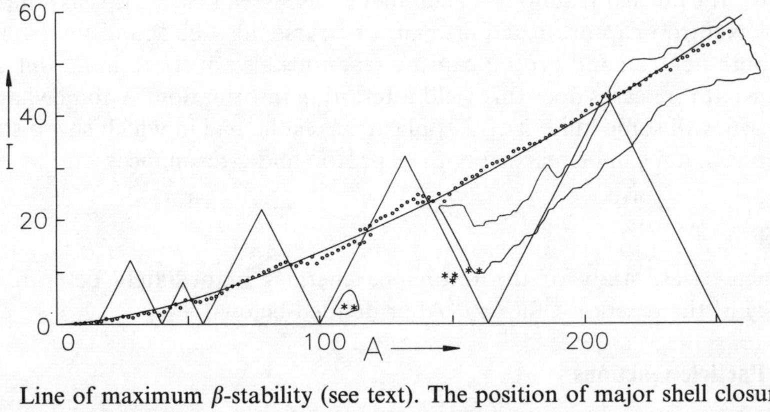

Fig. 2. Line of maximum β-stability (see text). The position of major shell closures is indicated by straight lines. Also shown: known cases of proton and α-radioactivity (crosses and regions, respectively).

isomeric states in the initial nuclide. They give useful information if the excitation energy of the isomer is known.

Proton, neutron and α-decay have also been observed for very short-lived states fed by β-decay (delayed proton (etc.) decay). In several cases, the study of the spectra of these delayed particles, in combination with other data, has yielded useful information on atomic masses.

VI. NUCLEAR REACTIONS

A. Capture reactions

The inverse process of neutron emission is neutron capture. Often, thermal neutrons (energy practically zero) are very easily absorbed by nuclides, and the energy $M(A,Z)+M_n-M(A+1,Z)$ then mostly appears as γ-rays. Measurements of such γ-ray spectra yield much useful and very precise information on nuclear masses. Of course, thermal capture of protons is not possible, but prolific proton capture can occur if the proton energy corresponds to a sharp level of the compound nucleus in the continuum; then a resonance is observed in the capture cross section. The mass of the final nucleus in the formed state is then

$$M'(A+1,Z+1)=M(A,Z)+M_H+\frac{ME_p}{M+M_p}\,\frac{2}{1+\sqrt{1+2MT/(M+M_p)^2}} \qquad (15)$$

(but for a small correction for electronic binding energies; the right hand part is the relativistic correction). Together with γ-ray measurements, as above, very useful information on the ground state masses can be derived. Whether these measurements determine atomic masses rather than nuclear ones is less directly clear than in decays. In fact, the above formula would not hold for very high energies of the incoming protons. For the energies in those cases of interest (and also for the nuclear reaction measurements discussed below), this assumption is valid. Neutron capture reactions can, of course, also be found in resonances, and both neutron and proton capture reactions also in continuum, but only in somewhat rare cases does this yield interesting information. A somewhat peculiar case with some quite useful applications is the one in which sharp states in the continuum can be reached both as proton and α-resonances, e.g.

^{27}Al + $\alpha \rightarrow {}^{31}$P
^{30}Si + $p \rightarrow {}^{31}$P

In such cases, study of the resonance energies immediately determines the energy of the reaction ^{30}Si$(p,\alpha)^{27}$Al as defined below.

B. Particle reactions

Reactions can occur in which the nuclei of two atoms (A,Z) and (a,z) interact to form two other atoms (A',Z') and (a',z'). Conservation laws then require that $A+a=A'+a'$ and $Z+z=Z'+z'$. (Pions and γ-rays may, for this discussion, be considered as particles with $a=0$ and the appropriate value of Z.) The reaction energy is defined as

$$Q = M(A,Z) + m(a,z) - M(A',Z') - m(a',z') \tag{16}$$

In all cases of interest, (A,Z) is a stationary target bombarded with ions of (a,z) with known (not too high) energy $e(a,z)$. One then measures the energy $e'(a',z')$ of the outgoing particles at a known angle with the incoming particles. From the conservation laws of energy and momentum, one can then calculate the recoil energy of the resulting atom $E'(A',Z')$. But for small(!) corrections for the binding energies of outer electrons,

$$Q = e'(a',z') + E'(A',Z') - e(a,z) \tag{17}$$

As in the cases of radioactive decay (which may be considered as special kinds of reactions), the reactions may lead to excited states in the nuclei. This can lead to an assignment of wrong values for the reaction energies. On the other hand, if the excitation energy is known, such a measurement still yields information on masses in nuclear ground states. Measurements of charged particle energies can be made in electric or magnetic spectrometers, or with scintillation or, better, semi-conductor detectors. The latter can also be used for γ-rays. Measurements of neutron energies are possible with time of flight methods; but for (γ, n) and (p, n) reactions, for example, another way exists: measurement of the reaction threshold T, that is the lowest energy for which such a (endothermic) reaction is possible. The advantage is that the determination of the numbers of low energy neutrons is easier than that of neutron energies.

$$Q = -\frac{M(A,Z)T}{M(A,Z)+m(a,z)}\,\frac{2}{1+\sqrt{1+2M(A,Z)T/[M(A,Z)-m(a,z)]^2}} \tag{18}$$

The reaction energy is the threshold in the centre of gravity coordinate system.

VII. OBTAINING MASS VALUES FROM EXTRAPOLATIONS

For several purposes, estimates of quantities derivable from atomic masses (like α- and β-decay energies, and proton and neutron instability) are desired. As mentioned above, many types of mass formulas have been developed that can be used to obtain mass values where no experimental data are available, and they are quite useful in many cases. But comparison with experimental data (see e.g. ref. 17) always shows the presence of systematic differences.

An extrapolation method then seems an attractive procedure. In earlier mass table work[1] such extrapolations were made starting from observed trends in four mass combinations chosen in such a way that the pairing energy did not cause staggering: S_{2n}, S_{2p} and Q^{--}, the last quantity being the sum of the decay energies of two successive β-decays. This approach has been quite successful for near extrapolations: in many cases the predicted mass values were found to agree with later observations.

A difficulty in this method is that the distances (in a plot of mass versus Z and N) between the four surfaces with different combinations of even and odd Z and N tend to diverge. This is avoided by using another possibility: an extrapolation of differences between empirical masses with a mass formula chosen, not to minimize the influence of more or less local effects, but expected to represent the overall effects reasonably well. The formula used for the pairing energies then acquires more importance. Fairly recently, a new formula for this effect was suggested[18], but recent investigations[19] cast some doubts on its validity. This method is under development.

For more remote extrapolations, one might expect that those based on theoretical results for certain combinations of neighboring nuclei, like the ones proposed by Garvey and Kelson[20], might be of advantage. For near extrapolations, their results are not bad, but for further ones the results are less than satisfactory[17].

For many cases with $A < 60$, extrapolations are proposed based on isobaric analog data. Theoretical predictions and experimental facts[21] show that multiplets of states with the same isospin T exist in all isobaric nuclei with $T_3 \leq T$ differing only in the value of T_3 and in mass. Their masses, according to theoretical expectations and, where checked, empirical data, obey a quadratic relation with T_3. If one of them is the ground state in the nucleus $|T_3| = +T$, another one is so for $T_3 = -T$ (an exception occurs for $A = 16$). Thus in those cases where the masses for a $T_3 = +T$ ground state and two excited members are known, that of the proton-rich $T_3 = -T$ ground state can be calculated. And, other than in using Garvey–Kelson-type relations, the result is based on more than just knowledge of other ground state masses. Yet the results have been used with some caution. Cases are known where the isobaric analog strength is distributed over some levels by a configuration interaction with lower-T levels with the same spin and parity; this may introduce uncertainty into the effective level energy. And, finally, the Thomas–Ehrman effect[22] is known to shift the levels of an unstable nuclide by non-easily estimated amounts.

REFERENCES

1. **Wapstra, A. H. and Audi, G.**, The 1983 atomic mass table, *Nucl. Phys. A* **432**, 1 (1985).
2. **Maripuu, S.**, 1975 mass predictions, *At. Data Nucl. Data Tables* **17**, 411 (1976).
3. **Haustein, P. E.**, 1986–1987 atomic mass predictions, *At. Data Nucl. Data Tables* **39**, 185 (1976).
4. **Bethe, H. A. and Backer, R. F.**, Nuclear physics, *Rev. Mod. Phys.* **8**, 82 (1936). **Weizsäcker, C. F. von,** *Z. Phys.* **6**, 431 (1936).
5. **Duckworth, H. E.**, Mass spectroscopy, Cambridge University Press, (1960).

6. **Smith, L. G. and Damm, C. C.**, Mass synchrometer, *Rev. Sci. Instrum.* **27**, 638 (1956).
7. **Van Dyck, Jr, R. S., Farnham, D. L., and Schwinberg, P. B.**, High precision Penning trap mass spectroscopy of the light ions, In *Nuclei Far From Stability/Atomic Masses and Fundamental Constants 1992, Inst. Phys. Conf. Ser. No 132,* Neugart, R. and Wöhr, A., editors, 3, IOP, Bristol (1993).
8. **Cornell, E. A., Weisskoff, R. M., Boyce, K. R., Flanagan, Jr, R. W., Lafyatis, G. P., and Pritchard, D. E.**, Single ion cyclotron resonance measurement of $M(CO^+) - M(N_2^+)$, *Phys. Rev. Lett.* **63**, 1674 (1989).
9. **Thibault, C., Klapisch, R., Rigaud, C., Poskanzer, A. M., Prieels, R., Lesard, L., and Reisdorf, W.**, Direct measurements of the masses of ^{11}Li and $^{26-32}$Na with an on-line mass spectrometer, *Phys. Rev. C* **12**, 644 (1975).
10. **Epherre, M., Audi, G., Thibault, C., Klapisch, R., Huber, G., Touchard, F., and Wollnik, H.**, Direct measurements of the masses of rubidium and cesium isotopes far from stability, *Phys. Rev. C* **19**, 1504 (1979).
11. **Audi, G., Epherre, M., Wapstra, A. H., and Bos, K.**, Masses of Rb, Cs and Fr isotopes, *Nucl. Phys. A* **378**, 443 (1982).
12. **Vieira, D. J., Wouters, J. M., Vaziri, K., Krauss, Jr, R. H., Wollnik, H., Butler, G. W., Wohn, F. K., and Wapstra, A. H.**, Direct mass measurements of neutron-rich light nuclei near N=20, *Phys. Rev. Lett.* **57**, 3253 (1986).
13. **Gillibert, A., Mittig, W., Bianchi, L., Consolo, A., Fernandez, B., Foti, A., Gastebois, J., Gregoire, C., Schutz, Y., and Stephan, C.**, New mass measurements far from stability, *Phys. Lett. B* **192**, 39 (1987).
14. **Sharma, K. S., Hagberg, E., Dyck, G. R., Hardy, J. C., Koslowsky, V. T., Schmeing, H., Barber, R. C., Yuan, S., Perry, W., and Watson, M.**, Masses of ^{103,104,105}In and ^{72,73}Br, *Phys. Rev. C* **44**, 2439 (1991).
15. **Stolzenberg, H., Becker, St., Bollen, G., Kern, G., Kluge, H.J., Otto, T., Savard, G., and Schweikard, L.**, Accurate mass determination of short-lived isotopes by a tandem Penning trap mass spectrometer, *Phy. Rev. Lett.* **65**, 3104 (1990).
16. **Rose, H. J. and Jones, G. A.**, A new kind of radioactivity, *Nature* **307**, 245 (1984).
17. **Audi, G. and Borcea, C.**, On the predictive power of nuclear models, In *Origin and Evolution of the Elements* Prantzos, N. *et al.*, editors, 421, Cambridge University Press, 1993.
18. **Jensen, A. S., Hansen, P. G., and Jonson, B.**, New mass relations and four nucleon correlations, *Nucl. Phys. A* **431**, 393 (1984).
19. **Wapstra, A. H.**, Improvement in the mass evaluation situation since 1985, In *Nuclei Far From Stability/Atomic Masses and Fundamental Constants 1992, Inst. Phys. Conf. Ser. No 132,* Neugart, R. and Wöhr, A., editors, 979, IOP, Bristol (1993).
20. **Garvey, G. T. and Kelson, I.**, New nuclidic mass relationship, *Phys. Rev. Lett.* **16**, 197 (1966).
21. **Wilkinson, D. J.**, *Isospin in nuclear physics,* North-Holland, Amsterdam, (1969).
22. **Thomas, R. G.**, On the determination of reduced widths from the one-level dispersion formula, *Phys. Rev.* **81**, 148 (1951).
Ehrman, J. B., On the displacement of corresponding levels of C^{13} and N^{13}, *Phys. Rev.* **81**, 412 (1951).

2

SHELL MODEL INTERPRETATION OF NUCLEAR MASSES

N. Zeldes

Table of contents

I. INTRODUCTION

This chapter describes an intuitive physical way to understand the empirical trends of nuclear masses using the shell model.

The nuclear ground state energy is the lowest eigenvalue of the hamiltonian matrix. In modern shell model calculations[1–4] it is arrived at by a numerical diagonalization of a large matrix. On the other hand, by using general considerations and basic properties of the empirical effective interaction of the shell model, one can obtain a simple analytical expression describing the dependence of the ground state energy on nucleon numbers.[5]

The derivation follows the order of the shell model approximations. One starts with a single configuration, using available explicit analytical results.[6–8] Then identical-nucleon and neutron–proton correlation terms[8–10] due to configuration interaction are added. When the latter terms are large the pure configuration results might get considerably distorted, but on the whole their basic features survive. The energy expression thus obtained has been used as a semi-empirical mass equation[5] with a high predictive power.[11]

After this Introduction the chapter comprises two parts: first empirical evidence on the masses is described. Then shell model results are summarized, and the properties of the effective interaction are used to obtain a ground state energy equation reproducing the empirical trends described in the first part.

II. EMPIRICAL TRENDS

A. Overall view (gross structure)

Fig. 1 shows the distribution of presently known nuclei in the N-Z plane (see also Fig. 2 of the chapter on atomic masses by A. H. Wapstra).

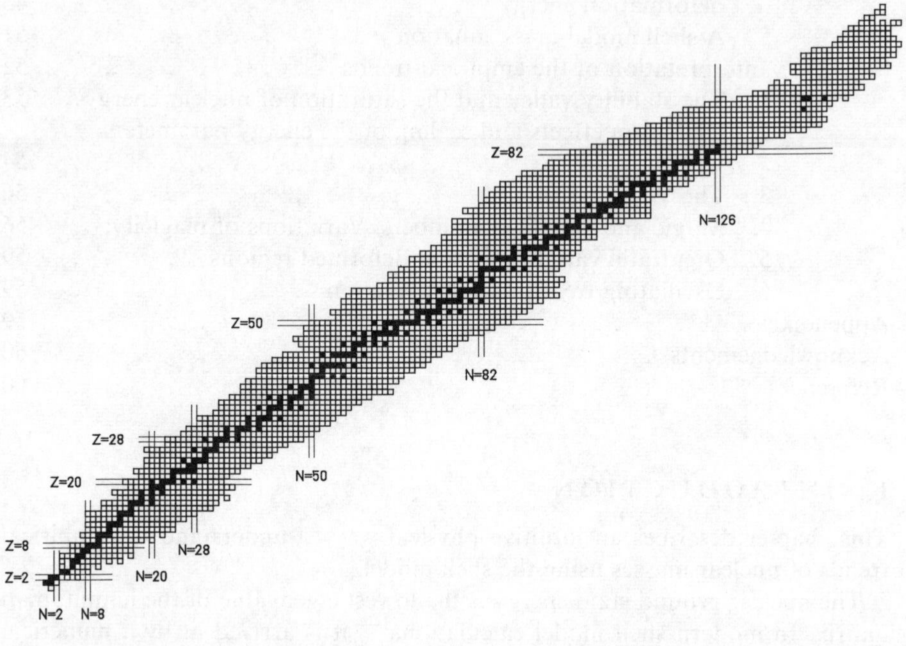

Fig. 1. Nuclear chart including major shell closures. Black squares represent nuclides found in nature. Taken from ref. 12.

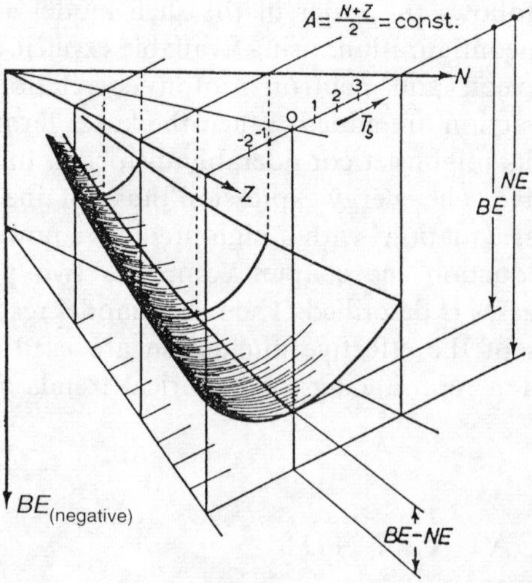

Fig. 2. Gross structure of the nuclear mass surface with isobaric mass parabolas. $NE = 7 - 9A$ MeV. Taken from ref. 13.

Fig. 2 shows the gross behaviour of the nuclear binding energy B as a function of nucleon numbers. The surface looks like a valley. The bottom corresponds to beta-stable nuclei, and the slopes to beta-unstable ones, decaying towards the bottom. The slope of the surface is approximately constant along the bottom, about 8 MeV/A, and it increases rapidly in the perpendicular direction, where the neutron–proton ratios deviate from their values at stability. This behaviour is like that of a macroscopic saturating system with composition-dependent energy, and it motivated the development of the liquid-drop model in the early days of nuclear physics.[14, 15]

Fig. 3 shows experimental B/A values of even-mass beta-most-stable isobars, in qualitative agreement with the liquid-drop model. On the other hand, there is a fine structure superposed on the above smooth trends. In particular, there are odd–even oscillations up to $A \approx 24$ (see insert), and there are kinks at ^{88}Sr, ^{118}Sn, ^{140}Ce and ^{208}Pb. (See also Fig. 1 of the chapter by Wapstra.) The fine structure of the mass surface is considered in the following sections II.B–F.

B. Odd–even effects

Closer scrutiny shows that there are four smooth mass surfaces corresponding to the four nuclear parity types. The even–even surface lies lowest and the odd–odd one highest, with an average separation distance of about 2.5 MeV. The two odd-A surfaces lie between the even-A ones, mostly nearer to the odd–odd surface, with a separation of a few hundred keV at most and with a changing relative position governed by the Suess–Jensen rule,[17] stating that on crossing magic numbers 28, 50 and $N = 82$ into a heavier mass region the corresponding odd-A surface becomes lower.*

The average separation distance between the odd-A and even–even surfaces decreases with A approximately like $12A^{-1/2}$ MeV[16] and it also seems to increase when $I = N - Z$ decreases (see, however, ref. 20).[18, 19] The additional lowering of the odd–odd surface towards the odd-A ones as compared to the separation of the even–even and odd-A surfaces decreases with A roughly like $20A^{-1}$ MeV.[16, 19] Local values of the separation distances from which variations with respect to A and I can be determined are given by appropriate third-order mass differences.[9, 19]

The odd–even splitting is illustrated in Figs. 4 and 5, showing isotopic, isotonic and isobaric sections of the mass surface in a non-diagonal shell region†. The vertical distance between the mass parabolas in the upper parts is

---

*The rule actually states that the corresponding pairing energy becomes smaller. The relation of pairing energy to odd–even separation distance is considered in section III.D.2.

†One distinguishes diagonal regions in the N-Z plane, where the major valence shells are the same for neutrons and protons, and non-diagonal regions, where they are different.

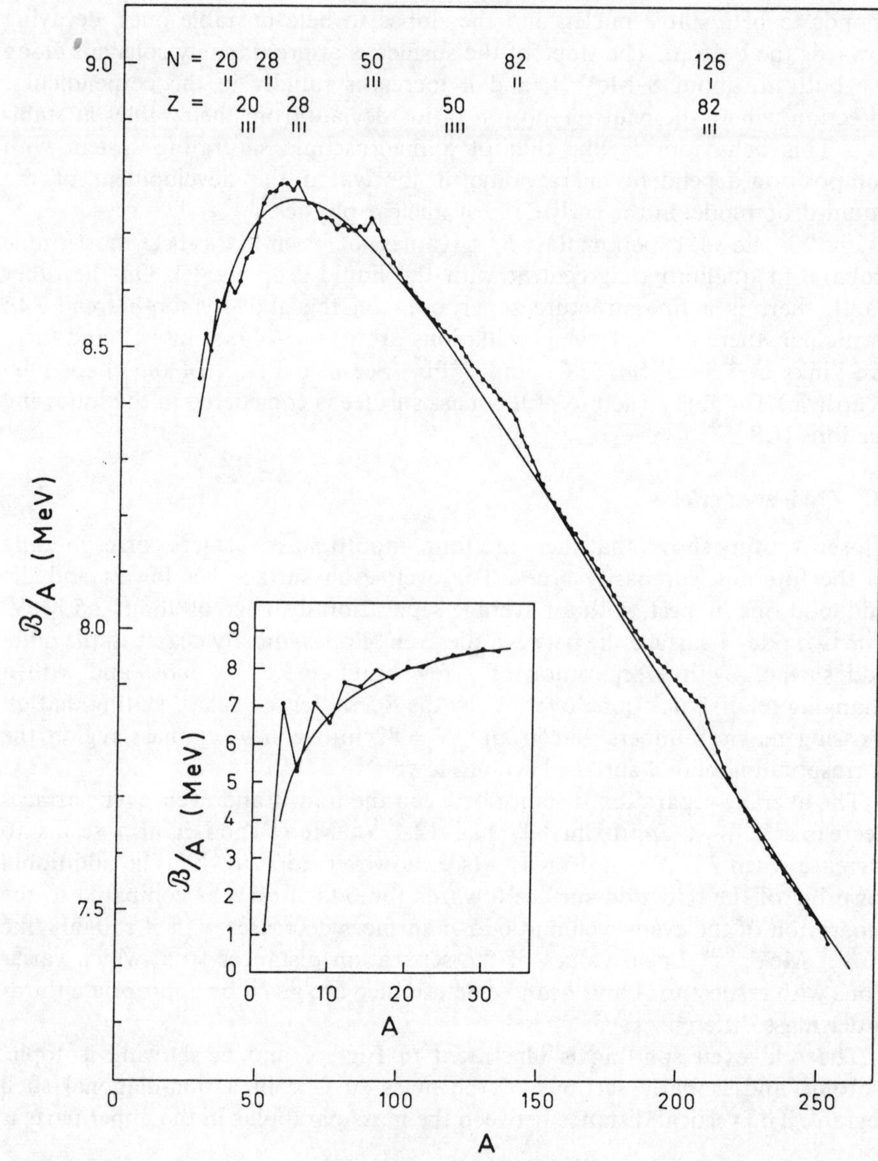

Fig. 3. *B/A* of even-*A* beta-most-stable nuclides. The smooth curve represents the liquid-drop mass formula. Taken from ref. 16.(A. Bohr/B.R. Mottelson, *Nuclear Structure*, *Volume 1* (pg. 168), © 1969 by W.A. Benjamin, Inc. Reprinted with permission of Addison-Wesley Publishing Company, Inc.)

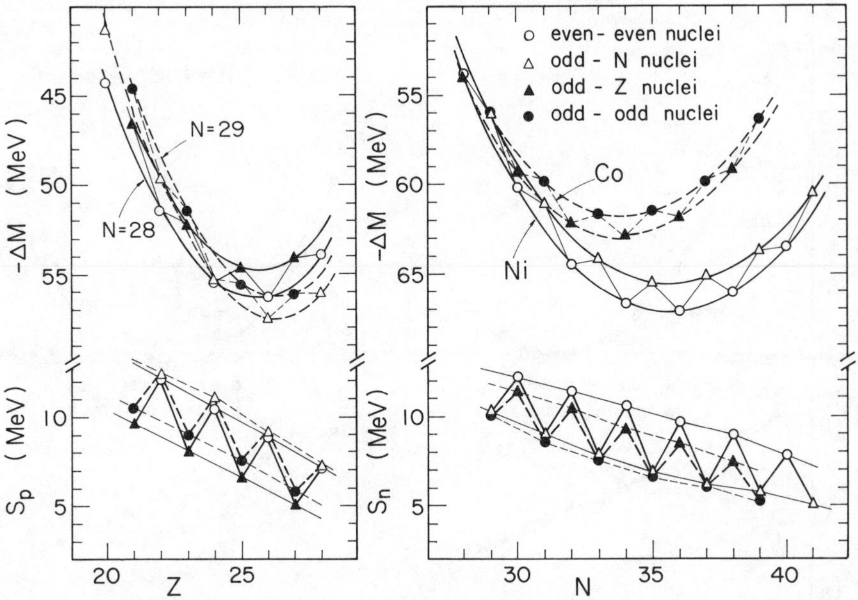

Fig. 4. Right part: isotopic mass parabolas (above) and corresponding neutron separation energies (below) for Co and Ni isotopes with $N \geq 28$. Left part: isotonic mass parabolas (above) and corresponding proton separation energies (below) for $N = 28$ and 29 isotones with protons in the $1f_{7/2}$ shell. Data from ref. 21.

doubled in the oscillation amplitude of the corresponding mass differences below. The $A = 99$ parabola splitting in Fig. 5 is in agreement with the Suess–Jensen rule. (See also Fig. 8 below.) Longer isodiapheric ($I =$ Const.) sections, demonstrating it directly, are shown in ref. 22.

We mention also that masses of light odd–odd nuclei with $Z = N$ are higher than the rest of the odd–odd surface. This is illustrated in Fig. 6, showing sections of the mass surface through $I = 0$ and $I = 2$ isodiapheres above and the corresponding deuteron separation energies below. The vertical distance between two isodiapheric lines in the upper part, which is doubled in the oscillation amplitude of the separation energies, is obviously larger for $I = 0$ than for $I = 2$ in the $1p$, $1d2s$ and also $1f_{7/2}$ shells (see also ref. 19).

The odd–even splitting is a most direct indication of the pairing property of the nuclear effective interaction (section III.B.2). The regular oscillations in light nuclei in Fig. 3 are a pairing effect.

The large splitting for even A was known in the early days of nuclear physics.[23,24] Then came the discovery of the smaller odd-A splitting,[25,17] and the additional lowering of the odd–odd surface (see also ref. 28).[26,27]

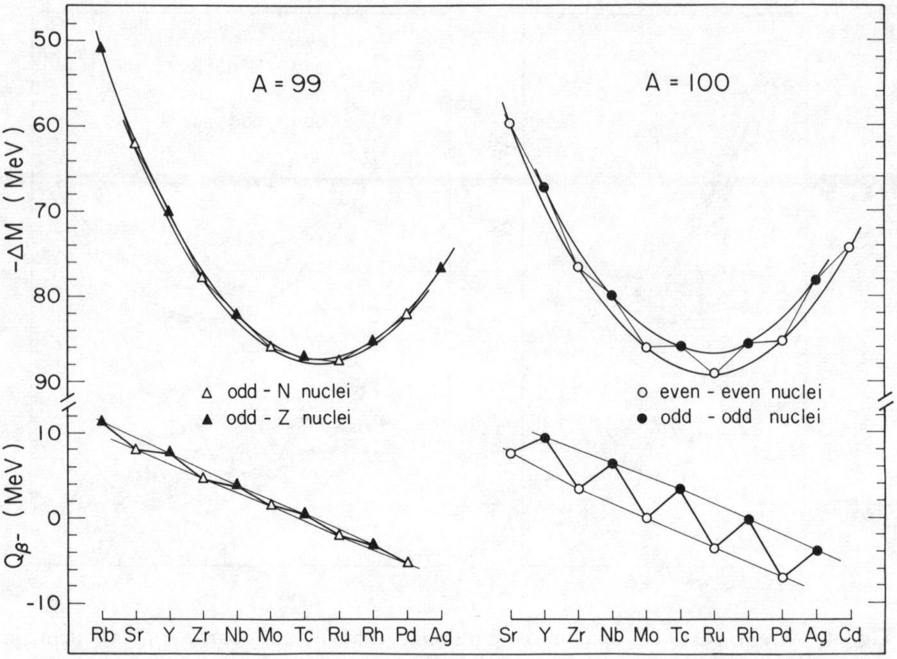

Fig. 5. Isobaric mass parabolas (above) and corresponding Q_{β^-} values (below) for $A = 99$ and 100 isobars. Data from ref. 21.

C. Slope discontinuity at $Z = N$: the Wigner term

Each of the four sheets of the mass surface has cusp lines with slope discontinuities. We consider the $Z = N$ line first.

A section of the mass surface through isobars of a given parity type in a diagonal region comprises two parabolic arcs with a cusp intersection at $Z = N$. This is demonstrated in Fig. 7, showing isobaric masses and Q_{β^-} values of $1f_{7/2}$ nuclei, after subtraction of Coulomb energies and neutron–proton mass differences. The sections are highly symmetric, demonstrating the charge symmetry of the effective nuclear interaction. (In actual nuclei, before subtraction, the Coulomb energy increases quadratically with Z, the minima of isobaric sections shift towards $Z < N$, and the $Z = N$ cusp line runs along the proton-rich slope of the stability valley.)

The existence of a cusp at $Z = N$ can also be demonstrated in other ways. One can plot isobaric mass residuals of a smooth mass equation as functions of I,[30] observing a spike at $I = 0$. Alternatively, one can plot second mass differences for nuclei lying on the same sheet of the mass surface. For a smooth surface these depend only weakly on N and Z. On the other hand, crossing a cusp line results in a corresponding spike. Such spikes at $Z = N$ are

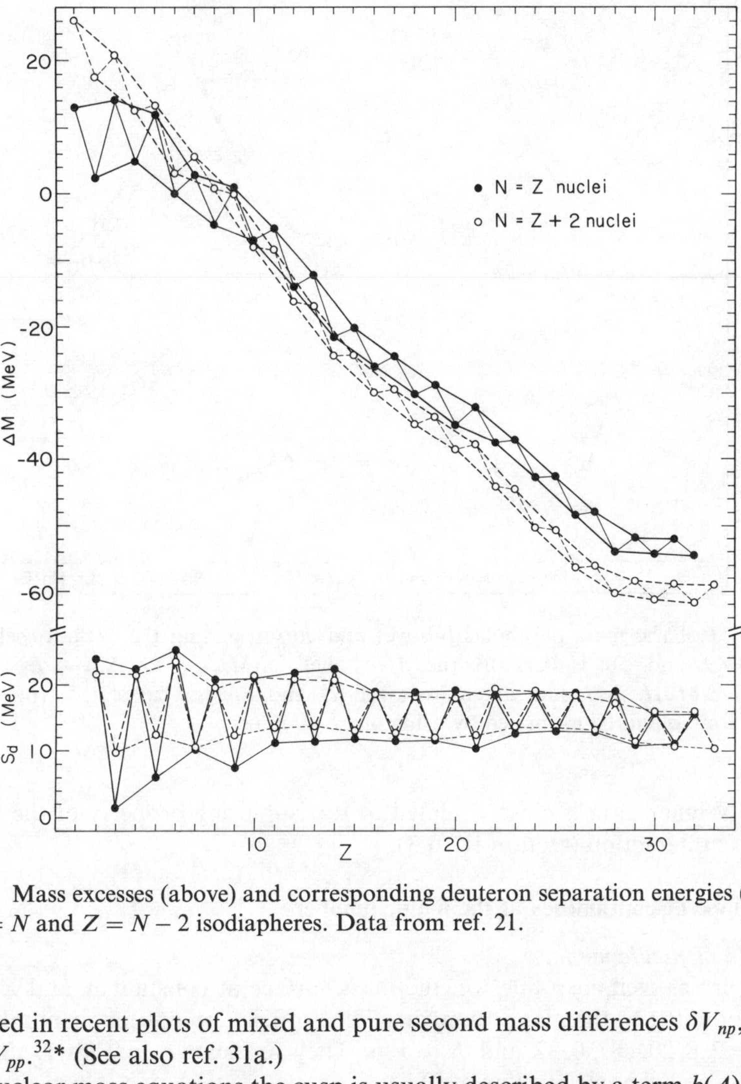

Fig. 6. Mass excesses (above) and corresponding deuteron separation energies (below) for $Z = N$ and $Z = N - 2$ isodiapheres. Data from ref. 21.

observed in recent plots of mixed and pure second mass differences δV_{np},[31] δV_{nn} and δV_{pp}.[32]* (See also ref. 31a.)

In nuclear mass equations the cusp is usually described by a term $b(A)T$ with a smoothly varying coefficient $b(A)$. The cusp at $Z = N$ is reproduced, since as a rule $T_{g.s.} = |T_z| = \frac{1}{2}|N - Z|$ (see section III.B.3). The T term is called the Wigner term, as Wigner first derived[33] a $T(T+4)$ term in the supermultiplet approximation. Both the course of the beta-stability line[34] and masses of $A < 70$ nuclei[19,35] indicate the superiority of $T(T+1)$ dependence, as in the *jj*-coupling shell model (section III.C.1). (See also ref. 36.)

*The δV_{nn} and δV_{pp} of ref. 32 are calculated from masses lying on two sheets of the mass surface rather than one. The results, though, are very similar.

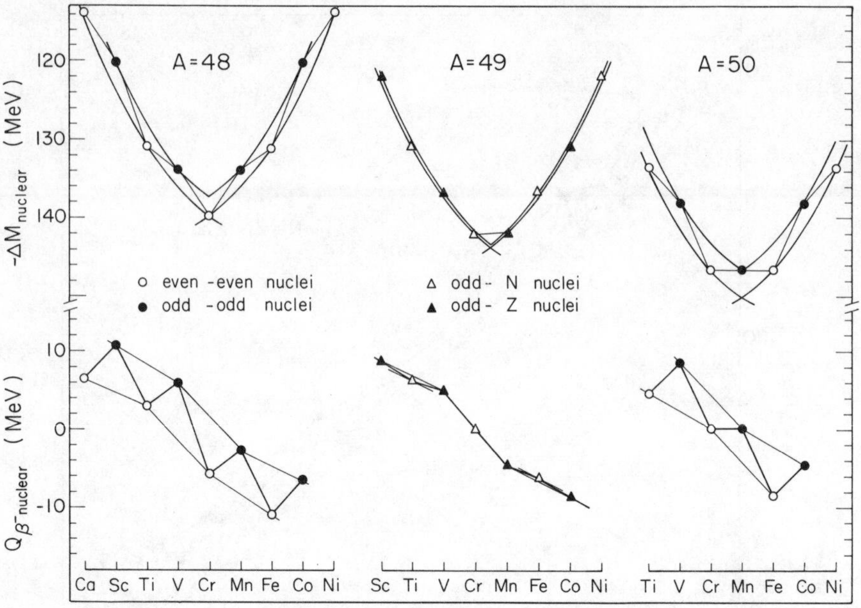

Fig. 7. Isobaric mass parabolas (above) and corresponding Q_{β^-} values (below) for $A = 48, 49$ and 50 isobars in the $1f_{7/2}$ shell. $\Delta M_{nuclear} = \Delta M - E_{Coulomb} - \frac{1}{2}(\Delta M_n - \Delta M_H)I$. ΔM data taken from ref. 21 and $E_{Coulomb}$ calculated from ref. 29. Values for Co and Ni estimated by using mirror symmetry.

The Wigner term is closely related to the isopairing property of the nuclear effective interaction (section III.B.3).

D. Slope discontinuities at the magic numbers

1. Major magic numbers

There are as well cusp lines on the mass surface at constant N and Z values corresponding to the magic numbers. The "official" major magic numbers are $N, Z = 2, 8, 20, 28, 50, 82$ and $N = 126$. They are shown in Fig. 1. (See also Fig. 2 of the chapter by Wapstra.)

Such slope discontinuities are illustrated in the upper parts of Fig. 8, showing the respective isotopic and isotonic sections of the mass surface for Sr and for $N = 66$ nuclei. Each section extends over two shell regions, with a discontinuous increase of slope when crossing the respective magic numbers $N = 50$ and $Z = 50$ into the heavier region. The lower parts show a discontinuous decrease at the magic number of the corresponding two-nucleon separation energy, $S_{2n} \approx -2(\partial E/\partial N)$ and $S_{2p} \approx -2(\partial E/\partial Z)$.

Fig. 9 shows S_{2n} systematics for $N = 79$ to 107. The $N = 82$ downwards discontinuity is universally observed for all elements.

Similar discontinuities are observed in other mass differences (which are

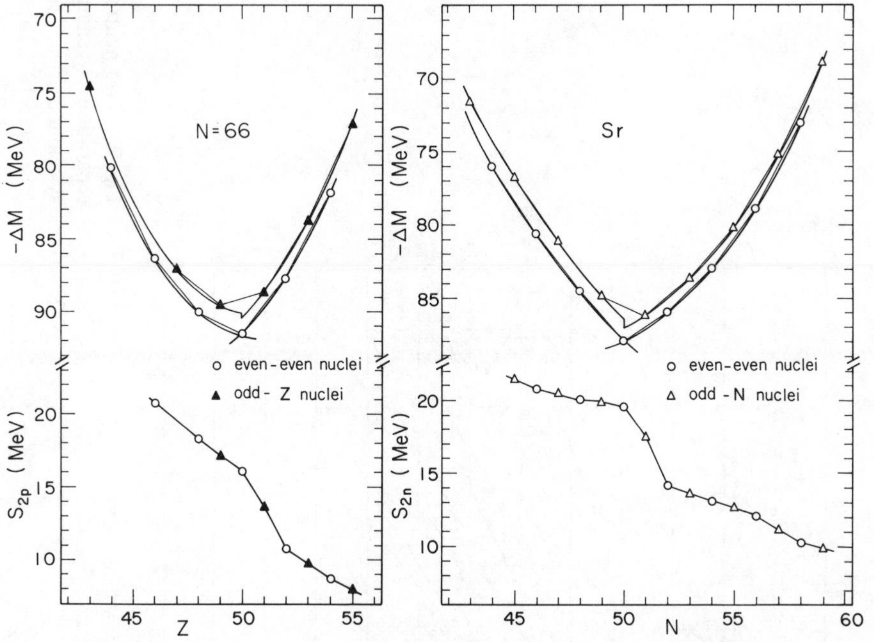

Fig. 8. Right part: isotopic mass parabolas (above) and corresponding S_{2n} values (below) for Sr isotopes. Left part: isotonic mass parabolas (above) and corresponding S_{2p} values (below) for $N = 66$ isotones. Data from ref. 21.

sums and differences of separation energies). Figs. 10 and 11 show Q_α systematics for $N = 79$ to 133. At $N = 82$ and 126 there are large upwards discontinuities of isotopic Q_α lines for all elements. Likewise, the vertical spacings of isotopic Q_α lines between $_{51}$Sb and $_{52}$Te and between $_{82}$Pb and $_{84}$Po are larger than for the other elements, universally for all N values.

The slope discontinuities of the mass surface at the magic numbers can also be demonstrated using the alternative ways mentioned in connection with the Wigner term. One can plot residuals of masses or of separation or decay energies for a smooth mass equation as functions of the corresponding nucleon number. This method was used to demonstrate the existence of magic numbers in the early 1950s.[38] Alternatively, the magic numbers are indicated by spikes in plots of pure second mass differences of nuclei of the same parity type.* [9,10,32]

The slope discontinuities of the mass surface result in discontinuous shifts at the magic numbers of the beta-stability line, which is a straight line segment in a given major shell region (see Fig. 2 of the chapter by Wapstra).[39–42,9]

The magic numbers constitute the most direct evidence for shell structure in

* See footnote in section II.C.

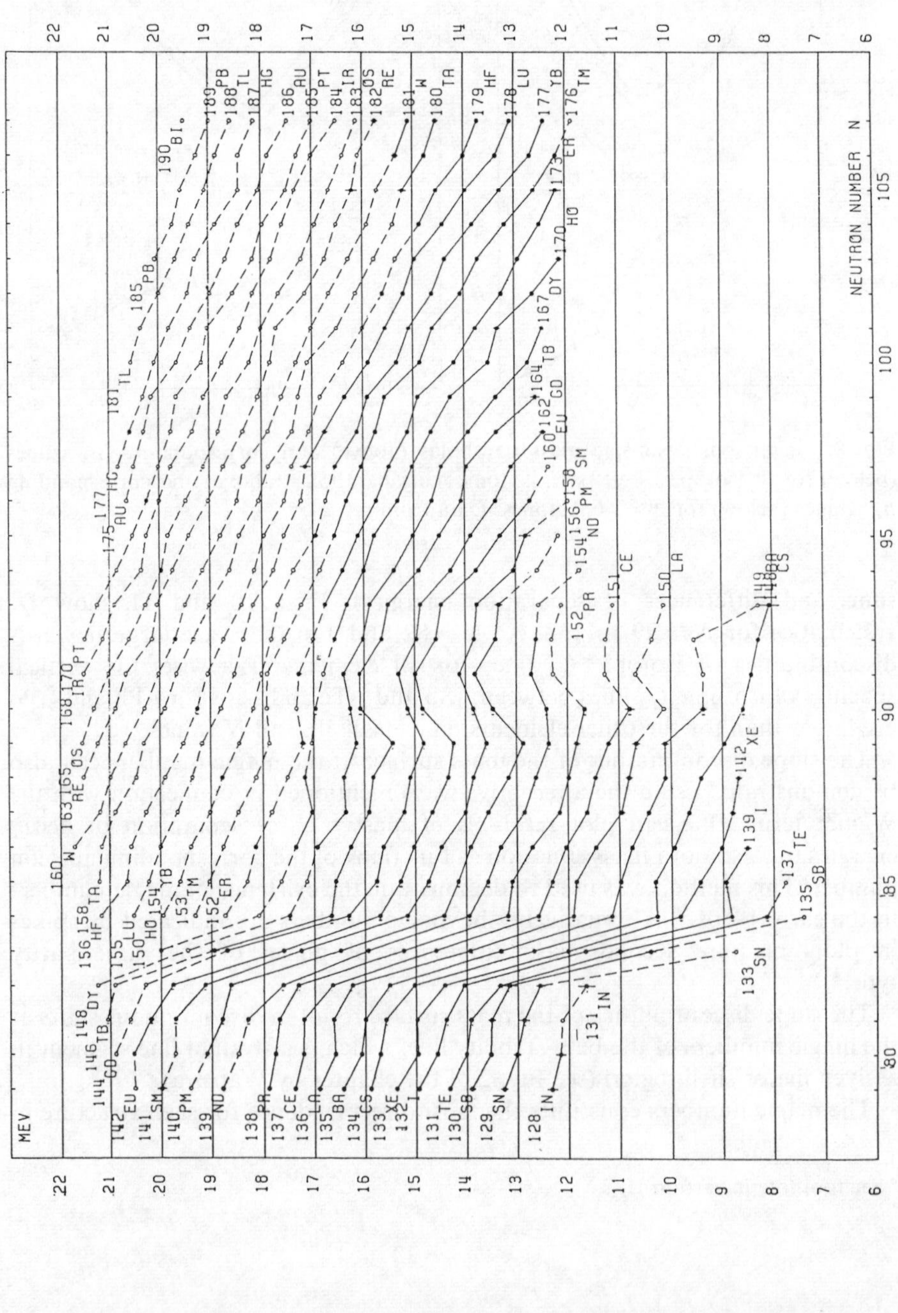

Fig. 9. Two-neutron separation energy systematics for $N = 79$ to 107. Taken from ref. 37.

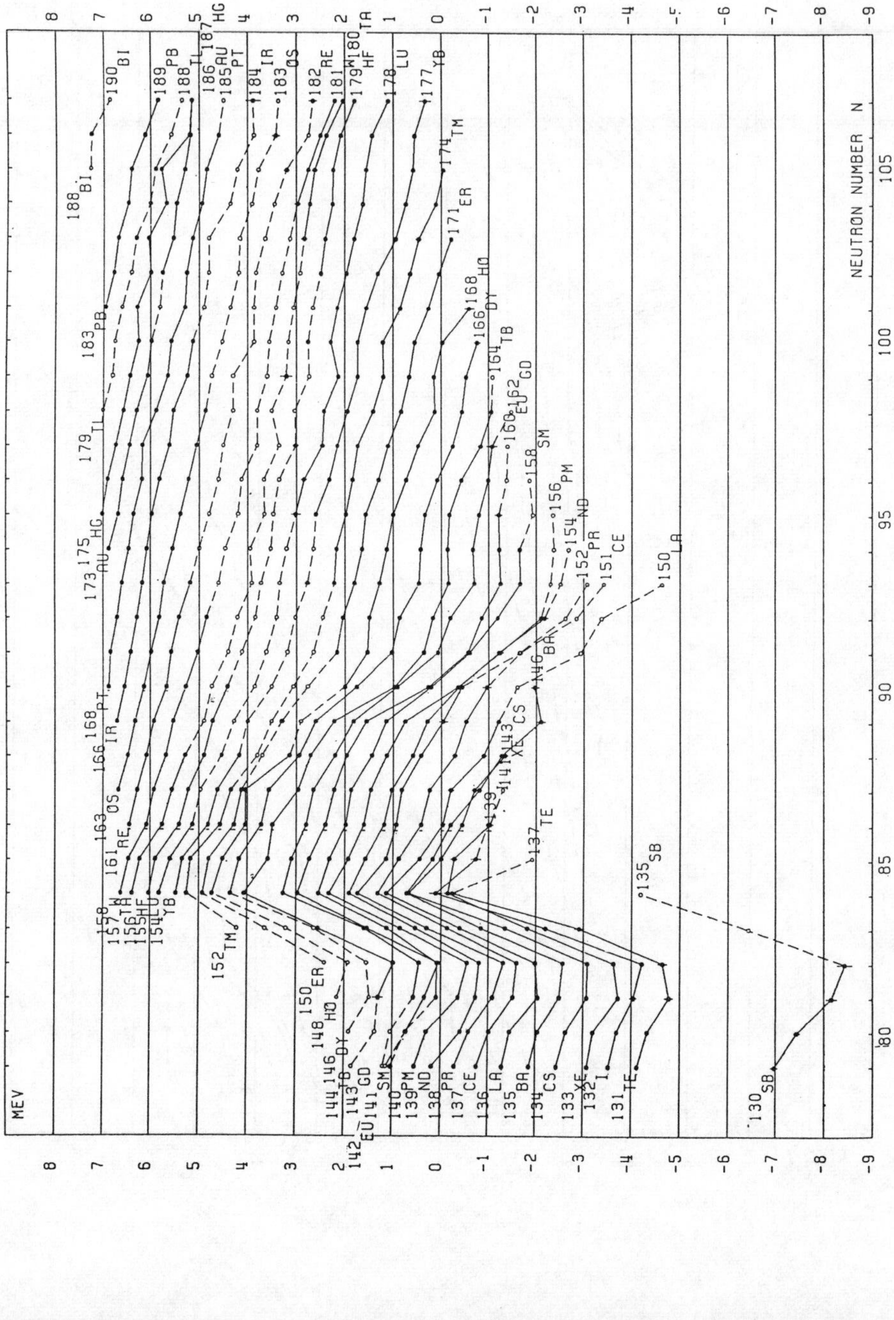

Fig. 10. Alpha decay energy systematics for $N = 79$ to 107. Taken from ref. 37.

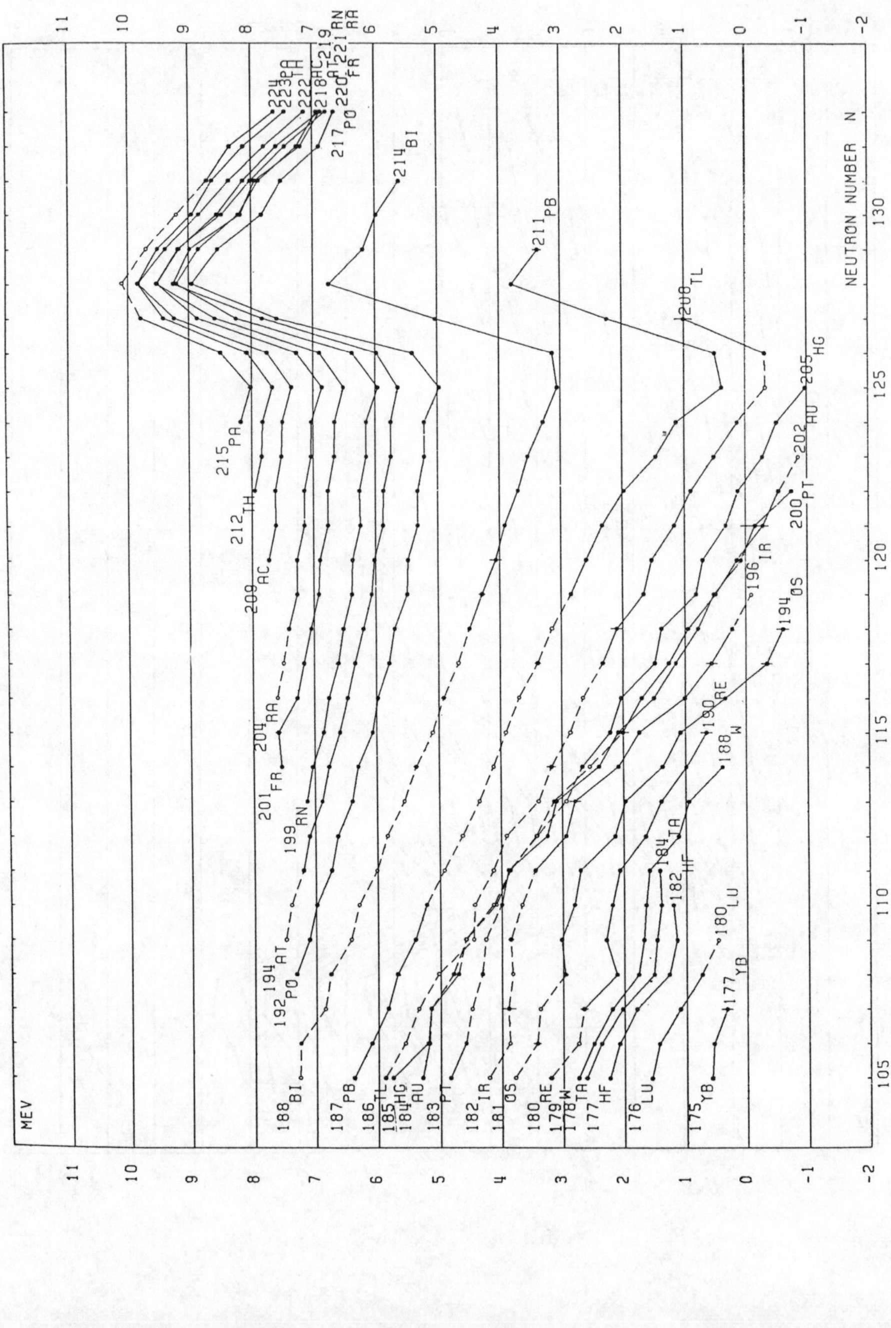

Fig. 11. Alpha decay energy systematics for $N = 105$ to 133. Taken from ref. 37.

nuclei. Their systematic exploration motivated[43] and pioneered[38] the development of the *jj*-coupling shell model and of earlier versions (see refs. 15, 38). The kinks in the B/A line in Fig. 3 are magic number effects.

2. Submagic numbers

At certain nucleon numbers there are slope discontinuities of the mass surface like at the magic numbers, but of a considerably smaller magnitude and of limited duration. These are called submagic numbers. Well-established submagic numbers in spherical nuclei are $N = 56$ and $Z = 40$ and 64.

In Fig. 12 there is an upwards discontinuity at $N = 56$ followed by a hump in isotopic Q_α lines of $_{38}$Sr, ($_{39}$Y), $_{41}$Nb and $_{42}$Mo (for $_{40}$Zr it starts at $N = 57$), but not for the other elements. The updated Fig. 18 of ref. 37a shows that the effect actually starts earlier, extending from $_{85}$Br through $_{92}$Mo. The magnitude of the discontinuity is much smaller than those at $N = 82$ and $N = 126$ in Figs. 10 and 11.

The submagic character of $Z = 40$ is illustrated in Fig. 12 too by the larger vertical spacing of isotopic Q_α lines between $_{39}$Y and $_{41}$Nb for $N \leq 59^\dagger$. Q_α systematics for $N = 27$ to 55 (Fig. 12 of ref. 37) show that this larger spacing continues down to $N = 48$, but not for lower N values. The magnitude of the increased spacing is considerably smaller than those observed above the $_{50}$Sn and $_{51}$Sb lines in Figs. 10 and 12 and above the $_{82}$Pb and $_{83}$Bi lines in Fig. 11.

Finally, the submagic character of $Z = 64$ is illustrated in Fig. 10, where there is an increased vertical spacing of isotopic Q_α lines between $_{64}$Gd and $_{66}$Dy for $81 \leq N \leq 87$. As for $Z = 40$ the effect is much smaller than at $Z = 50$ and 82.

Like the magic numbers, submagic numbers $N = 56$ and $Z = 40$ are respectively indicated by spikes in recent δV_{nn} and δV_{pp} plots.[32]

There are also submagic numbers in deformed nuclei, the most well known of which is $N = 152$, observed in S_{2n} and Q_α systematics (Figs. 6 and 16 of ref. 37). Submagicity of $Z = 66$ is revealed in Fig. 10 by a somewhat increased vertical spacing of isotopic Q_α lines between $_{66}$Dy and $_{68}$Er for $N \geq 91$. Sometimes[45] $N = 98$, 104 and 108 were proposed as submagic numbers in deformed nuclei, based on S_{2n} systematics.

3. Mutual support and weakening of magicities

In Fig. 11 the large spacing between the isotopic Q_α lines of $_{82}$Pb and $_{84}$Po associated with the magic character of $Z = 82$ is highest around $N = 126$ and

†Contrary to the case of the magic numbers $Z = 50$ and 82, where the increased spacing is between Z and $Z + 2$ (Figs. 12 and 11), the discontinuity associated with $Z = 40$ extends from $Z = 39$ to $Z = 41$ rather than from 40 to 42. This peculiarity is observed also in S_{2p} systematics (Fig.8 of ref. 37). A similar effect is seen in S_{2n} systematics at $N = 15$ rather than 16 (Fig. 13 below (see also ref. 37a and ref. 44)) and in Q_α systematics at both $N = 15$ and $Z = 15$ (Fig. 11 of ref 37 (see also ref. 37a)). These peculiarities are presumably[9] pairing effects due to the small pairing energies of the $2p_{1/2}$ and $2s_{1/2}$ subshells, as compared to the preceding subshells $2p_{3/2}$ or $1f_{5/2}$ and $1d_{5/2}$ (section III.D.2).

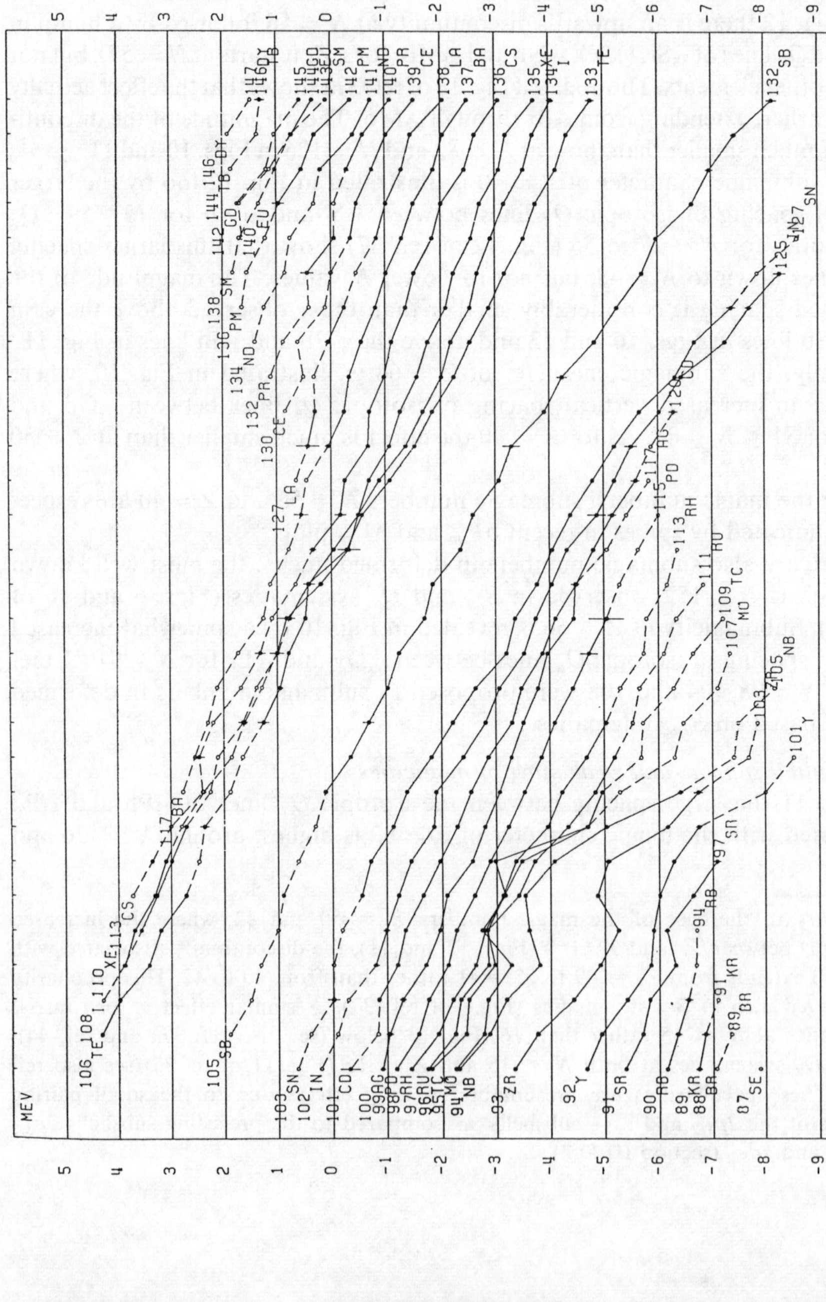

Fig. 12. Alpha decay energy systematics for $N = 53$ to 81. Taken from ref. 37.

decreases monotonously when $126 - N$ increases in the neutron-deficient isotopes. Likewise, the upwards discontinuity at $N = 126$ is highest for $_{82}Pb$, $_{83}Bi$ and $_{84}Po$, and decreases monotonously in both lighter and heavier elements when $|Z - 82|$ increases. The same effects are observed in separation energy systematics (see section III.D.4).

Thus, the magic character of $Z = 82$ seems to be strongest for $N = 126$ and vice versa. This effect is referred to as mutual support of magicities.[46] It is most conspicuous near ^{208}Pb, but it occurs near other doubly magic or submagic nuclei as well.[22,47] It occurs also in single-nucleon or single-hole excitation across shell and subshell gaps (see section III.D.4 and ref. 48).

In Fig. 11 the magicity of $Z = 82$, although weakened, persists all the way down to $N = 106$. Likewise, the magicity of $N = 126$ in Fig. 11, of $N = 82$ in Figs. 9 and 10 and of $Z = 50$ in Fig. 12 are universally valid. On the other hand, the smaller discontinuities associated with submagic numbers are completely washed out some distance away from the doubly magic or submagic nucleus where they are largest.

In lighter nuclei the variation of magicity with nucleon numbers is faster than in heavier ones, which makes even major magic numbers unstable. Thus the magicity of $N = 20$ has completely disappeared in exotic neutron-rich Na and Mg isotopes,[49] as can be seen in Fig. 13 and in updated versions.[50,37a] More drastically, the magicity of $N = 8$, which is very pronounced in $_6C$ isotopes, has completely disappeared in $_4Be$, where the ground state of ^{11}Be is $1/2+$.[51,22]

E. Quadratic variation in non-deformed regions

In a given non-diagonal shell region near its magic number boundaries each of the four sheets of the mass surface is well describable by a quadratic function of N and Z, with constant spacing between the four sheets. The same holds in a diagonal region on each side of the $Z = N$ line. Vertical sections of such a surface in any direction are quadratic mass parabolas, as in the upper parts of Figs. 4, 5, 7 and 8.

The quadratic variation of a single sheet is most directly demonstrated by linear variations with respect to N and Z of mass differences of nuclei lying on the same sheet, like double-nucleon separation energies, alpha, and double beta decay energies (Figs. 8–13 and others from refs. 37, 37a).

A direct demonstration of a constant spacing between two sheets is furnished by a constant-amplitude zig-zag oscillation of mass differences of nuclei lying alternatingly on the two sheets, like the single nucleon and deuteron separation energies and beta decay energies shown in the lower parts of Figs. 4, 5, 6 and 7.

(See also the very regular systems of parallelograms in the lower part of Fig. 17, below the magic number gaps.)

Starting in the 1950s,[52–55] linear systematics of various mass differences as in refs. 37, 37a have been widely used for predicting unknown masses by interpolation and short-distance extrapolation.[56–59,59a]

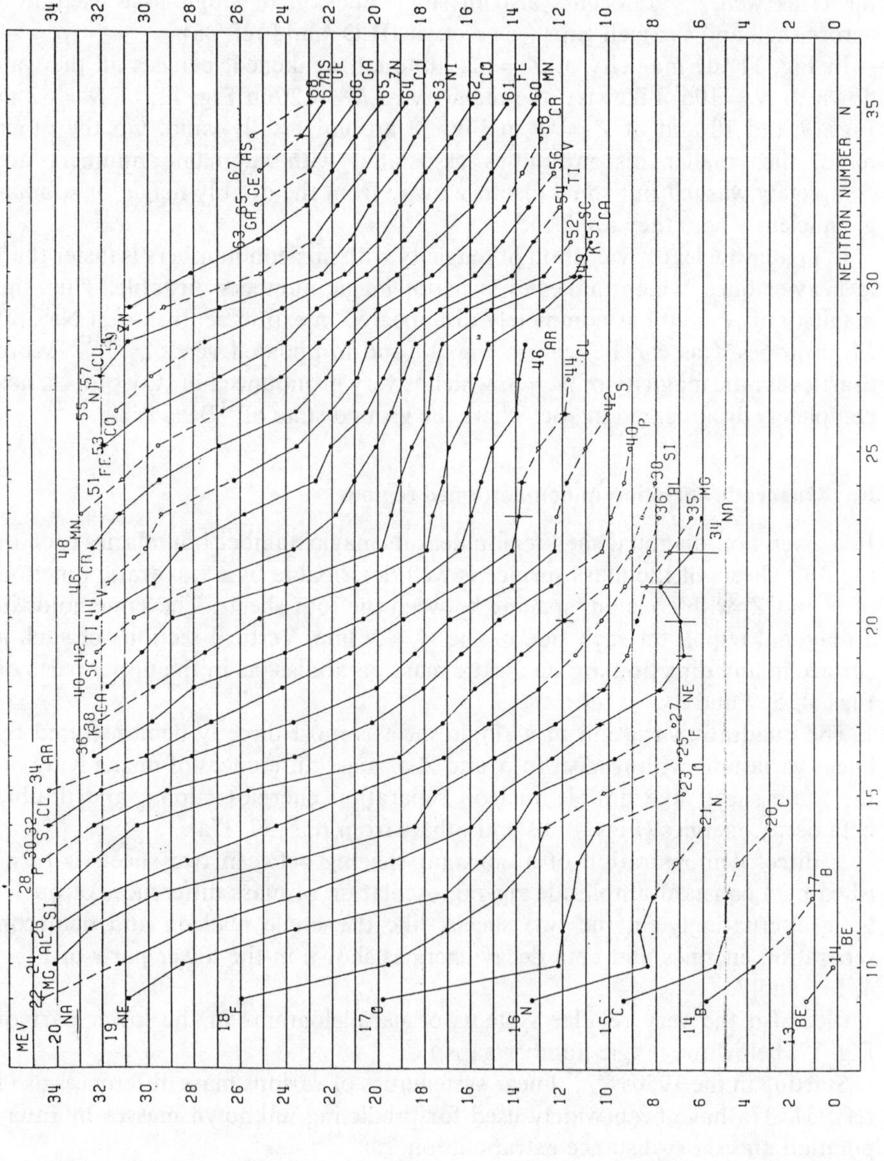

Fig. 13. Two-neutron separation energy systematics for $N = 9$ to 35. Taken from ref. 37.

F. Oscillating trends in deformed regions

Deeper inside major shell regions, where nuclei are deformed, there are oscillating trends superposed on the quadratic variation of the masses. These are reflected in the systematics of mass differences.

Most conspicuous are the humps observed in isotopic S_{2n} lines when there is an abrupt change from a spherical to a deformed shape. Such humps are observed in rare earth elements in Fig. 9 starting at $N = 89$, in elements near Zr in Fig. 3 of ref. 37 starting at $N = 59$ or 60 and in the heavy Na isotopes in Fig. 13 starting at $N = 19$ or 20. There are also smaller humps in Ra, Ac and Th lines in Fig. 6 of ref. 37, starting at $N = 132$. The ascending parts of these humps are essentially the only presently known places in the nuclear chart where S_{2n} increases rather than decreases when neutrons are added to the nucleus, reflecting the gain in binding energy accompanying deformation (see Fig. 1 of the chapter by Wapstra). Such humps are presently considered a fingerprint of deformation alongside the traditional ones considered in section III.B.5.

The energy gain accompanying deformation[60] and the hump at $N = 90$[61] were found in the 1950s. The humps at $N = 20$[62] and $N = 60$[63] were discovered later.

Milder oscillations superposed on linear decreasing trends are observed in S_{2n}, S_{2p} and Q_α lines in many places inside major shell regions and also on their boundaries (Figs. 9–13 and others from refs. 37, 37a).

Oscillations of the masses result in corresponding oscillations in plots of second mass differences,[9, 10, 31, 32] in particular in negative δV_{nn} values where deformation abruptly sets in.

The oscillations of the four sheets of the mass surface are independent, and their mutual distances oscillate as well around their average values, roughly symmetrically with respect to the middle of the shell region.[64, 9, 10]

III. SHELL MODEL INTERPRETATION

A. The shell model approach*

1. *The shell model approximations*

The shell model is a hierarchy of approximations based on perturbation theory. One separates the nuclear hamiltonian into two parts:

$$H = \sum t_i + \sum V_{ij} = H^0 + H' \tag{1}$$

$$H^0 = \sum h_i = \sum (t_i + U_i) \tag{2}$$

$$H' = \sum V_{ij} - \sum U_i \tag{3}$$

---

*A detailed exposition of shell model methods is given in ref. 6 (see also ref. 6a). Unreferenced results given in this and the following sections can be found there.

where t and V are respectively the single-nucleon kinetic energy and two-nucleon effective interaction, and U is an average nuclear potential with central and spin–orbit parts. The part H' is treated as a perturbation.

The zero-order hamiltonian H^0 corresponds to independent particle motion in the average potential U. The zero-order eigenstates $\Psi^{(0)}$ are Slater determinants built from single-nucleon wave functions φ_{nljm}†. The zero-order energies $E^{(0)}$, which are sums of single-nucleon energies ε_{nlj}, depend only on the configuration quantum numbers $n_i l_i j_i$ and are highly degenerate:

$$\varphi_{nljm}(\mathbf{r},\sigma) = \frac{R_{nlj}(r)}{r} \cdot Y_{ljm}(\theta,\varphi,\sigma) \tag{4}$$

$$h\varphi_{nljm} = \varepsilon_{nlj}\varphi_{nljm} \tag{5}$$

$$\Psi^{(0)} = \frac{1}{\sqrt{A!}} \det |\varphi_{n_1 l_1 j_1 m_1} \cdots \varphi_{n_A l_A j_A m_A}| \tag{6}$$

$$E^{(0)} = \sum \varepsilon_{nlj}. \tag{7}$$

In the ground state the nucleons occupy the lowest single-particle levels.

In the first-order approximation one diagonalizes H' separately in each degenerate configuration space. The eigenvector corresponding to the lowest eigenvalue $E^{(1)}$ is the ground state and its energy up to first order in the perturbation, $E^{(0)} + E^{(1)}$, is given by

$$E = \left\langle \sum t_i + \sum V_{ij} \right\rangle_{g.s.} = \langle H \rangle_{g.s.}. \tag{8}$$

Second- and higher-order terms in the perturbation expansion introduce configuration mixing.

When the valence major shell has several single-nucleon levels close in energy the low-lying many-nucleon configurations overlap and are largely mixed. In this quasi-degenerate case standard perturbation theory breaks down and the nuclear states and energies are obtained by diagonalizing H in the multi-configuration space of low-lying configurations. Eq. (8) holds in this case too. There has recently been much progress in large-basis shell model calculations.[1–4]

Interaction with high-lying configurations involving excitations across major shell gaps can be treated by perturbation theory.

2. Multipole expansions of the interaction for pure configurations

A charge-independent scalar two-body interaction in the J-T scheme can be written as $V_{12} = V^0_{12} + V^1_{12}(\mathbf{t}_1 \cdot \mathbf{t}_2)$, where V^0 and V^1 depend on the space and

† These are the single-nucleon wave functions used in the n–p scheme, where neutrons and protons are treated separately, as in non-diagonal shell regions and on the boundaries of diagonal regions. Eq. (6) applies to neutrons and to protons separately. Inside diagonal regions the J-T scheme with single-nucleon wave functions $\varphi_{nljm_t m_j}$ is more appropriate.

spin variables of the two nucleons. Each of them has in a given $j_1 j_2$ configuration* a multipole expansion

$$V_{12}^0 = \sum_k A^k(j_1 j_2)\left(\mathbf{u}_1^{(k)} \cdot \mathbf{u}_2^{(k)}\right) \tag{9}$$

$$V_{12}^1 = \sum_k B^k(j_1 j_2)\left(\mathbf{u}_1^{(k)} \cdot \mathbf{u}_2^{(k)}\right)(\mathbf{t}_1 \cdot \mathbf{t}_2) \tag{10}$$

where the coefficients $A^k(j_1 j_2)$ and $B^k(j_1 j_2)$ are products of radial and angular integrals depending on the subshell quantum numbers and the $\mathbf{u}^{(k)}$ are irreducible unit tensor operators of degree k with respect to $\mathbf{J}$, defined by irreducible matrix elements

$$\langle j \| u^{(k)} \| j' \rangle = \delta_{jj'}. \tag{11}$$

The two expansions can be unified to

$$V_{12} = \sum_{\kappa k} A^{\kappa k}(j_1 j_2)\left(\mathbf{u}_1^{(\kappa k)} \cdot \mathbf{u}_2^{(\kappa k)}\right) \tag{12}$$

with doubly irreducible unit tensors $\mathbf{u}^{(\kappa k)}$, of respective degrees κ (0 or 1) and k with respect to $\mathbf{T}$ and $\mathbf{J}$, defined by

$$\left\langle \frac{1}{2}\, j \,\| u^{(\kappa k)} \|\, \frac{1}{2}\, j' \right\rangle = \delta_{jj'} \tag{13}$$

and radial integrals $A^{\kappa k}$ given by

$$A^{0k}(j_1 j_2) = 2A^k(j_1 j_2) \tag{14}$$

$$A^{1k}(j_1 j_2) = \frac{3}{2}\, B^k(j_1 j_2). \tag{15}$$

For identical nucleons in two subshells $(\mathbf{t}_1 \cdot \mathbf{t}_2) = \frac{1}{4}$ and the multipole expansion (12) reduces to

$$V_{12}^{T=1} = \sum_k A^{k,T=1}(j_1 j_2)\left(\mathbf{u}_1^{(k)} \cdot \mathbf{u}_2^{(k)}\right) \tag{16}$$

with

$$A^{k,T=1}(j_1 j_2) = \frac{1}{2}\, A^{0k}(j_1 j_2) + \frac{1}{6}\, A^{1k}(j_1 j_2). \tag{17}$$

For a neutron and a proton subshell in the n–p scheme $(\mathbf{t}_1 \cdot \mathbf{t}_2) = -\frac{1}{4}$ and eq. (12) reduces to

---

*When talking about configurations we use a shortened notation j for the complete set of subshell quantum numbers nlj.

$$V_{12}^{np} = \sum_k A^{k,np}(j_1 j_2)\left(\mathbf{u}_1^{(k)} \cdot \mathbf{u}_2^{(k)}\right) \tag{18}$$

with

$$A^{k,np}(j_1 j_2) = \frac{1}{2} A^{0k}(j_1 j_2) - \frac{1}{6} A^{1k}(j_1 j_2). \tag{19}$$

Let n and p denote the respective numbers of neutrons and protons in a given subshell and $a = n + p$. For $j_1 = j_2$, using (12) and the identity

$$\sum_{i<j}\left(\mathbf{u}_i^{(\kappa k)} \cdot \mathbf{u}_j^{(\kappa k)}\right) = \frac{1}{2}\left[\left(\mathbf{U}^{(\kappa k)} \cdot \mathbf{U}^{(\kappa k)}\right) - \sum_i \mathbf{u}_i^{(\kappa k)^2}\right] \tag{20}$$

where $\mathbf{U}^{(\kappa k)} = \sum \mathbf{u}_i^{(\kappa k)}$ is the total $\mathbf{U}^{(\kappa k)}$ tensor of the subshell, one obtains for the energy matrix of the j^a configuration

$$\left\langle j^a xTJ \left| \sum_{i<j} V_{ij} \right| j^a x'TJ \right\rangle = \frac{1}{2}\sum_{\kappa k} A^{\kappa k}(jj) \times (-1)^{T+J} \times$$
$$\left[\sum_{x''T''J''}\left((-1)^{-T''-J''} \times \frac{\langle j^a xTJ \| U^{(\kappa k)} \| j^a x''T''J''\rangle\langle j^a x''T''J'' \| U^{(\kappa k)} \| j^a x'TJ\rangle}{(2T+1)(2J+1)}\right) - \frac{a}{4j+2}\delta xx'\right] \tag{21}$$

where x denotes quantum numbers needed additionally to $j^a TJM_T M_J$ for a complete specification of the states. The last term in the square brackets comes from the single-particle scalars $\mathbf{u}^{(\kappa k)^2}$ in eq. (20).

For the mutual interaction between two subshells one obtains

$$\left\langle j_1^{a_1}(x_1T_1J_1)j_2^{a_2}(x_2T_2J_2)TJ \left| \sum_{i_1 i_2} V_{i_1 i_2} \right| j_1^{a_1}(x_1'T_1'J_1')j_2^{a_2}(x_2'T_2'J_2')TJ \right\rangle =$$
$$(-1)^{T_1'+J_1'+T_2+J_2+T+J} \times \sqrt{(2T+1)(2J+1)} \times \left[\sum_{\kappa k} A^{\kappa k}(j_1 j_2) \times \right.$$
$$\left\langle j_1^{a_1} x_1 T_1 J_1 \left\| U_1^{(\kappa k)} \right\| j_1^{a_1} x_1' T_1' J_1' \right\rangle \left\langle j_2^{a_2} x_2 T_2 J_2 \left\| U_2^{(\kappa k)} \right\| j_2^{a_2} x_2' T_2' J_2' \right\rangle \times$$
$$\left.\begin{Bmatrix} T_1 & T_2 & T \\ T_2' & T_1' & \kappa \end{Bmatrix}\begin{Bmatrix} J_1 & J_2 & J \\ J_2' & J_1' & k \end{Bmatrix}\right]. \tag{22}$$

Analog expressions in the n–p scheme are obtained from (21) and (22) by suppressing the T-related parts, changing the $A^{\kappa k}$ to $A^{k,T=1}$ and $4j+2$ to $2j+1$ in eq. (21), and changing the $A^{\kappa k}$ to $A^{k,T=1}$ or $A^{k,np}$, as appropriate, in eq. (22).

3. Multipole expansions in multi-configuration spaces

In a multi-configuration space a charge-independent scalar interaction between two particles in two respective major shells has a multipole expansion[65,66,8] analogous to (12):

$$V_{12} = \sum_{\kappa k} \sum_{\alpha\beta\gamma\delta} A^{\kappa k}(j_\alpha j_\beta j_\gamma j_\delta)\left(\mathbf{u}_1^{(\kappa k)}(j_\alpha j_\gamma) \cdot \mathbf{u}_2^{(\kappa k)}(j_\beta j_\delta)\right) \tag{23}$$

with radial integrals $A^{\kappa k}(j_\alpha j_\beta j_\gamma j_\delta)$ and doubly irreducible generalized unit tensors $\mathbf{u}^{(\kappa k)}(j_\alpha j_\gamma)$ equivalent to coupled products of creation and annihilation operators in second quantization:

$$\left\langle \frac{1}{2} j \| u^{(\kappa k)}(j_\alpha j_\gamma) \| \frac{1}{2} j' \right\rangle = \delta_{jj_\alpha} \delta_{j' j_\gamma}. \tag{24}$$

The summation indices α, γ run independently over all subshells of the first major shell, and β, δ cover those of the second. The $\mathbf{u}^{(\kappa k)}$ are not hermitian but satisfy

$$\mathbf{u}^{(\kappa k)}(j_\alpha j_\gamma)^\dagger = (-1)^{j_\gamma - j_\alpha} \mathbf{u}^{(\kappa k)}(j_\gamma j_\alpha) \tag{25}$$

where † denotes hermitian conjugation. The radial integrals satisfy

$$A^{\kappa k}(j_\alpha j_\beta j_\gamma j_\delta) = (-1)^{j_\gamma - j_\alpha + j_\delta - j_\beta} A^{\kappa k}(j_\gamma j_\delta j_\alpha j_\beta)^* \tag{26}$$

where $*$ denotes complex conjugation.

For identical nucleons in the two major shells (23) reduces to

$$V_{12}^{T=1} = \sum_{k} \sum_{\alpha\beta\gamma\delta} A^{k,T=1}(j_\alpha j_\beta j_\gamma j_\delta)\left(\mathbf{u}_1^{(k)}(j_\alpha j_\gamma) \cdot \mathbf{u}_2^{(k)}(j_\beta j_\delta)\right) \tag{27}$$

where

$$\langle j \| u^{(k)}(j_\alpha j_\gamma) \| j' \rangle = \delta_{jj_\alpha} \delta_{j' j_\gamma} \tag{28}$$

and

$$A^{k,T=1}(j_\alpha j_\beta j_\gamma j_\delta) = \frac{1}{2} A^{0k}(j_\alpha j_\beta j_\gamma j_\delta) + \frac{1}{6} A^{1k}(j_\alpha j_\beta j_\gamma j_\delta). \tag{29}$$

For a neutron and a proton major shell in the *n–p* scheme (23) reduces to

$$V_{12}^{np} = \sum_{k} \sum_{\alpha\beta\gamma\delta} A^{k,np}(j_\alpha j_\beta j_\gamma j_\delta)\left(\mathbf{u}_1^{(k)}(j_\alpha j_\gamma) \cdot \mathbf{u}_2^{(k)}(j_\beta j_\delta)\right) \tag{30}$$

with

$$A^{k,np}(j_\alpha j_\beta j_\gamma j_\delta) = \frac{1}{2} A^{0k}(j_\alpha j_\beta j_\gamma j_\delta) - \frac{1}{6} A^{1k}(j_\alpha j_\beta j_\gamma j_\delta) \tag{31}$$

Eqs. (27) and (30) are the multi-configuration space analogs of (16) and (18).

The analog of eq. (21) when both particles are in the same major shell is

$$\left\langle zTJ \left| \sum_{i<j} V_{ij} \right| z'TJ \right\rangle = \frac{1}{2} \sum_{\kappa k} \sum_{\alpha\beta\gamma\delta} A^{\kappa k}(j_\alpha j_\beta j_\gamma j_\delta)(-1)^{T+J} \left[\sum_{z''T''J''} (-1)^{-T''-J''} \times \frac{\langle zTJ \| U^{(\kappa k)}(\alpha\gamma) \| z''T''J'' \rangle \langle z''T''J'' \| U^{(\kappa k)}(\beta\delta) \| z'TJ \rangle}{(2T+1)(2J+1)} - (-1)^{j_\alpha - j_\beta} \times \frac{a_\alpha}{4j_\alpha + 2} \times \delta_{\alpha\delta}\, \delta_{\beta\gamma}\, \delta_{zz'} \right] \tag{32}$$

where $\mathbf{U}^{(\kappa k)}(\alpha\gamma) \equiv \mathbf{U}^{(\kappa k)}(j_\alpha j_\gamma) = \sum \mathbf{u}_i^{(\kappa k)}(j_\alpha j_\gamma)$ is the total $\mathbf{U}^{(\kappa k)}(j_\alpha j_\gamma)$ tensor of the valence nucleons. The symbol z is introduced for the sake of compactness to denote the set of multi-subshell configuration quantum numbers $(j_1^{a_1} j_2^{a_2} \ldots)$ plus quantum numbers needed additionally to the configuration and to TJM_TM_J for a complete specification of the states. The last term in the square brackets comes from single-particle double scalars $\mathbf{u}_i^{(\kappa k)}(j_\alpha j_\gamma) \cdot \mathbf{u}_i^{(\kappa k)}(j_\beta j_\delta)$, as in (21).

The analog of eq. (22) is

$$\left\langle z_1 T_1 J_1, z_2 T_2 J_2, TJ \left| \sum_{i_1 i_2} V_{i_1 i_2} \right| z_1' T_1' J_1', z_2' T_2' J_2', TJ \right\rangle =$$
$$(-1)^{T_1' + J_1' + T_2 + J_2 + T + J} \times \sqrt{(2T+1)(2J+1)} \times \left[\sum_{\kappa k} \sum_{\alpha\beta\gamma\delta} A^{\kappa k}(j_\alpha j_\beta j_\gamma j_\delta) \times \right.$$
$$\left\langle z_1 T_1 J_1 \left\| U_1^{(\kappa k)}(\alpha\gamma) \right\| z_1' T_1' J_1' \right\rangle \left\langle z_2 T_2 J_2 \left\| U_2^{(\kappa k)}(\beta\delta) \right\| z_2' T_2' J_2' \right\rangle \times$$
$$\left. \begin{Bmatrix} T_1 & T_2 & T \\ T_2' & T_1' & \kappa \end{Bmatrix} \begin{Bmatrix} J_1 & J_2 & J \\ J_2' & J_1' & k \end{Bmatrix} \right]. \tag{33}$$

The analog equations for major shells in the n–p scheme are obtained from (32) and (33) as described after eq. (22).

4. *Particle–hole symmetries*

In the J–T scheme there is a one-to-one correspondence called particle–hole (p–h) conjugation between states $|j^a xTJM_TM_J\rangle$ of a nucleons in a j-subshell and states of $4j+2-a$ nucleons called a-hole states and denoted $|j^{-a}xTJ-M_T-M_J\rangle$. Conjugate states have the same xTJ and opposite M_TM_J values, and they multiply each other in the expansion of the unique antisymmetric $T=J=0$ state of the completely full subshell as a sum of coupled products of conjugate states[67]

$$|j^{4j+2}00\rangle = \binom{4j+2}{a}^{-1/2} \sum_{xTJ} \sqrt{(2T+1)(2J+1)} [|j^a xTJ\rangle \times |j^{-a} xTJ\rangle]^{T=0,\, J=0}. \tag{34}$$

Eq. (34) is a possible definition of p–h conjugation.

For identical nucleons in the n–p scheme there are analog relations, with states of $2j+1-n$ particles called n-hole states, the T quantum numbers suppressed, and $(4j+2)$ changed to $(2j+1)$ in eq. (34).

In a multi-configuration space there is similarly a one-to-one correspondence between particle states $|j_1^{a_1}(x_1T_1J_1)j_2^{a_2}(x_2T_2J_2)\ldots xTJM_TM_J\rangle$ and hole states $|j_1^{-a_1}(x_1T_1J_1)j_2^{-a_2}(x_2T_2J_2)\ldots xTJ-M_T-M_J\rangle$, where corresponding subshells are in subshell p–h conjugate states and corresponding subshell angular momenta and isospins are coupled in the same way. The unique antisymmetric $T=J=0$ state of the completely full major shell is here too a sum of coupled products of major shell particle–hole conjugate states as in (34).[8] Analog relations hold for a multi-configuration space of identical nucleons.

From eq. (34) and its analogs one obtains[67] relations between the reduced matrix elements of irreducible tensors in p–h conjugate configurations. For (34) the relation is

$$\langle j^{-a}xTJ\|T^{(\kappa k)}\|j^{-a}x'T'J'\rangle=(-1)^{\kappa+k+1}\langle j^{a}xTJ\|T^{(\kappa k)\dagger}\|j^{a}x'T'J'\rangle^{*} \qquad (\kappa,k)\neq(0,0) \qquad (35)$$

with $\mathbf{T}^{(\kappa k)}=\sum \mathbf{t}_i^{(\kappa k)}$. For a multi-configuration mixed major shell the corresponding relation is the same:[66,8]

$$\langle[-a]xTJ\|T^{(\kappa k)}\|[-a']x'T'J'\rangle=(-1)^{\kappa+k+1}\langle[a]xTJ\|T^{(\kappa k)\dagger}\|[a']x'T'J'\rangle^{*} \qquad (\kappa,k)\neq(0,0) \qquad (36)$$

where $[a]$ is a shortened notation for the configuration quantum number $(j_1^{a_1}j_2^{a_2}\ldots)$ and $[-a]$ denotes the hole configuration conjugate to $[a]$.

The analog relations for identical nucleons without isospin are obtained from (35) and (36) by suppressing κ and the T quantum numbers.

B. The nuclear ground state

1. *The empirical effective interaction*

The nuclear ground state and its energy (8) are essentially determined by the effective two-body interaction V.

Schiffer and collaborators[68,69] (see also[70,71]) made a systematic study of empirical effective two-nucleon interactions derived from two-nucleon spectra and ground state masses. They divided the empirical matrix elements $V_{j_1j_2J}$ by their centroid energy in order to eliminate radial-overlap and A-dependence effects, and plotted the normalized matrix elements as functions of the quasi-classical angle θ_{12} between $\mathbf{j}_1$ and $\mathbf{j}_2$, separately for all j^2 configurations and for all $j_1\neq j_2$ configurations.

Fig. 14 shows Schiffer's plots. The data in each plot nicely divide into two groups. For the j^2 configuration these are the $T=1$ (even J) and $T=0$ (odd J) groups. For $T=1$ there is only one strongly bound state with $\theta_{12}=180°$ $(J=0)$. On the other hand, for $T=0$ there is an inverted U-shape, with both the aligned $J=2j$ and the quasi-paired $J=1$ states strongly bound, and the

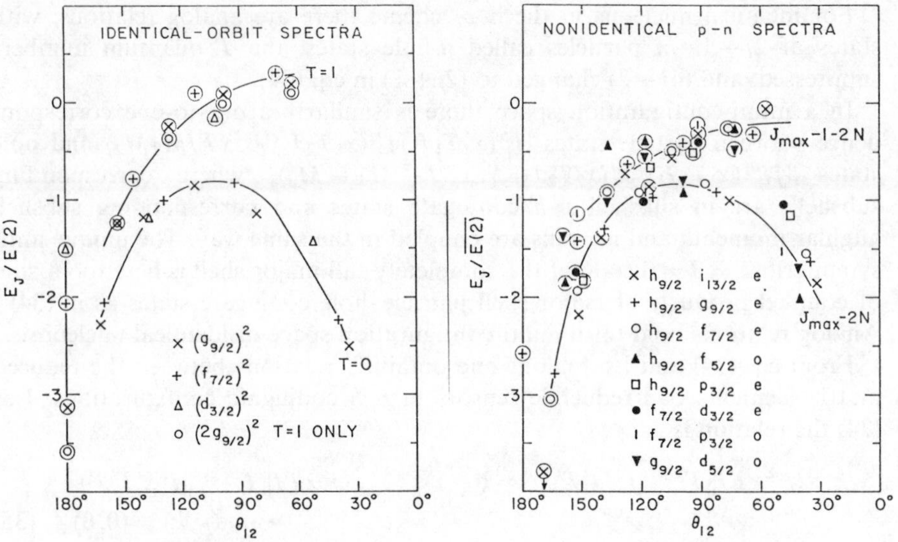

Fig. 14. Normalized empirical two-body matrix elements for identical orbits (left) and for non-identical orbits (right). Taken from ref. 68.

average $T = 0$ interaction is about an order of magnitude more binding than for $T = 1$.

For $j_1 \neq j_2$ in the n–p scheme the situation is similar: for odd $j_1 + j_2 + J$ the antialigned $J = |j_1 - j_2|$ state is the only strongly bound one, whereas for even $j_1 + j_2 + J$ there is the inverted U-shape with both the aligned $J = j_1 + j_2$ and the quasi-antialigned $J = |j_1 - j_2| + 1$ states strongly bound.

Finally, in the cases where $V_{j_1 j_2 TJ}$ matrix elements are available, the plot for $T = 0$ is similar to that in the n–p scheme, whereas the $V_{j_1 j_2 1J}$ are often unbinding.

These qualitative features are displayed as well by effective interactions used in multi-configuration shell model calculations.[72–75] They reflect the overall short-range attractive nature of the nuclear interaction. The smooth lines drawn through the n–p data points in Fig. 14 are from zero-range (delta-force) calculations.

2. The seniority scheme for identical nucleons

Since $|j^2 J = 0\rangle$ is the only strongly bound $T = 1$ state of the j^2 configuration one expects the ground state of the j^n configuration of identical nucleons to contain the highest possible number of $J_{12} = 0$ nucleon pairs. The amount of pairing is well defined in the seniority scheme, where the scalar two-body seniority operator q_{12} (first introduced by Racah[76] for electrons in LS-coupling) is defined by[77]

$$\langle j^2 JM|q|j^2 J'M'\rangle = (2j+1)\delta_{JJ'}\delta_{MM'}\delta_{J0}. \qquad (37)$$

A $|j^n vxJM\rangle$ state of seniority v is an antisymmetrized product of $\frac{1}{2}(n-v)$ $|j^2J=0\rangle$ pair states and a $|j^v xJM\rangle$ state of the v unpaired nucleons, of which components of lower seniority have been projected out.

For zero-range forces the identical-nucleon seniority is a good quantum number,[77] and the ground state has lowest seniority, namely $v=0$ for even n and $v=1$ for odd n. The empirical identical-nucleon effective interaction likewise seems[78–80] to satisfy the conditions under which the seniority is a good quantum number.[81]

For a configuration $j_n^n j_p^p$ the identical-nucleon energy $\sum V_{nn} + \sum V_{pp}$ is lowest when both neutrons and protons are in lowest-seniority states. However, in this case when either n or p is even (as in even–even and odd-mass nuclei) the corresponding J and hence the higher-multipole part of the neutron–proton interaction $\sum V_{np}$ vanishes (n–p analog of eq. (22)) and only the monopole ($k=0$) part remains. On the other hand, the tight neutron–proton binding in the aligned and antialigned states of the $j_1 j_2$ configuration (Fig. 14) is due to the higher multipoles, as the monopole interaction is J independent. Consequently, $\sum V_{np}$ can be made lower by relaxing the requirement of lowest seniority for identical nucleons and thereby letting the higher-multipole neutron–proton interaction contribute to the binding.

When the neutron and proton subshells are different the intra-subshell pairing interactions may be expected to win over and establish a lowest-seniority ground state. This is the basic assumption of the single-particle model of Mayer and Jensen.[38] On the other hand, for strongly overlapping neutron and proton orbitals the identical-nucleon pairing and neutron–proton aligning and antialigning interactions might have comparable magnitudes, resulting in seniority mixing. The more so, presumably, the higher the number np of interacting neutron–proton pairs is compared to the total number $a=n+p$ of valence nucleons.

The following relations are satisfied by irreducible tensors in the seniority scheme:[6]

$$\langle j^n xvJ \| T^{(k)} \| j^n x'v'J' \rangle = \langle j^v xvJ \| T^{(k)} \| j^v x'vJ' \rangle \delta_{vv'} \qquad k \text{ odd} \tag{38}$$

$$\langle j^n xvJ \| T^{(0)} \| j^n x'v'J' \rangle = n\sqrt{\frac{2J+1}{2j+1}} \langle j \| t^{(0)} \| j \rangle \delta_{xx'} \delta_{vv'} \delta_{JJ'} \tag{39}$$

3. *Seniority with isospin*

In a mixed subshell with charge-independent interactions the total isospin T is well defined. Since the $T=0$ effective interaction is lower than that for $T=1$ one expects a ground state with a high number of isopaired $T_{12}=0$ neutron–proton pairs, resulting in a low T value. As partly mentioned in section II.C, the lowest possible value, $T_{g.s.} = |T_z|$, is the empirical rule in all nuclei except heavier $Z=N$ odd–odd ones with $T_{g.s.} = 1$ (see below).

For mixed j^a configurations with isospin there is a J–T seniority scheme[82] which is a generalization of the one for identical nucleons. The seniority operator is the same:

$$\langle j^2 TJM_TM_J|q|j^2 T'J'M'_TM'_J\rangle = (2j+1)\delta_{TT'}\delta_{JJ'}\delta_{M_TM'_T}\delta_{M_JM'_J}\delta_{J0}. \qquad (40)$$

A state $|j^a xyvtTJ\rangle$ with seniority v and reduced isospin t (and additional quantum numbers x and y when required) is an antisymmetrized isospin-coupled product of $\frac{1}{2}(a-v)$ $|j^2 T=1\, J=0\rangle$ pair states and a $|j^v xT=tJ\rangle$ state of v unpaired nucleons, of which components of lower seniority have been projected out. For identical nucleons the T and t are superfluous, and J–T seniority is the same as identical-nucleon seniority.

For mixed subshells with isospin $T<\frac{a}{2}$ the seniority and reduced isospin are not exact quantum numbers whatever the range of the interaction. However, for zero-range forces they are approximately so,[83] and the ground state corresponds to the lowest possible seniority and isospin.[83,84] For even–even and odd-a subshells these are $T=|T_z|$ and $v=0$ and 1 respectively. On the other hand, for an odd–odd subshell the state with $v=0, T=|T_z|$ is forbidden by the Pauli Principle, and there is competition between $v=0, T=|T_z|+1$ and $v=2, T=|T_z|$. Moreover, in the latter case both $t=1$, even-J and $t=0$, odd-J values are allowed when $p\neq n$, whereas for $p=n$ the $t=1$, even-J possibility is excluded. These rules are obvious from the corresponding Young diagrams.

These rules of the seniority scheme are in agreement with experiment: all $Z\neq N$ odd–odd nuclei with a mixed valence subshell have $T_{g.s.}=|T_z|, J_{g.s.}$ even. For $Z=N$, odd–odd nuclei in the $1p$ and $2s1d$ shells have as a rule $T_{g.s.}=0, J_{g.s.}$ odd, and in the $1f_{7/2}$ shell and beyond they have $T_{g.s.}=1, J_{g.s.}=0$. (Exceptions are ^{34}Cl with $T_{g.s.}=1, J_{g.s.}=0$ and a 146 keV level with $T=0, J=3$, and ^{58}Cu with $T_{g.s.}=0, J_{g.s.}=1$ and a 203 keV level with $T=1, J=0$.) The competition between pairing and isopairing is discussed in ref. 85.

In spite of its general agreement with experimental ground state T and J values, the seniority scheme for mixed subshells is considerably violated. Presumably this has to do with the fact that for the empirical interaction the relative strengths of the V_{j^2J} with high odd-J values as compared to V_{j^20} are higher than for zero-range forces.[68] In particular, the ground state has mixed seniority and less pairing than for lowest seniority. Its largest component, though, is of lowest seniority.[86]

The violation of seniority in mixed subshells was found empirically in early shell model calculations.[87–89] Criteria for studying seniority-conserving (symplectic symmetry) properties of a given effective interaction are developed in refs. 81, 86.

The following relations hold:[6]

$$\langle j^a vtTJ \| T^{(0k)} \| j^a v't'T'J' \rangle = \sqrt{\frac{2T+1}{2t+1}} \langle j^v vT = tJ \| T^{(0k)} \| j^v vT = tJ' \rangle \delta_{vv'} \delta_{tt'} \delta_{TT'} \qquad k \text{ odd} \tag{41}$$

$$\langle j^a xyvtTJ \| T^{(00)} \| j^a x'y'v't'T'J' \rangle = a\sqrt{\frac{(2J+1)(2T+1)}{4j+2}} \langle j \| t^{(00)} \| j \rangle \delta_{xx'} \delta_{yy'} \delta_{vv'} \delta_{tt'} \delta_{TT'} \delta_{JJ'}. \tag{42}$$

4. Configuration interaction

When the distance between single-particle levels is comparable to their two-body interactions they fill simultaneously.[90,91] When the magnitude of the non-diagonal interaction matrix element connecting low-lying levels of neighbouring configurations is comparable to their distance apart there is configuration mixing. As a matter of fact, this is the situation in all heavier non-doubly-magic nuclei, washing out most submagic number effects as discussed in sections II.D.2 and 3. We consider configuration interaction in terms of the n–p scheme.

When there are n identical valence nucleons, as in semi-magic nuclei, the ground state is expected[92,93] to be a mixture of configurations in states of lowest overall seniority*, namely 0 and 1 respectively for even and for odd n. Such configurations, which differ in the distribution of $|j^2 J = 0\rangle$ identical-nucleon pairs in the various subshells, are obtained from one another by pair excitations between subshells. Their mixing is caused by large non-diagonal $\langle j^2 J = 0 | V | j'^2 J = 0 \rangle$ matrix elements. This kind of configuration mixing is called pairing correlations. (A recent review is given in ref. 94.)

In non-semi-magic nuclei the identical-nucleon interactions $\sum V_{nn}$ and $\sum V_{pp}$ each produce their own pairing correlations. However, the neutron–proton interactions $\sum V_{np}$ cannot contribute to the lowering of the energy by pairing correlations, as they are of single-nucleon character with respect to the separate neutron and proton variables.

On the other hand, $\sum V_{np}$ might have large non-diagonal matrix elements between low-lying configurations obtained from one another by single-nucleon excitations between subshells,[95,96] with identical-nucleon pair breaking and neutron–proton alignment and antialignment. The resulting configuration mixing is called neutron–proton correlations.

In nuclear structure both pairing and neutron–proton correlations play an essential role,[97,98] in particular in transitions from shell structure to collectivity. When both n and p (or the corresponding hole numbers after midshell) increase the number of low configurations which can mix increases fast (which makes it necessary to use larger vector spaces in shell model calculations of mid-major-shell nuclei as compared to semi-magic ones[1,2]). The relative

* By overall seniority one means the sum of seniorities of all subshells.

importance of neutron–proton correlations and their contribution to the binding energy presumably increase with np for a given a, reaching a maximum for $n \approx p$.

The high degree of configuration mixing in mixed major shells as compared to semi-magic nuclei was found empirically in early shell model calculations.[99,100] The importance of pairing[92,93,101] and of neutron–proton[95,96] correlations was respectively realized during the 1950s and early 1960s.

5. *Deformation*

Pure-configuration low-seniority states, as well as states obtained from them by pairing correlations, are essentially spherically symmetric. Thus the only plausible way to achieve deformation in the shell model is by configuration mixing with appropriate neutron–proton correlations. Consequently deformed nuclei are expected to be found inside major shell regions, whereas semi-magic nuclei should be spherical. This view has been emphasized since the early 1960s.[95] The crucial role of the neutron–proton interaction in producing deformation is presently common wisdom.[102–104a]

The above expectation is borne out by the facts. Fig. 15 shows a plot of the deformation parameter β_2 deduced from $B(E2)$ values of even–even nuclei. In the Figure semi-magic nuclei of either neutrons or protons correspond to local

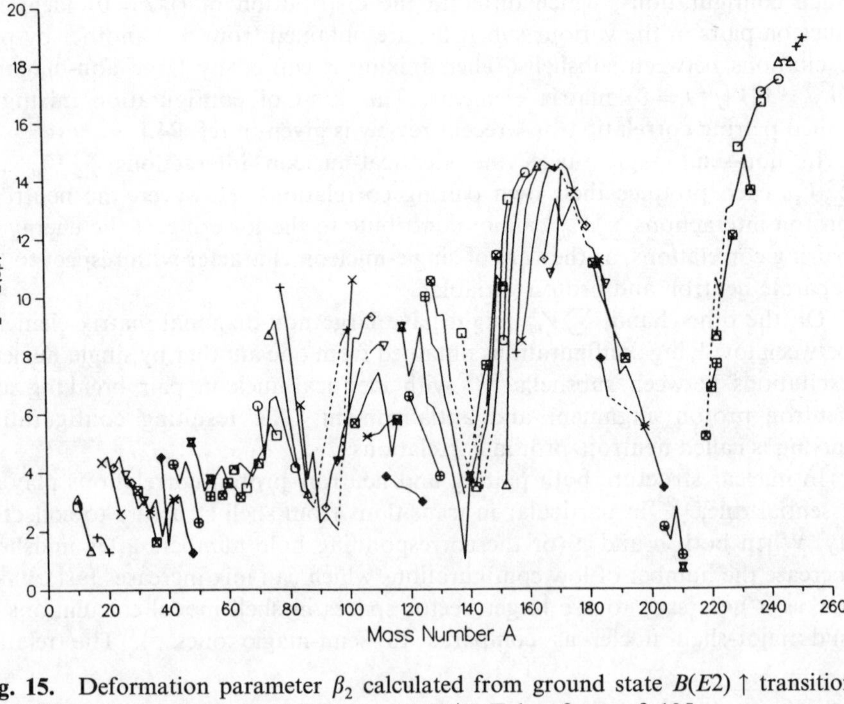

Fig. 15. Deformation parameter β_2 calculated from ground state $B(E2)\uparrow$ transition probabilities in even–even nuclei. $\beta_{2(s.p.)} = 1.59/Z$. Taken from ref. 105.

minima, and the deformation increases smoothly from the boundaries of shell regions towards saturation at their centres (see also Fig. 7 of ref. 106 and refs. 107–109).

Fig. 16 shows plots of a second fingerprint of deformation, the ratio $E(4_1^+)/E(2_1^+)$ of excitation energies of even–even nuclei in the heavy Zr and rare earth regions.

In the upper left part of the Figure the highest values, approaching the symmetric rigid rotor value of 3.33, occur for ^{100}Sr and ^{102}Zr with 62 neutrons,

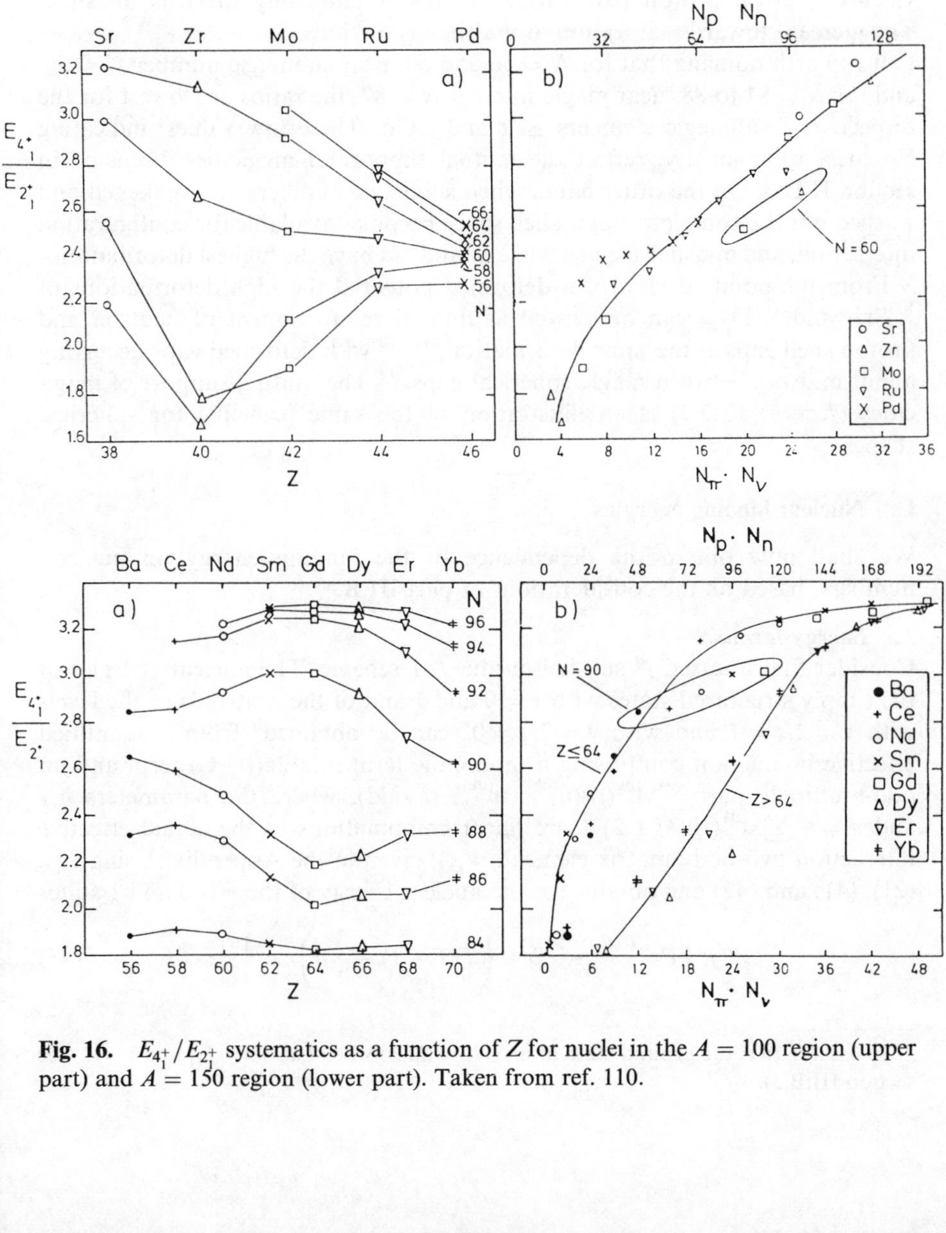

Fig. 16. $E_{4_1^+}/E_{2_1^+}$ systematics as a function of Z for nuclei in the $A = 100$ region (upper part) and $A = 150$ region (lower part). Taken from ref. 110.

and from the trend of the lines the ratios are expected to be higher for the $N = 64$ nuclei ^{102}Sr and ^{104}Zr and (perhaps even somewhat higher) for ^{104}Sr and ^{106}Zr with $N = 66$. These nuclei are nearest neighbours of $^{105}_{39}Y_{66}$, the central nucleus of the major shell region $50 \leq N \leq 82$, $28 \leq Z \leq 50$.

Likewise, in the lower left part the highest values, very close to 3.33, occur for ^{158}Sm, ^{160}Gd and ^{162}Dy with $N = 96$, and just below them their $N = 94$ isotopes. In these nuclei saturation has practically been achieved, and the ratio stays at about the same values up to ^{182}Hf and ^{184}W with $N = 110$, before dropping again for larger N and Z values (Fig. 18 of ref. 110).

On the right the same data are plotted as functions of the number N_pN_n of valence neutron–proton pairs, which increases uniformly towards midshell. The increase towards saturation at midshell is obvious.

It is worth noticing that for $N = 56$ and 58, near submagic number $N = 56$, and for $N = 84$ to 88, near magic number $N = 82$, the ratios are lowest for the respective Z-submagic elements $_{40}$Zr and $_{64}$Gd. These low values, indicating closeness to sphericity, reflect the mutual support of magicities discussed in section II.D.3. On the other hand, when submagic numbers are weakened and washed out the complete major shell space becomes available for configuration interaction, and midshell elements like Sr and Gd have the highest deformations.

From the point of view of a deformed potential the high deformations of $^{100}_{38}Sr_{62}$ and $^{162}_{66}Dy_{96}$ can be viewed as mutual reinforcement of neutron and proton shell gaps at the same deformation,[111,112] with deformed gaps occurring about midway between magic spherical gaps.[113] The mutual support of magicities (section II.D.3) is an illustration of the same principle for spherical shapes.

C. Nuclear binding energies

We shall now obtain the dependence of the binding energy on nucleon numbers, based on the considerations of part III.B.

1. *Energy formulas*

Consider first a mixed j^a subshell in the J–T scheme. The nuclear energies of the simply structured states with $v = 0$ and 1 and of the centroids of the levels with $v = 2, t = 1$ and with $v = 2, t = 0$* can be obtained[6] from a simplified effective interaction comprising a monopole term b, a $2\varepsilon(\mathbf{t}_1 \cdot \mathbf{t}_2)$ term and an odd-multipole part $\sum A^{0k}(jj)(\mathbf{u}_1^{(0k)} \cdot \mathbf{u}_2^{(0k)})$ (k odd), where the parameters b, ε and $\pi = -\sum A^{0k}(jj)(4j+2)^{-1}$ are linear combinations of the actual effective interaction two-body matrix elements V_{j^2J}, given in the Appendix. Using eqs. (21), (41) and (42) one obtains for the nuclear energy of the $v = 0$ and 1 states

$$W(j^aT) = \frac{a(a-1)}{2}b + \left(T(T+1) - \frac{3}{4}a\right)\varepsilon + \left[\frac{a}{2}\right]\pi \tag{43}$$

* We assume that the ground state of an odd–odd subshell has $T = |T_z|$ and $v = 2$ (see section III.B.3).

where the coefficient $[\frac{a}{2}]$ equals the number of $|j^2 J = 0\rangle$ pairs, namely $\frac{a}{2}$ for even a and $\frac{(a-1)}{2}$ for odd a. The ε and π are respectively called symmetry and pairing energies.

For the centroids of $v = 2$ levels one obtains

$$\overline{W(j^a T)} = \frac{a(a-1)}{2} b + \left(T(T+1) - \frac{3}{4} a \right) \varepsilon + \left[\delta_{t,1} \left(\frac{a}{2} - \frac{2j+1}{2j+2} \right) + \delta_{t,0} \left(\frac{a}{2} - \frac{2j+3}{2j+2} \right) \right] \pi. \quad (44)$$

Eqs. (43) and (44) are special cases of a general formula for centroids of levels with the same v and t, first derived by group theoretical methods (see also ref. 6a).[84] A derivation using fractional parentage coefficients without group theory is given in ref. 6.

For the lowest $v = 2$ level with a given t, which is lower than the centroid when there is more than one level in the multiplet, one can write

$$W(j^a T) = \frac{a(a-1)}{2} b + \left(T(T+1) - \frac{3}{4} a \right) \varepsilon + \frac{a}{2} \pi + \delta_{t,1} \kappa + \delta_{t,0} \lambda \quad (45)$$

with

$$\kappa = -\frac{2j+1}{2j+2} \pi + \kappa'$$

$$\lambda = -\frac{2j+3}{2j+2} \pi + \lambda' \quad (46)$$

where the κ' and λ' denote the corresponding lowering of the ground state energy as compared to the centroid. They depend in a complicated way on J, T and a.[114, 115] For short-range forces, though, they are expected to be small compared to the inter-centroid distance of $(j+1)^{-1} \pi$. In the following we ignore the J, T and a dependence of κ and λ.

Eqs. (43) and (45) can be unified to

$$W(j^a T) = \frac{a(a-1)}{2} b + \left(T(T+1) - \frac{3}{4} a \right) \varepsilon + \frac{a}{2} \pi + \frac{1-(-1)^a}{2} \left(-\frac{1}{2} \pi \right) + \frac{1-(-1)^{np}}{2} \left[(1 - \delta_{I,0}) \kappa + \delta_{I,0} \lambda \right]. \quad (47)$$

For n identical nucleons $T = \frac{1}{2} n$ and eq. (47) reduces to

$$W(j^n) = \frac{n(n-1)}{2} d + \left[\frac{n}{2} \right] \pi \quad (48)$$

with

$$d = b + \frac{1}{2} \varepsilon \quad (49)$$

This is like having a simplified equivalent interaction comprising only a monopole term d and odd-multipole terms. Eq. (48) is obtainable directly from such an interaction using the n–p scheme analog of eq. (21) and eqs. (38) and (39).

We now consider mutual nuclear interactions between two subshells. In even–even and odd-mass nuclei at least one subshell is in a $v = J = 0$ state and only the monopole–isomonopole $((\kappa, k) = (0,0))$ and monopole–isodipole $((\kappa, k) = (1,0))$ terms of eq. (12) contribute. Using eqs. (12) and (22) one obtains for the mutual interaction of two mixed subshells[8]

$$I(j_1^{a_1} T_1, j_2^{a_2} T_2, TJ) = a_1 a_2 b_{12} + 2\varepsilon_{12}(\mathbf{T}_1 \cdot \mathbf{T}_2) + \frac{1-(-1)^{a_1 a_2}}{2}\mu_{12} \tag{50}$$

where b_{12} and $2\varepsilon_{12}(\mathbf{t}_1 \cdot \mathbf{t}_2)$ are the respective $(\kappa, k) = (0,0)$ and $(\kappa, k) = (1,0)$ terms of eq. (12) and μ_{12} is the contribution of the higher multipoles to the interaction of the two odd subshells in an odd–odd nucleus. For the ground state μ_{12} is necessarily negative.

For two subshells of identical nucleons, where $(\mathbf{t}_1 \cdot \mathbf{t}_2) = \frac{1}{4}$, eq. (50) reduces to

$$I(j_1^{n_1} j_2^{n_2}) = n_1 n_2 I_{12}^{0,T=1} \tag{51}$$

where

$$I_{12}^{0,T=1} = b_{12} + \frac{1}{2}\,\varepsilon_{12} \tag{52}$$

is the identical-nucleon monopole term (eq. (16)). For a neutron and a proton subshell in the n–p scheme, where $(\mathbf{t}_1 \cdot \mathbf{t}_2) = -\frac{1}{4}$, eq. (50) reduces to

$$I(j_n^n j_p^p) = np I_{12}^{0,np} + \frac{1-(-1)^{np}}{2} I_{12}^{\prime np} \tag{53}$$

where

$$I_{12}^{0,np} = b_{12} - \frac{1}{2}\,\varepsilon_{12} \tag{54}$$

is the neutron–proton monopole term (eq. (18)). The $I_{12}^{\prime np}$ is the contribution of the higher multipoles in an odd–odd nucleus. It is negative for the ground state.

The monopole terms $b_{12}, \varepsilon_{12}, I_{12}^{0,T=1}$ and $I_{12}^{0,np}$ are linear combinations of the two-body matrix elements $V_{j_1 j_2 J}$, given in the Appendix. The respective higher multipole terms μ_{12} and $I_{12}^{\prime np}$ depend additionally on J, T, a_1, a_2 and on J, n, p. In the following we ignore this dependence, similarly to that of κ and λ (eq. (46)).

Finally we consider the Coulomb energy of the protons. In the n–p scheme it is a scalar interaction given by eqs. (48) and (51) with Coulomb coefficients d^c, π^c and $I^{0,T=1c}$ calculated for the Coulomb interaction $V_{12}^c = \frac{e^2}{r_{12}}$. On the other hand, in the J–T scheme it has isoscalar, isovector and isotensor parts[116,117] and its calculation is more involved (see also ref. 119).[114,118] It turns out, though, that the long-range Coulomb interaction is not very sensitive to total nucleon number and isospin coupling. For a single mixed subshell with isospin the main part with d^c is the same as in eq. (48), and the smaller part with π^c is smaller than (48) by less than $\frac{1}{2}\pi^c$ in all actual ground state subshells.[29] For two subshells with isospin and well-defined proton numbers the mutual Coulomb interaction is given by eq. (51). In the following we shall use the *n–p* scheme Coulomb energy expressions in both diagonal and non-diagonal regions.

2. Lowest-seniority energies of single-valence-subshell configurations

The total energy of the nucleus is the sum of the (properly) nuclear and Coulomb energies.

We consider first a nucleus in a diagonal region with a mixed valence configuration j^a outside a closed subshell core with $N_0 = \frac{1}{2}A_0$ neutrons and $Z_0 = \frac{1}{2}A_0$ protons. Adding to (47) E_0, the nuclear energy of the core, and ac, the kinetic energies and nuclear interactions with the core of the a valence nucleons, one obtains

$$E^{nuclear}(A_0 + a, T) = E_0 + ac + \frac{a(a-1)}{2}b + \left(T(T+1) - \frac{3}{4}a\right)\varepsilon + \frac{a}{2}\pi + \frac{1-(-1)^a}{2}\left(-\frac{1}{2}\pi\right) + \frac{1-(-1)^{np}}{2}[(1-\delta_{I,0})\kappa + \delta_{I,0}\lambda]. \quad (55)$$

The parameters E_0 and c are given by

$$E_0 = \sum_\alpha (4j_\alpha + 2)t_\alpha + \sum_\alpha W(j_\alpha^{4j_\alpha+2}) + \sum_{\alpha<\beta} I(j_\alpha^{4j_\alpha+2}, j_\beta^{4j_\beta+2}) \quad (56)$$

$$c = t + \sum_\alpha I(j_\alpha^{4j_\alpha+2}, j) \quad (57)$$

where the summation extends over all core *J–T* subshells j_α and j_β. The t_α and t are the respective single-nucleon kinetic energies in subshells j_α and j, the W's and I's in eq. (56) are respective internal and mutual potential energies of core subshells, and the I's in (57) are interactions of a valence nucleon with core subshells. The explicit expressions of the W's and I's in terms of the two-body interaction parameters are

$$W(j_\alpha^{4j_\alpha+2}) = (4j_\alpha+1)(2j_\alpha+1)b_\alpha - \frac{6j_\alpha+3}{2}\varepsilon_\alpha + (2j_\alpha+1)\pi_\alpha \quad (58)$$

$$I(j_\alpha^{4j_\alpha+2}, j_\beta^{4j_\beta+2}) = (4j_\alpha+2)(4j_\beta+2)b_{\alpha\beta} \quad (59)$$

$$I(j_\alpha^{4j_\alpha+2}, j) = (4j_\alpha+2)b_{\alpha a} \quad (60)$$

where $b_{\alpha a}$ denotes the monopole interaction b_{12} (eq. (50)) of a $j_\alpha j$ nucleon pair.

The nuclear energy of a nucleus in a non-diagonal region with a valence configuration $j_n^n j_p^p$ and a core with N_0 neutrons and Z_0 protons is similarly obtained by adding $E_0 + nc_n + pc_p$ to the sum of eq. (48) for the valence neutrons and protons and eq. (53):

$$E^{nuclear}(N_0+n, Z_0+p) = E_0 + nc_n + pc_p + \frac{n(n-1)}{2}d_n + \left[\frac{n}{2}\right]\pi_n + \frac{p(p-1)}{2}d_p + \left[\frac{p}{2}\right]\pi_p + npI^{0,np} + \frac{1-(-1)^{np}}{2}I'^{np} \quad (61)$$

with

$$E_0 = \sum_\alpha (2j_\alpha+1)t_\alpha + \sum_\alpha W(j_\alpha^{2j_\alpha+1}) + \sum_{\alpha<\beta} I(j_\alpha^{2j_\alpha+1}, j_\beta^{2j_\beta+1}) \quad (62)$$

$$c_{n(p)} = t_{n(p)} + \sum_\alpha I(j_\alpha^{2j_\alpha+1}, j_{n(p)}) \quad (63)$$

$$W(j_\alpha^{2j_\alpha+1}) = (2j_\alpha+1)j_\alpha d_\alpha + \frac{2j_\alpha+1}{2}\pi_\alpha \quad (64)$$

$$I(j_\alpha^{2j_\alpha+1}, j_\beta^{2j_\beta+1}) = (2j_\alpha+1)(2j_\beta+1)I_{\alpha\beta}^{0,T=1} \text{ for identical nucleon subshells} \quad (65)$$

$$I(j_\alpha^{2j_\alpha+1}, j_\beta^{2j_\beta+1}) = (2j_\alpha+1)(2j_\beta+1)I_{\alpha\beta}^{0,np} \text{ for a neutron and a proton subshell} \quad (66)$$

$$I(j_\alpha^{2j_\alpha+1}, j_{n(p)}) = (2j_\alpha+1)I_{\alpha n(p)}^{0,T=1} \text{ for a neutron (proton) subshell } j_\alpha \quad (67)$$

$$I(j_\alpha^{2j_\alpha+1}, j_{n(p)}) = (2j_\alpha+1)I_{\alpha n(p)}^{0,pn(np)} \text{ for a proton (neutron) subshell } j_\alpha. \quad (68)$$

The summation in eqs. (62)–(68) extends over all n–p subshells of the core.

Finally, the Coulomb energy is obtained from (48) as

$$E^{Coulomb}(N_0+n, Z_0+p) = E_0^c + pc^c + \frac{p(p-1)}{2}d^c + \left[\frac{p}{2}\right]\pi^c \quad (69)$$

with E_0^c and c^c obtained from the protonic part of eqs. (62)–(68), calculated for the Coulomb interaction with the summation extending over all proton subshells of the core.

One observes that both in diagonal and in non-diagonal regions a major part of the nuclear binding energy, owing to mutually interacting subshells, comes from monopole interactions.

The pairs of equations (55), (69) and (61), (69) give the respective lowest-seniority energies for nuclei with single valence subshells in diagonal and in non-diagonal regions. Eq. (55) was first proposed and used in the analysis of nuclear binding energies (only the centroids, eq. (44), for odd–odd nuclei) of $1p_{3/2}$ to $1f_{7/2}$ nuclei in ref. 120. The special case of both (55) and (61) for identical nucleons, which for semi-magic isotopes is

$$E(N_0+n, Z_0) = E_0 + nc + \frac{n(n-1)}{2}d + \left[\frac{n}{2}\right]\pi, \quad (70)$$

was first applied to the Ca isotopes and $N = 28$ isotones in the $1f_{7/2}$ shell in ref. 121. Eq. (70) is called the Talmi equation. Eq. (69) was first applied to

Coulomb energies of mirror nuclei in ref. 122. It is called the Carlson–Talmi equation.

In the modified shell model mass equations, both excluding[123,124] and including[125] odd–odd nuclei, the explicit lowest-seniority intra-subshell expressions of the first three lines of eq. (61) are replaced by general functions of n and p. The equations can be written

$$E(N_0+n, Z_0+p) = E_0 + U(Z_0+p) + W(N_0+n) + npI^{0,np} + \frac{1-(-1)^{np}}{2} I'^{np}, \qquad (71)$$

allowing for the admixture of higher seniorities but retaining only monopole neutron–proton interactions (the np term) in non-odd–odd nuclei.

3. Approximate lowest-seniority energies of multi-valence-subshell configurations

Most often nuclei have several simultaneously filling valence subshells with configuration mixing (section III.B.4). In this section we consider the energy of pure multi-subshell configurations. Mixing effects are discussed in section III.C.4.

In a nucleus in a diagonal region each subshell and each pair of subshells contribute to the nuclear energy an amount given respectively by (47) and (50). The corresponding contributions for a non-diagonal region are given by (48) and (51) (for identical nucleons) or (53) (for neutrons and protons). There are likewise several Coulomb energy contributions. The resulting energy expressions contain many intra-subshell and inter-subshell interaction parameters, with coefficients depending on the detailed subshell occupation numbers.

The expressions can be considerably simplified by making an approximation assuming equal corresponding parameters in neighbouring subshells*. For two mixed subshells in a diagonal region one assumes[8]

$$\begin{aligned} &c_1 = c_2 = c, && \pi_1 = \pi_2 = \pi, \\ &b_1 = b_2 = b_{12} = b, && \varepsilon_1 = \varepsilon_2 = \varepsilon_{12} = \varepsilon, \\ &\kappa_1 = \kappa_2 = -\pi + \mu_{12} = \kappa, && \lambda_1 = \lambda_2 = \lambda, \\ &c_1^c = c_2^c = c^c, && \pi_1^c = \pi_2^c = \pi^c, \\ &d_1^c = d_2^c = I_{12}^{0,T=1c} = d^c, && \end{aligned} \qquad (72)$$

where the last symbol in each extended equality relation denotes the common approximate value of the equated parameters. Similarly, for two neutron

*The approximations (72) and (73) are indicated when one considers separation or decay energies for specific single-nucleon levels. When networks of isotopic and isotonic lines for different nlj values are plotted in the same figure[9, 126–128] they nearly overlap, with a fine structure due to small differences in the locations and slopes of the lines. These locations and slopes are given by subshell parameters equated to one another in eqs. (72) and (73).

subshells j_1, j_2 and two proton subshells j_3, j_4 in a non-diagonal region one assumes[7]*

$$\begin{aligned} &c_1 = c_2 = c_n, \quad \pi_1 = \pi_2 = \pi_n, && c_3 = c_4 = c_p, \ \pi_3 = \pi_4 = \pi_p, \\ &d_1 = d_2 = I_{12}^{0,T=1} = d_n, && d_3 = d_4 = I_{34}^{0,T=1} = d_p, \\ &\quad I_{13}^{0,np} = I_{14}^{0,np} = I_{23}^{0,np} = I_{24}^{0,np} = I^{0,np}, && I_{13}^{lnp} = I_{14}^{lnp} = I_{23}^{lnp} = I_{24}^{lnp} = I^{lnp}, \\ &c_3^c = c_4^c = c^c, \ \pi_3^c = \pi_4^c = \pi^c, && d_3^c = d_4^c = I_{34}^{0,T=1c} = d^c. \end{aligned} \tag{73}$$

Corresponding terms can then be combined, using obvious identities like

$$\frac{a_1(a_1-1)}{2} + \frac{a_2(a_2-1)}{2} + a_1 a_2 = \frac{(a_1+a_2)(a_1+a_2-1)}{2},$$

$$T_1(T_1+1) + T_2(T_2+1) + 2(\mathbf{T}_1 \cdot \mathbf{T}_2) = T(T+1), \tag{74}$$

etc.

The resulting energy expressions for diagonal and non-diagonal regions are respectively identical in form with eqs. (55) and (69) and eqs. (61) and (69), with $a = \sum a_i, n = \sum n_i$ and $p = \sum p_i$. Thus in the approximation (72) and (73) of equal corresponding parameters in neighbouring subshells the energies of major shells can be calculated as if each comprises a single large subshell in a state of lowest seniority.

In the sequel we refer to the thus reinterpreted eqs. (55) and (69) and eqs. (61) and (69) as the respective approximate major-shell lowest-seniority equations in diagonal and in non-diagonal regions.

The above pure-configuration equations were derived for states of lowest overall seniority in the approximation (72) and (73). The same a and T dependence holds as well exactly for mixed-configuration states with lowest multi-subshell seniority in degenerate major shells[114, 129] and with $v = 0$ generalized seniority of identical nucleons.[130, 131]

The approximate major-shell lowest-seniority equation (61) was proposed and applied to the interpretation of nuclear energies in ref. 7. It is equivalent to Levy's empirical mass equation developed earlier on empirical grounds[132–134] and conceived as a manifestation of shell structure.[132] The validity of an approximation like (73) was indicated in ref. 135.

4. Seniority mixing, configuration interaction and deformation energy

As a rule mixed-configuration ground states are states with neither well-defined multi-subshell seniority in degenerate major shells nor well-defined generalized seniority. Consequently the approximate major-shell lowest-seniority equations have to be replaced by the lowest eigenvalue of the energy matrix (eq. (8)) in the multi-configuration space considered. The following considerations are based on ref. 8.

---

* The last and the first two lines of (73) are consequences of (72).

The energy matrix* is a sum of single-nucleon kinetic energies and two-body potential energies. The nuclear potential energy of a nucleus in a diagonal region is a sum of one-body monopole–isomonopole interactions (the last term in the square brackets in eq. (32) for $(\kappa, k) \neq (0{,}0)$), two-body monopole–isomonopole interactions (the terms with $(\kappa, k) = (0{,}0)$), and two-body higher-multipole interactions (the sum term in the square brackets with $(\kappa, k) \neq (0, 0)$). Likewise, the potential energy of a nucleus in a non-diagonal region has a single-neutron and single-proton monopole ($k = 0$) part, a two-body neutron–neutron, proton–proton and neutron–proton monopole part, and a neutron–neutron, proton–proton and neutron–proton higher-multipole part. The three parts can be identified with corresponding terms of the *n–p* scheme analogs of eqs. (32) and (33).

The kinetic energy and the monopole–isomonopole (in a diagonal region) and monopole (in a non-diagonal region) parts of the potential energy are real diagonal matrices which are constant in each configuration. The higher-multipole part for a nucleus in a diagonal region is a hermitian matrix which goes over into its complex conjugate under *J–T* major-shell *p–h* conjugation. This can be seen by using eqs. (32), (36), (25) and (26). Likewise, in a non-diagonal region the higher-multipole part undergoes complex conjugation under simultaneous *n–p* major-shell *p–h* conjugations of the neutrons and the protons, as seen when using the *n–p* scheme analogs of eqs. (32), (33), (36), (25) and (26). Additionally, in a non-diagonal region the higher-multipole parts of $\sum V_{nn}$ and $\sum V_{pp}$ and the odd-multipole part of $\sum V_{np}$ undergo complex conjugation under separate neutron and proton *p–h* conjugations. For common phase conventions the energy matrix is real, and the higher-multipole matrix is a *p–h* conjugation invariant.

Let us write

$$H = H_1 + H_2 + H_3 \tag{75}$$

$$H_1 = H^{low.sen.approx.} \tag{76}$$

$$H_2 = H - H^{low.sen.} \tag{77}$$

$$H_3 = H^{low.sen.} - H^{low.sen.approx.}. \tag{78}$$

Here $H^{low.sen.approx.}$ is a constant diagonal matrix equal everywhere to the approximate major-shell lowest-seniority eqs. (55) and (69) (in a diagonal region) or (61) and (69) (in a non-diagonal region). $H^{low.sen.}$ is a diagonal matrix equal in each configuration to the exact lowest-seniority energy of the configuration ($E_0 + \sum a_i c_i$ plus sums of intra-subshell (eq. (47)), inter-subshell

*We assume in the following that the Coulomb energy is well represented by the approximate major-shell lowest-seniority equation (69) and does not essentially affect the discussion. Studies of Coulomb displacement energies[136–138] indicate that there are deformation contributions too, but on the scale considered here they can be accounted for in an average way by a slight adjustment of the coefficients of eq. (69) in deformed regions.[29]

(eq. (50)) and Coulomb interactions in a diagonal region and $E_0 + \sum n_i c_i + \sum p_j c_j$ plus sums of intra-subshell (eq. (48)), inter-subshell (eqs. (51) and (53)) and Coulomb interactions in a non-diagonal region). The matrix H_2 comprises higher-multipole interactions only, as the contributions of the kinetic energies and the monopole and monopole–isomonopole interactions are the same in H and $H^{low.sen.}$. The diagonal matrix H_3 is equal in each configuration to the error introduced by using the approximation (72) and (73). It is presumably small as compared to H_1 and H_2 and will be treated as a perturbation.

Consider first the part $H_1 + H_2$. Its eigenvectors are those of H_2, and its eigenvalues are obtained from those of H_2 by adding them to H_1. Thus, neglecting H_3, the contribution of configuration interaction to the binding energy is given by the lowest eigenvalue of H_2.

Since the matrix H_2 is invariant under p–h conjugation its eigenvalues are the same in two p–h conjugate nuclei*, belonging to the same shell region and obtained from one another by the p–h transformation

$$n \longleftrightarrow (\delta_n - n), \quad p \longleftrightarrow (\delta_p - p), \tag{79}$$

where $\delta_n = \sum(2j_n + 1)$ and $\delta_p = \sum(2j_p + 1)$ are the total occupation numbers of the corresponding major shells. Consequently, the lowest eigenvalue can be written in a given shell region as a manifestly p–h symmetric expansion[8,5]

$$E^{conf.int.sym.}(N_0 + n, Z_0 + p) = \sum_{(k,l) \neq (0,0)} \varphi_{kl} \Phi_{kl}(n,p) + \sum_{k,l>0} \chi_{kl} X_{kl}(n,p) \tag{80}$$

in terms of p–h symmetric polynomials

$$\Phi_{kl}(n,p) = n^k(\delta_n - n)^k p^l(\delta_p - p)^l \tag{81}$$

$$X_{kl}(n,p) = n^k(\delta_n - n)^k p^l(\delta_p - p)^l \left(n - \frac{1}{2}\delta_n\right)\left(p - \frac{1}{2}\delta_p\right) \tag{82}$$

with separate expansion coefficients φ_{kl} and χ_{kl} for each nuclear parity type. In diagonal regions $\varphi_{kl} = \varphi_{lk}$ and $\chi_{kl} = \chi_{lk}$ due to the charge symmetry of the nuclear interaction. In semi-magic nuclei, along proton and neutron shell region boundaries, only the respective pure-neutron and pure-proton terms with φ_{k0} and φ_{0l} do not vanish. They are presumably mainly due to pairing correlations, whereas the contribution of neutron–proton correlations is included in the mixed terms. For completely empty and completely full major shells the complete expression (80) vanishes, as it should.

We now consider H_3, as well as interaction with configurations involving other major shells, as perturbations. These are not p–h symmetric, but like a general function of n and p they can be written as a sum of a p–h symmetric

---

*In such nuclei one can define p–h-conjugate schemes (section III.A.4) in which eq. (36) and its n–p scheme analog hold.

and a p–h antisymmetric part. The first part can be incorporated in the sum on the r.h.s. of (80) (but without the written restrictions on k and l). The second part can be written as a manifestly p–h antisymmetric expansion

$$E^{conf.int.antisym.}(N_0+n, Z_0+p) = \sum \psi_{kl}\Psi_{kl}(n,p) + \sum \omega_{kl}\Omega_{kl}(n,p) \tag{83}$$

in terms of p–h antisymmetric polynomials

$$\Psi_{kl}(n,p) = n^k(\delta_n - n)^k p^l(\delta_p - p)^l \left(n - \frac{1}{2}\delta_n\right) \tag{84}$$

$$\Omega_{kl}(n,p) = n^k(\delta_n - n)^k p^l(\delta_p - p)^l \left(p - \frac{1}{2}\delta_p\right) \tag{85}$$

separately for each nuclear parity type. In diagonal regions $\psi_{kl} = \omega_{lk}$.

The pure-neutron and pure-proton pairing correlation terms were added to the approximate major-shell lowest-seniority eqs. (61) and (69) in refs. 9,10. The complete expression (80) and (83) was added in refs. 8, 5. Similar terms were used earlier[139,140] (and also recently [141,142]) as shell- and deformation-correction terms in liquid-drop-type mass equations. The relation of eqs. (80) and (83) to deformation is considered in section III.D.6.

5. A shell model mass equation

Summarizing the results of part III.C we describe a shell model mass equation based on them.[8,5] The nuclear energy can be written as a sum of three parts:

$$E(N,Z) = E_{pair}(N,Z) + E_{def}(N,Z) + E_{Coul}(N,Z) \tag{86}$$

where the first part is

$$\begin{aligned} E_{pair}\left(\frac{A_0}{2}+n, \frac{A_0}{2}+p\right) &= \left(\frac{A_0}{A}\right)^{\lambda} \times \\ \Bigg\{E_0 + ac + \frac{a(a-1)}{2}b &+ \left(T(T+1) - \frac{3}{4}a\right)\varepsilon + \frac{a}{2}\pi + \\ \frac{1-(-1)^a}{2}\left(-\frac{1}{2}\pi\right) &+ \frac{1-(-1)^{np}}{2}[(1-\delta_{I,0})\kappa + \delta_{I,0}\lambda]\Bigg\}, \end{aligned} \tag{87}$$

eq. (55), in a diagonal region, and

$$\begin{aligned} E_{pair}(N_0+n, Z_0+p) &= \left(\frac{A_0}{A}\right)^{\lambda} \times \\ \Bigg\{E_0 + n\left(c_n + \frac{1}{2}\pi_n\right) &+ p\left(c_p + \frac{1}{2}\pi_p\right) + \\ \frac{n(n-1)}{2}d_n + \frac{p(p-1)}{2}d_p &+ npI^{0,np} + \\ \frac{1-(-1)^n}{2}\left(-\frac{1}{2}\pi_n\right) + \frac{1-(-1)^p}{2}\left(-\frac{1}{2}\pi_p\right) &+ \frac{1-(-1)^{np}}{2}I'^{np}\Bigg\}, \end{aligned} \tag{88}$$

eq. (61), in a non-diagonal region. The second part is

$$E_{def}(N_0+n, Z_0+p) = \left(\frac{A_0}{A}\right)^{\lambda} \times$$

$$\left\{\sum_{kl} [\varphi_{kl}\Phi_{kl}(n,p) + \chi_{kl}X_{kl}(n,p) + \psi_{kl}\Psi_{kl}(n,p) + \omega_{kl}\Omega_{kl}(n,p)] + \right.$$

$$\frac{1-(-1)^n}{2}\sum_{kl} [\bar{\varphi}^1_{kl}\Phi_{kl}(n,p) + \bar{\chi}^1_{kl}X_{kl}(n,p) + \bar{\psi}^1_{kl}\Psi_{kl}(n,p) + \bar{\omega}^1_{kl}\Omega_{kl}(n,p)] +$$

$$\frac{1-(-1)^p}{2}\sum_{kl} [\bar{\varphi}^2_{kl}\Phi_{kl}(n,p) + \bar{\chi}^2_{kl}X_{kl}(n,p) + \bar{\psi}^2_{kl}\Psi_{kl}(n,p) + \bar{\omega}^2_{kl}\Omega_{kl}(n,p)] +$$

$$\left.\frac{1-(-1)^{np}}{2}\sum_{kl} \left[\bar{\bar{\varphi}}_{kl}\Phi_{kl}(n,p) + \bar{\bar{\chi}}_{kl}X_{kl}(n,p) + \bar{\bar{\psi}}_{kl}\Psi_{kl}(n,p) + \bar{\bar{\omega}}_{kl}\Omega_{kl}(n,p)\right]\right\}, \quad (89)$$

eqs. (80) and (83), in both diagonal and non-diagonal regions. In a diagonal region the third and fourth lines of (89) are equal ($\bar{\varphi}^1_{kl} = \bar{\varphi}^2_{kl}$, etc.) in order to have one odd-A surface rather than two, and there are charge-symmetry constraints like those mentioned after eqs. (82) and (85). The third part is

$$E_{Coul}(N_0+n, Z_0+p) = \left(\frac{2Z_0}{A}\right)^{\lambda/3} \times$$

$$\left\{E^c_0 + pc^c + \frac{p(p-1)}{2}d^c + \left[\frac{p}{2}\right]\pi^c\right\}, \quad (90)$$

eq. (69).

The $(\frac{A_0}{A})^{\lambda}$ and $(\frac{2Z_0}{A})^{\lambda/3}$ are scaling factors. In refs. 8, 5 the values $\lambda = 1$ for nuclei beyond the $1p$ shell and $\lambda = 0$ for the $1p$ shell were adopted. For a nuclear radius which is constant in the $1p$ shell and varies according to $R = r_0 A^{1/3}$ in heavier nuclei[143,144] these are the respective A-dependences of the two-body radial integrals for zero-range and Coulomb forces for an infinite square well of radius R.[8] The estimate $\lambda = 1$ might be slightly too high, though. This point is addressed in section III.D.2.

The mass excess $\Delta M(N, Z)$ is obtained by adding to eq. (86) the sum of the nucleon mass excesses $N\Delta M_n + Z\Delta M_H$. The resulting mass equation was adjusted to masses and Coulomb displacement energies, using different sets of coefficients for different shell regions, and a mass table based on it has been given in ref. 5. Details of the adjustment process and a discussion of the resulting masses are given in the introduction to the table.

D. Interpretation of the empirical trends

The unscaled lowest seniority part of the mass equation, eqs. (87), (88) and (90), varies quadratically as a function of nucleon numbers throughout a shell region. On the other hand, terms of the configuration-interaction part, eq.

(89), are largest in the middle of the region and decrease towards the boundaries, where the mixed ($k,l > 0$) terms vanish altogether. The pure-neutron ($l = 0$) and pure-proton ($k = 0$) terms remain, but they too decrease towards the corners of the region, where the complete intra-major-shell configuration-interaction energy vanishes.

Thus the lowest-seniority quadratic approximation is expected to describe the data best around doubly magic nuclei, to deteriorate somewhat near semi-magic ones, and to be considerably distorted in midshell. This expectation comes out as well from the physical considerations of part III.B.

In sections III.D.1–5 the basic structure described in sections II.A–E is addressed in terms of the lowest-seniority part of the mass equation. The superposed oscillating trends of section II.F are considered in section III.D.6 in terms of the configuration-interaction part.

1. The stability valley and the saturation of nuclear energy[15,145,7]

Let us locally write the lowest-seniority equations for one sheet of the mass surface in a non-diagonal region, and also on one side of the $Z = N$ line in a diagonal region, as a quadratic equation*

$$E(n,p) = E_0 + E_n n + E_p p + \frac{1}{2}(E_{nn}n^2 + 2E_{np}np + E_{pp}p^2). \tag{91}$$

For eq. (91) the line of beta stability is a straight line. The components $(\Delta n, \Delta p)$ of a segment of this line connecting the points (n,p) and $(n + \Delta n, p + \Delta p)$ satisfy

$$(E_{nn}\Delta n + E_{np}\Delta p) - (E_{np}\Delta n + E_{pp}\Delta p) = 0. \tag{92}$$

The saturation requirement that the energy changes along the line at a constant rate leads to

$$(E_{nn}\Delta n + E_{np}\Delta p)\Delta n + (E_{np}\Delta n + E_{pp}\Delta p)\Delta p = 0. \tag{93}$$

The solution of eqs. (92) and (93) is[145]

$$\begin{aligned} E_{nn}\Delta n + E_{np}\Delta p &= 0 \\ E_{np}\Delta n + E_{pp}\Delta p &= 0. \end{aligned} \tag{94}$$

Eqs. (94) determine the ratios of the quadratic coefficients. Their signs are determined from the fact that isotopic and isotonic parabolas curve upwards (Figs. 2, 4 and 8). All together one has

$$E_{pp} : (-E_{np}) : E_{nn} = (\Delta n)^2 : \Delta n \Delta p : (\Delta p)^2 \tag{95}$$

and, since empirically $\Delta n \geq \Delta p$,

$$E_{pp} \geq -E_{np} \geq E_{nn} > 0. \tag{96}$$

---

*Local values of the quadratic coefficients E_{nn}, E_{np} and E_{pp} are given by the respective second mass differences δV_{nn}, δV_{np} and δV_{pp}[9,10,31,32] of nuclei lying on the same sheet.

For light nuclei with $\Delta n = \Delta p$ the inequalities become equalities. The empirical validity of (96) is obvious from $\delta V_{nn}, \delta V_{np}$ and δV_{pp} plots (see the previous footnote).

Neglecting the smooth A-scaling of the nuclear and Coulomb coefficients (eqs. (87), (88) and (90)) one has

$$E_{nn} = b + \frac{1}{2}\varepsilon, \quad E_{np} = b - \frac{1}{2}\varepsilon, \quad E_{pp} = b + \frac{1}{2}\varepsilon + d^c \tag{97}$$

in a diagonal region, and

$$E_{nn} = d_n, \quad E_{np} = I^{0,np}, \quad E_{pp} = d_p + d^c \tag{98}$$

in a non-diagonal region. The conditions (96) become

$$b + \frac{1}{2}\varepsilon + d^c \geq -b + \frac{1}{2}\varepsilon \geq b + \frac{1}{2}\varepsilon > 0 \tag{99}$$

from which one derives further

$$\varepsilon > 0, \quad b \leq 0, \quad d^c \geq 0 \tag{100}$$

in a diagonal region*, and

$$d_p + d^c \geq -I^{0,np} \geq d_n > 0 \tag{101}$$

in a non-diagonal region.

Evaluating the energy parameters in terms of the two-body matrix elements (see the Appendix) for the empirical and recent shell model effective interactions[69, 70, 72–75] and the Coulomb interaction[29] one finds that as a rule eqs. (99), (100) and (101) are satisfied. This is the shell model way of accounting for the gross liquid-drop-like features of nuclear masses. In particular, the course of the stability line is determined as a compromise between the tendency to maximize the number np of attracting ($b - \frac{1}{2}\varepsilon, I^{0,np} < 0$) neutron–proton pairs for a given total number of nucleons, and the tendency to minimize the repulsive Coulomb interaction ($d^c > 0$) by having less protons. The saturation along the stability line is achieved by the cancellation of attractive neutron–proton and repulsive identical-nucleon ($b + \frac{1}{2}\varepsilon, d > 0$) interactions. (The vanishing coefficient of a^2 along the stability line is given by $\frac{1}{2}(E_{nn}E_{pp} - E_{np}^2)/(E_{nn} + E_{pp} - 2E_{np})$.)

2. Odd–even effects and scaling of the energy parameters with A

The odd–even splitting of the mass surface is obvious from eqs. (87), (88) and (90). They represent four parallel sheets separated by parity-dependent pairing terms which we denote δ. Neglecting the scaling factors, the relative positions

* Strictly speaking, the considerations leading to eqs. (99) are valid only from the $1f_{7/2}$ shell onwards, where the stability line runs through real minima with $N > Z$ of isobaric mass parabolas (section II.C).

of the odd-A and odd–odd surfaces above the even–even surface are given by

$$\delta = \begin{cases} -\frac{1}{2}\pi - \frac{1}{2}\pi^c & \text{odd-}Z\text{ nuclei} \\ -\frac{1}{2}\pi & \text{odd-}N\text{ nuclei} \\ \kappa & \text{odd--odd } I \neq 0 \text{ nuclei} \\ \lambda & \text{odd--odd } I = 0 \text{ nuclei} \end{cases} \tag{102}$$

in a diagonal region, and

$$\delta = \begin{cases} -\frac{1}{2}\pi_p - \frac{1}{2}\pi^c & \text{odd-}Z\text{ nuclei} \\ -\frac{1}{2}\pi_n & \text{odd-}N\text{ nuclei} \\ -\frac{1}{2}\pi_p - \frac{1}{2}\pi_n + I'^{np} & \text{odd--odd nuclei} \end{cases} \tag{103}$$

in a non-diagonal region.

Evaluating the pairing energies, as given in the Appendix, for the empirical and current shell model effective interactions and the Coulomb interaction, and using the considerations of section III.C.1 for κ, λ and I'^{np}, one obtains the relations

$$\begin{aligned} &\pi, \pi_n, \pi_p < 0, \quad \pi^c > 0 \\ &\kappa + \pi \lesssim 0, \quad \lambda + \pi \gtrsim 0, \quad I'^{np} < 0. \end{aligned} \tag{104}$$

Thus the δ terms in eqs. (102) and (103) are positive and are arranged in increasing order of magnitude, accounting for the main odd–even effects of section II.B.

It is worth mentioning that the empirical splitting of the odd-A surface in a diagonal region is not fully accounted for by eq. (102), which predicts that the odd-Z surface will always be lower than the odd-N one. Instead, the splitting is charge symmetric (see Fig. 7 and ref. 146*).

We now consider the Suess–Jensen rule, stating that on crossing magic numbers 28, 50 and 82 into a heavier mass region the corresponding pairing energy decreases.

For zero-range forces the pairing energy is equal to the Slater radial integral $F^0(jj)$ multiplied by $\frac{1}{2}(2j+1)$,[147] which increases with j. For the empirical effective interaction too the magnitude of the pairing energy increases with j, as indicated by the large scatter of the $V_{j^2 0}$ ($\theta_{12} = 180°$) data points in Fig. 14, with the higher-j points lying lower. On crossing magic numbers 28, 50 and 82 the j of the filling subshell decreases ($f_{7/2} \to p_{3/2}$ or $f_{5/2}, g_{9/2} \to d_{5/2}$ or $g_{7/2}$ and $h_{11/2} \to f_{7/2}$ or $h_{9/2}$), resulting in reduced pairing energy.

Finally we address the A-scaling of the δ terms and the other energy parameters. The highest j value of a given major shell increases like $A^{1/3}$.[64] A variation like $A^{-1/2}$ of the pairing energies (II.B) implies that the radial integral

* A second failure of the lowest-seniority description in diagonal regions is mentioned in section III.D.5.

$F^0(jj)$ decreases like $A^{-\lambda}$ with $\lambda = 0.83$, rather than the value $\lambda = 1$ adopted for the nuclear energy parameters in ref. 5.

For nucleons in an infinite square well with zero-range interactions the radial integrals decrease like A^{-1} for a radius increasing like $A^{1/3}$.[147,8] For a corresponding harmonic oscillator potential they decrease like $A^{-1/2}$.[8] The $A^{-0.83}$ variation indicated by the empirical pairing energies is between the two, as expected for a nuclear potential intermediate between the square well and the harmonic oscillator.

In ref. 35 the best overall agreement between calculated and experimental masses from the $1d2s$ shell onwards was obtained for a scaling factor λ between 0.75 and 1. In ref. 70 the monopole components of the effective interaction were found to display an overall decrease like $A^{-0.75}$. All these findings indicate that for a uniform global power-law scaling a value of λ around 0.8 might do better than the value $\lambda = 1$ adopted in ref. 5.

This indication is further supported by the fact that the best global $A^{-\lambda/3}$ scaling of Coulomb radial integrals* for nuclei beyond the $1p$ shell is obtained[29] for $\lambda = 0.85$.

On the other hand, in large-basis shell model calculations in the $1d2s$[72,74] and $1f2p$[73] shells λ values of about 0.3, and more recently[73a] in the $1f2p$ shell a value of 0.5, have been adopted.

3. The Wigner term

Eq. (87) comprises a Wigner term εT. For $\varepsilon > 0$ (eq. (100)) and $T_{g.s.} = \frac{1}{2}|N - Z|$ (III.B.3) the slope discontinuity at $Z = N$ is reproduced. As already mentioned in section II.C, the complete $\varepsilon T(T+1)$ term reproduces quite well the course of the beta-stability line and masses of light nuclei.

4. Magic and submagic numbers. Variations of magicity

We consider magicity in terms of the n–p scheme. For the lowest-seniority equations (88) and (90) without the scaling factors the slopes in the N and Z directions of a given sheet of the mass surface are

$$\frac{\partial E}{\partial N} = \left(c_n + \frac{1}{2}\pi_n - \frac{1}{2}d_n\right) + nd_n + pI^{0,np} \tag{105}$$

$$\frac{\partial E}{\partial Z} = \left(c_p + c^c + \frac{1}{2}(\pi_p + \pi^c) - \frac{1}{2}(d_p + d^c)\right) + nI^{0,np} + p(d_p + d^c). \tag{106}$$

The coefficients are shell dependent. In particular, there is a discontinuous decrease of $|c_n|$ and $|c_p|$ on crossing a corresponding magic number due to the weaker interaction of a valence nucleon in the new outer shell with the inner core. This gives rise to the upwards slope discontinuities of the mass surface and the accompanying downwards discontinuities of separation energies at magic and submagic numbers (II.D.1 and 2).

---

*This scaling corresponds to the $A^{-\lambda}$ scaling of radial integrals for zero-range forces, both for the infinite square well and for the harmonic oscillator.[8]

We now consider the mutual support and weakening of magicities (II.D.3) in terms of nucleon separation energies.[148] The left lower part of Fig. 17 shows S_n systematics of odd-N Hg, Pb, Po and Rn isotopes. One observes that the magnitude of the $N = 126$ gap between the isotonic $N = 125$ and $N = 127$ lines is largest for $_{82}$Pb and decreases on both sides. Similarly, in the lower right part of the Figure the $Z = 82$ gap between the isotopic S_p lines of $_{81}$Tl and $_{83}$Bi is largest for $N = 126$ and decreases in the lighter isotopes.

The above variation of the neutron gap is directly related to a corresponding variation of the relative magnitude of slope of the $N = 125$ and $N = 127$ S_n lines. Before reaching Pb the slope is higher for $N = 125$, whereas after Pb the $N = 127$ slope is higher. Likewise, the increase of the proton gap up to $N = 126$ is a direct consequence of the higher slope of the Tl S_p line as compared to that of Bi.

The slopes just mentioned are given by the monopole interactions $I^{0,np}$ (eqs. (105) and (106)). The respective filling neutron and proton subshells just below and above ^{208}Pb are $\nu 3p_{1/2}\pi 3s_{1/2}$ and $\nu 2g_{9/2}\pi 1h_{9/2}$. The S_n lines indicate that

$$\left|I^{0,np}_{ps}\right| > \left|I^{0,np}_{gs}\right|, \qquad \left|I^{0,np}_{ph}\right| < \left|I^{0,np}_{gh}\right| \tag{107}$$

and the S_p lines indicate that

$$\left|I^{0,np}_{ps}\right| > \left|I^{0,np}_{ph}\right|. \tag{108}$$

Thus, as one would expect, when the neutron and proton are both in inner or both in outer subshells ((3p, 3s) or (2g, 1h)), and the overlap of their wave functions is high, their monopole interaction is stronger than when one is in an inner and the other in an outer subshell ((3p, 1h) or (2g, 3s)) and the overlap is poorer.

The left and right upper parts of Fig. 17 show the respective single-nucleon excitation energies across the $N = 126$ and $Z = 82$ gaps. The neutron and proton excitation energies are respectively highest for $_{82}$Pb and for $N = 126$. This is an illustration of the mutual support of magicities in excitation energies, mentioned in section II.D.3. It is accounted for too by eqs. (107) and (108).[148]

Gap variations due to equations like (107) and (108) are first-order variations. They are further augmented by configuration interaction, which increases as the gap decreases. When the number of nucleons of the other kind is high enough submagic gaps, and sometimes (in lighter nuclei) magic gaps too, can thus be completely obliterated. When np is large enough the ensuing neutron–proton correlations result in deformation, and new submagic numbers corresponding to a deformed potential appear (see also section III.B.5)*.

*Two recent mass models (see also ref. 148c)[148a, 148b] predict as well loss of proton magicity, accompanied by strong deformation, for exotic heavy isotopes of $_{50}$Sn and $_{82}$Pb, with respective midshell neutron numbers around 104 and 152.

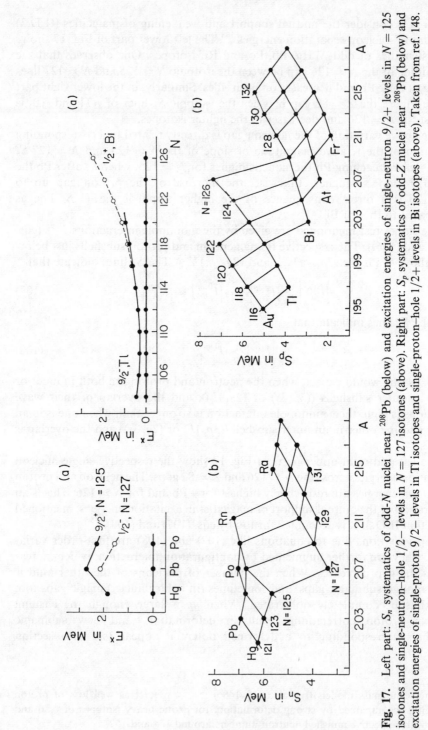

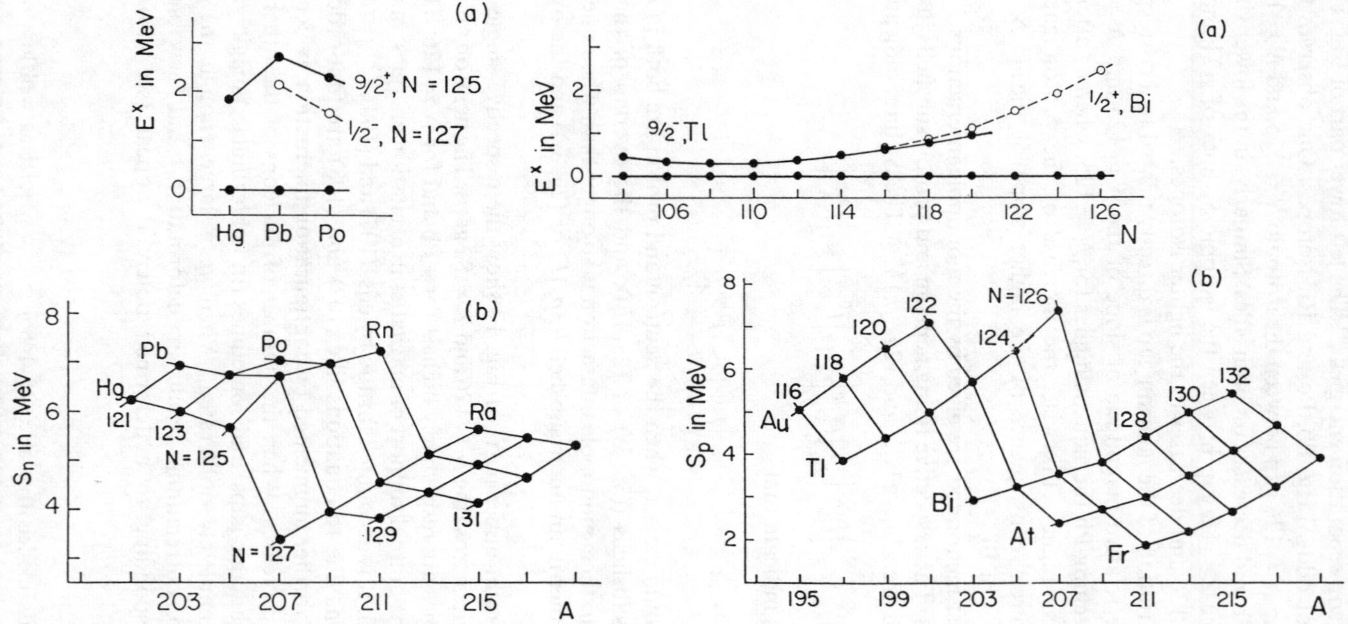

Fig. 17. Left part: S_n systematics of odd-N nuclei near ^{208}Pb (below) and excitation energies of single-neutron 9/2+ levels in $N = 125$ isotones and single-neutron–hole 1/2− levels in $N = 127$ isotones (above). Right part: S_p systematics of odd-Z nuclei near ^{208}Pb (below) and excitation energies of single-proton 9/2− levels in Tl isotopes and single-proton–hole 1/2+ levels in Bi isotopes (above). Taken from ref. 148.

The essential role of the neutron–proton monopole interaction in modifying single-nucleon gaps was recognized in the early 1950s.[92] The double role played by the neutron–proton interaction, of reducing semi-magic spherical gaps by its monopole part, and establishing neutron–proton correlations leading to deformation by its higher-multipole part, has recently been emphasized in refs. 149, 150.

5. Quadratic variation in non-deformed regions

As mentioned in section III.D.2, eqs. (87), (88) and (90) describe on each side of the $Z = N$ line in a given shell region four parallel quadratic surfaces corresponding to the four nuclear parity types. With the signs obtained for the quadratic coefficients (eqs. (100) and (101)) the quadratic mass parabolas and the linear systematics of mass differences near shell region boundaries (II.E) are accounted for.

It is worth mentioning that, like the odd-A splitting (III.D.2), the empirical quadratic variation in a diagonal region is not fully accounted for. Eq. (87) predicts the same E_{nn} and the same E_{pp} values (eq. (97)) on both sides of the $Z = N$ line. Instead, the empirical values of E_{nn} are higher for $N < Z$ than for $N > Z$, and in a charge-symmetric way the E_{pp} values are higher for $Z < N$ than for $Z > N$ (Fig. 6 of ref. 146).

6. Oscillating trends and deformation

The oscillating trends (section II.F) are accounted for by the configuration-interaction part (89) of the mass equation. In particular, the depression of the mass surface and the corresponding humps in separation energy lines in deformed regions result from the mixed terms representing mainly neutron–proton correlations. The milder oscillations of separation energy lines observed in semi-magic isotopes and isotones result from the pure-neutron and pure-proton terms representing mainly pairing correlations. The symmetric oscillations of the mutual distances between the four mass surfaces result from the oscillating pairing terms (the last three lines of eq. (89)).

The symmetric mixed polynomials $\Phi_{kl}(n,p)$ vanish on the boundaries of shell regions and increase towards their middle like the β_2 deformation parameter (Fig. 15). Sometimes[151, 138, 141, 142] (see also eqs. (67)–(69) of the chapter on nuclear deformations by W. Nazarewicz and I. Ragnarsson) the deformation has been parametrized in terms of the symmetric and antisymmetric polynomials of eqs. (81), (82), (84) and (85). Presumably the part E_{def}, eq. (89), can be viewed as an expansion of configuration-interaction energy in powers of the deformation.

APPENDIX

In the following we give the expressions[6, 6a, 8] of the energy parameters of section III.C.1 as linear combinations of the two-body matrix elements $V_{j_1 j_2 TJ}$.

The intra-subshell parameters are given by

$$b = \frac{(6j+5)\overline{V_2} + (2j+1)\overline{V_1} - 2V_0}{4(2j+1)} \tag{109}$$

$$\varepsilon = \frac{(2j+3)\overline{V_2} - (2j+1)\overline{V_1} - 2V_0}{2(2j+1)} \tag{110}$$

$$\pi = \frac{2j+2}{2j+1}(-\overline{V_2} + V_0) \tag{111}$$

$$d = \frac{(2j+2)\overline{V_2} - V_0}{2j+1} \tag{112}$$

where $\overline{V_2}$ and $\overline{V_1}$ are respective $(2J+1)$-averages of the V_{j^2TJ} for $T=1, v=2$ and $T=0, v=2$ (or, in terms of the n–p scheme, the V_{j^2J} for even $J \neq 0$ and for odd J). V_0 is the single matrix element V_{j^20}.

The inter-subshell parameters are

$$b_{12} = \frac{3}{4}\overline{V_{T=1}} + \frac{1}{4}\overline{V_{T=0}} \tag{113}$$

$$\varepsilon_{12} = \frac{1}{2}(\overline{V_{T=1}} - \overline{V_{T=0}}) \tag{114}$$

$$I_{12}^{0,T=1} = \overline{V_{T=1}} \tag{115}$$

$$I_{12}^{0,np} = \frac{1}{2}(\overline{V_{T=1}} + \overline{V_{T=0}}) \tag{116}$$

where $\overline{V_{T=1}}$ and $\overline{V_{T=0}}$ are respective $(2J+1)$-averages of the matrix elements $V_{j_1j_2T=1,J}$ and $V_{j_1j_2T=0,J}$ (or, in terms of the n–p formalism, the respective antisymmetrized and symmetrized matrix elements $V_{j_1j_2J}$).

ACKNOWLEDGEMENTS

I thank Georges Audi for ref. 21, Alex Brown for the TBLC8 matrix elements quoted in ref. 73a, and Ernst Roeckl for ref. 12.

REFERENCES

1. **McGrory, J.B. and Wildenthal, B.H.,** *Annu. Rev. Nucl. Part. Sci.,* 30, 383, 1980.
2. **Brown, B.A. and Wildenthal, B.H.,** *Annu. Rev. Nucl. Part. Sci.,* 38, 29, 1988.
3. **Brown, B.A., Warburton, E.K., and Wildenthal, B.H.,** in *Exotic Nuclear Spectroscopy,* McHarris, W.C., Ed., Plenum Press, New York, 1990, 295.
4. **Brown, B.A.,** *Nucl. Phys.,* A522, 221c, 1991.
5. **Liran, S. and Zeldes, N.,** *At. Data Nucl. Data Tables,* 17, 431, 1976.
6. **de-Shalit, A. and Talmi, I.,** *Nuclear Shell Theory,* Academic Press, New York, 1963.

6a. **Talmi, I.,** *Simple Models of Complex Nuclei,* Harwood Academic Publishers, Chur, 1993.

7. **Zeldes, N.,** *Nucl. Phys.,* 7, 27, 1958.
8. **Liran, S.,** *Calculation of Nuclear Masses in the Shell Model,* Ph.D. Thesis, Jerusalem, (in Hebrew), unpublished, 1973.

9. **Zeldes, N., Gronau, M., and Lev, A.,** *Nucl. Phys.*, 63, 1, 1965.
10. **Zeldes, N., Grill, A., and Simievic, A.,** *Mat. Fys. Skr. Dan. Vidensk. Selsk.*, 3, No. 5, 1967.
11. **Haustein, P.E.,** in *Atomic Masses and Fundamental Constants 7,* Klepper, O., Ed., THD-Schriftenreihe Wissenschaft und Technik, Bd. 26, Darmstadt, 1984, 413.
12. **Pfeng, E.,** GSI Darmstadt, personal communication from E. Roeckl, July 1994.
13. **Cappeller, V.,** in *Nuclear Masses and their Determination,* Hintenberger, H., Ed., Pergamon Press, London, 1957, 27.
14. **v. Weizsäcker, C.F.,** *Z. Phys.*, 96, 431, 1935.
15. **Bethe, H.A. and Bacher, R.F.,** *Rev. Mod. Phys.*, 8, 82, 1936.
16. **Bohr, A. and Mottelson, B.R.,** *Nuclear Structure,* Vol. 1, Benjamin, New York, 1969.
17. **Suess, H.E. and Jensen, J.H.D.,** *Ark. Fys.*, 3, 577, 1951.
18. **Vogel, P., Jonson, B., and Hansen, P.G.,** *Phys. Lett.*, 139B, 227, 1984.
19. **Jensen, A.S., Hansen, P.G., and Jonson, B.,** *Nucl. Phys.*, A431, 393, 1984.
20. **Wapstra, A.H.,** in *Nuclei Far From Stability/Atomic Masses and Fundamental Constants 1992,* Neugart, R. and Wöhr, A., Eds., Institute of Physics Publishing, Bristol, 1993, 979.
21. **Wapstra, A.H. and Audi, G.,** Interim atomic mass evaluation, personal communication from G. Audi, February 1992.
22. **Kravtsov, V.A. and Skachkov, N.N.,** *Nucl. Data,* A1, 491, 1966.
23. **Heisenberg, W.,** *Z. Phys.*, 78, 156, 1932.
24. **Gamow, G.,** *Z. Phys.*, 89, 592, 1934.
25. **Glueckauf, E.,** *Proc. Phys. Soc. London,* 61, 25, 1948.
26. **Ghoshal, S.N. and Saxena, A.N.,** *Proc. Phys. Soc. London,* A69, 293, 1956.
27. **Way, K.,** in ref. 13, p. 39.
28. **Kravtsov, V.A.,** *JETP,* 41, 1852, 1961.
29. **Ashktorab, K., Elitzur, S., Jänecke, J., Liran, S., and Zeldes, N.,** *Nucl. Phys.*, A517, 27, 1990.
30. **Myers, W.D. and Swiatecki, W.J.,** *Nucl. Phys.*, 81, 1, 1966.
31. **Brenner, D.S., Wesselborg, C., Casten, R.F., Warner, D.D., and Zhang, J.-Y.,** *Phys. Lett.*, B243, 1, 1990.
31a. **Van Isacker, P., Warner, D.D., and Brenner, D.S.,** *Phys. Rev. Lett.*, 74, 4607, 1995.
32. **Zamfir, N.V. and Casten, R.F.,** *Phys. Rev.*, C43, 2879, 1991.
33. **Wigner, E.,** *Phys. Rev.*, 51, 947, 1937.
34. **Tseng, C.Y., Cheng, T.S., and Yang, F.C.,** *Nucl. Phys.*, A334, 470, 1980.
35. **Zeldes, N., Novoselsky, A., and Taraboulos, A.,** *Kinam,* 4, 459, 1982.
36. **Jänecke, J.,** *Nucl. Phys.*, 73, 97, 1965.
37. **Bos, K., Audi, G., and Wapstra, A.H.,** *Nucl. Phys.*, A432, 140, 1985.
37a. **Borcea, C., Audi, G., Wapstra, A.H., and Favaron, P.,** *Nucl. Phys.*, A565, 158, 1993.
38. **Mayer, M.G. and Jensen, J.H.D.,** *Elementary Theory of Nuclear Shell Structure,* Wiley, New York, 1955.
39. **Coryell, C.D.,** *Annu. Rev. Nucl. Sci.* 2, 305, 1953.
40. **Bouchez, R., Robert, J., and Tobailem, J.,** *J. Phys. Rad.*, 14, 281, 1953.
41. **Kumar, K. and Preston, M.A.,** *Can. J. Phys.*, 33, 298, 1955.

42. **Dewdney, J.W.,** *Nucl. Phys.,* 43, 303, 1963.
43. **Mayer, M.G.,** *Phys. Rev.,* 74, 235, 1948.
44. **Wouters, J.M., Kraus, Jr., R.H., Vieira, D.J., Butler, G.W., and Löbner, K.E.G.,** *Z. Phys.,* A331, 229, 1988.
45. **Barber, R.C., Meredith, J.O., Southon, F.C.G., Williams, P., Barnard, J.W., Sharma, K., and Duckworth, H.E.,** *Phys. Rev. Lett.,* 31, 728, 1973.
46. **Schmidt, K.-H., Faust, W., Münzenberg, G., Clerc, H.-G., Lang, W., Pielenz, K., Vermeulen, D., Wohlfarth, H., Ewald, H., and Güttner, K.,** *Nucl. Phys.,* A318, 253, 1979.
47. **Schmidt, K.-H. and Vermeulen, D.,** in *Atomic Masses and Fundamental Constants 6,* Nolen, Jr., J.A. and Benenson, W., Eds., Plenum Press, New York, 1980, 119.
48. **Zeldes, N.,** in *4th Int. Conf. on Nuclei Far From Stability,* Hansen, P.G. and Nielsen, O.B., Eds., CERN 81-09, Geneva, 1981, 93.
49. **Détraz, C. and Vieira, D.J.,** *Annu. Rev. Nucl. Part. Sci.,* 39, 407, 1989.
50. **Vieira, D.J., Zhou, X.G., Tu, X.L., Wouters, J.M., Seifert, H.L., Löbner, K.E.G., Zhou, Z.Y., Lind, V.G., and Butler, G.W.,** in *Nuclear Shapes and Nuclear Structure at Low Excitation Energies,* Vergnes, M., Sauvage, J., Heenen, P.H., and Duong, H.T., Eds., Plenum Press, New York, 1992, 365.
51. **Talmi, I. and Unna, I.,** *Phys. Rev. Lett.,* 4, 469, 1960.
52. **Perlman, I., Ghiorso, A., and Seaborg, G.T.,** *Phys. Rev.,* 77, 26, 1950.
53. **Sengupta, S.,** *Z. Phys.,* 134, 413, 1953.
54. **Way, K. and Wood, M.,** *Phys. Rev.,* 94, 119, 1954.
55. **Glass, R.A., Thompson, S.G., and Seaborg, G.T.,** *J. Inorg. Nucl. Chem.,* 1, 3, 1955.
56. **Wapstra, A.H., Audi, G., and Hoekstra, R.,** *Nucl. Phys.,* A432, 185, 1985.
57. **Wapstra, A.H., Audi, G., and Hoekstra, R.,** *At. Data Nucl. Data Tables,* 39, 281, 1988.
58. **Borcea, C., Audi, G., and Duflo, J.,** in *Nuclei Far From Stability/Atomic Masses and Fundamental Constants 1992,* Neugart, R. and Wöhr, A., Eds., Institute of Physics Publishing, 1993, 59.
59. **Borcea, C. and Audi, G.,** *CSNSM-92.38*, Orsay, 1992.
59a. **Audi, G., Wapstra, A.H., and Dedieu, M.,** *Nucl. Phys.,* A565, 193, 1993.
60. **Hogg, B.G. and Duckworth, H.E.,** *Can. J. Phys.,* 32, 65, 1954.
61. **Johnson, Jr., W.H. and Nier, A.O.,** *Phys. Rev.,* 105, 1014, 1957.
62. **Thibault, C., Klapisch, R., Rigaud, C., Poskanzer, A.M., Prieels, R., Lessard, L., and Reisdorf, W.,** *Phys. Rev.,* C12, 644, 1975.
63. **Epherre, M., Audi, G., Thibault, C., Klapisch, R., Huber, G., Touchard, F., and Wollnik, H.,** *Phys. Rev.,* C19, 1504, 1979.
64. **Nemirovsky, P.E. and Adamchuk, Yu. V.,** *Nucl. Phys.,* 39, 551, 1962.
65. **Racah, G. and Stein, J.,** *Phys. Rev.,* 156, 58, 1967.
66. **French, J.B.,** in *Many Body Description of Nuclear Structure and Reactions,* Bloch, C., Ed., Academic Press, New York, 1966, 278.
67. **Racah, G.,** *Phys. Rev.,* 62, 438, 1942.
68. **Schiffer, J.P.,** *Ann. Phys. NY,* 66, 798, 1971.
69. **Schiffer, J.P. and True, W.W.,** *Rev. Mod. Phys.,* 48, 191, 1976.
70. **Daehnick, W.W.,** *Phys. Rep.,* 96, 317, 1983.
71. **Molinari, A., Johnson, M.B., Bethe, H.A., and Alberico, W.M.,** *Nucl. Phys.,* A239, 45, 1975.

72. **Brown, B.A., Richter, W.A., Julies, R.E., and Wildenthal, B.H.,** *Ann. Phys. NY,* 182, 191, 1988.
73. **Richter, W.A., Van der Merwe, M.G., Julies, R.E., and Brown, B.A.,** *Nucl. Phys.,* A523, 325, 1991.
73a. **Van der Merwe, M.G., Richter, W.A., and Brown, B.A.,** *Nucl. Phys.,* A579, 173, 1994, and personal communication from B.A. Brown.
74. **Warburton, E.K. and Brown, B.A.,** *Phys. Rev.,* C46, 923, 1992.
75. **Ji, X. and Wildenthal, B.H.,** *Phys. Rev.,* C37, 1256, 1988.
76. **Racah, G.,** *Phys. Rev.,* 63, 367, 1943.
77. **Racah, G. and Talmi, I.,** *Physica,* 18, 1097, 1952.
78. **Talmi, I. and Unna, I.,** *Nucl. Phys.,* 19, 225, 1960.
79. **Lanford, W.A.,** *Phys. Lett.,* 30B, 213, 1969.
80. **Lawson, R.D.,** *Z. Phys.,* A303, 51, 1981.
81. **French, J.B.,** *Nucl. Phys.,* 15, 393, 1960.
82. **Flowers, B.H.,** *Proc. R. Soc. London,* A212, 248, 1952.
83. **Edmonds, A.R. and Flowers, B.H.,** *Proc. R. Soc. London,* A214, 515, 1952.
84. **Racah, G.,** in *L. Farkas Memorial Volume,* Farkas, A. and Wigner, E.P., Eds., Research Council of Israel, Jerusalem, 1952, 294.
85. **Zeldes, N. and Liran, S.,** *Phys. Lett.,* 62B, 12, 1976.
86. **Ginocchio, J.N.,** *Nucl. Phys.,* 63, 449, 1965.
87. **Edmonds, A.R. and Flowers, B.H.,** *Proc. R. Soc. London,* A215, 120, 1952.
88. **Lawson, R.D.,** *Phys. Rev.,* 124, 1500, 1961.
89. **Lawson, R.D. and Zeidman, B.,** *Phys. Rev.,* 128, 821, 1962.
90. **Klinkenberg, P.F.A.,** *Rev. Mod. Phys.,* 24, 63, 1952.
91. **Zeldes, N.,** *Nucl. Phys.,* 2, 1, 1956.
92. **de-Shalit, A. and Goldhaber, M.,** *Phys. Rev.,* 92, 1211, 1953.
93. **Mottelson, B.R.,** in *The Many Body Problem,* Dunod, Paris, 1959, 283.
94. **Allaart, K., Boeker, E., Bonsignori, G., Savoia, M., and Gambhir, Y.K.,** *Phys. Rep.,* 169, 209, 1988.
95. **Talmi, I.,** *Rev. Mod. Phys.,* 34, 704, 1962.
96. **Unna, I.,** *Phys. Rev.,* 132, 2225, 1963.
97. **Bohr, A. and Mottelson, B.R.,** *Nuclear Structure,* Vol. 2, Benjamin, Reading, MA, 1975.
98. **Casten, R.F.,** *Nuclear Structure from a Simple Perspective,* Oxford University Press, New York, 1990.
99. **Elliott, J.P. and Flowers, B.H.,** *Proc. R. Soc. London,* A229, 536, 1955.
100. **Redlich, M.G.,** *Phys. Rev.,* 99, 1427, 1955.
101. **Bohr, A., Mottelson, B.R., and Pines, D.,** *Phys. Rev.,* 110, 936, 1958.
102. **Åberg, S., Flocard, H., and Nazarewicz, W.,** *Annu. Rev. Nucl. Part. Sci.,* 40, 439, 1990.
103. **Wood, J.L., Heyde, K., Nazarewicz, W., Huyse, M., and Van Duppen, P.,** *Phys. Rep.,* 215, 101, 1992.
104. **Kirchuk, E., Federman, P., and Pittel, S.,** *Phys. Rev.,* C47, 567, 1993.
104a. **Werner, T.R., Dobaczewski, J., Guidry, M.W., Nazarewicz, W., and Sheikh, J.A.,** *Nucl. Phys.,* A578, 1, 1994.
105. **Raman, S., Malarkey, C.H., Milner, W.T., Nestor, Jr., C.W., and Stelson, P.H.,** *At. Data Nucl. Data Tables,* 36, 1, 1987.
106. **Raman, S., Nestor, Jr., C.W., and Bhatt, K.H.,** *Phys. Rev.,* C37, 805, 1988.
107. **Casten, R.F., Heyde, K., and Wolf, A.,** *Phys. Lett.,* B208, 33, 1988.

108. **Zhang, J.-Y., Casten, R.F., and Brenner, D.S.,** *Phys. Lett.,* B227, 1, 1989.
109. **Feng, D.H., Wu, C.-L., Guidry, M., and Li, Z.-P.,** *Phys. Lett.,* B205, 156, 1988.
110. **Casten, R.F.,** *Nucl. Phys.,* A443, 1, 1985.
111. **Hamilton, J.H., Ramayya, A.V., Maguire, C.F., Piercey, R.B., Bengtsson, R., Möller, P., Nix, J.R., Zhang, J.-Y., Robinson, R.L., and Frauendorf, S.,** *J. Phys. G: Nucl. Phys.,* 10, L87, 1984.
112. **Hamilton, J.H.,** in *Treatise on Heavy-Ion Science, Vol. 8, Nuclei Far From Stability,* Bromley, D.A., Ed., Plenum Press, New York, 1989, 3.
113. **Myers, W.D. and Swiatecki, W.J.,** *Ark. Fys.,* 36, 343, 1967.
114. **Hecht, K.T.,** *Nucl. Phys.,* A102, 11, 1967.
115. **Hemenger, R.P. and Hecht, K.T.,** *Nucl. Phys.,* A145, 468, 1970.
116. **MacDonald, W.M.,** *Phys. Rev.,* 101, 271, 1956.
117. **Wigner, E.P.,** in *Proc. Robert A. Welch Found. Conf. on Chemical Research. I. The Structure of the Nucleus,* Milligan, W.O., Ed., The Robert A. Welch Foundation, Houston, 1958, 67.
118. **Hecht, K.T.,** *Nucl. Phys.,* A114, 280, 1968.
119. **Jänecke, J.,** in *Isospin in Nuclear Physics,* Wilkinson, D.H., Ed., North-Holland, Amsterdam, 1969, 297.
120. **Talmi, I. and Thieberger, R.,** *Phys. Rev.,* 103, 718, 1956.
121. **Talmi, I.,** *Phys. Rev.,* 107, 326, 1957.
122. **Carlson, B.C. and Talmi, I.,** *Phys. Rev.,* 96, 436, 1954.
123. **Jelley, N.A., Cerny, J., Stahel, D.P., and Wilcox, K.H.,** *Phys. Rev.,* C11, 2049, 1975.
124. **Davids, C.N.,** *Phys. Rev.,* C13, 887, 1976.
125. **Hotchkis, M.A.C.,** *Studies of Light Neutron-Rich Nuclei,* Ph.D. Thesis, Canberra, unpublished, 1984.
126. **Everling, F.,** *Nucl. Phys.,* 36, 228, 1962.
127. **Kienle, P., Vienna, K., Zahn, U., and Weckermann, B.,** *Z. Phys.,* 176, 226, 1963.
128. **Zeldes, N.,** in *Atomic Masses and Fundamental Constants 4,* Sanders, J.H. and Wapstra, A.H., Eds., Plenum Press, London, 1972, 245.
129. **Kerman, A.K., Lawson, R.D., and Macfarlane, M.H.,** *Phys. Rev.,* 124, 162, 1961.
130. **Talmi, I.,** *Nucl. Phys.,* A172, 1, 1971.
131. **Shlomo, S. and Talmi, I.,** *Nucl. Phys.,* A198, 81, 1972.
132. **Levy, H.B.,** *UCRL-4588,* Livermore, 1955.
133. **Levy, H.B.,** *UCRL-4713,* Livermore, 1956.
134. **Levy, H.B.,** *Phys. Rev.,* 106, 1265, 1957.
135. **Thieberger, R. and de-Shalit, A.,** *Phys. Rev.,* 108, 378, 1957.
136. **Jänecke, J., Aarts, E.H.L., Drentje, A.G., Harakeh, M.N., and Gaarde, C.,** *Nucl. Phys.,* A394, 39, 1983.
137. **Comay, E. and Jänecke, J.,** *Nucl. Phys.,* A410, 103, 1983.
138. **Jänecke, J. and Comay, E.,** *Phys. Lett.,* 140B, 1, 1984.
139. **Kümmel, H., Mattauch, J., Thiele, W., and Wapstra, A.H.,** in *Nuclidic Masses,* Johnson, Jr., W.H., Ed., Springer, Vienna, 1964, 42.
140. **Kümmel, H., Mattauch, J.H.E., Thiele, W., and Wapstra, A.H.,** *Nucl. Phys.,* 81, 129, 1966.
141. **Johansson, S.A.E. and Spanier, L.,** in ref. 11, p. 429.
142. **Spanier, L. and Johansson, S.A.E.,** *At. Data Nucl. Data Tables,* 39, 259, 1988.
143. **Collard, H.R., Elton, L.R.B., and Hofstadter, R.,** *Nuclear Radii,* Springer, Berlin, 1967.

144. **de Vries, H., de Jager, C.W., and de Vries, C.,** *At. Data Nucl. Data Tables,* 36, 495, 1987.
145. **Fuchs, K.,** *Proc. Cambridge Philos. Soc.,* 35, 242, 1939.
146. **Zeldes, N.,** in ref. 47, p. 129.
147. **Mayer, M.G.,** *Phys. Rev.,* 78, 22, 1950.
148. **Zeldes, N., Dumitrescu, T.S., and Köhler, H.S.,** *Nucl. Phys.,* A399, 11, 1983.
148a. **Aboussir, Y., Pearson, J.M., Dutta, A.K., and Tondeur, F.,** *Nucl. Phys.,* A549, 155, 1992.
148b. **Möller, P., Nix, J.R., Myers, W.D., and Swiatecki, W.J.,** *At. Data Nucl. Data Tables,* 59, 185, 1995.
148c. **Möller, P.and Nix, J.R.,** *J.Phys.G: Nucl. Part. Phys.,* 20, 1681, 1994.
149. **Heyde, K., Van Isacker, P., Casten, R.F., and Wood, J.L.,** *Phys. Lett.,* 155B, 303, 1985.
150. **Heyde, K., Jolie, J., Moreau, J., Ryckebusch, J., Waroquier, M., Van Duppen, P., Huyse, M., and Wood, J.L.,** *Nucl. Phys.,* A466, 189, 1987.
151. **Jänecke, J.,** *Phys. Lett.,* 103B, 1, 1981.

3

PENNING TRAP MASS SPECTROMETER

G. Bollen and H.-J. Kluge

Table of contents

I. INTRODUCTION

It has been proved that Penning traps are efficient tools for high-accuracy mass determinations. With these devices the mass ratios of a number of stable or long-lived light particles, as for example electron and positron, electron and proton, proton and antiproton, helium-3 and tritium or the molecules N_2 and CO, have been determined with unsurpassed accuracy.[1–5] So far, only one spectrometer has been set up for the determination of the mass of radioactive nuclei. This is the tandem Penning trap mass spectrometer ISOLTRAP,[6–8] installed at the on-line mass separator ISOLDE at CERN in Geneva. Up to now it has been used to determine mass values of more than 70 radioactive isotopes of alkali and alkali earth elements with accuracies of $\delta m/m \approx 10^{-7}$ and resolving powers of $R = m/\Delta m$ exceeding 1 million.[7–11]

The shortest half-life of an isotope investigated is $T_{1/2} = 1.8$ s (^{142}Cs). The high resolving power of a Penning trap mass spectrometer enables one to discriminate between the mass of an isotope in its ground and isomeric state.[12]

The aim of this chapter is to discuss briefly the principles of a Penning trap and its specific requirements when investigating radioactive nuclei. The set-up and performance of the ISOLTRAP spectrometer are described in more detail as it represents the prototype for mass measurements of radioactive isotopes.

II. PRINCIPLE OF TECHNIQUE

A. Eigen frequencies of ions in a Penning trap and the cyclotron resonance

The mass determination in a Penning trap is based on the determination of the cyclotron frequency

$$\omega_c = q/m \cdot B$$

of ions with mass m and charge q in a magnetic field of strength B. The confinement of charged particles is achieved by an electrode configuration which creates an axially symmetric electrostatic quadrupole field; this in turn prevents the ions from escaping along the magnetic field lines. In general this configuration (Fig. 1) consists of three electrodes, two endcaps and one ring electrode which follow the shape of the equipotential surfaces of the quadrupole potential. The surfaces are defined by the equations

$$z^2 = z_0^2 + \frac{\rho^2}{2} \tag{1}$$

for the endcaps and

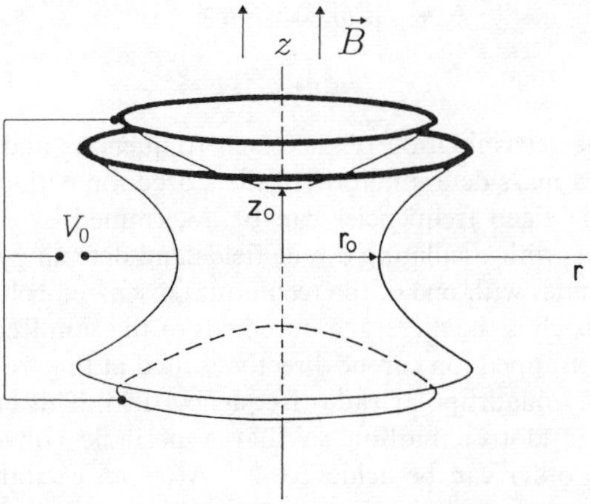

Fig. 1. Basic configuration of a Penning trap, consisting of two endcaps and a ring electrode.

$$z^2 = \frac{1}{2} \cdot (\rho^2 - \rho_0^2) \tag{2}$$

for the ring electrode, where ρ_0 is the inner radius of the ring and $2 \cdot z_0$ is the spacing between the endcaps. A characteristic trap dimension d can be defined by

$$d^2 = \frac{1}{2} \cdot \left(z_0^2 + \frac{\rho_0^2}{2} \right). \tag{3}$$

Owing to the additional electric force the ion motion is not a pure cyclotron motion with frequency ω_c but a superposition of three harmonic eigenmotions: a slow drift around the trap axis called magnetron motion with frequency ω_-, a modified cyclotron motion with frequency ω_+ and an axial oscillation with frequency ω_z.[13] These eigen frequencies follow the relation $\omega_+ \gg \omega_z \gg \omega_-$ (for $V_0 \ll \frac{1}{2} \frac{q}{m} (B \cdot d)^2$) and are determined by

$$\omega_+ = \frac{\omega_c}{2} + \sqrt{\frac{\omega_c^2}{4} - \frac{\omega_z^2}{2}} \tag{4}$$

$$\omega_- = \frac{\omega_c}{2} - \sqrt{\frac{\omega_c^2}{4} - \frac{\omega_z^2}{2}} \tag{5}$$

$$\omega_z = \sqrt{\frac{q \cdot V_0}{m \cdot d^2}}, \tag{6}$$

where V_0 is the potential difference between the ring electrode and endcaps.

With respect to high-accuracy mass measurements important relationships between the radial frequencies of an ion in a pure quadrupole field and a homogeneous magnetic field are

$$\omega_c = \omega_+ + \omega_- \tag{7}$$

$$\omega_c = \omega_+ + \omega_z^2/(2\omega_+) \tag{8}$$

$$\omega_c^2 = \omega_+^2 + \omega_-^2 + \omega_z^2. \tag{9}$$

The independent determination of the eigen frequencies and the use of these equations allow a mass determination of the stored ion with only the magnetic field known. The eigen frequencies can be determined by driving the corresponding motions with oscillating dipole fields and detecting the change of the motional amplitudes with one of the techniques discussed below. However, the straightest approach is the direct measurement of the sum frequency $\omega_+ + \omega_-$. The motion of a trapped ion can be directly excited at this frequency by means of an azimuthally quadrupolar radio frequency field. This excitation couples magnetron and cyclotron motion so that a periodic conversion from one motion into the other can be achieved.[14,15] After an excitation time T_{RF} an initially pure magnetron motion is transformed into a pure cyclotron motion. In general, this transformation leads to a large change in radial energy (since $\omega_+ \gg \omega_-$) which can be detected. The width of the cyclotron resonance curve

$\Delta\nu(FWHM) \simeq 0.9/T_{RF}$ is determined by the Fourier limit of the driving RF field switched on for a time T_{RF}. For example, a singly charged ion with mass number $A = 100$ in a field of $B = 6$ T has $\nu_c \simeq 1$ MHz. Exciting it for $T_{RF} = 900$ ms yields a linewidth of $\Delta\nu \approx 1$ Hz corresponding to a resolving power of $R = 10^6$.

B. Injection of ions into traps and ion cooling

In order to perform mass determinations on unstable isotopes delivered by an on-line mass separator a mechanism is needed to transfer these ions into the trap. Two possibilities exist for capturing an ion from the outside world in a Penning trap. Either the trapping potential is simply turned on when the ion has moved into the trap or the ion loses kinetic energy when passing through it, i.e. by collisions in a buffer gas. Both schemes are applied in the two traps of the ISOLTRAP spectrometer, and will be discussed in more detail in section III.A and III.B.

The most important task of cooling in traps is the decrease of the motional amplitudes of the stored ion. Small amplitudes are needed in order to reduce the effects of trap imperfections (see section II.D) on the ion motion. Buffer gas cooling is the standard technique applied in Paul traps. In the case of the Penning trap the use of buffer gas is only possible if the resulting increase of the ions magnetron motion (which has negative energy) can be prevented. This can be achieved by a coupling of both radial motions by a cyclotron excitation at $\omega_c = \omega_+ + \omega_-$ (see section III.A).[15,16]

Other basic cooling techniques are resistive cooling, laser cooling and cooling with cold charged particle clouds. With respect to mass measurements on short-lived heavy isotopes these techniques are not applicable. In the case of resistive cooling, the motional energy of the ions is dissipated in a cooled external circuit via the image currents induced in the trap electrodes.[17] This technique works well only for light particles with high motional frequencies. Sub-Kelvin temperatures can be reached via laser cooling, but optical transitions of appropriate wave length are required and therefore the applicability is limited to some particular candidates.[18,19] Cooling with cold clouds of electrons, which cool themselves by synchrotron radiation, has successfully been applied in the antiproton experiment, but cooling of ions with positrons has not yet been demonstrated.[20]

C. Principle of the cyclotron resonance detection

There exist a number of schemes for the detection of the eigen frequencies of a trapped ion. Possibilities are the narrow- or broad-band detection of the image currents induced in the trapping electrodes. But both schemes are in general not applicable to unstable isotopes. Narrow-band detection requires tuned circuits which cannot be changed quickly for the investigation of a new isotope.[1–5,21] Broad-band detection (commonly used in FT-ICR experiments) requires a large number of stored ions for a detectable signal.[22] Another technique, as applied in the case of the ISOLTRAP spectrometer, makes use of the conversion of radial energy into axial energy in a magnetic field gradient.[23] After excitation of their radial motion by the RF field (see above) the ions are gently kicked out of the trap, leave the homogeneous part of the magnetic field and drift along the magnetic field lines to an ion detector placed in the weak fringe field of the magnet. The radial energy gained by the excitation is converted into axial energy in the inhomogeneous part of the magnetic field. Therefore the measurement of the drift time as a function of the applied radio frequency yields a resonance curve with a minimum at $\omega_c = \omega_+ + \omega_-$, corresponding to a maximum of gained radial energy.

D. Systematic errors

For an accuracy of 10^{-7} for the mass values it is important to find and understand sources of systematic errors caused by shifts of the cyclotron resonance. These frequency shifts can firstly arise from the trap itself and are due to deviations of the electric trapping field from a pure axial quadrupole, magnetic field inhomogeneities and misalignment of the trap relative to the magnetic field axis.[13,14] By careful design and construction of the spectrometer and use of a superconducting magnet with a field homogeneity of $\Delta B/B < 10^{-7}$ over the trapping region these sources can be kept low. The temporal stability of the magnetic field has to be sufficiently high $((\delta B/\delta t)/B < 10^{-7}/h)$ during the time needed to measure a series of cyclotron resonances. Commercial NMR magnets fulfil these requirements.

Another source of systematic errors at a level of $\delta m/m \approx 10^{-7}$ are shifts due to the ion–ion interaction of simultaneously stored ions with different masses. Besides the isotopes in the ground state, isomers and isotopes of neighbouring elements are produced in nuclear reactions. In the cases of mass differences of $\Delta E > h \cdot \Delta\nu_c(FWHM)$, contaminant ions can be removed from the trap by selectively exciting them to large cyclotron orbits by a dipole RF field at ω_+. Should this technique not be applicable, one is forced to work with a low number of stored ions. This diminishes the effects of the ion–ion interaction but increases the measuring time.

III. CONSTRUCTION OF A PENNING TRAP MASS SPECTROMETER

As an example of a Penning trap mass spectrometer for short-lived nuclei,[6–8] the experimental set-up of the ISOLTRAP spectrometer and its performance will be discussed. ISOLTRAP was first installed at the on-line mass separator ISOLDE-2 and then at the PS-Booster ISOLDE at CERN/Geneva. This chapter describes the configuration of ISOLTRAP as it was realized until end of 1993. A brief summary of subsequent, still on-going modifications can be found in ref. 8.

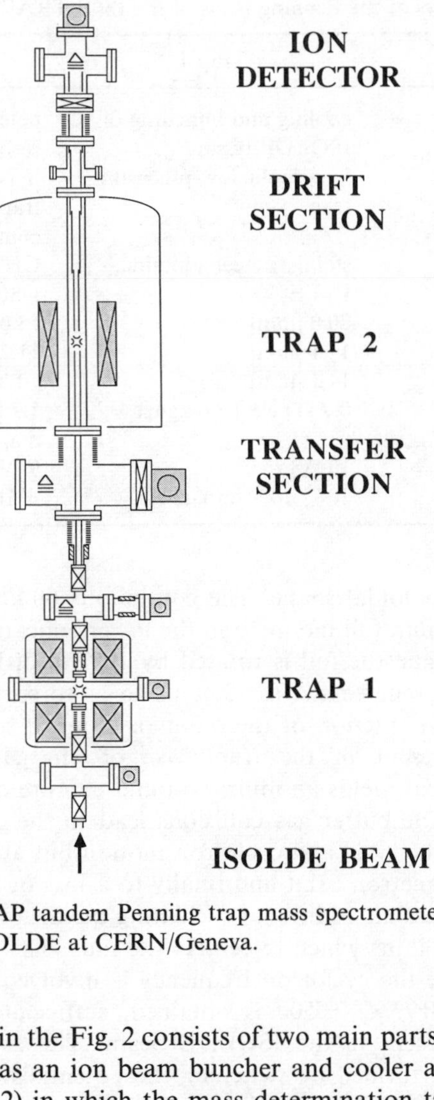

Fig. 2. The ISOLTRAP tandem Penning trap mass spectrometer, installed at the on-line mass separator ISOLDE at CERN/Geneva.

The set-up shown in the Fig. 2 consists of two main parts, the lower Penning trap (trap 1) acting as an ion beam buncher and cooler and the upper high-precision trap (trap 2) in which the mass determination takes place. Table 1 gives an overview of the properties of these traps. A multi-stage differential pumping system allows the use of buffer gas in trap 1 and assures a vacuum of $p < 10^{-8}$ in the chamber containing trap 2. Such a low value is essential for reaching narrow cyclotron resonances and through this a high resolving power of the spectrometer.

A. The cooler and buncher trap

This trap (trap 1) is placed in a vacuum chamber mounted in a 0.7 T electromagnet. It is a low-precision trap and consists of two endcaps and a ring

TABLE 1 Parameters of the Penning traps of the ISOLTRAP spectrometer

		Trap 1	Trap 2
Purpose		cooling and bunching of ISOLDE beam	determination of cyclotron resonance
Trap design		hyperbolic low-precision trap	hyperbolic high-precision trap with four compensation electrodes
Materials		stainless steel, alumina, PTFE	OHFC copper (gold plated), glass ceramics
Dimensions	ρ_0	20.0 [mm]	13.0 [mm]
	z_0	14.1 [mm]	11.2 [mm]
	d	14.1 [mm]	10.2 [mm]
Magnetic field	B	0.7 [T] electromagnet	5.7 [T] superconducting magnet
Trapping voltage	V_0	10 [V]	9 [V]
Vacuum	p	10^{-4}–10^{-3} [mbar] He	$< 10^{-8}$ [mbar]

electrode of hyperboloidal shape. The continuous 60 keV ISOLDE beam is collected on a rhenium foil mounted in the lower endcap of this trap. After a certain collection time the foil is rotated by 180° in order to face the inside of the trap and current heated for 500 ms so as to surface-ionize the accumulated material. A fraction of these ions is stopped by collisions with the He buffer gas present in the trap. Use of He gas at a pressure of $p = (10^{-4}–10^{-3})$ mbar yields an ionization and capture efficiency of $\varepsilon > 10^{-4}$ for alkali atoms. The buffer gas collisions lead to the desirable decrease of the amplitudes of the axial and cyclotron motion but at the same time to an increase of the magnetron orbit and finally to a loss of the ions. In order to avoid the last process an additional oscillating quadrupole field at frequency ω_c is applied for 200 ms which re-centers the ions. This cooling technique is mass selective since the cyclotron frequency is involved.[15,16] A mass resolution of $m/\Delta m$ $(FWHM)$ = 500 is obtained, sufficient to remove isotopic contaminations present in the ISOLDE beam. The extraction of the ions from the trap is performed by switching the potentials at the trapping electrodes to give a nearly constant electric field from one endcap to the other ($E = 1$ V/cm). After ejection, the ion bunch is accelerated to 1 keV by means of an electrostatic lens system for transport to the second trap. The total time needed to release the ions from the foil, to cool and finally to eject them, is less than 1 second.

B. The ion transport

The ion bunch has to be delivered as efficiently as possible from the cooler and buncher trap to the high-precision trap. Special attention has to be paid to the injection of the ions into the fringe field of the magnet. If this injection is not performed correctly, the ions will transform part of their axial energy into radial energy. This is not tolerable since the resonance detection scheme used in the ISOLTRAP spectrometer, as discussed in section II.C, is based on the detection of the change of the radial energy caused by the RF excitation.

The ion transport system consists of three electrostatic lenses plus a number of deflection plates. The last lens focuses the ions into the magnetic field in such a way that the ions pick up as little radial energy as possible. In total four sets of adjustable diaphragms allow the control of the path of the ions. At the entrance to the precision trap the ion bunch is retarded to an energy of about 1 eV. When arriving at the precision trap the bunch is captured in flight by pulsing the lower endcap of the trap.[24] Transfer efficiencies between the two traps close to 100% have been achieved.

C. The high-precision trap and the resonance detection

The high-precision trap shown in Fig. 3 is constructed from oxygen-free copper and is gold plated. Glass ceramics are used for insulation. All parts are machined as thin as possible in order to minimize magnetic field inhomogeneities due to the susceptibility of the materials. Deviations from the ideal quadrupole field are compensated by correction electrodes installed between the ring electrode and the endcaps and at the holes required in the

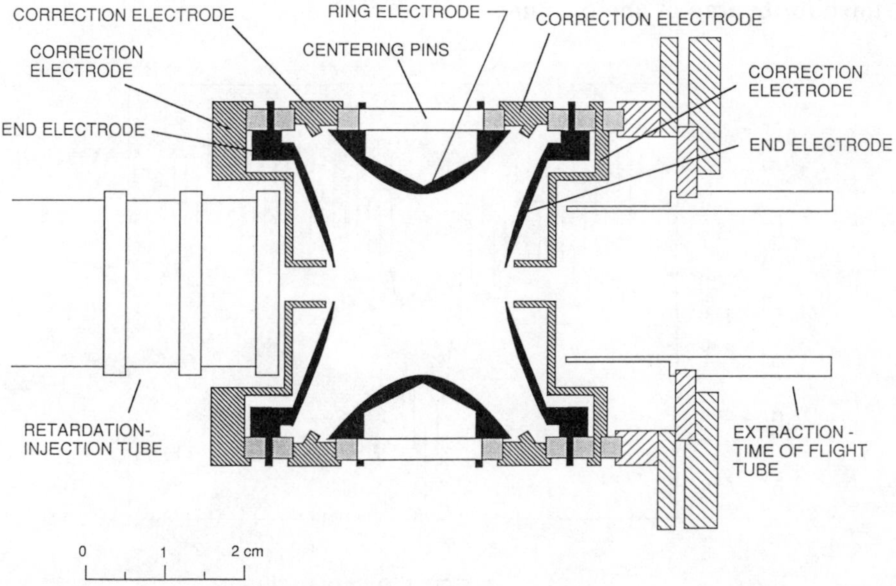

Fig. 3. The high-precision trap with four compensation electrodes used in the ISOLTRAP spectrometer.

endcaps for the injection and ejection of ions. The azimuthal quadrupole field for the excitation of the ions motion is created by splitting the ring electrode into four quadrants. The trap itself is placed off-center with respect to the axis of the magnetic field so as to capture the ions at a magnetron orbit of $\rho \approx 1$ mm radius. The first step carried out in the trap is the removal of unwanted ions (i.e. isobars) by selectively driving their motion with a dipole RF field at the corresponding ω_+ frequency. For this purpose two of the four ring segments are used. Subsequently, the remaining ions are excited by the azimuthal quadrupole RF field created by all four ring segments for a period of typically $T_{RF} = 900$ ms. For the detection of the gained radial energy, the ions are extracted as short pulses from the trap. This is accomplished by lowering the trapping potential slowly and by applying short voltage pulses to the upper endcap and correction electrode. The extracted ion bunches travel through a drift section consisting of a number of electrodes and finally arrive at a multi-channel plate ion detector, used for the determination of their time of flight after extraction. As discussed above, the cyclotron resonance curve is obtained from this time of flight as a function of the applied radio frequency. As an example, Fig. 4 shows a resonance curve obtained for ^{120}Cs ($T_{1/2} = 64$ s). As expected from the excitation time of $T_{RF} = 1.8$ s used in this case, a Fourier-limited line-width of $\Delta\nu_c(FWHM) = 0.5$ Hz is achieved, corresponding to a resolving power of $R = 1.5$ million. The time and the total number of detected ions required for the measurement of such a resonance are typically < 10 min and of the order of 1000, respectively. Typically, 20–40 radioactive ions are stored in the trap at any one time.

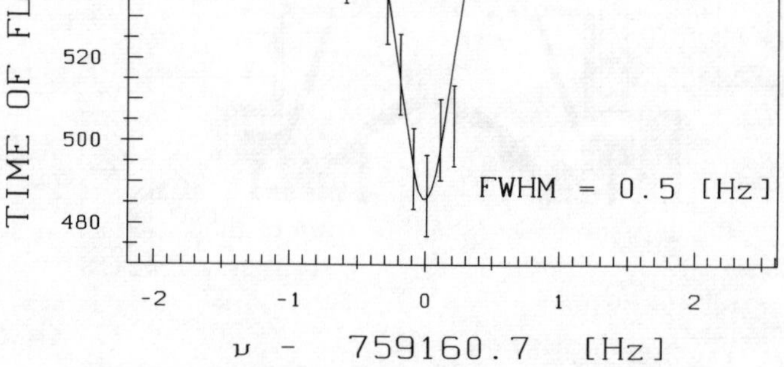

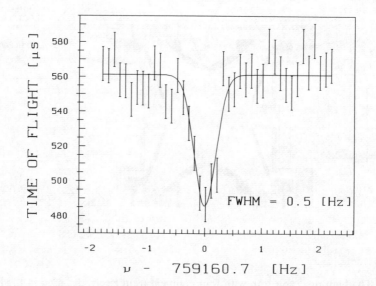

Fig. 4. Cyclotron resonance of ^{120}Cs obtained with the ISOLTRAP spectrometer.

IV. PERFORMANCE

A. Calibration

The calibration of the magnetic field is performed via the determination of the cyclotron resonance of a stable isotope with a well-known mass.[25] This measurement is carried out frequently in order to follow a possible drift or to detect sudden changes of the magnetic field. The time dependence of the magnetic field between two calibration measurements is assumed to be linear and is taken into account in the evaluation of the mass values for the unstable isotope measured in between.

B. Accuracy

The accuracy of the system has been checked by measuring the well-known mass of stable isotopes to high statistical precision. In particular, numerous measurements have been performed on ^{85}Rb, ^{87}Rb and ^{133}Cs whose masses are known to an accuracy of a few keV.[25] A single determination of the cyclotron resonance frequency of these stable isotopes has a typical fit error of some parts in 10^{-8}. Repeating the measurements several times one obtains an instrumental error below 10^{-8}. Taking ^{85}Rb as a reference isotope and comparing the determined masses with the tabulated ones, agreement within 3×10^{-8} is observed for ^{87}Rb and better than 1×10^{-7} for ^{133}Cs. Hence it can be concluded that calibration errors are less than 10^{-7} within a mass range of up to $\Delta A/A \approx 60\%$ for $A = 85$. In on-line runs, where less time can be devoted to each isotope, the accuracy is determined by the limited statistics available for each isotope and the accuracy with which the various parameters can be controlled. Therefore a systematic error of 10^{-7} is assigned to each cyclotron resonance frequency determination which includes possible calibration errors. This is a conservative estimate for all possible errors other than statistical uncertainties.

The high resolving power of a Penning trap mass spectrometer allows the resolution of the cyclotron resonances of ground and isomeric states in the case of $\Delta E = E_{isomer} - E_{groundst.} > h \cdot \Delta\nu(FWHM)$. Figure 5 shows as an example the cyclotron resonance of both the ground ($T_{1/2} = 17.7$ min) and isomeric state ($T_{1/2} = 5.7$ min) of ^{84}Rb (top) and the cyclotron resonance ^{84g}Rb (bottom).[12] For the latter mass determination, the measurement was started after the isomeric state of the ^{84}Rb sample collected on the foil had already decayed. Since the masses of the ground and isomeric state can be distinguished, the uncertainty of the population of the two states in the nuclear reaction has no influence on the accuracy of the mass determination. This is an important advantage of the Penning trap in comparison to mass spectrometers of low resolving power where the isomeric contamination may lead to large errors or requires additional γ-spectroscopy in order to determine the respective population.

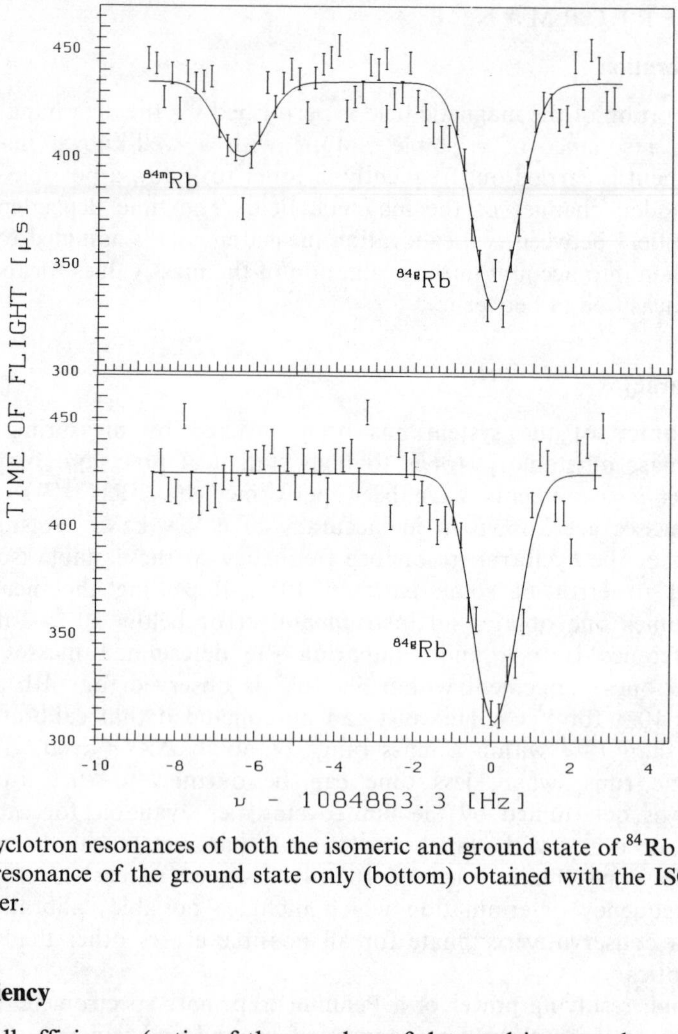

Fig. 5. Cyclotron resonances of both the isomeric and ground state of ^{84}Rb (top) and cyclotron resonance of the ground state only (bottom) obtained with the ISOLTRAP spectrometer.

C. Efficiency

The overall efficiency (ratio of the number of detected ions to the number of ions implanted in the foil) of the ISOLTRAP spectrometer is in the order of 10^{-4} for alkali atoms and results from the following partial efficiencies:

1. The release and ionization efficiency of the ions implanted in the rhenium foil in trap 1. This efficiency depends strongly on the ionization energy of the investigated element and the temperature of the ionizer. It is about unity for alkali atoms and typically one order less for the alkali earth elements.
2. The capture efficiency in trap 1. Only a small fraction (10^{-4} – 10^{-3}) of the ions leaving the foil lose enough kinetic energy in the buffer gas to be trapped.
3. The transfer efficiency from trap 1 to trap 2. This efficiency was found to be close to 100%.

4. The efficiency of the multi-channel plate ion detector, which is assumed to be about 30%.

A significant improvement can be achieved, if the ions enter the first trap already as an ion bunch. Present developments, either using a pulsed laser for heating the surface ionizer or applying an additional Paul trap as a buncher for the ISOLDE beam, will considerably improve the total efficiency.[8,26,27] A further increase of the efficiency can be achieved if a long cooler trap with higher buffer gas pressure is used. Such a trap has been built and has recently been incorporated in the ISOLTRAP set-up.[8]

D. Applicability

Presently, the ISOLTRAP mass spectrometer can only be applied to mass measurements of isotopes of those elements which are surface ionizable. By adding a third trap (Paul trap), the ion beam of an on-line isotope separator can be retarded and captured in flight. Positive and negative ions of an energy of 60 keV have recently been captured with an efficiency of $\varepsilon \approx 0.2\%$. An increase in efficiency can be anticipated with minor improvements.[26,27] All isotopes available at on-line isotope facilities with sufficient yields would then be accessible to mass measurements with this triple-trap mass spectrometer.

As shown above, the limit in respect to half-life is of the order of $T_{1/2} = 1$ s for an accuracy of $\delta m/m = 10^{-7}$ and for the case of an isotope of mass number $A = 100$. Since the half-life limit scales with the mass number and inversely with the desired accuracy, even mass measurements of short-lived isotopes like, for example, ^{11}Li ($T_{1/2} = 9$ ms) could be performed with an accuracy of $\delta m \approx 10$ keV.

REFERENCES

1. **Schwinberg, P.B., Van Dyck, R.S., Jr, and Dehmelt, H.G.,** Trapping and thermalization of positrons for Geonium experiments, *Phys. Lett. A,* **81**, 119 (1981).
2. **Van Dyck, R.S., Jr, Moore, F.L., Farnham, D.L., and Schwinberg, P.B.,** Improved measurement of the proton-electron mass ratio, *Bull. Am. Phys. Soc.,* **31**, 244 (1986).
3. **Gabrielse, G., Fei, X., Orozco, L.A., Tjoelker, R.L., Haas, J., Kalinowsky, H., Trainer, T., and Kells, W.,** Thousandfold improvement in the measured antiproton mass, *Phys. Rev. Lett.,* **65**, 1317 (1990).
4. **Van Dyck Jr., R.S., Farnham, D.L., and Schwinberg, P.B.,** Tritium-helium-3 mass difference using the Penning trap mass spectroscopy, *Phys. Rev. Lett.,* **19** 2888 (1993).
5. **Cornell, E.A., Weisskopf, R.M., Boyce, K.R., Flanagan, R.W., Jr, Lafatis, G.P., and Pritchard, D.E.,** Single ion cyclotron resonance measurement of $M(CO^+)/M(N_2^+)$, *Phys. Rev. Lett.,* **63**, 1674 (1989).
6. **Bollen, G., Hartmann, H., Kluge, H.-J., König, M., Otto, Th., Savard, G., and Stolzenberg, H.,** Towards a"perfect" Penning trap mass spectrometer for unstable isotopes, *Phys. Scr.,* **46**, 581 (1992).

7. **Kluge, H.-J. and Bollen, G.**, ISOLTRAP: A Tandem Penning Trap Mass Spectrometer for Radioactive Isotopes, *Hyperfine Interactions,* **81**, 15 (1993).
8. **Bollen, G.**, Mass Determination of Radioactive Isotopes with the ISOLTRAP Spectrometer at ISOLDE /CERN, Proc. 91th Nobel Symposium on "Trapped Charged Particles and Fundamental Physics", Lysekil, Sweden, 1994, *Physica Scripta,* (1995) in print.
9. **Stolzenberg, H., Becker, St., Bollen, G., Kern, F., Kluge, H.-J., Otto, Th., Savard, G., Schweikhard, L., Audi, G., and Moore, R.B.,** Accurate mass determination of short-lived isotopes by a tandem Penning-trap mass spectrometer, *Phys. Rev. Lett.,* **65**, 3104 (1990).
10. **Bollen, G., Kluge, H.-J., Otto, Th., Savard, G., Schweikhard, L., Stolzenberg, H., Audi, G., Moore, R.B., and Rouleau, G.,** Mass determination of francium and radium isotopes by a Penning trap mass spectrometer, *J. Mod. Opt.,* **39**, 257 (1992).
11. **Otto, Th., Bollen, G., Kluge, H.-J., Savard, G., Schweikhard, L., Stolzenberg, H., Audi, G., Moore, R.B., Rouleau, G., Szerypo, J., The ISOLDE Collaboration,** Penning trap mass spectrometry of neutron-deficient Rb and Sr isotopes, *Nucl. Phys. A,* **567**, 281 (1994).
12. **Bollen, G., Kluge, H.-J., König, M., Otto, Th., Savard, G., Stolzenberg, H., Moore, R.B., Rouleau, G., and Audi, G.,** Resolution of nuclear ground and isomeric states by a Penning trap mass spectrometer, *Phys. Rev. C,* **46**, R2140 (1992).
13. **Brown, L.S., and Gabrielse, G.,** Geonium theory: physics of a single electron or ion in a Penning trap, *Rev. Mod. Phys.,* **58**, 233 (1986).
14. **Bollen, G., Moore, R.B., Savard, G., and Stolzenberg, H.,** The accuracy of heavy-ion mass measurements using time of flight ion cyclotron resonance in a Penning trap. *J. Appl. Phys.,* **68**, 4355 (1990).
15. **König, M., Bollen, G., Kluge, H.-J., Otto, T., and Szerypo, J.**, Quadrupole Excitation of Stored Ion Motion at the True Cyclotron Frequency, *Int. J. Mass Spec. Ion. Processes,* **142**, 95 (1995).
16. **Savard, G., Becker, St., Bollen, G., Kluge, H.-J., Moore, R.B., Otto, Th., Schweikhard, L., Stolzenberg, H., and Wiess, U.,** A new cooling technique for heavy ions in a Penning trap, *Phys. Lett. A,* **158**, 247 (1991).
17. **Dehmelt, H.G.,** Radiofrequency spectroscopy of stored ions. II. Spectroscopy, *Adv. At. Mol. Phys.,* **5**, 109 (1969).
18. **Itano, W.M., and Wineland, D.J.** Laser cooling of ions in harmonic and Penning traps, *Phys. Rev. A,* **25**, 35 (1982).
19. **Wineland, D.J., Itano, W.M., Bergquist, J.C., and Huelt, R.G.,** Laser-cooling limits and single-ion spectroscopy, *Phys. Rev. A,* **36**, 2220 (1987).
20. **Gabrielse, G., Fei, X., Orozco, L.A., Tjoelker, R.L., Haas, J., Kalinowsky, H., Trainor, T., and Kells, W.,** Cooling and slowing of trapped antiprotons below 100 meV, *Phys. Rev. Lett.,* **63**, 1360 (1989).
21. **Wineland, D.J., and Dehmelt, H.G.,** Principles of the stored ion calorimeter *J. Appl. Phys.,* **46**, 919 (1975).
22. **Marshall, A.G., and Schweikhard, L.,** Fourier-transform ion cyclotron resonance mass spectrometry: technique developments, *Int. J. Mass Spectrom. Ion Processes,* **37**, 118 (1992).
23. **Graeff, G., Kalinowsky, H., and Traut, J.,** A direct determination of the proton-electron mass ratio, *Z. Phys. A,* **297**, 35 (1980).
24. **Schnatz, H., Bollen, G., Dabkiewicz, P., Egelhof, P., Kalinowsky, H., Schweikhard,**

L., Stolzenberg, H., and Kluge, H.-J., In-flight capture of ions into a Penning trap, *Nucl. Instrum. Methods A,* **251**, 17 (1986).

25. **Audi, G. and Wapstra, A.H.**, The 1993 atomic mass evaluation (I). Atomic mass table, *Nucl. Phys. A,* **565**, 1 (1993).
26. **Moore, R.B., and Rouleau, G.,** In-flight capture of an ion beam in a Paul trap, *J. Mod. Opt.,* **39**, 361 (1992).
27. **Moore, R.B., Lunney, M.D.N., Rouleau, G., and Savard, G.,** The manipulation of ions using electromagnetic traps, *Phys. Scr.,* **46**, 569 (1992).

4

NUCLEAR DEFORMATIONS

W. Nazarewicz and I. Ragnarsson

Table of contents

I. INTRODUCTION

In specific regions of the nuclear periodic chart, large quadrupole moments are observed and the low-lying excitations have a rotational character. These general features are understood if the nuclei in question are assumed to have a stable deformation, i.e. a non-spherical distribution of the nuclear matter. The need to introduce nuclear deformation becomes even more evident when considering the coexisting (isomeric) nuclear states and the fission process. While fission is a dynamical process, the nucleus can also be *trapped* in a long-lived very elongated state. Such superdeformed configurations were found experimentally in the actinides (fission isomers) and more recently in lighter nuclei such as the rare-earths (superdeformed high spin states).

There are many ways to describe the phenomenon of nuclear deformation. One can use a geometric language of shapes, shell effects, and closed classical trajectories.[1] Another closely related picture comes from a spontaneous symmetry-breaking mechanism (Jahn–Teller effect) caused by the multipole–multipole interaction which leads to a strong coupling between degenerate single-particle states.[2] The actual presence of a static deformation depends crucially on a delicate balance between the symmetry-violating particle–vibration coupling and the symmetry-restoring pairing force, which tries to push the system towards the region of highest density of states.

The models explicitly constructed in the *laboratory frame* such as the spherical shell model based on the rotationally-invariant many-body Hamiltonian do not really need the notion of nuclear deformation. However, nuclear deformations can be extracted *in a model-dependent way* from calculated moments and transition rates and then related to specific effective interactions. For instance, in the spherical shell model the quadrupole deformation can be attributed to the long-ranged proton–neutron quadrupole–quadrupole force.[3]

In this chapter, we mainly concentrate on ground state deformations. The axial quadrupole degree of freedom corresponding to a reflection-symmetric stretching preserving axial symmetry is then most important. Other types of deformation can be introduced without affecting the nuclear properties in any major way as long as the axial and reflection symmetry are preserved. New features show up if one and/or the other of these symmetries are broken. It is still disputed in which way permanent triaxial or reflection-asymmetric shapes are realized in atomic nuclei. In any case, these nuclear collective degrees of freedom are important and it is interesting to consider such stable deformations.

A detailed discussion of nuclear deformations certainly goes beyond the scope of this chapter. Our intention is to give the reader a short summary of basic theoretical concepts and experimental techniques, and present different examples of calculations compared with experimental data. For a more complete discussion on specific subjects we would like to refer the reader to recent publications: ref. 4 (review of nuclear shapes in the mean field theory) and ref. 5 (review on shape coexistence). The material is organized as follows. Parametrizations of deformed nuclear shapes are discussed in section II. In section III the methods of extracting nuclear deformations from measured moments are discussed. Section IV reviews different experimental techniques that enable us to extract information on nuclear deformations. Experimental systematics are discussed in section V. Microscopic models for nuclear deformations are presented in section VI. Finally, section VII contains examples of calculations compared with the data.

II. PARAMETRIZATION OF NUCLEAR SHAPE

Thanks to the saturation property of the nuclear forces the nucleonic density, $\rho(\vec{r})$, is roughly constant $[\rho(\vec{r}) \approx \rho_\text{o}]$ inside the nucleus ($r < R \approx 1.2A^{1/3}$ fm) and quickly drops to zero in a rather narrow region of $\Delta r \sim 1$ fm which defines the nuclear surface $\Sigma(\vec{r})$. Since the surface is not sharp, the associated shape must be introduced through a specific contour in the nucleonic density distribution, usually the midpoint between the maximum value and zero:

$$\rho(\vec{r})_\Sigma = \frac{1}{2}\rho_\text{o}. \tag{1}$$

A. Multipole expansions. Deformations $\alpha_{\lambda\mu}$ and β_λ

At ground state deformations, the nuclear shape is usually parametrized in terms of a spherical harmonic (multipole) expansion. The length of the radius vector pointing from the origin to the surface is given by[6]

$$R(\Omega;\bar{\alpha}) = C(\bar{\alpha})r_\text{o}A^{1/3}\left[1+\sum_{\lambda=1}^{\lambda_\text{max}}\sum_{\mu=-\lambda}^{\lambda}\alpha^*_{\lambda\mu}Y_{\lambda\mu}(\Omega)\right], \tag{2}$$

where $\Omega \equiv (\theta,\varphi)$ and $(\bar{\alpha})$ denotes the set of deformation parameters $\alpha_{\lambda\mu}$. The requirement that the radius is real imposes the condition

$$(\alpha_{\lambda\mu})^* = (-)^\mu \alpha_{\lambda-\mu}, \tag{3}$$

In many cases one assumes that the shape is invariant with respect to the three symmetry planes. This leads to

$$\alpha_{\lambda\mu} = \alpha_{\lambda\,-\mu} \text{ and } \alpha_{\lambda\mu} = 0 \text{ for odd } \lambda \text{ and/or odd } \mu\,. \tag{4}$$

To make sure that the total volume enclosed by Σ, V_Σ, is conserved (i.e. independent of nuclear deformation) the constant $C(\bar{\alpha})$ is introduced through the requirement

$$\int_{V_\Sigma} d^3\vec{r} = \frac{4}{3}\,\pi R_o^3\,. \tag{5}$$

In the expansion of the nuclear radius, eq. (2), a term $\alpha_0 Y_{00}$ is sometimes introduced instead of $C(\bar{\alpha})$. Because $Y_{00} = 1/\sqrt{4\pi}$, it is straightforward to rewrite the corresponding equation to find

$$\begin{aligned} &1 + \alpha_0(\bar{\alpha})\sqrt{1/4\pi} \rightarrow C(\bar{\alpha})\,,\\ &\alpha_{\lambda\mu}/\left[1 + \alpha_0(\bar{\alpha})\sqrt{1/4\pi}\right] \rightarrow \alpha_{\lambda\mu}\,. \end{aligned} \tag{6}$$

The three dipole deformations, $\alpha_{1\pm1}$ and α_{10}, are given by the condition that fixes the center of mass at the origin of the body-fixed frame

$$\int_{V_\Sigma} \vec{r} d^3\vec{r} = 0\,. \tag{7}$$

For the shapes axially symmetric with respect to the z-axis all deformation parameters with $\mu \neq 0$ disappear. The remaining deformation parameters $\alpha_{\lambda 0}$ are usually called β_λ

$$\beta_\lambda \equiv \alpha_{\lambda 0}. \tag{8}$$

Nuclear shapes with quadrupole ($\lambda = 2$), octupole ($\lambda = 3$), and hexadecapole ($\lambda = 4$) deformations are shown in Fig. 1.

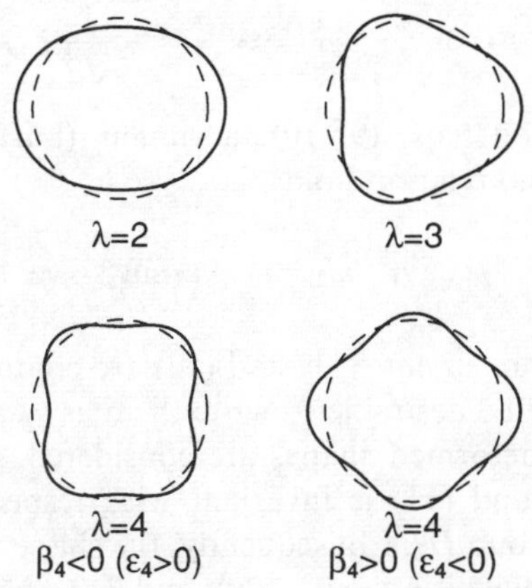

Fig. 1. Nuclear shapes with quadrupole ($\lambda = 2$), octupole ($\lambda = 3$), and hexadecapole ($\lambda = 4$) deformations.

B. Deformations $\varepsilon_{\lambda\mu}$

In an alternative approach, the shape can be defined through the shape of the average potential. This approach is naturally accommodated if the modified oscillator (MO) potential is used. The spatial part of the mo potential is given in terms of an expansion in Legendre polynomials[7, 8],

$$V_{MO} = \frac{1}{2}\hbar C(\bar{\varepsilon})\mathring{\omega}_0 \cdot \rho_t^2 \left\{ 1 - \frac{2}{3}\sqrt{\frac{4\pi}{5}}\left[\varepsilon_{20} Y_{20} - \sqrt{\frac{3}{8}}\varepsilon_{22}(Y_{22} + Y_{2-2}) \right] + 2\sum_{\lambda} \varepsilon_\lambda P_\lambda \right\}, \tag{9}$$

where the equipotential surfaces, $V_{MO} = \text{const.}$, define the shape:

$$\rho_t^2 = \frac{\rho_0^2}{C(\bar{\varepsilon})}\left\{ 1 - \frac{2}{3}\sqrt{\frac{4\pi}{5}}\left(\varepsilon_{20} Y_{20} - \sqrt{\frac{3}{8}}\varepsilon_{22}(Y_{22} + Y_{2-2}) \right) + 2\sum_{\lambda} \varepsilon_\lambda P_\lambda \right\}^{-1/2}. \tag{10}$$

In these equations, the constant $C(\bar{\varepsilon})$ is defined through the relation (5). The nuclear radius is denoted by ρ_t because it is defined in the stretched[9, 10] coordinate system, as discussed below.

C. Quadrupole deformations

The quadrupole deformations ($\lambda = 2$) are usually the most important of the shape parameters. Three of the five quadrupole deformation parameters determine the orientation of the nucleus in space, and may be related to the Euler angles by means of the three conditions for the principal axes of the quadrupole tensor; i.e. $\alpha_{21} = \alpha_{12} = \alpha_{22} - \alpha_{2-2} = 0$. The remaining two quadrupole deformations, α_{20} and α_{22}, are usually written in terms of the polar coordinates β_2 and γ[2, 6]

$$\alpha_{20} = \beta_2 \cos\gamma, \quad \alpha_{22} = \alpha_{2-2} = -\frac{1}{\sqrt{2}}\beta_2 \sin\gamma. \tag{11}$$

In the ε-parametrization, eqs. (9, 10), a similar (but not identical due to conventional definitions) relation holds, i.e.

$$\varepsilon_{20} = \varepsilon_2 \cos\gamma, \quad \varepsilon_{22} = \frac{2}{\sqrt{3}}\varepsilon_2 \sin\gamma, \quad \varepsilon_2 \equiv \varepsilon. \tag{12}$$

The triaxial deformations in eqs. (11) and (12) are commonly denoted by γ in both although it would be desirable to write them as γ_β and γ_ε.

If only quadrupole-deformed shapes are considered, the surface Σ defined by means of eqs. (2) and (11) is invariant with respect to transformations induced by the point group D_2. Consequently, the shape is axial if $\beta_2 > 0$ and γ is a multiple of 60° (prolate for $\gamma = 0°$, 120°, and 240°; oblate for $\gamma = 60°$, 180°, and 300°). If γ is not a multiple of 60°, the shape is triaxial. Thus, for quadrupole nuclear shapes, it is enough to consider only one sector, e.g. $0° \leq \gamma \leq 60°$;

other sectors are obtained by a simple relabelling of the axes. For axial deformations, oblate shapes can also be achieved with $\beta_2 < 0$ (or $\varepsilon < 0$) and $\gamma = 0°$. In eqs. (11) and (12), the signs are chosen so that for oblate shape at $\gamma = 60°$ the x-axis is the symmetry axis (Lund convention).

If only the ε and γ parameters are considered in eq. (9), V_{MO} becomes equivalent to an anisotropic harmonic oscillator potential defined by means of the three frequencies ω_x, ω_y, ω_z. The surface (10) becomes an ellipsoidal surface with principal axis proportional to $(1/\omega_x)$, $(1/\omega_y)$, and $(1/\omega_z)$. The coefficients of eq. (9) are chosen so that for $\gamma = 0°$, the ratio between the symmetry axis and the perpendicular axis becomes $(1 + \varepsilon)$ to the lowest order. Eqs. (9, 10) are expressed in the stretched coordinate system which has been introduced to remove the coupling between different N-shells. The stretched (transformed, "t") coordinates, namely $\xi = x_t$, $\eta = y_t$, $\zeta = z_t$, are related to the normal coordinates (x, y, z) by means of the simple scaling transformation, $(x_t)_i = x_i\sqrt{M\omega_i/\hbar}$. Consequently, "half"of the deformation comes from the scaling ("stretching") of the coordinate system, and another "half"from an explicit deformation within the stretched system. The angles entering $Y_{\lambda\mu}(\theta_t, \varphi_t)$ and $P_\lambda(\cos\theta_t)$ in eqs. (9, 10) are defined in the stretched system.

The harmonic oscillator potential can also be defined in the spherical coordinates[9], in which case the factor $(2\varepsilon/3)$ in eqs. (9, 10) is replaced by a factor $(4\delta/3)$; i.e. the quadrupole deformation coordinate is labelled δ instead of ε. In this case, the "full" deformation is expressed in the spherical coordinate system. Since the parameters $(\varepsilon, \gamma_\varepsilon)$ and (δ, γ_δ) describe ellipsoidal shape, it is possible to find an exact correspondence as given, e.g. in ref. 11. Other terms, $2\sum\delta_\lambda P_\lambda$, are then introduced in a similar way as in eqs. (9, 10). However, one should note that the shapes generated by the ε_λ and δ_λ $(\lambda \geq 3)$ deformations are different.

In many applications, only axially symmetric quadrupole deformation $(\gamma = 0°)$ is considered. The equipotential surface of eq. (10) has a spheroidal shape. Spheroidal shapes are often expressed in β-parameters in which case the value of β is chosen so that the axis ratio becomes the same as in eq. (2).

D. Relationship between ε_λ and β_λ

There is no unique way to relate one set of deformation parameters to another. As discussed, e.g. in ref. 12, a standard method of comparing shapes is to require the equality between multipole moments. For shapes with smooth curvatures (no wiggles) the connection between different parameter sets is rather well-defined. For example, technical details on the transformation between the β and ε parametrizations are given in ref. 12, and the relationship between the corresponding quadrupole and hexadecapole parameters is shown in Fig. 2. Similar figures have been published, for example, in refs. 7, 13 where, contrary to Fig. 2, no higher multipoles have been considered when mapping the shapes. Note also that there is an error in the sign of ε_4 in ref. 7.

Since very elongated (superdeformed) shapes are also discussed in this

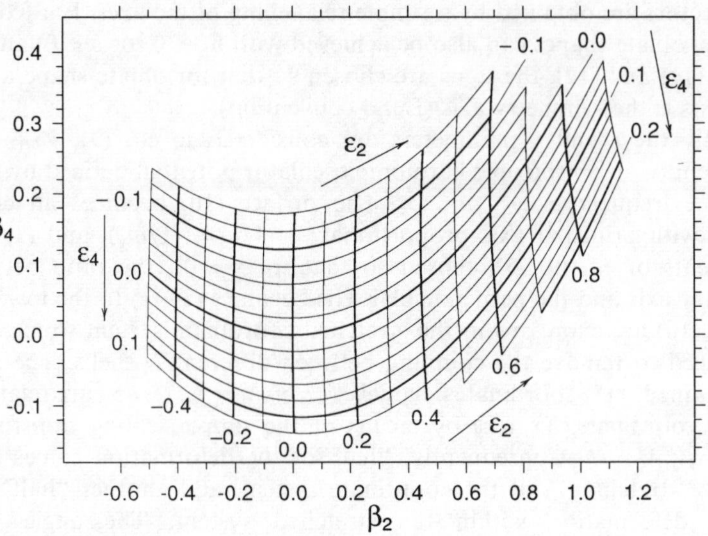

Fig. 2. The transformation diagram between $(\varepsilon(\equiv \varepsilon_2), \varepsilon_4)$ and (β_2, β_4). Deformation parameters β_λ with $\lambda \leq 12$ were considered in the fit, where the Q_λ moments with $\lambda = 2$, 4, 6, 8, 10, and 12 were required to be the same. Note that (i) the transformation between ε and β_2 is rather insensitive to hexadecapole deformations, (ii) the shapes along the $\varepsilon_4 = 0$ and $\beta_4 = 0$ lines are very different, and (iii) ε_4 and β_4 have opposite signs.

chapter, it is worth pointing out some specific features of the hexadecapole parameters which become especially evident at large deformations. For a fixed quadrupole deformation, we may define an average fission valley by minimizing the liquid drop energy with respect to the hexadecapole deformation. Then, excluding very light nuclei, this valley falls somewhere in the middle of the $\beta_4 = 0$ and $\varepsilon_4 = 0$ lines of Fig. 2. The average fission trajectory can then be described by either a positive β_4 or a positive ε_4. However, the roles of the hexadecapole deformations are very different in both cases. In the ε parametrization, the positive ε_4 introduces a tendency for "necking" while in the β parametrization, the positive β_4 tends to remove a rather strong necking which is present at large β_2 deformations. Other examples are the superdeformed high-spin states around ^{152}Dy which have spheroidal-like shapes (with $\varepsilon_2 = 0.55$–0.60 or $\beta_2 = 0.58$–0.62) with *small* ε_4 ($\varepsilon_4 = 0.02$–0.04)[14] but *large* β_4 ($\beta_4 = 0.10$–0.12).[15]

E. Other deformation parameters

There exist some variants of the ε or δ parametrizations, namely the η deformations[16] and the ν deformations[17]. Both of these describe the same shapes as those given by the δ deformations discussed above and exact relationships exist between these different parameters. Specific examples are listed in the Appendix A of ref. 17. For ε values in the range of –0.3 to 0.6, we give in Table 1 a translation between several standard elongation parameters. This comparison, where only one (quadrupole) deformation degree of freedom is considered explicitly, is certainly somewhat crude, but still provides a good idea of the general scaling between different deformations. For more details on how to relate different spheroidal parametrizations we refer the reader to ref. 20.

When discussing large deformations and fission, the straightforward expansions in spherical harmonics are not very appropriate. For axial systems the (c, h, α) cylindrical parametrization of nuclear shape has been introduced[21,22]

$$(x^2 + y^2)/C^2 = \begin{cases} (1 - z^2/C^2)(A + Bz^2/C^2 + \alpha z/C) & \text{for } B \geq 0, \\ (1 - z^2/C^2)\,[(A + \alpha z/C)\exp{(Bc^3 z^2/C^2)}] & \text{for } B < 0, \end{cases} \tag{13}$$

where C is determined by the volume conservation condition, $B = 2h + (c - 1)/2$, and $A = 1/c^3 - B/5$. The parameter c describes the elongation of the system, h is the neck coordinate, and α is an asymmetry parameter. Ellipsoidal shapes come out as a special case. A different functional dependence is used for diamond-like shapes ($\varepsilon_4 < 0$) than for "tin-can" shapes ($\varepsilon_4 > 0$). The liquid drop valley is approximately described by $h = 0$.

Recently, the (c, h) parametrization has been written in terms of new deformations β and r.[19] For ellipsoidal shape, β is defined from the ratio of the nuclear axes and r is a measure of the nuclear surface in the (x, y)-plane for

Table 1 Relationship between different elongation parameters. In all cases (except for the β_2 parametrization) the shape is spheroidal which means that an exact relation exists. The parameters δ and ν are associated with a modified oscillator potential expressed in spherical coordinates (see e.g. refs. 9, 11 and ref. 17, respectively). The β_2 values are calculated by requiring the equality of the Q_{20} and Q_{40} surface multipole moments.[12] Different assumptions for the hexadecapole deformation indicate the approximate character of the relation between ε and β_2. The β parameter (used, for example, in refs. 18, 19) corresponds to the spheroidal shape with the same major-to-minor axis ratio $a_\parallel/a_\perp$ as defined by β_2. The axis ratio is given in the last column.

ε	δ	ν	$\beta_2(\beta_4 = 0)$	$\beta_2(\varepsilon_4 = \varepsilon^2/6)$	$\beta_2(\varepsilon_4 = 0)$	β	$a_\parallel/a_\perp$
−0.3	−0.3088	−0.2877	−0.308	−0.307	−0.306	−0.2882	0.750
−0.2	−0.2048	−0.1942	−0.207	−0.207	−0.207	−0.1982	0.824
−0.1	−0.1014	−0.0984	−0.104	−0.104	−0.104	−0.1022	0.906
0.0	0.0	0.0	0.0	0.0	0.0	0.0	1.0
0.1	0.0981	0.1018	0.107	0.107	0.107	0.1093	1.107
0.2	0.1916	0.2076	0.217	0.217	0.217	0.2265	1.231
0.3	0.2794	0.3185	0.332	0.330	0.329	0.3523	1.375
0.4	0.3605	0.4353	0.452	0.447	0.444	0.4878	1.545
0.5	0.4342	0.5596	0.581	0.568	0.561	0.6341	1.750
0.6	0.5000	0.6931	0.723	0.695	0.680	0.7927	2.000

$z = 0$ ($r = 0$ at scission, $r = 1$ for ellipsoidal shapes, and $r > 1$ for diamond-like shapes).

A different parametrization has been used in ref. 23. It is based on the observation that the liquid drop fission valley is quite accurately described by Cassinian ovals [similar to the (c, h) family of shapes with $h = 0$] with the elongation parameter denoted as ε. Deviations from such shapes are then expressed by means of an expansion in Legendre polynomials. The corresponding deformation parameters are denoted by α_m where, for example, α_4 plays a similar role as h in the (c, h) parametrization. Both (c, h) and (α_m) parametrizations are capable of describing the bifurcation of the surface into two parts.

The three-quadratic surface parametrization of ref. 24 is sometimes used at large elongations. It is, however, not very suitable to describe nuclear ground state deformations.

F. Triaxial deformations

The most important deformation that breaks axial symmetry is the quadrupole degree of freedom γ. If $\gamma \neq 0$ and higher multipolarity deformations are considered, it becomes necessary to include also triaxial deformations with $\mu \neq 0$ and $\lambda > 2$. Consider, for example, an axial (β_2, β_4) shape symmetric around the z-axis. If $\gamma \neq 0$, the shape becomes triaxial. If $\gamma \to 60°$, the quadrupole deformations generate a shape which becomes axially symmetric around the x-axis. On the other hand, the hexadecapole term being axially symmetric around the z-axis breaks the axial symmetry of the shape. It is possible, however, to parametrize the hexadecapole deformations in such a way that the resulting shape transforms in the same way as the quadrupole shape.[25] The hexadecapole tensor can, for example, be constructed as a product of two quadrupole tensors.[26] Triaxial hexadecapole deformations along these lines were considered in refs. 27–30. A generalized parametrization of the hexadecapole deformation has been proposed by Rohoziński and Sobiczewski.[25] In the presence of the three symmetry planes, the hexadecapole tensor can be parametrized by means of three parameters, β_4, γ_4, and δ_4:

$$\alpha_{40} = \beta_4\left(\sqrt{\frac{7}{12}}\cos\delta_4 + \sqrt{\frac{5}{12}}\sin\delta_4\cos\gamma_4\right), \tag{14}$$

$$\alpha_{42} = -\frac{1}{\sqrt{2}}\,\beta_4\sin\delta_4\sin\gamma_4\,, \tag{15}$$

$$\alpha_{44} = \beta_4\left(\sqrt{\frac{5}{24}}\cos\delta_4 - \sqrt{\frac{7}{24}}\sin\delta_4\cos\gamma_4\right), \tag{16}$$

where $0 \leq \delta_4 \leq \pi$ and $0 \leq \gamma_4 \leq \pi/3$.

A parametrization of the octupole tensor has recently been proposed in refs. 31, 32. For recent calculations involving triaxial octupole deformations, see refs. 33, 34.

III. NUCLEAR MOMENTS AND SHAPE DEFORMATIONS

In this section the methods of extracting nuclear deformations from the data are discussed. Since the nuclear deformations are not observables themselves, the information on the nuclear shape usually comes from measured moments of the charge (or mass) density

$$Q_{\lambda\mu}(i \to f) = \left(\frac{16\pi}{2\lambda+1}\right)^{1/2} \langle f|M(\lambda\mu)|i\rangle, \tag{17}$$

where

$$M(\lambda\mu;\vec{r}) = f_{\lambda\mu}(r)\, Y_{\lambda\mu}(\Omega). \tag{18}$$

These are either static moments ($i = f$) or transition moments ($i \neq f$).

The radial form factor, $f_{\lambda\mu}(r)$, in eq. (18) has a particularly simple form for the electric multipole operators

$$f_{\lambda\mu} = e\left[\frac{1}{2} - t_3 + (-1)^{\lambda}\, \frac{Z}{A^{\lambda}}\right] r^{\lambda}, \tag{19}$$

where $t_3 = +1/2$ ($-1/2$) for neutrons (protons) and the (Z, A)-dependent term represents the recoil effect of the nuclear core (especially important for dipole transitions).[35]

If the nucleus is well deformed, the measured moments can be expressed in terms of the intrinsic moments defined in the intrinsic (body-fixed) frame.[1] For instance, the relation between the measured spectroscopic (static) quadrupole moment, $Q_{20}^{(s)}$, and the (static) intrinsic quadrupole moment of the axial nucleus, Q_d, is given by

$$Q_{20}^{(s)} = \left(\frac{16\pi}{5}\right)^{1/2} \langle I|M(E2)|I\rangle = \frac{3\langle K^2\rangle - I(I+1)}{(I+1)(2I+3)}\, Q_d, \tag{20}$$

where K is the projection of the total spin on the intrinsic nuclear axis and $Q_d = Q_{20}$. (Analogous expressions exist[1] for higher moments.) For the axial system, the transition intrinsic quadrupole moment, Q_t, can be deduced from the reduced gamma transition probability within a rotational band [eq. (38)]

$$B(E2; KI_i \to KI_f) = \frac{5}{16\pi}\, Q_t^2\, \langle I_i K 20|I_f K\rangle^2, \tag{21}$$

where $Q_t = Q_{20}$.

Nuclear deformations are extracted from the measured intrinsic moments in a model dependent way through

$$Q_{\lambda\mu} = \left(\frac{16\pi}{2\lambda+1}\right)^{1/2} \int \rho_{\bar{\beta}}(\vec{r}) M(\lambda\mu; \vec{r}) d^3\vec{r}, \tag{22}$$

where the average nuclear (charge) density, $\rho_{\bar{\beta}}(\vec{r})$, is assumed to be a simple function of deformation parameters. Often, a sharp uniform density distribution is assumed:

$$\rho_{\bar{\beta}}(\vec{r}) = \begin{cases} \rho_o & \text{if } \vec{r} \in V_\Sigma \\ 0 & \text{otherwise.} \end{cases} \tag{23}$$

For the density (23), the intrinsic electric moment (22) is given by [neglecting the recoil term in eq. (19)]

$$Q_{\lambda\mu} = \left(\frac{16\pi}{2\lambda+1}\right)^{1/2} \frac{\rho_o}{\lambda+3} \int d\Omega R^{\lambda+3}(\Omega;\bar{\alpha}) Y_{\lambda\mu}(\Omega). \tag{24}$$

If the surface is defined through eq. (2) and deformations are small (one can neglect second-order contributions in α),

$$Q_{\lambda\mu} = \frac{3Ze(r_o A^{1/3})^\lambda}{\sqrt{(2\lambda+1)\pi}} \alpha_{\lambda\mu}. \tag{25}$$

There are also other forms of $\rho_{\bar{\beta}}(\vec{r})$ frequently used, e.g. the Fermi distribution which accounts for the thickness of the nuclear surface.[1]

If the nuclear shape is triaxial, the relations (20) and (21) do not hold. For small quadrupole deformations, the static intrinsic quadrupole moment, Q_d, and the transition quadrupole moment, Q_t, are[36]

$$Q_d = Q_{20} \frac{\sin(\gamma + 30^\circ)}{\sin(30^\circ)}, \; Q_t = Q_{20} \frac{\cos(\gamma + 30^\circ)}{\cos(30^\circ)}, \tag{26}$$

where Q_{20} is given by eq. (25).

Below are given some useful expressions that relate the intrinsic quadrupole moment to the associated quadrupole deformations (sharp density distribution is assumed):

(i) The uniformly charged ellipsoid with an axis ratio $a_\parallel / a_\perp$:

$$Q_{20} = \frac{2}{5} r_o^2 Z A^{2/3} [(a_\parallel/a_\perp)^2 - 1] (a_\parallel/a_\perp)^{-2/3}; \tag{27}$$

(ii) The δ parametrization:

$$Q_{20} = \frac{4}{5} Z r_o^2 A^{2/3} \left(\delta_2 + \frac{2}{3}\delta_2^2 + \frac{25}{33}\delta_4^2 - 2\delta_2\delta_4 \right) + O(\delta^3); \tag{28}$$

(iii) The ε parametrization:

$$Q_{20} = \frac{4}{5} Z r_o^2 A^{2/3} \left[\varepsilon_2 \left(1 + \frac{1}{2}\varepsilon_2\right) + \frac{25}{33}\varepsilon_4^2 - \varepsilon_2\varepsilon_4 \right] + O(\varepsilon^3); \tag{29}$$

(iv) The (β_2, β_4) parametrization (for an exact expression, see ref. 15):

$$Q_{20} = \frac{3}{\sqrt{5\pi}} Zr_o^2 A^{2/3} \left(\beta_2 2 + \frac{2}{7}\sqrt{\frac{5}{\pi}}\beta_2^2 + \frac{20}{77}\sqrt{\frac{5}{\pi}}\beta_4^2 + \frac{12}{7\sqrt{\pi}}\beta_2\beta_4 \right) + O(\beta^3). \quad (30)$$

IV. EXPERIMENTAL TECHNIQUES

This section contains a summary of basic experimental methods to measure nuclear deformations. Only a very brief review of the relevant experimental methods is given below.

A. Coulomb excitation

Coulomb excitation of atomic nuclei (i.e. an inelastic scattering process in which a charged bombarding particle excites the target nucleus via the electromagnetic field only) is one of the oldest ways to probe nuclear shape. Within the semiclassical approach[37, 38] Coulomb excitation is governed by a system of coupled equations for time-dependent transition amplitudes:

$$\frac{da_i(t)}{dt} = \sum_{\lambda\mu j} f_{\lambda\mu}(t)\langle j\|M(\lambda)\|i\rangle a_j(t). \quad (31)$$

By integrating eq. (31) and performing the χ^2 fit to the Coulomb-excitation cross sections, one can deduce the values of the $\langle j\|M(\lambda)\|i\rangle$ matrix elements. The contemporary Coulomb-excitation codes allow for the determination of hundreds of matrix elements ($E1, E2, E3, E4, E5, E6, M1, M2$) from several thousand data points obtained from many independent experiments.[39, 40]

Recent developments in the field of Coulomb excitation make it possible to measure practically all the $E2$ matrix elements for the low-lying nuclear states in stable nuclei from heavy-ion-induced multiple Coulomb-excitation data. This information can then be employed in the sum rules method[40, 41] for extracting the collective shape parameters. By analogy with the usual quadrupole shape parameters (β_2, γ) given in eq. (11), the $M(E2)$ operators can be related to the two quantities Q and δ:

$$M(E2, 0) = Q\cos\delta, \quad M(E2, 2) = M(E2, -2) = -\frac{1}{\sqrt{2}} Q\sin\delta. \quad (32)$$

By making the $I = 0$ tensor products of several $E2$ operators, e.g.

$$[M(E2) \times M(E2)]_0 = \frac{1}{\sqrt{5}} Q^2, \quad [[M(E2) \times M(E2)]_2 \times M(E2)]_0$$

$$= -\sqrt{\frac{2}{35}} Q^3 \cos 3\delta, \quad (33)$$

one can extract the $E2$ invariants in a model-independent way. For instance,

$$\langle I_i|[M(E2)\times M(E2)]_0|I_i\rangle = \frac{(1)^{2I_i}}{\sqrt{2I_i+1}}\sum_{I_j}\langle I_i\|M(E2)\|I_j\rangle\,\langle I_j\|M(E2)\|I_i\rangle$$

$$\begin{Bmatrix} 2 & 2 & 0 \\ I_i & I_i & I_j \end{Bmatrix}. \tag{34}$$

Knowing the experimental values of the reduced matrix elements of the $E2$ operator one can determine the distribution of the intrinsic quadrupole moment in the body-fixed frame. Examples of such analysis can be found in ref. 42. Representative examples of Coulomb-excitation experiments can be found in ref. 43 (observation of $E4$ transitions in ^{152,154}Sm), ref. 44 (investigation of quadrupole shapes in ^{182,184,186}W), and ref. 45 (angular momentum dependence of $E1$, $E2$, and $E3$ matrix elements in ^{148}Nd).

A recent interesting development is the use of polarized heavy ions. By measuring the tensor analyzing powers T_{20}, one can determine the electric polarizability and the quadrupole moment of the aligned heavy ion.[46,47]

B. Strongly interacting probes

Inelastic hadron scattering experiments at energies above the Coulomb barrier are a useful tool for the determination of nuclear deformation parameters.[48] Moreover, hadronic probes (such as α particles, protons, or neutrons) interact with protons and neutrons, thus yielding information on mass deformations.

Measured angular distributions of differential cross sections are analyzed using the coupled-channel or distorted wave Born approximation formalism with a deformed optical potential, which usually is of a Woods–Saxon type and contains a real part, an imaginary part, and a spin–orbit term. The deformation lengths of the real part of the optical potential can be related to the deformations of the nuclear matter density using a theorem by Satchler.[49,50]

The similarity of the nuclear charge-deformation parameters, β_λ^C, as measured by Coulomb excitation and by inelastic electron scattering, with the potential or matter deformations, β_λ^N, inferred from strong interaction probes, suggests that, at least for deformed nuclei, the charge deformation appears very close to the potential or matter deformation.[51–54] There exists a simple scaling law that relates β_λ^N and β_λ^C.[57] For the systematic trend of deviations between β_λ^N and β_λ^C, see ref. 58.

For additional information, see ref. 55 [nuclear $\lambda = 2$, 4 deformations in the rare earth nuclei from (α, α′)], ref. 51 [nuclear $\lambda = 2$, 4 deformations in ^{182,184,186}W from (α, α′)], ref. 56 (the Fourier–Bessel method for extracting density distributions from inelastic α scattering), ref. 52 [nuclear $\lambda = 2$, 4 deformations in the *sd* nuclei from (n,n′); review on multipole moments of the nuclear potential], refs. 53, 54 (nuclear $\lambda = 2$, 4, 6 deformations in well-deformed rare earth nuclei from the polarized proton scattering; comparison between different folded potentials and different reactions), ref. 59 (nuclear $\lambda = 2$, 4, 6 deformations in ^{176}Yb and ^{182}W from inelastic proton scattering;

comparison between different experimental methods; comparison between different folded potentials and different reactions), ref. 60 [nuclear $\lambda = 2, 4, 6$ deformations in ^{168}Er from (α, α') and (d,d')], and ref. 61 [nuclear $\lambda = 2, 4$ deformations in the *sd* nuclei from (α, α'); comparison between folding potential and Woods–Saxon potential].

C. Electron scattering

Electron scattering provides very precise information on the shape of the nucleus by measuring the charge and current transition densities. In the plane wave Born approximation formalism, the differential cross section for electroexcitation of a discrete nuclear electric transition of multipolarity λ is proportional to the form factor

$$F_\lambda^C(q) = \sqrt{\frac{2I_f+1}{2I_i+1}} \int_0^\infty \rho_\lambda(r) j_\lambda(qr) r^2 dr, \tag{35}$$

which is the Fourier–Bessel transform of the multipole component λ of the transition charge density

$$\rho_\lambda(r) = \int \langle \Psi_f \| \rho(\Omega) Y_\lambda(\Omega) \| \Psi_i \rangle d\Omega. \tag{36}$$

By measuring the cross section as a function of the momentum transfer q, one can deduce the electric form factors and, by reversing eq. (35), the charge density ρ_λ. In other words, the electron scattering experiments provide detailed information on the multipole distribution of charge density, and hence nuclear deformations.[62–66] Representative examples are ref. 67 ($\lambda = 2, 4, 6$ moments in deformed rare earth and actinide nuclei), ref. 68 (combined analysis of (e,e') and muonic X-ray data for ^{152}Sm), ref. 69 (ground state and transition densities for transitional Os and Pt isotopes with $A = 188, 196$), and ref. 70 (ground state and transition densities with $\lambda = 2, 4, 6$ for ^{152}Sm).

D. Lifetime measurements

From lifetimes of the excited nuclear states one can extract information on transition matrix elements of multipole operators and, indirectly, nuclear deformations. The probability for a γ-transition with an energy E_γ is

$$T(\lambda) = \frac{8\pi(\lambda+1)}{\lambda[(2\lambda+1)!!]^2} \frac{1}{\hbar} \left(\frac{E_\gamma}{\hbar c}\right)^{2\lambda+1} B(\lambda), \tag{37}$$

where

$$B(\lambda; I_i \to I_f) = (2I_1+1)^{-1} |\langle f \| M(\lambda) \| i \rangle|^2 \tag{38}$$

is the reduced transition probability. For electric transitions $M(\lambda, \mu)$ is given by eqs. (18)–(19).

Short lifetimes (from ~0.5 ps to ~0.01 ps) of excited (high-spin) states can be determined from the analysis of γ-ray lineshapes using the Doppler shift attenuation method (DSAM)[71]. The residual heavy ions produced by the heavy-ion-induced fusion–evaporation reaction have an initial velocity v/c of the order of a few percent. During the γ-decay process, the excited ions are slowed down in the thick target backing. When the lifetime of an excited state is comparable to the time of the slowing down, the corresponding γ-rays are Doppler shifted thus leading to broadened lineshapes. By simulating the slowing-down process and making the lineshape analysis (taking into account sidefeeding from unknown states), one can extract the lifetimes of the excited (high-spin) states. Finally, from known lifetimes the transition moments can be deduced. Owing to various experimental uncertainties and systematic errors, the accuracy of the DSAM can approach 10–30%. Representative examples of the DSAM analysis are discussed in ref. 72 (lifetimes within the superdeformed band in ^{192}Hg), ref. 73 (lifetimes of high-spin states in Yb nuclei), ref. 74 (lifetime measurements of dipole collective bands in neutron-deficient Pb nuclei).

A powerful method for determining short lifetimes ($\tau < 10^{-11}$ s) of excited nuclear states is the gamma-ray-induced Doppler broadening (GRID) technique.[75] The main idea is to measure the Doppler-shifted γ-rays emitted by the excited states populated by thermal neutron capture, (n,γ). Here, the recoil velocities are about three orders of magnitude lower than in the DSAM method. For recent developments, see ref. 76 (lifetimes in ^{168}Er).

The recoil distance method (RDM) also makes use of the Doppler-shifted γ-rays. In this method, a foil is placed at some (varied) distance from the target. The fraction of γ-rays emitted from nuclei stopped (slowed down) in the foil versus those emitted in flight gives the lifetime of the excited state. The range of lifetimes that can be studied using the RDM varies from $\sim$1 ns to $\sim$1 ps. For representative examples, see ref. 77 (lifetimes of high-spin states in ^{160,161}Yb), ref. 78 (lifetimes of high-spin states in ^{156,157,158}Dy), ref. 79 (lifetime measurements in ^{128,130,132}Nd).

For lifetimes of long-lived (isomeric) states, the standard electronic (pulse beam) techniques are used.[80] Here, the lifetimes are deduced from time spectra relative to the beam pulse obtained with gates on individual γ-rays. Lifetimes can also be directly deduced from the partial width for γ-decay, Γ_γ. The latter quantity can be measured using, for example the nuclear resonance fluorescence method (resonant photon scattering).[81]

E. Hyperfine effects in muonic atoms

The negative muon can serve as a sensitive probe of nuclear charge distributions.[82–84] The main part of the muon–nucleus interaction is the Coulomb potential

$$V = -e^2 \int \frac{\rho(\vec{r})}{|\vec{r}_\mu - \vec{r}|}\, d^3 r, \tag{39}$$

where $\rho(\vec{r})$ is the charge density of a nucleus. By making a multipole expansion of $\rho(\vec{r})$ [or by assuming $\rho(\vec{r})$ in the form of a deformed Fermi distribution], one can extract moments of the nuclear charge distribution by analyzing the X-rays emitted by a muonic atom (typical values of the muonic hyperfine splitting in the uranium $3d$ states are about 30 keV and 1 keV for $E2$ and $E4$ splitting, respectively). Of course, in the detailed calculations it is necessary to take into account the coupling between the muonic levels and the low-lying nuclear states (nuclear polarization), magnetic interaction, radiative corrections, etc.

It should be pointed out that this method allows for the determination of sign, magnitude, and radial distribution of the intrinsic nuclear multipole moments. There exists a rather rich literature on charge distribution analysis by means of muonic atoms, see e.g. refs. 85–90. In recent studies[91,92] the analysis of muonic K, L, M, and N X-rays allowed for a precise determination of $E2$ matrix elements and intrinsic quadrupole and hexadecapole moments in a number of nuclei. Typical precision of the method for the quadrupole and hexadecapole moments is about 1% and 5%, respectively, i.e. it is as least twice that available from other techniques (e.g. Coulomb excitation).

F. Hyperfine techniques

The electric quadrupole interaction between a nucleus with an electric quadrupole moment, eQ, and the electric field gradient (EFG) at the nuclear site, $eq = \partial^2 V/\partial z^2$, leads to a hyperfine splitting of nuclear levels with energies

$$E_m = \hbar\omega_Q \left\{ m^2 - \frac{1}{3} I(I+1) \right\}, \tag{40}$$

where the quadrupole splitting parameter, ω_Q, is given by

$$\hbar\omega_Q = \frac{3e^2qQ}{4I(2I-1)}. \tag{41}$$

[In eq. (40) the asymmetry parameter was put to zero.] There exists a variety of different experimental methods based on the analysis of interaction (40). In addition to old techniques such as Mössbauer spectroscopy or radiofrequency resonance techniques (e.g. the nuclear magnetic resonance or nuclear quadrupole resonance method), new methods based on low temperature nuclear orientation or collinear laser spectroscopy have been developed. A review of various methods can be found in refs. 93–94. A table of nuclear quadrupole moments of ground state and metastable states ($T_{1/2} > 1$ min) is given in ref. 95. For representative examples of experiments, see ref. 96 (quadrupole moments of light thallium isotopes from collinear fast atom beam laser spectroscopy), ref. 97 (quadrupole moments of radium isotopes from collinear fast atom beam laser spectroscopy), and ref. 98 (the quadrupole moment in ^{186}Au from nuclear magnetic resonance).

The time differential perturbed angular distribution technique (TDPAD) exploits the quadrupole interaction (40) of the quadrupole moment of a high-spin isomer with the EFG in non-cubic crystals (see ref. 99). By measuring the time dependence of gamma anisotropy, one can determine $|Q|$. In order to measure the sign of Q one can combine TDPAD with the tilted multifoil (TMF) technique which leads to the nuclear spin polarization via magnetic hyperfine coupling.[100] Representative examples of the TDPAD or TMF/TDPAD analysis are found in ref. 101 (quadrupole moments of the 12^+ isomers in ^{188,190}Hg), ref. 102 (quadrupole moments of high-spin isomers in ^{144,147}Gd), ref. 103 (quadrupole moments of Rn isomers), and ref. 104 (the quadrupole moment of the $K = 25$ isomer in ^{182}Os). Another recently developed technique, based on the same basic principle as TDPAD, is level mixing spectroscopy – a method very suited to studies of high-spin isomers (see recent works on isomers in At and Fr[105] and Tl and Bi[106]).

G. Nuclear radii; *E*0 moments

The mean square nuclear radius (monopole moment) is a sensitive probe of nuclear deformation. Assuming the standard multipole expansion of the nuclear radius vector in eq. (2), the expectation value of the $E0$ moment can be approximated by

$$\langle r^2 \rangle \approx \langle r^2 \rangle_\circ \left(1 + \frac{5}{4\pi} \sum_\lambda \langle \beta_\lambda^2 \rangle \right), \qquad (42)$$

where $\langle r^2 \rangle_\circ = \frac{3}{5} r_\circ^2 A^{2/3}$ (for a recent semiempirical formula, see ref. 107). The quantity $\sum_\lambda \langle \beta_\lambda^2 \rangle$ reflects the deformation susceptibility of the nucleus.[108] For well-deformed nuclei with large quadrupole deformations it will be close to β_2^2. Consequently, by analyzing the deviations of experimental $\langle r^2 \rangle$ values from those provided by the spherical droplet model formula, one can extract the information on nuclear deformation. It is also seen from eq. (42) that an isotope shift, $\delta\langle r^2 \rangle$, is proportional to $\langle r^2 \rangle$ and the change in $\sum_\lambda \langle \beta_\lambda^2 \rangle$ – thus providing information on relative deformation changes between isotopes.[108]

For the systematics of nuclear isotope shifts, see ref. 109. The influence of deformation on $\delta\langle r^2 \rangle$ is discussed, for example in ref. 110.

V. SYSTEMATICS

Since global nuclear properties, such as masses, radii, moments of inertia, or deformations, usually change rather smoothly with proton and neutron number, valuable information can be obtained by analyzing systematic trends.

A. Quadrupole deformations

For well-deformed nuclei the intrinsic quadrupole moment Q_{20} can be expressed via the $B(E2)\uparrow = B(E2; 0^+_{g.s.} \rightarrow 2^+_1)$ transition rate by means of the strong coupling formula (21)

$$Q_{20} = \left[\frac{16\pi}{5}\frac{B(E2)}{e^2}\right]^{1/2}. \tag{43}$$

The quantity $B(E2)\uparrow$ can be related to the γ-ray lifetime

$$\tau_{E2} = \tau(1+\alpha) = (40.82 \times 10^{13})E_{2_1^+}^{-5}[B(E2)\uparrow/e^2b^2]^{-1} \tag{44}$$

where α is the total internal conversion coefficient, the lifetimes τ_{E2} and τ are expressed in ps, $E_{2_1^+}$ in keV, and $B(E2)\uparrow$ in e^2b^2. Equations (25), (43), and (44) provide a way to extract quadrupole deformation from measured lifetimes or $B(E2)$ rates. For example, for small deformations

$$\beta_2 \approx -7\sqrt{\frac{\pi}{80}} + \sqrt{\frac{49\pi}{80} + \frac{7\pi Q_{20}}{6eZr_\circ^2 A^{2/3}}} \tag{45}$$

[see eq. (30)]. There exists a number of phenomenological formulae which relate γ-ray lifetimes to $E_{2_1^+}$, Z, and A. One of the oldest ones is the expression developed by Grodzins:[111,112]

$$\tau_{E2} = (2.74 \pm 0.91) \times 10^{13} E_{2_1^+}^{-4} Z^{-2} A. \tag{46}$$

Another formula based on the global systematics has been introduced in refs. 113, 114. It is given by

$$\tau_{E2} = (1.25 \pm 0.50) \times 10^{14} E_{2_1^+}^{-(4.00 \pm 0.03)} Z^{-2} A^{0.69 \pm 0.05}. \tag{47}$$

or

$$B(E2)\uparrow \approx 3.27 E_{2_1^+}^{-1.0} Z^2 A^{-0.69}. \tag{48}$$

In refs. 114, 115 the authors disçuss many phenomenological expressions for $B(E2)\uparrow$ based on local[116, 117] or regional[118, 119] systematics.

A survey of experimental quadrupole deformations deduced from $0^+_{g.s.} \to 2^+_1$ transitions is contained in a recent compilation.[114] The phenomenological expressions for quadrupole deformation can be found in, for example, ref. 120 [eq. (67) below] and ref. 121.

B. Octupole deformations

In the limit of strong coupling the octupole deformation, β_3, is related to the reduced transition probability $B(E3)$ from the ground state to the first 3^- state:

$$\beta_3 = \frac{4\pi}{ZR_\circ^3}\left[\frac{B(E3)\uparrow}{e^2}\right]^{1/2}, \tag{49}$$

where the value of $B(E3)$ can be deduced from the partial mean lifetime for $E3$ γ-ray emission to the ground state:

$$\tau_{E3} = (1.226 \times 10^{32})\, E_{3_1^-}^{-7}[B(E3)\uparrow/e^2b^3]^{-1}. \tag{50}$$

In eq. (50) no correction has been made for internal conversion since it is expected to be insignificant in the considered energy range. The systematics of experimental octupole deformations deduced from $0^+_{g.s.} \rightarrow 3^-_1$ transitions can be found in refs. 122, 115.

C. Higher deformations

The systematics of experimental β_4 or β_6 deformations can be found in refs. 23, 28, 59, 120, 124. For phenomenological expressions, see eqs. (68), (69) below.

VI. MICROSCOPIC DESCRIPTIONS

Various theoretical methods to describe nuclear deformations are reviewed in this section. The main emphasis is put on the mean field approach in which the deformation concept appears naturally.

A. Mean field approach

The concept of nuclear deformation appears inherently in the mean field approach in which the average potential and, consequently, the total energy is a functional of the nucleonic density. Depending on proton or neutron number, the self-consistent density may be spherical or deformed. For a detailed discussion of the mean field approach we refer the reader to ref. 125.

1. Hartree–Fock method

The average nuclear mean field, in which nucleons move as independent particles, can be obtained from the Hartree–Fock (HF) theory. The starting point is the general two-body Hamiltonian

$$H = \sum_{ij} t_{ij} c_i^+ c_j + \frac{1}{4} \sum_{ijkl} \bar{v}_{ijkl} c_i^+ c_j^+ c_l c_k, \tag{51}$$

where $\bar{v}_{ijkl}$ is the antisymmetrized matrix element of a two-body effective (usually density-dependent) force. The A-body wave function is approximated by a Slater determinant whose orbitals are determined by a minimization of the total energy. This variation leads to an eigenvalue problem which defines both the single-particle orbitals $\{\Phi_i,\ i = 1 \ldots A\}$ and the single-particle energies e_i,

$$(t + \Gamma)\, \Phi_i = e_i \Phi_i. \tag{52}$$

The resulting self-consistent HF potential

$$\Gamma_{ij} = \sum_{kl} \bar{v}_{ikjl}\, \rho_{lk} \tag{53}$$

depends on the orbitals through the density matrix ρ.

The pairing (particle–particle) components of the effective force can be approximated in the framework of the BCS theory, or they can be treated on the same footing as the particle–hole interactions through the Hartree–Fock–Bogolyubov (HFB) theory. The solution of the HF(B) equations gives the binding energy of the nucleus at the local minimum. Sometimes, however, one wants to calculate the total energy as a function of one or several collective parameters q_i, such as deformations. Consequently, one is interested in a wave function which minimizes the total energy under the constraint that certain one-body operators $\hat{Q}_i$ (defining the collective fields) have fixed expectation values, $\langle \hat{Q}_i \rangle = q_i$. By solving the constrained HF(B) equations one can find the whole multidimensional *potential energy surface* (PES), $E_{HFB}(q_1, q_2, \ldots, q_{max})$. Sometimes more than one minimum exists in the PES. In such cases the phenomenon of *shape coexistence* is expected.[5]

2. Multipole–multipole forces

Multipole deformations are directly related to a special class of effective interactions, namely the separable multipole–multipole forces:

$$V_{QQ} = -\frac{1}{2}\sum_{\lambda=2}^{\lambda_{max}} \sum_{\mu=-\lambda}^{\lambda} \kappa_{\lambda\mu} Q^{+}_{\lambda\mu} \cdot Q_{\lambda\mu} \tag{54}$$

where

$$Q^{+}_{\lambda\mu} = \sum_{ij} \langle i | f_{\lambda\mu}(r) Y_{\lambda\mu}(\Omega) | j \rangle \, c^{+}_{i} c_j \tag{55}$$

is a multipole operator of multipolarity $\lambda\mu$. For the short-ranged part of the effective interaction the multipole expansion leads to the multipole pairing forces. A simple model for deformation in heavy nuclei consists of a multipole–multipole force and a monopole pairing force, e.g. the pairing-plus-quadrupole model.[126–128] In this chapter, however, the discussion will be rather focused on surface deformations.

The multipole–multipole Hamiltonian can be written as

$$H = \sum_i \varepsilon_i c^{+}_{i} c_i + V_{QQ}, \tag{56}$$

where the first term on the right hand side is the spherical shell-model potential and V_{QQ} is given by eq. (54). The associated HF Hamiltonian is

$$H_{HF} = \sum_i \varepsilon_i \, c^{+}_{i} c_i - \sum_{\lambda\mu} q_{\lambda\mu} Q^{+}_{\lambda\mu}, \tag{57}$$

where, according to eq. (53),

$$q_{\lambda\mu} = \kappa_{\lambda\mu} \langle Q_{\lambda\mu} \rangle \tag{58}$$

is the self-consistent shape deformation. With fixed deformation parameters $q_{\lambda\mu}$ the H_{HF} Hamiltonian (57) is equivalent to a deformed shell model Hamiltonian (see below).

3. Deformed shell model
As early as 1955, Nilsson developed the deformed shell model (DSM) and demonstrated[9] that nuclear deformations could be predicted by considering the sum of single-particle energies $\sum_i e_i$ of a phenomenological average potential designed to reproduce the characteristics of the mean field potential[9, 129, 130] (for the inclusion of pairing see ref. 131). The DSM is an approximation to the HF approach. Here, the HF field Γ is not determined self-consistently from eq. (53), but is assumed to be of the form of some phenomenological *average potential*, V_{DSM}, which contains a central part, a spin–orbit term, and a Coulomb potential for protons. All these terms depend explicitly on a set of external deformation parameters, $\alpha_{\lambda\mu}$, defining the nuclear surface. The first DSM potential was introduced by Nilsson[9] who employed the anisotropic harmonic oscillator potential [eq. (9)] modified by a spin–orbit term and an $\vec{\ell}^2$ term simulating the effect of a flattened well bottom. Other commonly used average potentials are the Woods–Saxon potential[132, 133] and the folded-Yukawa potential.[134]

4. Shell correction approach
The DSM cannot be used to predict binding energies. This is because the DSM energy differs from the full HF energy by the two-body interaction term. On the other hand, it is well known that with a good accuracy, binding energies are accounted for by the classical model of the liquid drop.[135]

The two approaches are merged into the so-called macroscopic–microscopic method [also called the shell correction method or Nilsson–Strutinsky (NS) approach] developed by Swiatecki, and Strutinsky and collaborators (see refs. 21, 136, 137 and references therein). The main assumption of the NS approach is that the total energy of a nucleus can be composed of two parts,

$$E = E^{macr} + E^{shell} \tag{59}$$

where E^{macr} is the macroscopic energy (smoothly depending on the number of nucleons and thus associated with the "uniform" distribution of single-particle orbitals), and E^{shell} is the shell-correction term fluctuating with particle number (reflecting the non-uniformities of the single-particle level distribution, i.e. shell effects). The macroscopic part, E^{macr}, is usually replaced by the corresponding liquid drop (or droplet) model value, whilst the shell correction term is calculated using the deformed independent-particle model. The relationship between the macroscopic–microscopic method and the HF approach is given by the so-called Strutinsky energy theorem.[138]

B. Shell model

In the shell model (see, e.g. refs. 139–141) the nuclear many-body wave function is expanded in Slater determinants built from single-particle orbits defined in the laboratory frame. Thus, no intrinsic frame is introduced and it is not necessary to specify any nuclear shape or deformation.

The dimensions of the shell model matrices grow very fast with the number of active orbitals. Therefore, in the heavy region, it is only for nuclei with a few particles outside closed shells that it has been possible to perform meaningful shell model calculations. These nuclei are essentially spherical and, therefore, of little interest in the present context. For *sd*-shell nuclei on the other hand, it is possible to do complete calculations covering a full shell, namely $d_{5/2}$, $s_{1/2}$, and $d_{3/2}$, which is gradually filled when the number of protons or neutrons increases in the range of 8–20. Furthermore, as some of these nuclei show typical features of intrinsic deformation, it seems reasonable to test the collective properties of the shell model in this region.

The occupation of single-j subshells was analyzed in ref. 142 and subsequently, for the nucleus ^{24}Mg, compared[11] with the occupations of the Nilsson–Strutinsky model. An interesting feature is how the deformation or collectivity changes with increasing spin, especially if the band structures survive to the maximal spins (band terminations). Based on a shell model analysis, it has been argued[143] that the bands in the *sd*-shell lose their collectivity for spins around $I = 8\hbar$ which could be compared with the maximum spin, $I_{max} = 12\hbar$ in ^{24}Mg and $I_{max} = 14\hbar$ in ^{28}Si. As only the *sd*-shell is usually included in the basis, the coupling to other shells is taken care of by effective interactions and charges. These are fitted at low spins, but it seems somewhat unclear if the same effective parameters are optimal also at higher spins. Consequently, it still remains to be investigated whether the shell model gives a reliable description of collectivity at high spins. Some of those questions will certainly be resolved by the large-scale shell model calculations, such as the $4\hbar\omega$ shell model calculations for ^{16}O of ref. 144.

As mentioned above, for heavier nuclei the full spherical shell model calculations become virtually impossible because of the enormous dimension of the configuration space involved. A possible way out of this dilemma is the variational approach to the shell model represented by MONSTER and VAMPIR calculations.[145–148] In these configuration-mixing calculations, the basic building blocks are the general symmetry-projected HFB wave functions, and the truncation scheme is based on the variational method. It has been demonstrated[148] that such variational shell model calculations can be performed for medium-mass and heavy systems. In practice, however, it turns out that to get "acceptable" convergence for heavy nuclei, the associated matrices are still too large for today's computers but may be accommodated by the next generation of computers. Another problem, typical of the shell model approach, is the difficulty of finding an effective interaction for the heavier systems.

C. Algebraic models

The nuclear models based on group theoretical techniques are usually constructed in the *laboratory frame*, i.e. their Hamiltonians are rotationally invariant. The link between the algebraic and geometric frameworks can be provided by the intrinsic state formalism. The main idea is to separate the

Hamiltonian into collective and intrinsic parts, $H = H_{coll} + H_{int}$, where the intrinsic Hamiltonian depends on deformation parameters.

In the Elliott model[149] the many-body Hamiltonian (containing the shell model quadrupole–quadrupole interaction) is an SU(3) scalar. Consequently its eigenstates can be labeled by means of the quantum numbers λ and μ. The highest-weight state corresponds to the *intrinsic* state of a rotational band. The intrinsic quadrupole moments of an intrinsic state are given, in units of $\hbar/m\mathring{\omega}_0$, by

$$\langle Q_{20}\rangle = 2\lambda + \mu, \quad \langle Q_{22}\rangle = \sqrt{\frac{3}{2}}\mu. \tag{60}$$

By introducing quadrupole deformation parameters (β_2, γ) by means of eq. (32), one can find a direct relationship between the intrinsic shape parameters (β_2, γ) and the (λ and μ) labels of SU(3):

$$\beta_2 \sim \sqrt{\lambda^2 + \mu^2 + \lambda\mu} \qquad \tan\gamma = -\sqrt{3}\,\frac{\mu}{2\lambda + \mu}. \tag{61}$$

A similar expression is obtained if one compares the quadrupole invariants (33).[150] By means of eq. (61) the deformation concept can be given an algebraic interpretation. In particular, the above relations have been used in the framework of the pseudo-SU(3) model or the pseudo-symplectic model.[151, 152]

It has been demonstrated that the mean field approximation to an algebraic model can be realized via the coherent-state theory (see e.g. ref. 153). The trial wave functions are the coherent states $|\eta\rangle$ of the dynamical symmetry group of the system. They define the potential energy $E(\eta) = \langle\eta|H|\eta\rangle$.

The interacting boson model (IBM) is based on monopole, s^+ ($J^\pi = 0^+$), and quadrupole, d^+ ($J^\pi = 2^+$), phenomenological bosons. In the IBM-1 there is no distinction between protons and neutrons; the two types of bosons (namely, proton bosons and neutron bosons) are distinguished in the IBM-2. There exists extensive literature on the geometrical interpretation of the IBM.[150, 154] An intrinsic state for the IBM-2 is a coherent state

$$|N_\pi N_\nu; \alpha\rangle = (N_\pi! N_\nu!)^{-1/2}[B_\pi^+(\alpha^{(\pi)})]^{N_\pi}[B_\nu^+(\alpha^{(\nu)})^{N_\nu}|0\rangle, \tag{62}$$

where $|0\rangle$ is the boson vacuum and B_ρ^+ ($\rho = \pi, \nu$) is a hybrid-boson operator given by

$$B_\rho^+ \alpha^{(\rho)}) = N\left(s_\rho^+ + \sum_\mu \alpha_\mu^{(\rho)} d_{\rho\mu}^+\right). \tag{63}$$

By considering the time-even condensate, eq. (3), one can parametrize eq. (63) in terms of two intrinsic shape variables β_ρ, γ_ρ and three Euler angles of spatial rotation, Ω_ρ. The energy surface of H depends on the seven intrinsic variables: the four quadrupole deformation parameters β_π, β_ν, γ_π, γ_ν, and the three Euler angles of relative orientation of the proton and neutron condensates. In the fermion dynamical symmetry model (FDSM) the coherent states are also given by eqs. (62)–(63).[153, 155] However, in the case of the FDSM the s^+ and d^+ bosons are replaced by the monopole and quadrupole pair operators S^+ and D^+ which are constructed explicitly from the fermion single-particle operators. For a critical discussion of the predictive power of algebraic models for the $B(E2)$ rates, see refs. 155, 156.

D. Cluster models

For strongly deformed nuclei, cluster models give an intuitive picture of the nuclear shape. For example, superdeformed nuclei with an axis ratio of 2:1 might be viewed as two touching spheres.[157, 158] In the light nuclei strongly deformed configurations have been treated as clusters, especially as built from α particles. In refs. 159, 160 a description of ^{20}Ne in terms of an $\alpha + {}^{16}$O bi-molecule was suggested. This picture was extended in ref. 161 to describe the close-lying 0^+ and 2^+ states in ^{8}Be, ^{12}C, ^{16}O, ^{20}Ne, and ^{24}Mg as built from α particles in a row (see ref. 162 for a discussion of various cluster configurations in light nuclei). Ikeda *et al.*[163] suggested that cluster configurations would appear near the threshold energy for decay into the fragments. They made an attempt to systematize the α-clustering phenomenon by means of the so-called Ikeda diagram.[164]

The natural model for describing the large α widths is the α-cluster model[165, 166] (see e.g. refs. 167, 168 for recent developments) and in the LCCO (Linear Combination of Cluster Orbits) model.[169] In these models, the clusters are fixed at definite points, but relative motion between the clusters can be introduced through the generator coordinate method (GCM); see ref. 170 for a review.

The occurrence of clustering can also be understood from calculations based on the mean field approach where many strongly deformed configurations can be identified with α clusters. These deformed structures with favored shell energy are alternatively described as closed configurations in the harmonic oscillator model, i.e. configurations where the axes ratio (or rather the oscillator frequency ratio) is given by small integer numbers.[8] For these light nuclei, the realistic nuclear potential is similar to that of the harmonic oscillator. Therefore, it is not surprising that states which show cluster character may also be identified in the HF calculations (e.g. refs. 171–173). The occupation of shell model orbitals which result when the two spherical or deformed systems are combined has been systematized by Harvey[174] (see also recent reviews[175, 176]). Harvey's prescription can be understood by considering the quanta in the three Cartesian directions of the harmonic oscillator wave functions or, equivalently, from the number of nodes in deformed orbitals. For symmetric combinations, these numbers remain unchanged in the perpendicular direction while they must increase in the parallel direction so that the Pauli principle is obeyed (for the relevant two-centre shell model calculations, see ref. 177). For example, if two ^{8}Be nuclei built from α particles are combined in a linear arrangement, we end up in the ^{16}O configuration with a 4:1 axis ratio which can also be described as four α particles in a row. The most spectacular state of this kind is probably the six α-particle chain configuration of ^{24}Mg for which evidence has recently been published[178] (see also ref. 179).

Similar arguments can then be used when combining heavier clusters to form superdeformed shapes. With three equal clusters in a row, we obtain hyperdeformed nuclear states with a 3:1 axis ratio. Such a state has been calculated for ^{48}Cr in the NS approach[180] and, more recently, in the cluster model.[181] This hyperdeformed state could be viewed as built either from three spherical ^{16}O nuclei or from two deformed ^{24}Mg nuclei. The latter scenario might be supported by ^{24}Mg + ^{24}Mg scattering experiments[182] where observed resonances might be associated with the calculated hyperdeformed states in ^{48}Cr.

VII. CALCULATIONS

This section contains examples of various calculations of nuclear deformations. The examples are presented along with relevant experimental data.

A. Global and regional calculations of equilibrium deformations

In Tables 2–4 we list references to some calculations on nuclear deformation which are systematic in the sense that they cover essentially all nuclei (Table 2), the whole region(s) (Table 3), or at least several nuclei (Table 4). Even though we are aware that this compilation is far from complete, we believe it could be helpful for further study of the literature. The most complete calculations are probably those based on the NS approach with the folded-Yukawa average potential by Möller and Nix, first published in ref. 183. The calculated equilibrium quadrupole and hexadecapole moments are discussed in ref. 184 where figures of $\varepsilon, \varepsilon_4$ deformation parameters as a

Table 2 Calculations of ground state shapes in various mass regions. In the last column, parentheses indicate that this degree of freedom is only partly included, while square brackets indicate that the results are expressed in these parameters although they do not enter the original calculations.

Ref.	Model	Nuclei	Deformations/Moments
186	NS	$Z, N \geq 20$	$\varepsilon, \varepsilon_4, [\beta_2, \beta_4]$
183, 184, 185	NS	$Z, N \geq 8$	$\varepsilon, \varepsilon_4, [Q_{20}, Q_{40}]$
187	NS + C	$Z = 34–84$	ε, γ
188	NS	$Z, N \geq 8$ (selected nuclei)	$\varepsilon, \varepsilon_4, (\varepsilon_6)$
189	DDM	$A = 12–240$	Q_{20}
190	ETFSI	$A \geq 36$	c, h

C: Cranking approximation; DDM: Dynamic Deformation Model; ETFSI: Extended Thomas–Fermi Strutinsky Integral; NS: Nilsson–Strutinsky approach.

Table 3 Regional calculations of ground state shapes in one/few mass regions. For other explanations, see caption to Table 2.

Ref.	Model	Nuclei	Deformations/Moments
27	NS + BH	$20 \leq Z \leq 38$	$\varepsilon, \varepsilon_4, \gamma$
192	DS	$28 \leq Z \leq 50, 50 \leq N \leq 82$	ε, γ
193	NS	$Z = 34$–46	$c, h, (\beta_2, \gamma)$
191	EDM	$Z = 32$–40	ε
194	NS	$Z = 36$–46	ν_2, γ
195	NS	$Z = 40$–62	$\varepsilon, \varepsilon_4$
196, 197	DS	$50 \leq Z \leq 82, 50 \leq N \leq 82$	ε, γ
127, 198	PPQ (+ BH)	$Z = 50$–82, $N = 82$–126	β, γ
199	NS	$Z = 62$–78, 90–100	$\beta_2, \beta_4, [Q_{20}, Q_{40}]$
200	NS	$Z = 68$–108	$\varepsilon, \varepsilon_4, [\beta_2, \beta_4]$
201	NS	$Z = 50$–82	$\varepsilon, \varepsilon_4$
202	NS	$Z = 52$–66	β_2, β_4, γ
203	NS	$Z = 56$–82	$c, h, (\beta_2, \beta_4), \beta, \gamma$
204	NS + C	$Z = 58$–74	β_2, β_4
124	NS + BH	$Z = 60$–72	$\beta_2, \beta_4, [Q_{20}, Q_{40}]$
205	PPQ	$Z = 62$–72	$Q_{20}, Q_{22}, [\beta, \gamma]$
206	NS + GCM	$Z = 60$–72	$\varepsilon, \varepsilon_4, [Q_{20}, Q_{40}]$
207	NS	$Z = 63$–75	$\varepsilon, \varepsilon_4$
208	NS	$Z = 63$–70	$\beta_2, \beta_4, [Q_{20}, Q_{40}]$
28	NS	$Z = 62$–76	$\beta_2, \beta_4, [Q_{20}, Q_{40}]$
209	NS	$Z = 62$–79	β_2, β_4
7	NS	$Z = 62$–72, 92–100	$\varepsilon, \varepsilon_4, [(\beta_2, \beta_4, \beta_6)]$
210	NS	$Z \geq 52, A \geq 120$	β_2, β_4
211	NS + C	$Z = 88$–96	β_2, β_4
212	PPQ + C	$Z = 90$–98	β_2, β_4
213	NS	$Z = 92$–102	$\varepsilon, \varepsilon_4, \varepsilon_6, [Q_{60}]$

BH: Bohr Hamiltonian; C: Cranking approximation; EDM: Energy Density Method; GCM: Generator Coordinate Method; PPQ: Pairing + Quadrupole Model; NS: Nilsson–Strutinsky approach; DS: Deformed Shell model.

function of Z and N are exhibited and discussed. Similar diagrams, based on the most recent calculations by Möller *et al.*,[185] are presented in Figs. 3 and 4.

Figure 3 shows the absolute values of ε, i.e. there is no distinction between prolate and oblate shapes. In fact, the vast majority of deformed nuclei have prolate shapes. The oblate ground state deformations can be found only in very few regions, e.g. in the Ge and Se nuclei around ^{72}Se, in the Hg isotopes, or in the heavy Os and Pt isotopes.

Global calculations of nuclear shapes at high angular momentum were published by Åberg[187] who employed the cranked NS approach with the MO potential. The recent large-scale calculations of masses and deformations by Pearson *et al.*[190] are based on the extended Thomas–Fermi Strutinsky integral

Table 4 Local calculations of ground state shapes in one/few mass regions. For other explanations, see caption to Table 2

Ref.	Model	Nuclei	Deformations/ Moments
214	HF	$N = Z = 8[2]20$	Q_{20}, Q_{22}
215	HF	$N = Z = 4[2]14$	$Q_{20},, Q_{40}, [\beta_2, \beta_4]$
216	HF	$N = Z = 8[2]20$, ^{152}Sm, ^{170}Er	$Q_{20}, [\beta_2]$
217	HF	$N = Z = 6, 10, 12, 14, 16, 18$	$[\beta_2, \beta_4, \beta_6]$
8	NS	$N = Z = 6[2]22$	$\varepsilon, \varepsilon_4, \gamma, (\varepsilon_3, \varepsilon_{33})$
218	HF + BCS	$^{54-58}$Fe, $^{58-64}$Ni, $^{64-70}$Zn, $Z = 60–70$ (7 nuclei), ^{238}U, ^{252}Cf	Q_{20}, Q_{40}
219	HFB + GCM	Fe, Ni, and Zn	$Q_{20}, [\beta_2]$
220	HFB	^{58}Fe, ^{62}Zn, ^{74}Ge, ^{132}Ba, ^{134}Ce, ^{186}W	β_2, γ
221	HFB	$^{98-110}$Mo	
222	HFB	$^{94-102}$Zr, $^{92-106}$Mo	Q_{20}
223	HFB	$^{88-100}$Sr, $^{90-102}$Zr, $^{96-106}$Mo	$Q_{20}, [\beta_2]$
224	NS	Zr isotopes	$\varepsilon, \varepsilon_4$
225	HF + BCS	Kr, Sr, Zr, and Mo isotopes	Q_{20}, γ
226	HF	$^{126, 134, 140, 150}$Ce	$Q_{20}, [\beta_2, \beta_4]$
227	HF	$Z = 62–76$ (9 nuclei)	$Q_{20}, [\beta_2, \beta_4]$
228	HF + BCS	$^{92-110}$Cd, $^{132-142}$Sm	$Q_{20}, Q_{22}, [\delta, \gamma]$
229	RMF	^{160}Gd, $^{168, 170}$Er	$Q_{20}, [\beta_2]$
230	HF + BCS	Os, Pt, and Hg isotopes	$Q_{20}, [\beta_2, \beta_4]$

BCS: BCS pairing; GCM: Generator Coordinate Method; HF: Hartree–Fock; HFB: Hartree–Fock–Bogolyubov; NS: Nilsson–Strutinksy approach; RMF: Relativistic Mean Field.

method, a modified version of the energy density method.[191] All the above methods are variants of the *static* mean field approach. The correlations (mainly of the quadrupole character) are taken into account by the dynamic deformation model (DDM) by Kumar,[189] who constructs the collective wave functions by means of the collective Bohr Hamiltonian.

The "regional" calculations of equilibrium deformations are summarized in Table 3. Most works are based on the NS approach and concentrate on deformed nuclei from the light ($28 < N < 50$) and heavy ($50 < N < 82$) "zirconium" region ($28 < Z < 50$), the rare earth region ($50 < Z < 82$, $82 < N < 126$), and actinides ($Z > 82$, $N > 126$). Finally, some representative examples of "local" calculations are contained in Table 4. Our intention here is to present calculations which, because of their computational complexity, cannot (or could not at the time they were performed) be applied on a large scale. The use of the words "numerically complex" is a qualified one. With new, powerful, massively parallel computers, it will certainly be possible (and desirable) to perform "regional" or even "global" calculations in the framework of, for example, the HFB + GCM method.

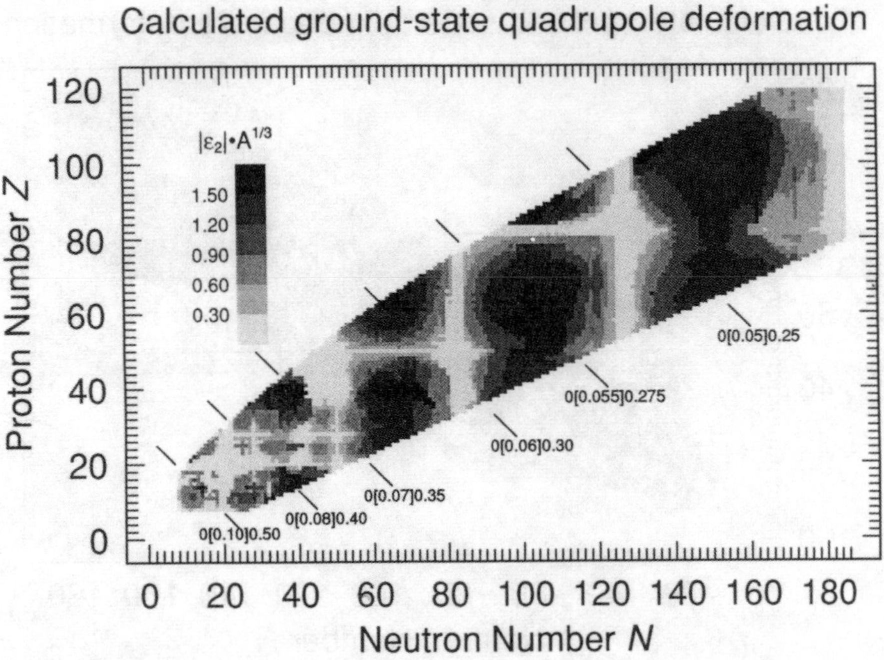

Fig. 3. Contour plot of ground state ε_2 deformations according to the Nilsson–Strutinsky calculations of ref. 185. As typical ground state ε_2 values tend to decrease approximately as $A^{-1/3}$ with increasing mass, the plotted values are multiplied by $A^{1/3}$ to get similar amplitudes for all values of A. Corresponding variations in ε_2 values are indicated for some different A values. For example, the dark shading for a few nuclei in the middle of the rare earth region indicates that their calculated ground state ε_2 deformations are larger than ~ 0.275.

B. Particle number dependence of multipole moments

From shell model arguments, it follows that the total intrinsic moment of even multipolarity λ should change sign $(\frac{\lambda}{2}-1)$ times within a major shell.[231] One can understand this behavior in terms of the single-j shell ($j \gg 1$) model. The matrix element of the multipole operator $Q_{\lambda 0}$ in the state $|jm\rangle$ is given by

$$\langle jm|Q_{\lambda 0}|jm\rangle = \langle jm\lambda 0|jm\rangle \frac{\langle j\|Q_{\lambda 0}\|j\rangle}{\sqrt{2j+1}}. \tag{64}$$

For $j \gg \lambda$ one can approximate the Clebsch–Gordan coefficient in eq. (64) by its asymptotic value

$$\langle jm\lambda 0|jm\rangle \approx P_\lambda(\cos\theta), \tag{65}$$

where $\cos\theta = m/j$. Assuming that the j-shell is slightly split by the quadrupole deformation, the relative position of the Fermi level within the shell is given by

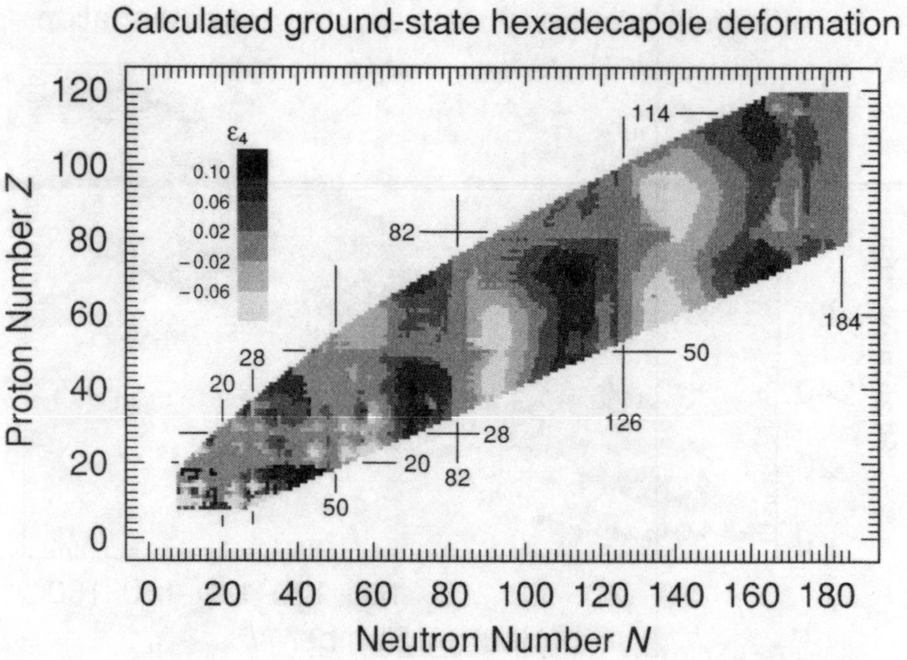

Fig. 4. Contour plot of ground state ε_4 deformations according to the Nilsson–Strutinsky calculations of ref. 185. Closed shells are marked.

$\xi = \cos\theta_\circ$ [ξ is a partial filling of the valence shell, i.e. $\xi = 0$ (1) corresponds to the bottom (top) of the shell]. The expectation value of the total multipole moment, $q_{\lambda 0}$, is thus proportional to

$$q_{\lambda 0} \propto \int_0^{\xi} P_\lambda(\cos\theta)\, d(\cos\theta) \propto \sin^2\theta_\circ P'_\lambda(\cos\theta_\circ). \tag{66}$$

It follows from eq. (66) that $q_{\lambda 0}$ vanishes for $\xi = 0$ and 1, and it changes sign $(\frac{\lambda}{2} - 1)$ times within the shell. Consequently, it is expected that β_4 should change sign once around the mid-shell, β_6 should change sign twice within the shell, and so on.

Based on the results of ref. 231 Jänecke suggested[120] a simple parametrization of nuclear deformations:

$$\beta_2 = \beta_2^m \xi_p(1 - \xi_p)(2 - \xi_p)\xi_n(1 - \xi_n)(2 - \xi_n) + \beta_2^\circ, \tag{67}$$

$$\beta_4 = \beta_4^m \xi_p(1 - \xi_p)(2 - \xi_p)\xi_n(1 - \xi_n)(2 - \xi_n)(7\xi_{np}^2 - 14\xi_{np} + 4.9), \text{ and} \tag{68}$$

$$\beta_6 = \beta_6^m \xi_p(1 - \xi_p)(2 - \xi_p)\xi_n(1 - \xi_n)(2 - \xi_n)[33(\xi_{np} - 1)^4 - 30(\xi_{np} - 1)^2 + 1], \tag{69}$$

where ξ_n (ξ_p) is the partial filling of the neutron (proton) valence shell and $\xi_{np} \equiv \frac{1}{2}(\xi_n + \xi_p)$. The parameters β_λ^m and β_2° are obtained by a least-squares fit to the data. For instance, for the rare earth nuclei [$\xi_n = (N - 82)/(126 - 82)$, $\xi_p = (Z - 50)/(82 - 40)$] the recommended parameters[120] are $\beta_2^m = 1.721$, $\beta_2^\circ = 0.082$, $\beta_4^m = 0.504$, and $\beta_6^m = 0.0215$.

C. Quadrupole and hexadecapole deformations

The typical behavior of the calculated ground state equilibrium deformations β_2 and β_4 is presented in Fig. 5, representative of deformed rare earth nuclei. As discussed in section VII.B, the quadrupole deformation reaches its maximum in the middle of the shell. Figure 6 shows the systematics of the experimental values of the quadrupole deformations in the rare earths and the actinides. They are compared to the trend given by eq. (67). Systematic deviations from the empirical formula seen for the neutron-deficient platinum isotopes can be explained in terms of the interaction between well-deformed and weakly deformed configurations.

In general, calculations give a rather close reproduction of experimental quadrupole moments. A representative example of these calculations is displayed in Fig. 7. The experimental deformations in even–even Yb and Hf isotopes are compared to the predictions of the NS approach. It seems that the deviations from experiment are usually caused by neglecting dynamical correlations in transitional nuclei by the static mean field calculations. There are

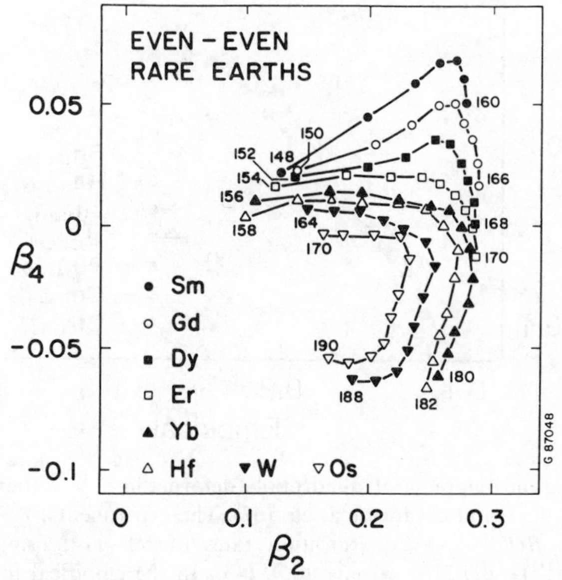

Fig. 5. Calculated equilibrium ground state deformations (β_2–β_4) for doubly even rare earth nuclei (from ref. 209).

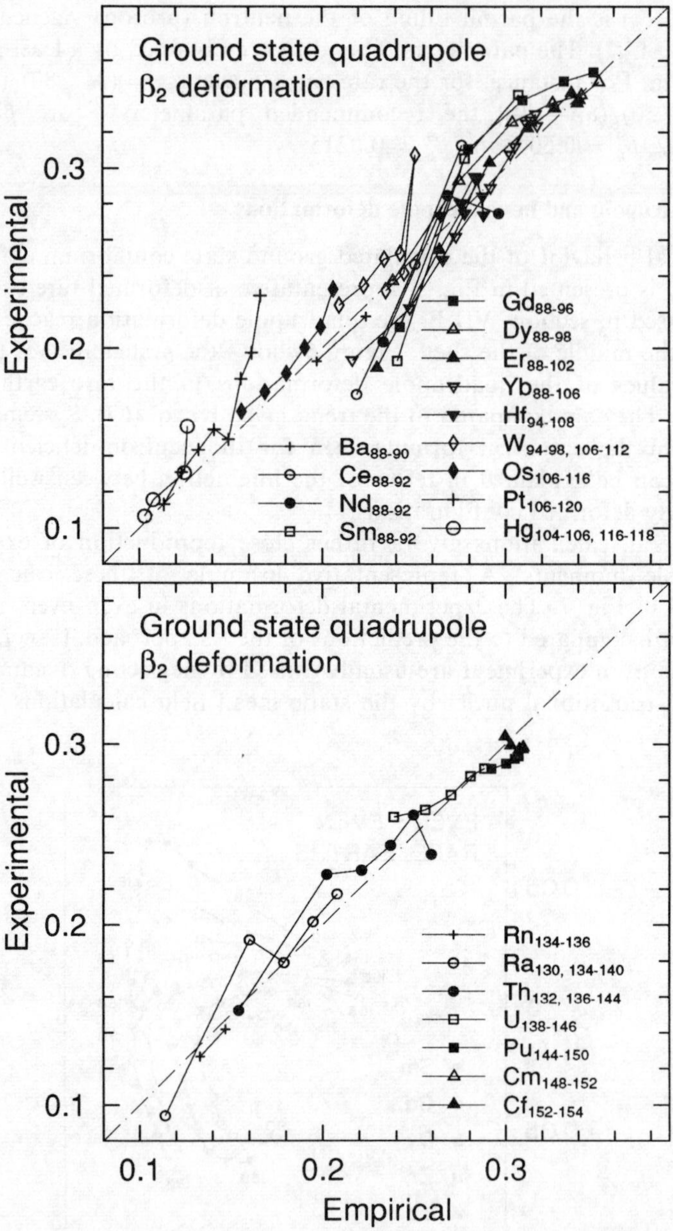

Fig. 6. Correlation diagrams of quadrupole deformations β_2 obtained from experiment and from the empirical formula, eq. (67). The experimental β_2 values are calculated from the $B(E2; 0^+ \rightarrow 2^+)$ transition rates of ref. 114, using the expression $\beta_2 = (4\pi/3Zr_\circ^2 A^{2/3})\sqrt{B(E2)/e}$, cf. eqs. (25), (43). In the empirical formula, the parameters have been slightly modified compared with ref. 120; namely, $\beta_2^m = 1.90$, $\beta_2^\circ = 0.082$ for the rare earths and $\beta_2^m = 1.721$, $\beta_2^\circ = 0.055$ for the actinides.

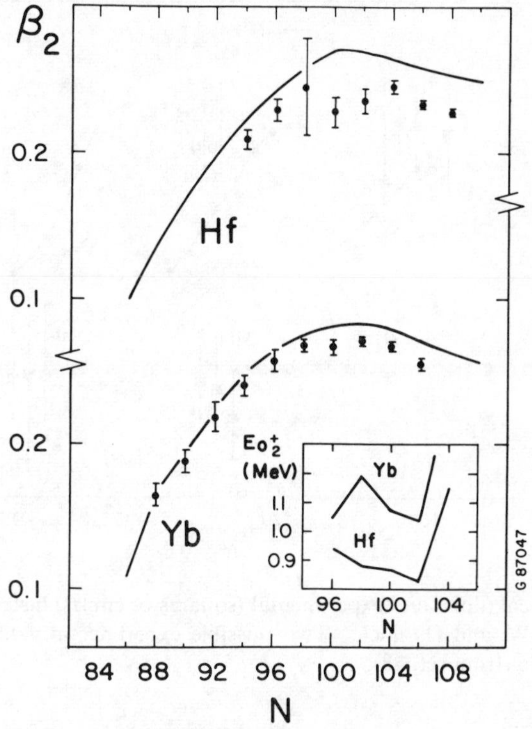

Fig. 7. Theoretical and experimental quadrupole deformations for Yb and Hf nuclei. The inset shows the energy of the first excited 0^+ state in the $N = 96$–104 Yb and Hf nuclei (from ref. 209).

also other sources of discrepancies, such as possible configuration mixing with excited states (see inset in Fig. 7), the centrifugal stretching effect leading to deformation changes between 0^+ and 2^+ states, or the influence of higher deformations (e.g. β_4).

The hexadecapole deformations are positive (i.e. β_4 positive which means ε_4 negative, see section II.D) in the first half of the shell, and are negative in the second half of the shell (see Fig. 5). In most cases, calculations give a fairly good description of hexadecapole moments. There are, however, some exceptions. An important example of the discrepancy between experiment and theory is the long-standing problem of the so-called hexadecapole anomaly at the border of the rare earth region. Figure 8 shows the comparison between calculated and experimental hexadecapole moments in well-deformed rare earth nuclei. If good agreement between experiment and theory is obtained for the hexadecapole moments of Er and Yb isotopes, there is a large deviation for the heavy Hf, W, and Os isotopes. The experimental hexadecapole deformations in this nuclei are around $\beta_4 = -0.2$, which is by a factor of two larger than the calculated values. In addition,

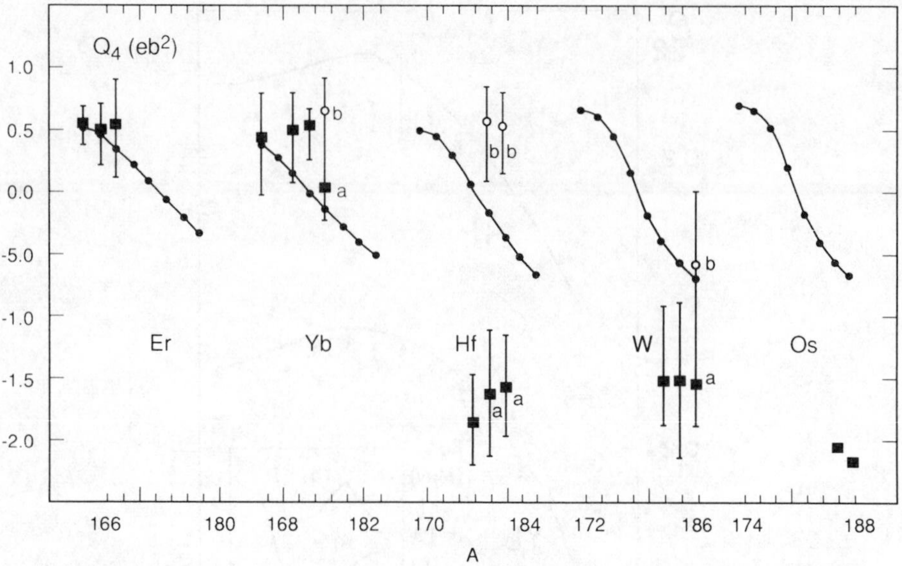

Fig. 8. Theoretical (dots) and experimental (squares or circles) hexadecapole moments for Er, Yb, Hf, W, and Os nuclei. Two possible experimental values (a) and (b) are shown if available (from ref. 28).

there are large differences in the nuclear and Coulomb β_4 deformations.[232] The nuclei with the larger deviations lie at the border of the deformed region, and are expected to be γ-soft or even γ-unstable. Therefore, the contribution from non-axial degrees of freedom or dynamical correlations should be taken into account.[28] There is some experimental evidence from α scattering studies[60] for the non-axial hexadecapole deformations, α_{42} and α_{44}, around ^{168}Er. (For the systematic behaviour of the experimental $q_{4\mu}$ values, see fig. 9 of ref. 233.)

D. Higher-order moments

Experimentally, very little is known about higher-order multipole moments. The first determination of the $\lambda = 6$ (hexacontatetrapole) nuclear moment was done by Hendrie *et al.*[234] by means of (α, α') scattering. For other experimental works, see refs. 53, 54, 60, 67, 70, 235–238.

The first theoretical analysis of equilibrium β_6 deformations[7] reproduced the experimental trend qualitatively. It was pointed out[239] that the inclusion of a $\lambda = 6$ field in the average potential is important at very elongated shapes, e.g. at fission barriers (see also ref. 240). Recently, the equilibrium values of β_6 deformation have been computed[241–244] for nuclei in the heavy-Ba and Ra–Th regions, where reflection-asymmetric deformations are important. On the average, for light actinides with $130 \leq N \leq 140$, β_6 is positive (around 0.02) whilst it is negative (around –0.01) for Xe–Sm nuclei with $86 \leq N \leq 94$. The inclusion of odd-multipolarity deformations may, in some cases, significantly influence the equilibrium value of β_6[243,245] Möller *et al.*[185,246] took into account the $\lambda = 6$ deformation in their large-scale mass calculations. The largest correction to the mass due to ε_6, about –1.3 MeV, was calculated near ^{252}Fm. (For the role of β_6 in the actinide region see also refs. 213, 247.)

There are almost no experimental data available on higher *static* moments with $\lambda > 6$. Barlett *et al.*[248] attempted to deduce the value of β_8 in ^{176}Yb through (p,p′) scattering, but this result is very uncertain as it depends strongly on various model assumptions.

E. Reflection-asymmetric shapes

A coupling between intrinsic states of opposite parity is produced by the long-ranged octupole–octupole residual interaction. In some cases the mixing is so strong that the nucleus appears to acquire stable octupole deformation in the body-fixed frame. For normally-deformed systems the condition for strong octupole coupling occurs for particle numbers associated with the maximum $\Delta N = 1$ interaction between the intruder subshell (ℓ,j) and the normal parity subshell ($\ell - 3, j - 3$). The regions of nuclei with strong octupole correlations correspond to particle numbers near 34 ($g_{9/2} \leftrightarrow p_{3/2}$ coupling), 56 ($h_{11/2} \leftrightarrow d_{5/2}$ coupling), 88 ($i_{13/2} \leftrightarrow f_{7/2}$ coupling), and 134 ($j_{15/2} \leftrightarrow g_{9/2}$ coupling). That is, the tendency towards octupole deformation occurs just above closed shells.[249,250] Indeed, for the Ra–Th ($Z \sim 88$, $N \sim 134$) and Ba–Sm ($Z \sim 56$, $N \sim 88$) nuclei, the features of stable octupole deformation (namely, low-lying negative-parity states, parity doublets, alternating parity bands with enhanced $E1$ transitions) have been established (see refs. 4, 251, 252).

1. Octupole deformations

Several mean field calculations predict reflection instability for nuclei around ^{222}Th and ^{146}Ba (see Table 5). Figure 9 displays the contour plot of ground state ε_3 deformations according to the Nilsson–Strutinsky calculations of ref. 185. It is seen that the reflection-asymmetric shapes are predicted in nuclei with particle numbers around 56, 88, and 134.

All calculations yield similar results for stable octupole deformations, but give slightly different predictions for the height of the octupole barrier. The calculated octupole minima are usually very shallow with octupole barriers varying between 0.5 and 2 MeV, depending on the model. Consequently, dynamical fluctuations are expected to play a significant role. Some attempts towards including dynamic corrections by means of parity projection or Gaussian overlap approximation have been made in refs. 243, 244, 253, 254. Other candidates for the strong octupole–quadrupole coupling can be found in the medium-mass nuclei, i.e. in the $A \sim 70$ mass region,[33,255]] in the nuclei around ^{96}Zr,[256] and around ^{112}Ba.[257]

Table 5 Systematic calculations of reflection-asymmetric shapes around ^{222}Th (actinides) and ^{146}Ba (lanthanides)

Ref.	Model	Nuclei	Deformations/Moments
258	M + P	actinides	$\varepsilon, \varepsilon_3, \varepsilon_4$
249	NS	actinides	$\varepsilon, e_3, \varepsilon_4$
250	NS	actinides, lanthanides	$\beta_2, \beta_3, \beta_4$
259	NS	actinides	ν_2, ν_3, ν_4
260	HF + BCS	actinides	Q_{20}, Q_{30}
261	HFB	actinides	Q_{20}, Q_{30}
262	NS + C	actinides	$\beta_2, \beta_3, \beta_4$
242	NS	actinides, lanthanides	$\beta_2, \beta_3, \beta_4, \beta_5, \beta_6, \beta_7, \beta_8$
263	NS + PR	actinides	$\beta_2, \beta_3, \beta_4$
264	HFB	actinides	Q_{20}, Q_{30}
245	NS	lanthanides	$\beta_2, \beta_3, \beta_4, \beta_5, \beta_6, \beta_7$
264	M + P	actinides	$\nu_2, \nu_3, \nu_4, \nu_5, \nu_6$
243	HFB	actinides	Q_{20}, Q_{30}
244	HFB	lanthanides	Q_{20}, Q_{30}
265	NS	actinides	$\beta_2, \beta_3, \beta_4, \beta_5, \beta_6, \beta_7$
266	NS + C	lanthanides	$\beta_2, \beta_3, \beta_4$

BCS: BCS pairing; C: Cranking approximation; HF: Hartree–Fock; HFB: Hartree–Fock–Bogolyubov; M + P: Multipole + Pairing Hamiltonian; NS: Nilsson–Strutinsky approach; PR: Particle + Rotor.

2. *Dipole deformations*

In the presence of low-lying octupole excitations, a large isovector $E1$ moment may arise in the intrinsic frame due to a shift between the center of charge and the center of mass. Such a static dipole moment manifests itself by very enhanced electric dipole transitions between opposite parity members of quasi-molecular rotational bands. As pointed out in refs. 267, 268, the $E1$ moment results from an interplay between the strongly attractive proton–neutron force represented by the symmetry energy term, and the Coulomb force.

For an axially-deformed system, the nuclear electric dipole moment is given by

$$D_\circ = e\frac{N}{A}\langle z_p\rangle - e\frac{Z}{A}\langle z_n\rangle, \tag{70}$$

where $\langle\ldots\rangle$ denotes the expectation value in the *intrinsic* state. If the intrinsic parity is broken by, for example, the presence of static reflection-asymmetric moments, then, generally, $N\langle z_p\rangle \neq Z\langle z_n\rangle$ and $D_\circ \neq 0$. Nuclei from the regions of low-energy octupole collectivity indeed exhibit very enhanced $E1$ transitions. Dipole moments extracted from measured $B(E1)$ rates are unusually large and of the order of 0.1–0.3 efm.[269] An expression for the *macroscopic* contribution to the intrinsic dipole moment has been derived in ref. 270 based

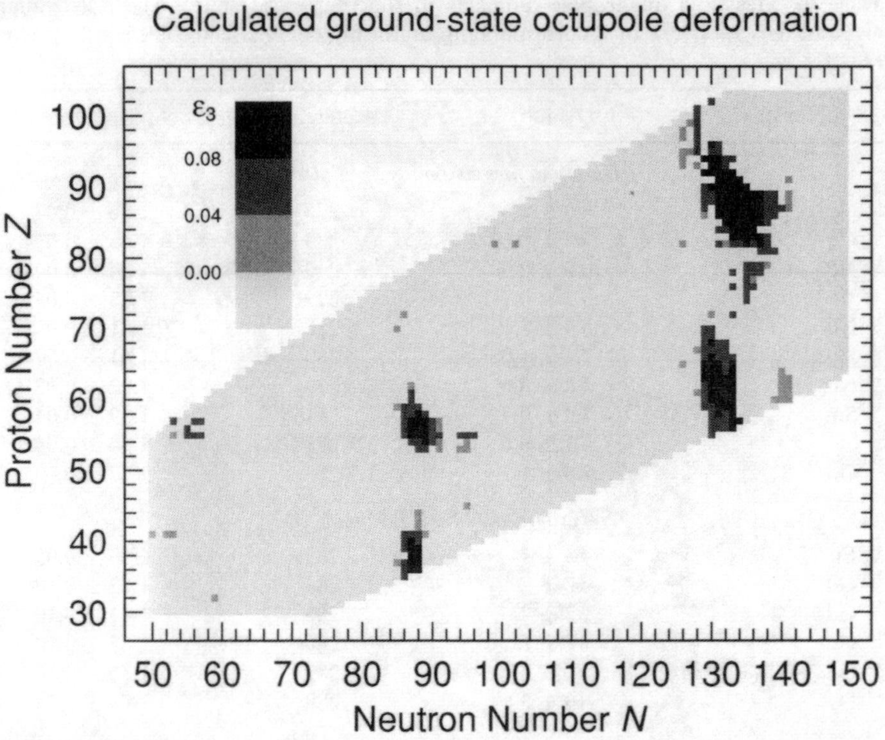

Fig. 9. Contour plot of ground state ε_3 deformations according to the Nilsson–Strutinsky calculations of ref. 185. The light-grey shading indicates nuclei which are calculated to be stable towards octupole deformation, i.e. $\varepsilon_3 = 0$. Essentially all light nuclei shown in Figs. 3 and 4 (but absent in the present figure) are predicted to have reflection-symmetric shapes.

on the two-fluid droplet model. It consists of two terms. The first term represents essentially the *redistribution* effect and reflects the fact that electric charge tends to move towards regions of the surface with large curvature. The second term represents the contribution to the dipole moment arising from the *neutron skin*. Since both contributions have opposite signs, the presence of the neutron skin leads to a reduction of the macroscopic contribution to the intrinsic dipole moment.

The *microscopic* contribution to the intrinsic dipole moment was introduced in ref. 271, where an approach based on the ns method was formulated. The resulting reflection-asymmetric deformed shell model was able to explain many properties of low-lying $E1$ modes in the transitional nuclei near ^{146}Ba and ^{224}Th.[272, 269] Parity-projected HF calculations with Gogny interaction were able to reproduce the $B(E1)$ rates in the Ra–Th[243] and Ba[244] nuclei.

Table 6 Measured quadrupole moments in superdeformed states. The last column gives an axis ratio c/a of a corresponding uniformly charged ellipsoid, eq. (27), with $r_\circ = 1.2$ fm

Nucleus	Q_{20} (eb)	Reference	c/a
	Highly deformed bands around ^{132}Ce		
^{131}Ce	6.0 ± 0.6	277	1.36 ± 0.04
^{132}Ce	7.5 ± 0.6	278	1.45 ± 0.03
^{133}Nd	6.0 ± 0.7	279	1.34 ± 0.04
	6.7 ± 0.7	279	1.38 ± 0.04
^{135}Nd	5.4 ± 1.0	280	1.31 ± 0.07
	7.4 ± 1.0	280	1.41 ± 0.07
^{137}Nd	4.0 ± 0.5	279	1.22 ± 0.03
^{135}Sm	7.0 ± 0.7	281	1.39 ± 0.04
	6.2 ± 0.6	281	1.34 ± 0.04
^{139}Gd	7.0 ± 1.5	282	1.37 ± 0.09
	Superdeformed bands around ^{152}Dy		
^{143}Eu	13 ± 1	283	1.69 ± 0.05
^{146}Gd	12 ± 2	284	1.62 ± 0.10
^{146}Gd*	8 ± 2	285	1.41 ± 0.10
^{149}Gd	17 ± 2	286	1.87 ± 0.10
^{150}Gd	17 ± 3	287	1.86 ± 0.10
^{152}Dy	19 ± 3	288	1.93 ± 0.14
	Superdeformed bands around ^{192}Hg		
^{190}Hg	18 ± 3	289	1.62 ± 0.10
^{191}Hg	20 ± 2	290, 291	1.69 ± 0.07
^{192}Hg	20 ± 2	72	1.69 ± 0.08
	20 ± 3	292	1.69 ± 0.10
^{194}Pb	21 ± 2	293	1.70 ± 0.10
	Fission isomers in actinides		
^{236}U	32 ± 5	273	1.84 ± 0.13
^{238}U	29 ± 3	294	1.75 ± 0.08
^{236}Pu	37^{+14}_{-8}	295	1.95 ± 0.30
^{239}Pu	36 ± 4	296	1.91 ± 0.10
^{240}Am	33 ± 2	297	1.83 ± 0.03

F. Superdeformation

Superdeformed high-spin bands, together with fission isomers, are the most spectacular examples of shape coexistence. The lifetimes of levels in the superdeformed band yield transition quadrupole moments which correspond to a large quadrupole deformation ($0.45 \lesssim \beta_2 \lesssim 0.65$). In heavy nuclei, three regions of superdeformed (SD) shapes have been established: fission isomers in actinides ($92 \leq Z \leq 97$, $141 \leq N \leq 151$), high-spin SD states around ^{152}Dy

Table 7 Calculations of superdeformed shapes related to (possible) collective high-spin configurations in various mass regions. For other explanations, see caption to Table 2

Ref.	Model	Nuclei	Deformations/Moments
298	NS	Pt–Po	β_2, β_4
299	NS	Pt–Pb	$\varepsilon_2, \varepsilon_4$
203	NS	Os–Hg	β_2, β_4
300	HF + BCS	Pt	Q_{20}
301, 302	NS + C	rare earths	$\varepsilon, \gamma\,(\varepsilon_4)$
303, 304	NS + C	rare earths	β, γ
305	NS	rare earths	$\varepsilon, \gamma\,(\varepsilon_4)$
240, 188	NS + C	$Z = 76$–88	$\varepsilon, \varepsilon_4\,(\gamma)$
157	NS + C	rare earths	$\varepsilon, \gamma\,(\varepsilon 3, \varepsilon 4)$
187	NS + C	$A \geq 40$	$\varepsilon, \gamma\,(\varepsilon_4)$
306	NS + C + D	rare earths	β_2, β_4, γ
30	NS + C + D	Yb, Hf, ^{185}Au	$\varepsilon, \varepsilon_4, \gamma$
307	NS + C	rare earths	ν_2, ν_4
14	NS + C + D	rare earths	$\varepsilon, \varepsilon_4$
323	NS + C	Pt–Pb	ν_2, ν_4
15	NS + P + C	Sm–Er	β_2, β_4
308	HFB	Hg	Q_{20}, Q_{22}
309	HF + BCS	Hg	Q_{20}, Q_{22}
310	HFB	Hg	Q_{20}
311	NS + P + C	Gd–Dy	$\varepsilon_2, \varepsilon_4$
312	NS + P + C	Hg	β_2, β_4
313	NS + C	Hg	$\varepsilon_2, \varepsilon_3\,(\varepsilon_4)$
314	NS + C	Dy, Hg	$\beta_2, \beta_3\,(\beta_4)$
316	HF + BCS + GCM	^{194}Hg	$Q_{20}, B(E2)$
315	NS + P + C	Os–Pt	β_2, β_4, γ
317	RMF + C	^{80}Sr, ^{152}Dy	Q_{20}
318	NS + P + C	Au–Ra	$\beta_2, \beta_3, \beta_4$
319	NS + C	Dy, Hg	$\beta_2, \alpha_{3\mu}, \beta_4$
33	NS + C	Hg	$\beta_2, \alpha_{3\mu}, \beta_4, \gamma$
320	HF + BCS + GCM	Hg	Q_{20}, Q_{30}
321, 322	NS + C	Gd–Dy	$Q_{20}, \varepsilon, \varepsilon_4$

BCS: BCS pairing; C: Cranking approximation; D: Diabetic approach; GCM: Generator Coordinate Method; HF: Hartree–Fock; HFB: Hartree–Fock–Bogolyubov; NS: Nilsson–Strutinsky approach; P: Pairing; RMF: Relativistic Mean Field.

($63 \leq Z \leq 66$, $80 \leq N \leq 87$), and SD bands around ^{192}Hg ($80 \leq Z \leq 82$, $109 \leq N \leq 116$). An impressive experimental and theoretical effort has been devoted to exploring the underlying physics. These investigations have opened up a new exciting field of nuclear superdeformed spectroscopy.

A detailed discussion of SD configurations goes beyond the scope of this chapter. Therefore, we refer the reader to the reviews in refs. 13, 273 (fission isomers) and refs. 4, 274, 275 (high-spin SD states), where many properties of SD structures are discussed.

The measured quadrupole moments in nuclei are displayed in Table 6. Table 6 also contains experimental quadrupole moments in well-deformed intruder bands around ^{132}Ce (for representative calculations see ref. 276). Table 7 contains a survey of calculations of deformations in SD states. The agreement between experimental and calculated deformations is usually found to be good. Figure 10 shows the measured quadrupole transition moment in the SD band in ^{192}Hg. The theoretical prediction of the cranked NS approach with the Woods–Saxon average potential and pairing agrees well with the data. Another example is presented in Fig. 11, which shows the shape trajectories in the $(\varepsilon, \varepsilon_4)$-plane calculated in the MO model for the yrast SD bands in ^{145}Gd–^{153}Dy. In this mass range, the crucial orbitals which are filled with increasing particle number originate from the high-N shells, $N = 6$ and 7 for neutrons, and $N = 6$ for protons. They all have positive quadrupole moments and lead to increasing quadrupole deformation.

ACKNOWLEDGEMENTS

The authors would like to thank P. Möller for discussions and communication of his unpublished results. Useful discussions with C. Baktash, J. Draayer, J. Garrett, R. Robinson, R. Ronningen, and Z. Szymański are gratefully acknowledged. We also wish to express our sincere thanks to S. Madison for

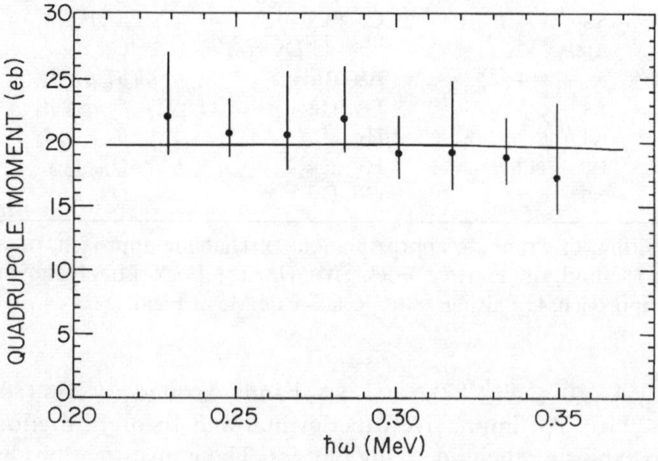

Fig. 10. Measured quadrupole transition moment, Q_t, in the SD band in ^{192}Hg. The theoretical prediction of the cranked NS approach with the Woods–Saxon average potential and pairing is indicated by the solid line (from ref. 72).

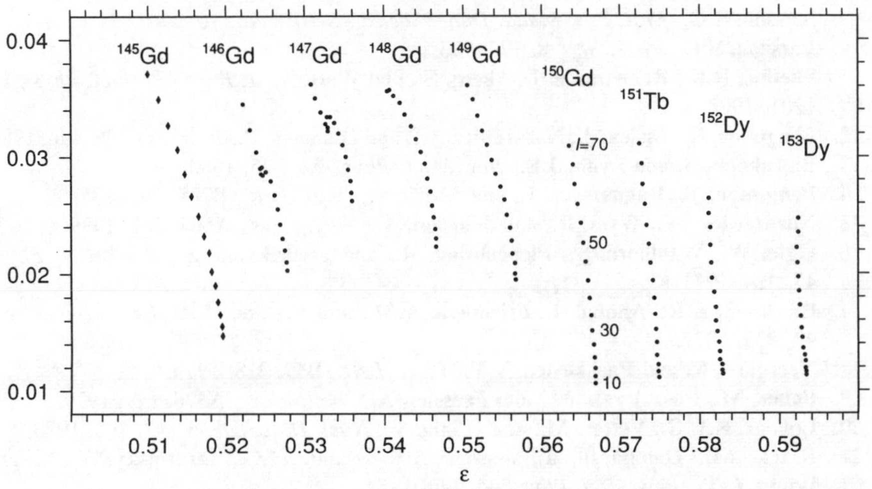

Fig. 11. Shape trajectories in the $(\varepsilon, \varepsilon_4)$-plane calculated for the yrast sd bands in ^{145}Gd–^{153}Dy in the NS method with the cranked MO potential. Deformations are given for approximate spins $I = 10[4]70$ as indicated for ^{150}Gd. In all cases, ε_4 increases with increasing spin. The configurations are those assigned[14,322] to the observed SD bands except for the nucleus ^{145}Gd where no SD band has been clearly established experimentally.

drafting the figures. The Joint Institute for Heavy Ion Research has as member institutions the University of Tennessee, Vanderbilt University, and the Oak Ridge National Laboratory; it is supported by the members and by the U.S. Department of Energy (DOE) through Contract No. DE–FG05–87ER40361 with the University of Tennessee. This work was also partly supported by the Swedish Natural Science Research Council (NFR), DOE through Contract No. DE-F605-93ER40770, and by the Polish Committee for Scientific Research under Contract No. 2P03B 034 08.

REFERENCES

1. **Bohr, A. and Mottelson, B.R.**, *Nuclear Structure*, vol. 2, Benjamin, New York, 1975.
2. **Hill, D.L. and Wheeler, J.A.**, *Phys. Rev.*, **89**, 1102, 1953.
3. **de-Shalit, A. and Goldhaber, M.**, *Phys. Rev.*, **92**, 1211, 1953.
4. **Åberg, S., Flocard, H., and Nazarewicz, W.**, *Annu. Rev. Nucl. Part. Sci.*, **40**, 439, 1990.
5. **Wood, J.L., Heyde, K., Nazarewicz, W., Huyse, M., and van Duppen, P.**, *Phys. Rep.*, **215**, 101, 1992.
6. **Bohr, A.**, *Mat. Fys. Medd. Dan. Vidensk. Selsk.*, No. 14, **26**, 1952.
7. **Nilsson, S.G., Tsang, C.F., Sobiczewski, A., Szyma'nski, Z., Wycech, S., Gustafson, C., Lamm, I.-L., Möller, P., and Nilsson, B.**, *Nucl. Phys.*, **A131**, 1, 1969.
8. **Leander, G.A. and Larsson, S.E.**, *Nucl. Phys.*, **A239**, 93, 1975.

9. **Nilsson, S.G.**, *Mat. Fys. Medd. Dan. Vidensk. Selsk.*, No. 16, **29**, 1955.
10. **Larsson, S.E.**, *Phys. Scr.*, **8**, 17, 1973.
11. **Sheline, R.K., Ragnarsson, I., Åberg, S., and Watt, A.**, *J. Phys. G: Nucl. Phys.*, **14**, 1201, 1988.
12. **Bengtsson, R., Dudek, J., Nazarewicz, W., and Olanders, P.**, *Phys. Scr.*, **39**, 196, 1989.
13. **Bjørnholm, S. and Lynn, J.E.**, *Rev. Mod. Phys.*, **52**, 725, 1980.
14. **Bengtsson, T., Ragnarsson, I., and Åberg, S.**, *Phys. Lett.*, **B208**, 39, 1988.
15. **Nazarewicz, W., Wyss, R., and Johnson, A.**, *Nucl. Phys.*, **A503**, 285, 1989.
16. **Ogle, W., Wahlborn, S., Piepenbring, R., and Fredriksson, S.**, *Rev. Mod. Phys.*, **43**, 424, 1971.
17. **Chasman, R.R., Ahmad, I., Friedman, A.M., and Erskine, J.R.**, *Rev. Mod. Phys.*, **49**, 833, 1977.
18. **Neergård, K. and Pashkevich, V.V.**, *Phys. Lett.*, **B59**, 218, 1975.
19. **Faber, M., Poszajczak, M., and Faessler, A.**, *Nucl. Phys.*, **A326**, 129, 1979.
20. **Löbner, K.E.G., Vetter, M., and Hönig, V.**, *Nucl. Data Tables*, **A7**, 495, 1970.
21. **Brack, M., Damgaard, J., Jensen, A.S., Pauli, H.C., Strutinsky, V.M., and Wong, C.Y.**, *Rev. Mod. Phys.*, **44**, 320, 1972.
22. **Nix, J.R.**, *Annu. Rev. Nucl. Sci.*, **22**, 65, 1972.
23. **Pashkevich, V.V.**, *Nucl. Phys.*, **A169**, 275, 1971.
24. **Nix, J.R.**, *Nucl. Phys.*, **A130**, 241, 1969.
25. **Rohoziński, S.G. and Sobiczewski, A.**, *Acta Phys. Pol.*, **B12**, 1001, 1981.
26. **Rohoziński, S.G.**, Report INR 1520/VII/PH/B, Warszawa, unpublished, 1974.
27. **Larsson, S.E., Leander, G.A., Ragnarsson, I., and Alenius, N.G.**, *Nucl. Phys.*, **A261**, 77, 1976.
28. **Nazarewicz, W. and Rozmej, P.**, *Nucl. Phys.*, **A369**, 396, 1981.
29. **Nazarewicz, W., Dudek, J., Bengtsson, R., Bengtsson, T., and Ragnarsson, I.**, *Nucl. Phys.*, **A435**, 397, 1985.
30. **Bengtsson, T. and Ragnarsson, I.**, *Nucl. Phys.*, **A436**, 14, 1985.
31. **Rohoziński, S.G.**, J. Phys. G: *Nucl. Phys.*, **16**, L173, 1990.
32. **Hamamoto, I., Zhang, X.Z., and Xie, H.-X.**, *Phys. Lett.*, **B257**, 1, 1991; erratum: **B278**, 511.
33. **Skalski, J.**, *Phys. Rev.*, **C43**, 140, 1991.
34. **Skalski, J.**, *Phys. Lett.*, **B274**, 1, 1992.
35. **Bohr, A. and Mottelson, B.R.**, *Mat. Fys. Medd. Dan. Vidensk. Selsk.*, No. 16, **27**, 1953.
36. **Ring, P., Hayashi, A., Hara, K., Emling, H., and Grosse, E.**, *Phys. Lett.*, **B110**, 423, 1982.
37. **Alder, K., Bohr, A., Huus, T., Mottelson, B.R., and Winther, A.**, *Rev. Mod. Phys.*, **28**, 432, 1956.
38. **Alder, K. and Winther, A.**, *Electromagnetic Excitation, Theory of Coulomb Excitation with Heavy Ions*, North-Holland, Amsterdam, 1975.
39. **Czosnyka, T., Cline, D., and Wu, C.Y.**, *Bull. Am. Phys. Soc.*, **28**, 745, 1983.
40. **Cline, D.**, *Annu. Rev. Nucl. Part. Sci.*, **36**, 683, 1986.
41. **Kumar, K.**, *Phys. Rev. Lett.*, **28**, 249, 1972.
42. **Cline, D.**, in *Nuclear Structure 1985*, Broglia, R., Hagemann, G.B., and Herskind, B., Eds., Elsevier, Amsterdam, 1985, 313.
43. **Stephens, F.S., Diamond, R.M., Glendenning, N.K., and de Boer, J.**, *Phys. Rev. Lett.*, **24**, 1137, 1970; **Stephens, F.S., Diamond, R.M., and de Boer, J.**, *Phys. Rev. Lett.*, **27**, 1151, 1971.

44. **Kulessa, R., Bengtsson, R., Bohn, H., Emling, H., Faestermann, T., von Feilitsch, F., Grosse, E., Nazarewicz, W., Schwalm, D., and Wollersheim, H.J.**, *Phys. Lett.*, **B218**, 421, 1989.
45. **Ibbotson, R., White, C.A., Czosnyka, T., Kotliński, B., Butler, P.A., Clarkson, N., Cunningham, R.A., Cline, D., Devlin, M., Helmer, K.G., Hoare, T.H., Hughes, J.R., Kavka, A.E., Jones, G.D., Poynter, R.J., Regan, P., Vogt, E.G. Wadsworth, R., Watson, D.L., and Wu, C.Y.**, *Proc. Int. Conf. on Nuclear Structure at High Angular Momentum, Ottawa, 1992,* vol. 2, Proceedings, AECL 10613, 434, 1992.
46. **Fick, D.**, *Annu. Rev. Nucl. Part. Sci.*, **31**, 53, 1981.
47. **Fick, D.**, *J. Phys. (Paris), Colloq.*, **C6**, 265, 1990.
48. **Bernstein, A.M.**, *Adv. Nucl. Phys.*, **3**, 325, 1969.
49. **Satchler, G.R.**, *J. Math. Phys.*, **13**, 1118, 1972.
50. **Mackintosh, R.S.**, *Nucl. Phys.*, **A266**, 379, 1976.
51. **Baker, F.T., Scott, A., Styles, R.C., Kruse, T.H., Jones, K., and Suchannek, R.**, *Nucl. Phys.*, **A351**, 63, 1981.
52. **Haouat, G., Lagrange, Ch., de Swiniarski, R., Dietrich, F., Delaroche, J.P., and Patin, Y.**, *Phys. Rev.*, **C30**, 1795, 1984.
53. **Ichihara, T., Sakaguchi, H., Nakamura, M., Noro, T., Ohtani, F., Sakamoto, H., Ogawa, H., Yosoi, M., Ieiri, M., Isshiki, N., and Kobayashi, S.**, *Phys. Rev.*, **C29**, 1228, 1984.
54. **Ogawa, H., Sakaguchi, H., Nakamura, M., Noro, T., Sakamoto, H., Ichihara, T., Yosoi, M., Ieiri, M., Isshiki, N., Takeuchi, Y., and Kobayashi, S.**, *Phys. Rev.*, **C33**, 834, 1986.
55. **Lee, I.Y., Saladin, J.X., Holden, J., O'Brien, J., Baktash, C., Bemis, Jr., C., Stelson, P.H., McGowan, F.K., Milner, W.T., Ford, Jr., J.L.C., Robinson, R.L., and Tuttle, W.**, *Phys. Rev.*, **C12**, 1483, 1975.
56. **Rebel, H., Pesl, R., Gils, H.J., and Friedman, E.**, *Nucl. Phys.*, **A368**, 61, 1981.
57. **Hendrie, D.L.**, *Phys. Rev. Lett.*, **31**, 478, 1973.
58. **Madsen, V.A. and Brown, V.R.**, *Phys. Rev. Lett.*, **52**, 176, 1984.
59. **Lay, B.G., Banks, S.M., Spicer, B.M., Shute, G.G., Officer, V.C., Ronningen, R.M., Crawley, G.M., Anantaraman, N., and DeVito, R.P.**, *Phys. Rev.*, **C32**, 440, 1985.
60. **Govil, I.M., Fulbright, H.W., Cline, D., Wesołwski, E., Kotliński, B., Bäcklin, A., and Gridnev, K.**, *Phys. Rev.*, **C33**, 793, 1986.
61. **Fritze, J., Neu, R., Abele, H., Hoyler, F., Staudt, G., Eversheim, P.D., Hinterberger, F., and Müther, H.**, *Phys. Rev.*, **C43**, 2307, 1991.
62. **Heisenberg, J.**, *Adv. Nucl. Phys.*, **12**, 61, 1981.
63. **Bertozzi, W.**, *Nucl. Phys.*, **A374**, 109c, 1982.
64. **Heisenberg, J., and Blok, H.P.**, *Annu. Rev. Nucl. Part. Sci.*, **33**, 569, 1983.
65. **Frois, B.**, in *Nuclear Structure 1985,* Broglia, R., Hagemann, G.B., and Herskind, B., Eds., Elsevier, Amsterdam, 1985, 25.
66. **Heisenberg, J.**, in *Nuclear Structure 1985,* Broglia, R., Hagemann, G.B., and Herskind, B., Eds., Elsevier, Amsterdam, 1985, 229.
67. **Cooper, T., Bertozzi, W., Heisenberg, J., Kowalski, S., Turchinetz, W., Williamson, C., Cardman, L., Fivozinsky, S., Lightbody, Jr., J., and Penner, S.**, *Phys. Rev.*, **C13**, 1083, 1976.
68. **Reuter, W., Shera, E.B., Wohlfahrt, H.D., and Tanaka, Y.**, *Phys. Lett.*, **B124**, 293, 1983.

69. **Boeglin, W., Egelhof, P., Sick, I., Cavedon, J.M., Frois, B., Goutte, D., Phan, X.H., Platchkov, S.K., Williamson, S., and Girod, M.**, *Phys. Lett.*, **B186**, 285, 1987.
70. **Phan, X.H., Andersen, H.G., Cardman, L.S., Cavedon, J.-M., Clemens, J.-C., Frois, B., Girod, M., Gogny, D., Goutte, D., Grammaticos, B., Hofmann, R., Huet, M., Leconte, P., Platchkov, S.K., Sick, I., and Williamson, S.E.**, *Phys. Rev.*, **C38**, 1173, 1988.
71. **Fossan, D.B. and Warburton, E.K.**, in *Nuclear Spectroscopy and Reactions*, part C, Cerny, J., Ed., Academic Press, New York, 1974, 307.
72. **Moore, E.F., Janssens, R.V.F., Ahmad, I., Carpenter, M.P., Fernandez, P.B., Khoo, T.L., Ridley, S.L., Wolfs, F.L.H., Ye, D., Beard, K.B., Garg, U., Drigert, M.W., Benet, Ph., Daly, P.J., Wyss, R., and Nazarewicz, W.**, *Phys. Rev. Lett.*, **64**, 3127, 1990.
73. **Lisle, J.C., Clarke, D., Chapman, R., Khazaie, F., Mo, J.N., Hübel, H., Schmitz, W., Theine, K., Garrett, J.D., Hagemann, G.B., Herskind, B., and Schiffer, K.**, *Nucl. Phys.*, **A520**, 451c, 1990.
74. **Wang, T.F., Henry, E.A., Becker, J.A., Kuhnert, A., Stoyer, M.A., Yates, S.W., Brinkman, M.J., Cizewski, J.A., Macchiavelli, A.O., Stephens, F.S., Deleplanque, M.A., Diamond, R.M., Draper, J.E., Azaiez, F.A., Kelly, W.H., Korten, W., Rubel, E., and Akovali, Y.A.**, *Phys. Rev. Lett.*, **69**, 1737, 1992.
75. **Börner, H.G.**, J. Phys. G: *Nucl. Phys.*, **14**, Supplement, S143, 1988.
76. **Börner, H.G., Jolie, J., Robinson, S.J., Krusche, B., Piepenbring, R., Casten, R.F., Aprahamian, A., and Draayer, J.P.**, *Phys. Rev. Lett.*, **66**, 691, 1991.
77. **Fewell, M.P., Johnson, N.R., McGowan, F.K., Hattula, J.S., Lee, I.Y., Baktash, C., Schutz, Y., Wells, J.C., Riedinger, L.L., Guidry, M.W., and Pancholi, S.C.**, *Phys. Rev.*, **C31**, 1057, 1985.
78. **Emling, H., Grosse, E., Kulessa, R., Schwalm, D., and Wollersheim, H.J.**, *Nucl. Phys.*, **A419**, 187, 1984.
79. **Moscrop, R., Campbell, M., Gelletly, W., Goettig, L., Lister, C.J., and Varley, B.J.**, *Nucl. Phys.*, **A499**, 565, 1989.
80. **Löbner, K.E.G.**, in *The Electromagnetic Interactions in Nuclear Spectroscopy*, Hamilton, W.D., Ed., North-Holland, Amsterdam, 1975, 173.
81. **Metzger, F.R.**, *Prog. Nucl. Phys.*, **7**, 53, 1959.
82. **Wheeler, J.A.**, *Phys. Rev.*, **92**, 812, 1953.
83. **Wilets, L.**, *Mat. Fys. Medd. Dan. Vidensk. Selsk.*, **29**, No. 3, 1954.
84. **Jacobsohn, B. A.**, *Phys. Rev.*, **96**, 1637, 1954.
85. **Wu, C.S., and Wilets, L.**, *Annu. Rev. Nucl. Sci.*, **19**, 527, 1969.
86. **Hufner, J., Scheck, F., and Wu, C.S.**, in *Muon Physics*, Hughes, V.M., and Wu, C.S., Eds., Academic Press, New York, 1977.
87. **Hitlin, D., Bernow, S., Devons, S., Duerdoth, I., Kast, J.W., Macagno, E.R., Rainwater, J., Wu, C. S., and Barrett, R.C.**, *Phys. Rev.*, **C1**, 1184, 1970.
88. **Wagner, L.K., Shera, E.B., Rinker, G.A., and Sheline, R.K.**, *Phys. Rev.*, **C16**, 1549, 1977.
89. **Yamazaki, Y., Shera, E.B., Hoehn, M.V., and Steffen, R.M.**, *Phys. Rev.*, **C18**, 1474, 1978.
90. **Tanaka, Y., Steffen, R.M., Shera, E.B., Reuter, W., Hoehn, M.V., and Zumbro, J.D.**, *Phys. Rev.*, **C29**, 1830, 1984.
91. **Zumbro, J.D., Shera, E.B., Tanaka, Y., Bemis, Jr., C.E., Naumann, R.A., Hoehn, M.V., Reuter, W., and Steffen, R.M.**, *Phys. Rev. Lett.*, **53**, 1888, 1984.

92. **Zumbro, J.D., Naumann, R.A., Hoehn, M.V., Reuter, W., Shera, E.B., Bemis, Jr., C.E., and Tanaka, Y.**, *Phys. Lett.*, **B167**, 383, 1986.
93. *Low Temperature Nuclear Orientation*, Stone, N.J. and Postma, H., Eds., North-Holland, Amsterdam, 1986.
94. *On Line Nuclear Orientation–1*, Proc. First Int. Conf. on On-Line Nuclear Orientation (OLNO–1), Stone, N.J. and Rikovska, J., Eds., Baltzer AG, Basel, 1988; *Hyp. Int.*, **43**, 1988.
95. **Raghavan, P.**, *At. Data Nucl. Data Tables,* **42**, 189, 1989.
96. **Bounds, J.A., Bingham, C.R., Carter, H.K., Leander, G.A., Mlekodaj, R.L., Spejewski, E.H., and Fairbank, Jr., W.M.**, *Phys. Rev.*, **C36**, 2560, 1987.
97. **Neu, W., Neugart, R., Otten, E.-W., Passler, G., Wendt, K., Fricke, B., Arnold, E., Kluge, H.J., Ulm, G., and the ISOLDE Collaboration**, *Z. Phys.,* **D11**, 105, 1989.
98. **Hinfurtner, B., Hagn, E., Zech, E., Eder, R., the NICOLE Collaboration, and the ISOLDE Collaboration**, *Phys. Rev. Lett.,* **67**, 812, 1991.
99. **Broude, C., Bendahan, J., Dafni, E., Goldring, G., Hass, M., Lahmer-Naim, E., and Rafailovich, M.H.**, in *The Variety of Nuclear Shapes,* Garrett, J.D. *et al.*, Eds., World Scientific, Singapore, 1988, 292.
100. **Dafni, E., Hass, M., Bertschat, H.H., Broude, C., Davidovsky, F., Goldring, G., and Lesser, P.M.S.**, *Phys. Rev. Lett.,* **50**, 1652, 1983.
101. **Dracoulis, G.D., Lönnroth, T., Vajda, S., Kistner, O.C., Rafailovich, M.H., Dafni, E., and Schatz, G.**, *Phys. Lett.,* **B149**, 311, 1984.
102. **Dafni, E., Bendahan, J., Broude, C., Goldring, G., Hass, M., Naim, E., Rafailovich, M.H., Chasman, C., Kistner, O.C., and Vajda, S.**, *Nucl. Phys.,* **A443**, 135, 1985.
103. **Berger, A., Grawe, H., Mahnke, H.-E., Dafni, E., Hass, M., Naim, E., and Rafailovich, M.H.**, *Phys. Lett.,* **B182**, 11, 1986.
104. **Broude, C., Hass, M., Goldring, G., Alderson, A., Ali, I., Cullen, D. M., Fallon, P., Hanna, F., Roberts, J.W., and Sharpey-Schafer, J.F.**, *Phys. Lett.,* **B264**, 17, 1991.
105. **Hardeman, F., Scheveneels, G., Neyens, G., Nowen, R., and Coussement, R.**, *Hyp. Int.,* **59**, 13, 1990.
106. **Scheveneels, G., Hardeman, F., Neyens, G., and Coussement, R.**, *Phys. Rev.,* **C43**, 2560, 1991.
107. **Wesołwski, E.**, *J. Phys. G: Nucl. Phys.,* **11**, 909, 1985.
108. **King, W.H.**, *Isotope Shifts in Atomic Spectra,* Plenum Press, New York, 1984.
109. **Aufmuth, P., Heilig, K., and Steudel, A.**, *At. Data Nucl. Data Tables,* **37**, 455, 1987.
110. **Ahmad, S.A., Klempt, W., Neugart, R., Otten, E.W., Reinhard, P.-G., Ulm, G., Wendt, K., and the ISOLDE Collaboration**, *Nucl. Phys.,* **A483**, 244, 1988.
111. **Grodzins, L.**, *Phys. Lett.,* **2**, 88, 1962.
112. **Goldhaber, A.S. and Scharff-Goldhaber, G.**, *Phys. Rev.,* **C17**, 1171, 1978.
113. **Raman, S., Nestor, Jr., C.W., and Bhatt, K.H.**, *Phys. Rev.,* **C37**, 805, 1988.
114. **Raman, S., Nestor, Jr., C.W., Kahane, S., and Bhatt, K.H.**, *At. Data Nucl. Data Tables,* **42**, 1, 1989.
115. **Raman, S., Nestor, Jr., C.W., Kahane, S., and Bhatt, K.H.**, *Phys. Rev.,* **C43**, 556, 1991.
116. **Ross, C.K. and Bhaduri, R.K.**, *Nucl. Phys.,* **A196**, 369, 1972.
117. **Patnaik, R., Patra, R., and Satpathy, L.**, *Phys. Rev.,* **C12**, 2038, 1975.
118. **Hamamoto, I.**, *Nucl. Phys.,* **73**, 225, 1965.
119. **Casten, R.F.**, *Nucl. Phys.,* **A443**, 1, 1985.

120. **Jänecke, J.**, *Phys. Lett.*, **B103**, 1, 1981.
121. **Wesołowski, E.**, *Acta Phys. Pol.*, **B15**, 559, 1984.
122. **Spear, R.H.**, *At. Data Nucl. Data Tables,* **42**, 55, 1989.
123. **Ronningen, R.M., Melin, R.C., Nolen, Jr., J.A., Crawley, G.M., and Bemis, Jr., C.E.**, *Phys. Rev. Lett.,* **47**, 635, 1981.
124. **Rozmej, P.**, *Nucl. Phys.,* **A445**, 495, 1985.
125. **Ring, P. and Schuck, P.**, *The Nuclear Many-Body Problem,* Springer-Verlag, New York, 1980.
126. **Kisslinger, L.S. and Sorensen, R.A.**, *Rev. Mod. Phys.,* **35**, 853, 1963.
127. **Kumar, K. and Baranger, M.**, *Nucl. Phys.,* **A110**, 529, 1968.
128. **Baranger, M. and Kumar, K.**, *Nucl. Phys.,* **A122**, 241, 1968.
129. **Mottelson, B.R. and Nilsson, S.G.**, *Phys. Rev.,* **99**, 1615, 1955.
130. **Moszkowski, S.A.**, *Phys. Rev.,* **99**, 803, 1955.
131. **Bès, D.R. and Szymański, Z.**, *Nucl. Phys.,* **28**, 42, 1961.
132. **Damgaard, J., Pauli, H.C., Pashkevich, V.V., and Strutinsky, V.M.**, *Nucl. Phys.,* **A135**, 432, 1969.
133. **Ćwiok, S., Dudek, J., Nazarewicz, W., Skalski, J., and Werner, T.**, *Comput. Phys. Commun.,* **46**, 379, 1987.
134. **Möller, P., and Nix, J.R.**, *Nucl. Phys.,* **A361**, 117, 1981.
135. **Myers, W.D. and Swiatecki, W.J.**, *Ann. Phys. (New York),* **55**, 395, 1969.
136. **Swiatecki, W.J.**, in *Proc. 2nd Int. Conf. on Nuclidic Masses,* Johnson, Jr., W.H., Ed., Springer-Verlag, Vienna, 1964, 58.
137. **Strutinsky, V.M.**, *Nucl. Phys.,* **A95**, 420, 1967.
138. **Strutinsky, V.M.**, *Nucl. Phys.,* **A218**, 169, 1974.
139. **Whitehead, R.R., Watt, A., Cole, B.J., and Morrison, I.**, in Adv. *Nucl. Phys.,* vol. 9, Baranger, M. and Vogt, E., Eds., Plenum Press, New York, 1977, 123.
140. **Wildenthal, B.H.**, *Prog. Part. Nucl. Phys.,* **11**, 5, 1983.
141. **Heyde, K.**, *The Nuclear Shell Model,* Springer-Verlag, Berlin, Heidelberg, 1990.
142. **Watt, A., Kelvin, D., and Whitehead, R.R.**, *Phys. Lett.,* **B63**, 388, 1976.
143. **Watt, A., Kelvin, D., and Whitehead, R.R.**, *J. Phys. G: Nucl. Phys.,* **6**, 35, 1980.
144. **Haxton, W.C. and Johnson, C.**, *Phys. Rev. Lett.,* 65, 1325, 1990.
145. **Schmid, K.W., Grümmer, F., and Faessler, A.**, *Phys. Rev.,* **C29**, 291, 1984.
146. **Schmid, K.W., Grümmer, F., Kyotoku, M., and Faessler, A.**, *Nucl. Phys.,* **A452**, 493, 1986.
147. **Schmid, K.W. and Grümmer, F.**, *Rep. Prog. Phys.,* **50**, 731, 1987.
148. **Schmid, K.W.**, in *High-Spin Physics and Gamma-Soft Nuclei,* Saladin, J.X., Sorensen, R.A., and Vincent, C.M., Eds., World Scientific, Singapore, 106, 1991.
149. **Elliott, J.P.**, *Proc. R. Soc.,* **A245**, 128, 1958; ibid., 562, 1958.
150. **Castaños, O., Draayer, J.P., and Leschber, Y.**, *Z. Phys.,* **A329**, 33, 1988.
151. **Rowe, D.J.**, *Rep. Prog. Phys.,* **48**, 1419, 1985.
152. **Draayer, J.P.**, *Nucl. Phys.,* **A520**, 259c, 1990.
153. **Zhang, W.M., Feng, D.H., and Gilmore, R.**, *Rev. Mod. Phys.,* **62**, 867, 1990.
154. **Leviatan, A. and Kirson, M.W.**, *Ann. Phys. (New York),* **201**, 13, 1990.
155. **Zhang, W.M., Wu, C.L., Feng, D.H., Ginocchio, J.N., and Guidry, M.W.**, *Phys. Rev.,* **C38**, 1475, 1988.
156. **Bhatt, K.H., Nestor, Jr., C.W., and Raman, S.**, *Phys. Rev.,* **C46**, 164, 1992.
157. **Bengtsson, T., Faber, M.E., Leander, G., Möller, P., Płoszajczak, M., Ragnarsson, I., and Åberg, S.**, *Phys. Scr.,* **24**, 200, 1981.
158. **Nazarewicz, W. and Dobaczewski, J.**, *Phys. Rev. Lett.,* **68**, 154, 1992.

159. **Dennison, D.M.**, *Phys. Rev.,* **57**, 454, 1940.
160. **Dennison, D.M.**, *Phys. Rev.,* **96**, 378, 1954.
161. **Morinaga, H.**, *Phys. Rev.,* 101, 254, 1956.
162. **Sheline, R.K. and Wildermuth, K.**, *Nucl. Phys.,* **21**, 196, 1960.
163. **Ikeda, K., Takigawa, N., and Horiuchi, H.**, *Prog. Theor. Phys. Suppl. Extra Number,* 464, 1968.
164. **Horiuchi, H., Ikeda, K., and Suzuki, Y.**, *Suppl. Prog. Theor. Phys.,* **52**, 89, 1972.
165. **Brink, D.M.**, *Proc. Int. School of Physics,* 'Enrico Fermi' 1965, Bloch, C., Ed., Academic Press, New York, 1966, 247.
166. **Buck, B., Dover, C.B., and Vary, J.P.**, *Phys. Rev.,* **C11**, 1803, 1975.
167. **Descouvemont, P. and Baye, D.**, *Phys. Rev.,* **31**, 2274, 1985.
168. **Funck, C., Grund, B., and Langanke, K.**, Z. Phys., **A334**, 1, 1989.
169. **Abe, Y., Hiura, J., and Tanaka, H.**, *Prog. Theor. Phys.,* **46**, 352, 1971.
170. **Arima, A., Horiuchi, H., Kubodera, K., and Takigawa, N.**, *Adv. Nucl. Phys.,* vol. 5, Baranger, M. and Vogt, E., Eds., Plenum Press, New York, 1972, 345.
171. **Flocard, H., Heenen, P.H., Krieger, S.J., and Weiss, M.S.**, *Nucl. Phys.,* **A391**, 285, 1982.
172. **Provoost, D., Grümmer, F., Goeke, K., and Reinhard, P.-G.**, *Nucl. Phys.,* **A431**, 139, 1984.
173. **Strayer, M.R., Cusson, R.Y., Umar, A.S., Reinhard, P.-G., Bromley, D.A., and Greiner, W.**, *Phys. Lett.,* **B135**, 261, 1984.
174. **Harvey, M.**, USDERA Report ORO-4856-26, in *Proc. 2nd Int. Conf. on Clustering Phenomena in Nuclei,* College Park, 1975, 549.
175. **Rae, W.D.M.**, *Int. J. Mod. Phys.,* **A3**, 1343, 1988.
176. **Cseh, J. and Scheid, W.**, *J. Phys. G: Nucl. Phys.,* **18**, 1419, 1992.
177. **Chandra, H. and Mosel, U.**, *Nucl. Phys.,* **A298**, 151, 1978.
178. **Wuosmaa, A.H., Betts, R.R., Back, B.B., Freer, M., Glagola, B.G., Happ, Th., Henderson, D.J., Wilt, P., and Bearden, I.G.**, *Phys. Rev. Lett.,* **68**, 1295, 1992.
179. **Merchant, A.C. and Rae, W.D.M.**, *Phys. Rev.,* **C46**, 2096, 1992.
180. **Betts, R.R. and Rae, W.D.M.**, Inst. Phys. Conf. Ser., **86**, Durell, J.L., Irvine, J.M., and Morrison, G.C., Eds., 1986, 189.
181. **Rae, W.D.M. and Merchant, A.C.**, *Phys. Lett.,* **B279**, 207, 1992.
182. **Wuosmaa, A.H., Zürmuhle, R.W., Kutt, P.H., Pate, S.F., Saini, S., Halbert, M.L., and Hensley, D.C.**, *Phys. Rev.,* **C41**, 2666, 1990.
183. **Möller, P. and Nix, J.R.**, At. Data *Nucl. Data Tables,* **26**, 165, 1981.
184. **Bengtsson, R., Möller, P., Nix, J.R., Zhang, J.-Y.**, *Phys. Scr.,* **29**, 402, 1984.
185. **Möller, P., Nix, J.R., Myers, W.D., and Swiatecki, W.J.**, *At. Data Nucl. Data Tables,* **59**, 185, 1995.
186. **Seeger, P.A. and Howard, W.M.**, *Nucl. Phys.,* **A238**, 491, 1975.
187. **Åberg, S.**, *Phys. Scr.,* **25**, 23, 1982.
188. **Ragnarsson, I. and Sheline, R.K.**, *Phys. Scr.,* **29**, 385, 1984.
189. **Kumar, K.**, *Nuclear Models and the Search for Unity in Nuclear Physics,* Bergen; Universitetsforlaget, 1984.
190. **Pearson, J.M., Aboussir, Y., Dutta, A.K., Nayak, R.C., Farine, M., and Tondeur, F.**, *Nucl. Phys.,* **A528**, 1, 1991.
191. **Tondeur, F.**, *Nucl. Phys.,* **A359**, 278, 1981.
192. **Arseniev, D.A., Sobiczewski, A., and Soloviev, V.G.**, *Nucl. Phys.,* **A139**, 269, 1969.
193. **Faessler, A., Galonska, J.E., Götz, U., and Pauli, H.C.**, *Nucl. Phys.,* **A230**, 302, 1974.

194. **Chasman, R.R.**, *Z. Phys.,* **A339**, 111, 1991.
195. **Ragnarsson, I.**, in *Proc. Int. Conf. on the Properties of Nuclei far from the Region of Beta-Stability,* vol. 2, CERN 70-30, Leysin, Switzerland, 1970, 847.
196. **Arseniev, D.A., Sobiczewski, A., and Soloviev, V.G.**, *Nucl. Phys.,* **A126**, 15, 1969.
197. **Arseniev, D.A., Malov, L.A., Pashkevich, V.V., Sobiczewski, A., and Soloviev, V.G.**, *Sov. J. Nucl. Phys.,* **8**, 514, 1969.
198. **Kumar, K. and Baranger, M.**, *Nucl. Phys.,* **A122**, 273, 1968.
199. **Gareev, F.A., Ivanova, S.P., and Pashkevich, V.V.**, *Sov. J. Nucl. Phys.,* **11**, 667, 1970.
200. **Möller, P., Nilsson, S.G., and Nix, J.R.**, *Nucl. Phys.,* **A229**, 292, 1974.
201. **Ragnarsson, I., Sobiczewski, A., Sheline, R.K., Larsson, S.E., and Nerlo-Pomorska, B.**, *Nucl. Phys.,* **A233**, 329, 1974.
202. **Kern, B.D., Mlekodaj, R.L., Leander, G.A., Kortelahti, M.O., Zganjar, E.F., Braga, R.A., Fink, R.W., Perez, C.P., Nazarewicz, W., and Semmes, P.B.**, *Phys. Rev.,* **C36**, 1514, 1987.
203. **Götz, U., Pauli, H.C., Alder, K., and Junker, K.**, *Nucl. Phys.,* **A192**, 1, 1972.
204. **Werner, T.R. and Dudek, J.**, *At. Data Nucl. Data Tables,* **50**, 179, 1992.
205. **Kumar, A. and Gunye, M.R.**, *J. Phys. G: Nucl. Phys.,* **8**, 975, 1982.
206. **Nerlo-Pomorska, B., Pomorski, K., Brack, M., and Werner, E.**, *Nucl. Phys.,* **A462**, 252, 1987.
207. **Nielsen, B.S. and Bunker, M.E.**, *Nucl. Phys.,* **A245**, 376, 1975.
208. **Alikov, B.A., Zuber, K., Pashkevich, V.V., and Tsoi, E.G.**, *Izv. Akad. Nauk SSSR, Ser. Fiz.,* **48**, 875, 1984.
209. **Nazarewicz, W., Riley, M.A., and Garrett, J.D.**, *Nucl. Phys.,* **A512**, 61, 1990.
210. **Kahane, S., Raman, S., and Dudek, J.**, *Phys. Rev.,* **C40**, 2282, 1989.
211. **Dudek, J., Nazarewicz, W., and Szymański, Z.**, *Phys. Rev.,* **C26**, 1708, 1982.
212. **Egido, J.L. and Ring, P.**, *Nucl. Phys.,* **A423**, 93, 1984.
213. **Böning, K. and Sobiczewski, A.**, *Acta Phys. Pol.,* **B14**, 287, 1983.
214. **Žofka, J. and Ripka, G.**, *Nucl. Phys.,* **A168**, 65, 1971.
215. **Abgrall, Y., Morand, B., and Caurier, E.**, *Nucl. Phys.,* **A192**, 372, 1972.
216. **Vautherin, D.**, *Phys. Rev.,* **C7**, 296, 1973.
217. **Grin, Yu.T. and Kochetov, A.B.**, *Sov. J. Nucl. Phys.,* **18**, 145, 1974.
218. **Negele, J. W. and Rinker, G.**, *Phys. Rev.,* **C15**, 1499, 1977.
219. **Girod, M. and Reinhard, P.-G.**, *Nucl. Phys.,* **A384**, 179, 1982.
220. **Girod, M. and Grammaticos, B.**, *Phys. Rev. Lett.,* **40**, 361, 1978.
221. **Federman, P. and Pittel, S.**, *Phys. Lett.,* **B77**, 29, 1978.
222. **Khosa, S.K., Tripathi, P.N., and Sharma, S.K.**, *Phys. Lett.,* **B119**, 257, 1982.
223. **Kumar, A. and Gunye, M.R.**, *Phys. Rev.,* **C32**, 2116, 1985.
224. **Heyde, K., Moreau, J., and Waroquier, M.**, *Phys. Rev.,* **C29**, 1859, 1984.
225. **Bonche, P., Flocard, H., Heenen, P.H., Krieger, S.J., and Weiss, M.S.**, *Nucl. Phys.,* **A443**, 39, 1985.
226. **Flocard, H., Quentin, P., Kerman, A.K., and Vautherin, D.**, *Nucl. Phys.,* **A203**, 433, 1973.
227. **Flocard, H., Quentin, P., and Vautherin, D.**, *Phys. Lett.,* **B46**, 304, 1973.
228. **Redon, N., Meyer, J., Meyer, M., Quentin, P., Bonche, P., Flocard, H., and Heenen, P.-H.**, *Phys. Rev.,* **C38**, 550, 1988.
229. **Gambhir, Y.K. and Ring, P.**, *Phys. Lett.,* **B202**, 5, 1988.
230. **Sauvage-Letessier, J., Quentin, P., and Flocard, H.**, *Nucl. Phys.,* **A370**, 231, 1981.
231. **Bertsch, G.F.**, *Phys. Lett.,* **B26**, 130, 1967.

232. **Ronningen, R.M., Baker, F.T., Scott, A., Kruse, T.H., Suchannek, R., Savin, W., and Hamilton, J.H.**, *Phys. Rev. Lett.,* **40**, 364, 1981.
233. **Govil, I.M., Fulbright, H.W., and Cline, D.**, *Phys. Rev.,* **C36**, 1442, 1987.
234. **Hendrie, D.L., Glendenning, N.K., Harvey, B.G., Jarvis, O.N., Duhm, H.H., Saudinos, J., and Mahoney, J.**, *Phys. Lett.,* **B26**, 127, 1968.
235. **Aponick, Jr., A.A., Chesterfield, C.M., Bromley, D.A., and Glendenning, N.K.**, *Nucl. Phys.,* **A159**, 367, 1970.
236. **Hendrie, D.L., Harvey, B.G., Meriwether, J.R., Mahoney, J., Faivre, J.-C., and Kovar, D.G.**, *Phys. Rev. Lett.,* **30**, 571, 1973.
237. **Moss, J.M., Terrien, Y.D., Lombard, R.M., Brassard, C., Loiseaux, J.M., and Resmini, F.**, *Phys. Rev. Lett.,* **26**, 1488, 1971.
238. **Ronningen, R.M., Crawley, G.M., Anantaraman, N., Banks, S.M., Spicer, B.M., Shute, G.G., Officer, V.C., Wastell, J.M.R., Devins, D.W., and Friesel, D.L.**, *Phys. Rev.,* **C28**, 123, 1983.
239. **Möller, P. and Nilsson, B.**, *Phys. Lett.,* **B31**, 171, 1970.
240. **Ragnarsson, I.**, in *Future Directions in Structure of Nuclei Far From Stability*, Hamilton, J.H., Spejewski, E.H., Bingham, C.R., and Zganjar, E.F., Eds., North-Holland, Amsterdam, 1989, 367.
241. **Chasman, R.R.**, *Phys. Lett.,* **B175**, 254, 1986.
242. **Sobiczewski, A., Patyk, Z., Ćwiok, S., and Rozmej, P.**, *Nucl. Phys.,* **A485**, 16, 1988.
243. **Egido, J.L. and Robledo, L.M.**, *Nucl. Phys.,* **A494**, 85, 1989.
244. **Egido, J.L. and Robledo, L.M.**, *Nucl. Phys.,* **A518**, 475, 1990.
245. **Ćwiok, S. and Nazarewicz, W.**, *Nucl. Phys.,* **A496**, 367, 1989.
246. **Möller, P., Nix, J.R., Myers, W.D., and Swiatecki, W.J.**, LBL Preprint, 1991.
247. **Patyk, Z. and Sobiczewski, A.**, *Nucl. Phys.,* **A533**, 132, 1991.
248. **Barlett, M.L., McGill, J.A., Ray, L., Barlett, M.M., Hoffmann, G.W., Hintz, N.M., Kyle, G.S., Franey, M.A., and Blanpied, G.**, *Phys. Rev.,* **C22**, 1168, 1980.
249. **Leander, G.A., Sheline, R.K., Möller, P., Olanders, P., Ragnarsson, I., and Sierk, A.J.**, *Nucl. Phys.,* **A388**, 452, 1982.
250. **Nazarewicz, W., Olanders, P., Ragnarsson, I., Dudek, J., Leander, G.A., Möller, P., and Ruchowska, E.**, *Nucl. Phys.,* **A429**, 269, 1984.
251. **Rohoziński, S.G.**, Rep. Prog. Phys., **51**, 541, 1988.
252. **Nazarewicz, W.**, *Nucl. Phys.,* **A520**, 333c, 1990.
253. **Bonche, P.**, in *The Variety of Nuclear Shapes*, Garrett, J.D., *et al.*, Eds., World Scientific, Singapore, 1988, 302.
254. **Robledo, L.M., Egido, J.L., Nerlo-Pomorska, B., and Pomorski, K.**, *Phys. Lett.,* **B201**, 409, 1988.
255. **Ennis, P.J., Lister, C.J., Gelletly, W., Price, H.G., Varley, B.J., Butler, P.A., Hoare, T., Ćwiok, S., and Nazarewicz, W.**, *Nucl. Phys.,* **A535**, 392, 1992.
256. **Mach, H., Nazarewicz, W., Kusnezov, D., Moszyński, M., Fogelberg, B., Hellström, M., Spanier, L., Gill, R.L., Casten, R.F., and Wolf, A.**, *Phys. Rev.,* **C41**, R2469, 1990.
257. **Skalski, J.**, *Phys. Lett.,* **B238**, 6, 1990.
258. **Chasman, R.R.**, *Phys. Lett.,* **B96**, 7, 1980.
259. **Chasman, R.R.**, *J. Phys. (Paris), Colloq.,* **C6**, 167, 1984.
260. **Bonche, P., Heenen, P.H., Flocard, H., and Vautherin, D.**, *Phys. Lett.,* **B175**, 387, 1986.

261. **Robledo, L.M., Egido, J.L., Berger, J.F., and Girod, M.**, *Phys. Lett.*, **B187**, 223, 1987.
262. **Nazarewicz, W., Leander, G.A., and Dudek, J.**, *Nucl. Phys.*, **A467**, 437, 1987.
263. **Leander, G.A. and Chen, Y.S.**, *Phys. Rev.*, **C37**, 2744, 1988.
264. **Chasman, R.R.**, *Phys. Lett.*, **B219**, 232, 1989.
265. **Ćwiok, S. and Nazarewicz, W.**, *Nucl. Phys.*, **A529**, 95, 1991.
266. **Nazarewicz, W. and Tabor, S.L.**, *Phys. Rev.*, **C45**, 2226, 1992.
267. **Strutinsky, V.M.**, *Atomnaya Energiya*, **1**, 150, 1956; *J. Nucl. Energy*, **4**, 523, 1957.
268. **Bohr, A. and Mottelson, B.R.**, *Nucl. Phys.*, **9**, 687, 1959.
269. **Butler, P.A. and Nazarewicz, W.**, *Nucl. Phys.*, **A533**, 249, 1991.
270. **Dorso, C.O., Myers, W.D., and Swiatecki, W.J.**, *Nucl. Phys.*, **A451**, 189, 1986.
271. **Leander, G.A.**, in *AIP Conf. Proc.*, **125**, American Institute of Physics, New York, 1985, 125.
272. **Leander, G.A., Nazarewicz, W., Bertsch, G.F., and Dudek, J.**, *Nucl. Phys.*, **A453**, 58, 1986.
273. **Metag, V., Habs, D., and Specht, H.J.**, *Phys. Rep.*, **65**, 1, 1980.
274. **Twin, P.J.**, *Nucl. Phys.*, **A522**, 13c, 1991.
275. **Janssens, R.V.F. and Khoo, T.L.**, *Annu. Rev. Nucl. Part. Sci.*, **41**, 321, 1991.
276. **Wyss, R., Nyberg, J., Johnson, A., Bengtsson, R., and Nazarewicz, W.**, *Phys. Lett.*, **B215**, 211, 1988.
277. **He, Y., Godfrey, M.J., Jenkins, I., Kirwan, A.J., Nolan, P.J., Mullins, S.M., Wadsworth, R., and Love, D.J.G.**, *J. Phys. G: Nucl. Phys.*, **16**, 657, 1990.
278. **Kirwan, A.J., Ball, G.C., Bishop, P.J., Godfrey, M.J., Nolan, P.J., Thornley, D.J., Love, D.J.G., and Nelson, A.H.**, *Phys. Rev. Lett.*, **58**, 467, 1987.
279. **Mullins, S.M., Jenkins, I., He, Y.-J., Kirwan, A.J., Nolan, P.J., Hughes, J.R., Wadsworth, R., and Wyss, R.A.**, *Phys. Rev.*, **C45**, 2683, 1992.
280. **Diamond, R.M., Beausang, C.W., Macchiavelli, A.O., Bacelar, J.C., Burde, J., Deleplanque, M.A., Draper, J.E., Duyar, C., McDonald, R.J., and Stephens, F.S.**, *Phys. Rev.*, **C41**, R1327, 1990.
281. **Regan, P.H., Wyss, R., Wadsworth, R., Fossan, D.B., He, Y.-J., Hughes, J.R., Jenkins, I., Ma, R., Metcalfe, M.S., Mullins, S.M., Nolan, P.J., Paul, E.S., Poynter, R.J., and Xu, N.**, *Phys. Rev.*, **C42**, R1805, 1990.
282. **Paul, E.S., Forbes, S.A., Fossan, D.B., Gizon, J., Hughes, J.R., Mullins, S.M., Metcalfe, M.S., Nolan, P.J., Poynter, R.J., Regan, P.H., Smith, G., and Wadsworth, R.**, *J. Phys. G: Nucl. Phys.*, **18**, 121, 1992.
283. **Forbes, S.A., Mullins, S.M., Nolan, P.J., Paul, E.S., Clark, R.M., Regan, P.H., Wadsworth, R., Atac, A., Hagemann, G.B., Herskind, B., Nyberg, J., Piiparinen, M.J., Dewald, A., Boehm, G., and Kruecken, R.**, in *Proc. Int. Conf. on Nuclear Structure at High Angular Momentum, Ottawa, 1992*, vol. 1, Contributions, AECL 10613, 1992, 65.
284. **Hebbinghaus, G., Strähle, K., Rzaca-Urban, T., Balabanski, D., Gast, W., Lieder, R.M., Schnare, H., Urban, W., Wolters, H., Ott, E., Theuerkauf, J., Zell, K.O., Eberth, J., von Brentano, P., Alber, D., Maaier, K.H., Schmitz, W., Beck, E.M., Hübel, H., Bengtsson, T., Ragnarson, I., and Åberg, S.**, *Phys. Lett.*, **B240**, 311, 1990.
285. **Strähle, K., Rzaca-Urban, T., Hebbinghaus, G., Lieder, R.M., Balabanski, D., Gast, W., Schnare, H., Urban, W., von Brentano, P., Dewald, A., Eberth, J., Ott, E., Theuerkauf, J., Wolters, H., Zell, K.O., Alber, D., Maier, K.H., Beck, E.M., Hübel, H., and Schmitz, W.**, in *Proc. Int. Conf. on Nuclear Structure at*

High Angular Momentum, Ottawa, 1992, vol. 1, Contributions, AECL 10613, 1992, 15.

286. **Haas, B., Taras, P., Flibotte, S., Banville, F., Gascon, J., Cournoyer, S., Monaro, S., Nadon, N., Prévost, D., Thibault, D., Johansson, J.K., Tucker, D.M., Waddington, J.C., Andrews, H.R., Ball, G.C., Horn, D., Radford, D.C., Ward, D., Pierre, C.St., and Dudek, J.**, *Phys. Rev. Lett.*, **60**, 503, 1988.
287. **Fallon, P., Alderson, A., Ali, I., Cullen, D.M., Forsyth, P.D., Riley, M.A., Roberts, J.W., Sharpey-Schafer, F. F., Twin, P. J., Bentley, M. A., and Bruce, A.M.**, *Phys. Lett.*, **B257**, 269, 1991.
288. **Bentley, M.A., Ball, G.C., Cranmer-Gordon, H.W., Forsyth, P.D., Howe, D., Mokhtar, A.R., Morrison, J.D., Sharpey-Schafer, J.F., Twin, P.J., Fant, B., Kalfas, C.A., Nelson, A.H., Simpson, J., and Sletten, G.**, *Phys. Rev. Lett.*, **59**, 2141, 1987.
289. **Drigert, M.W., Carpenter, M.P., Janssens, R.V.F., Moore, E.F., Ahmad, I., Fernandez, P., Khoo, T.L., Wolfs, F.L.H., Bearden, I.G., Benet, Ph., Daly, P.J., Garg, U., Reviol, W., Ye, D., and Wyss, R.**, *Nucl. Phys.*, **A530**, 452, 1991.
290. **Moore, E.F., Janssens, R.V.F., Chasman, R.R., Ahmad, I., Khoo, T.L., Wolfs, F.L.H., Ye, D., Beard, K.B., Garg, U., Drigert, M.W., Benet, Ph., Grabowski, Z.W., and Cizewski, J.A.**, *Phys. Rev. Lett.*, **63**, 360, 1989.
291. **Carpenter, M.P., Bingham, C.R., Courtney, L.H., Janzen, V.P., Larabee, A.J., Liu, Z.-M., Riedinger, L.L., Schmitz, W., Bengtsson, R., Bengtsson, T., Nazarewicz, W., Zhang, J.-Y., Johansson, J.K., Popescu, D.G., Waddington, J.C., Baktash, C., Halbert, M.L., Johnson, N.R., Lee, I.Y., Schutz, Y.S., Nyberg, J., Johnson, A., Wyss, R., Dubuc, J., Kajrys, G., Monaro, S., Pilotte, S., Honkanen, K., Sarantites, D.G., and Haenni, D.R.**, *Nucl. Phys.*, **A513**, 125, 1990.
292. **Lee, I.Y., Baktash, C., Cullen, D.M., Garrett, J.D., Johnson, N.R., McGowan, F.K., Winchell, D.F., and Yu, C.H.**, *Phys. Rev.*, **C50**, 2602, 1994.
293. **Willsau, P., Hübel, H., Korten, W., Azaiez, F., Deleplanque, M.A., Diamond, R.M., Macchiavelli, A.O., Stephens, F.S., Kluge, H., Hannachi, F., Bacelar, J.C., Becker, J.A., Brinkman, M.J., Henry, E.A., Kuhnert, A., Wang, T.F., Draper, J.A., and Rubel, E.**, Z. Phys., **A344**, 351, 1993.
294. **Ulfert, G., Metag, V., Habs, D., and Specht, H.J.**, *Phys. Rev. Lett.*, **42**, 1596, 1979.
295. **Metag, V. and Sletten, G.**, *Nucl. Phys.*, **A282**, 77, 1977.
296. **Habs, D., Metag, V., Specht, H.J., and Ulfert, G.**, *Phys. Rev. Lett.*, **38**, 387, 1977.
297. **Bemis, Jr., C.E., Beene, J.R., Young, J.P., and Kramer, S.D.**, *Phys. Rev. Lett.*, **43**, 1854, 1979.
298. **Rastopchin, E.M., Smirenkin, G.N., and Pashkevich, V.V.**, in *Proc. Workshop on Microscopic Theories of Superdeformation in Heavy Nuclei at Low Spin*, Lyon, 1990; V.V. Pashkevich, Preprint JINR (Dubna) P4-4383, 1969.
299. **Tsang, C.F. and Nilsson, S.G.**, *Nucl. Phys.*, **A140**, 275, 1970.
300. **Cailliau, M., Letessier, J., Flocard, H., and Quentin, P.**, *Phys. Lett.*, **B46**, 11, 1973.
301. **Bengtsson, R., Larsson, S.E., Leander, G.A., Möller, P., Nilsson, S.G., Åberg, S., and Szymański, Z.**, *Phys. Lett.*, **B57**, 301, 1975.
302. **Andersson, C.G., Larsson, S.E., Leander, G.A., Möller, P., Nilsson, S.G., Ragnarsson, I., berg, S., Bengtsson, R., Dudek, J., Nerlo-Pomorska, B., Pomorski, K., and Szymański, Z.**, *Nucl. Phys.*, **A268**, 205, 1976.
303. **Neergråd, K. and Pashkevich, V.V.**, *Phys. Lett.*, **B59**, 218, 1975.
304. **Neergrd, K., Pashkevich, V.V., and Frauendorf, S.**, *Nucl. Phys.*, **A262**, 61, 1976.
305. **Ragnarsson, I., Nilsson, S.G., and Sheline, R.K.**, *Phys. Rep.*, **45**, 1, 1978.

306. **Dudek, J. and Nazarewicz, W.**, *Phys. Rev.,* **C31**, 298, 1985.
307. **Chasman, R.R.**, *Phys. Lett.,* **B187**, 219, 1987.
308. **Girod, M., Delaroche, J.P., Gogny, D., and Berger, J.F.**, *Phys. Rev. Lett.,* **62**, 2452, 1989.
309. **Bonche, P., Krieger, S.J., Quentin, P., Weiss, M.S., Meyer, J., Meyer, M., Redon, N., Flocard, H., and Heenen, P.-H.**, *Nucl. Phys.,* **A500**, 308, 1989.
310. **Delaroche, J.P., Girod, M., Libert, J., and Deloncle, I.**, *Phys. Lett.,* **B232**, 145, 1989.
311. **Shimizu, Y.R., Vigezzi, E., and Broglia, R.A.**, *Nucl. Phys.,* **A509**, 80, 1990.
312. **Riley, M.A., Cullen, D.M., Alderson, A., Ali, I., Fallon, P., Forsyth, P.D., Hanna, F., Mullins, S.M., Roberts, J.W., Sharpey-Schafer, J.F., Twin, P.J., Poynter, R., Wadsworth, R., Bentley, M.A., Bruce, A.M., Simpson, J., Sletten, G., Nazarewicz, W., Bengtsson, T., and Wyss, R.**, *Nucl. Phys.,* **A512**, 178, 1990.
313. **Höller, J. and Åberg, S.**, *Z. Phys.,* **A336**, 363, 1990.
314. **Dudek, J., Werner, T.R., and Szymański, Z.**, *Phys. Lett.,* **B248**, 235, 1990.
315. **Wyss, R., Satuła, W., Nazarewicz, W., and Johnson, A.**, *Nucl. Phys.,* **A511**, 324, 1990.
316. **Bonche, P., Dobaczewski, J., Flocard, H., Heenen, P.-H., and Meyer, J.**, *Nucl. Phys.,* **A510**, 466, 1990.
317. **Koepf, W. and Ring, P.**, *Nucl. Phys.,* **A511**, 279, 1990.
318. **Satuła, W., Ćwiok, S., Nazarewicz, W., Wyss, R., and Johnson, A.**, *Nucl. Phys.,* **A529**, 289, 1991.
319. **Dudek, J.**, in *High-Spin Physics and Gamma-Soft Nuclei*, Saladin, J.X., Sorensen, R.A., and Vincent, C.M., Eds., World Scientific, Singapore, 1991, 146.
320. **Bonche, P., Krieger, S.J., Weiss, M.S., Dobaczewski, J., Flocard, H., and Heenen, P.-H.**, *Phys. Rev. Lett.,* **66**, 876, 1991.
321. **Haas, B., Janzen, V.P., Ward, D., Andrews, H.R., Radford, D.C., Prévost, D., Kuehner, J.A., Omar, A., Waddington, J.C., Drake, T.E., Galindo-Uribarri, A., Zwartz, G., Flibotte, S., Taras, P., and Ragnarsson, I.**, *Nucl. Phys.,* **A561**, 251, 1993.
322. **Ragnarsson, I.**, *Nucl. Phys.,* **A557**, 167c, 1993.
323. **Chasman, R.R.**, *Phys. Lett.,* **B219**, 227, 1989.

5

NUCLEAR STABILITY, HADRON, AND CLUSTER EMISSION

Dorin N. Poenaru and Walter Greiner

Table of contents

I. INTRODUCTION

The present chapter is devoted to the problem of nuclear stability against various *spontaneous* decay modes, in which a *parent* nucleus A, Z in its ground state is split into nuclei A_1, Z_1 and A_2, Z_2

$$^{A}Z \rightarrow {}^{A_1}Z_1 + {}^{A_2}Z_2 \tag{1}$$

in a way that conserves the hadron numbers $A = A_1 + A_2$, $Z = Z_1 + Z_2$. The heavy fragment A_1, Z_1 is called *daughter* and the light one *emitted cluster*, or heavy ion, or hadron (if it is a proton or a neutron), etc. Consequently we also use, alternatively, the subscripts $_{d,e}$ instead of $_{1,2}$. The emitted fragment gives the name of the corresponding radioactivity; it can be a proton (p), two-proton (2p), neutron (n), ^{4}He (α-particle), ^{12}C, ^{14}C, ^{20}O, ^{23}F, ^{24}Ne, ^{28}Mg, ^{32}Si, light fission fragment, etc.

There are also two-step processes, *beta-delayed particle emissions*, in which, after the population of an (usually isobaric analog) excited state by β-decay (β^-, β^+ or electron capture) of a *precursor*, a neutral (n, 2n, 3n, 4n) or a charged particle (p, 2p, 3p, d, t, α, light fission fragment, etc.) is rapidly emitted, the half-life being determined by *beta*-decay.

We shall briefly discuss several processes, spanning a wide range of mass and charge asymmetry, with the mass and atomic numbers of the emitted particle in the range $(1, A/2)$ and $(0, Z/2)$, respectively.

Preliminary information about these decay modes can be obtained by calculating, from the known masses M_i, the released energy

$$Q = [M - (M_1 + M_2)]c^2 \tag{2}$$

where c is the light velocity. In this way we can see if the process is energetically allowed ($Q > 0$). In the absence of any energy loss for fragment deformation and excitation, as in cold fission phenomena, one can find the kinetic energy of the small fragment

$$E_{k2} = QA_1/A \tag{3}$$

which can be experimentally determined.

Another important measurable quantity is the partial half-life, T, of the parent nucleus against this decay mode. The emission rate is inversely proportional to it. The half-life can be calculated within an adequate dynamical theoretical model. In principle, any parent nucleus for which the released energy is a positive quantity, $Q > 0$, can decay by fragment emission, but practically there is a severe selection imposed by the available techniques, requiring a short-enough half-life and a large-enough branching ratio with respect to the main disintegration mode (e.g. α-decay). We shall give below some examples of the technical achievements.

In fission the mass asymmetry parameter $\eta = (A_1 - A_2)/A$ takes relatively low values. One can guess that the macroscopic collective motion[1] should prevail in such a complex rearrangement process. Thus, a convenient way to simplify the nuclear many-body problem is to consider only a small number of suitably chosen macroscopic parameters, as the degrees of freedom describing the nuclear shape.[2] Even within this approach it is difficult to choose a shape parametrization simultaneously taking into account all the important collective coordinates (fragment separation distance or nuclear elongation, necking-in, deformations, non-axiality, mass and charge asymmetry, etc.).

A rich variety of nuclear decay modes have been recently discovered and intensively studied.[3–5] The phenomenological liquid drop model (LDM), assuming exclusively a collective motion, applied since 1939, has only been partly successful in describing fission phenomena. It fails to explain the ground state nuclear deformations, clearly observed in various regions of the nuclear chart, the asymmetric distribution of the fission mass fragments, the phenomenon of fission isomerism, the intermediate structure resonances in fission cross-sections, etc. In 1966 it was understood that even in this highly collective process one has to take into account the single–particle behavior of the nucleon motion. In 1967 Strutinsky[6] introduced his famous macroscopic-microscopic model, by adding to the LDM deformation energy the shell and pairing corrections calculated on the basis of a single-particle deformed shell model (like the Nilsson model, the two-center shell model, etc.). This hybrid theory has been developed[7] and succeeded in giving a satisfactory explanation of the fission properties mentioned above.

Significant progress has been achieved with the development of the asymmetric two-center shell model[8] and the theory of fragmentation,[9] in order to calculate the mass yield very often measured in fission. New, very asymmetric, fission modes have also been predicted by using this approach.[10] Cold fusion[11] with the reaction partner ^{208}Pb has been of practical importance for the synthesis[12,13] of the heaviest elements 107 to 111.

Fission models, extended to very large mass asymmetry, have been applied since 1980 to predict *cluster radioactivities* (see the review papers[14–16] and references therein). In this case the spontaneously emitted light fragment is a small nucleus (like ^{14}C, ^{24}Ne, etc.) heavier than an α particle, but lighter than the lightest fission fragment. Wrong conclusions have been drawn in other papers published before 1980, including speculations since 1924 concerning the abundance of some gases (like N and Ne) in uranium ores.

Four years later, the experiment performed in 1984 by Rose and Jones[17] on the spontaneous emission of ^{14}C from ^{223}Ra gave the first confirmation of our conclusions, based on penetrability calculations, that ^{14}C should be the most probable emitted cluster from such a nucleus. Before any other model was published, we estimated, within our analytical superasymmetric fission (ASAF) model,[18,19] the half-lives for more than 150 decay modes, including all cases experimentally confirmed during 1984–1994.

Up to now there have been successful measurements[20–23] on ^{14}C, ^{20}O, ^{23}F, $^{24-26}$Ne, ^{28,30}Mg, and ^{32,34}Si cluster decays of the following transfrancium nuclei: ^{221}Fr, $^{221-224,226}$Ra, ^{225}Ac, ^{228,230}Th, ^{231}Pa, $^{232-234}$U, ^{236,238}Pu, and (preliminary) ^{242}Cm. The shorter half-life, $T \simeq 10^{11}$ s, corresponds to the ^{14}C radioactivity of the ^{222}Ra parent nucleus, leading to a double magic daughter ^{208}Pb.

There is strong competition with α-decay. The largest branching ratio with respect to α-decay, $b = T_\alpha/T$, is about $10^{-9.2}$ and has been observed in the ^{14}C radioactivity of ^{223}Ra. With the high sensitivity of solid state track detectors, it was possible to measure a branching ratio as low as $10^{-16.25}$ for Mg emission from ^{238}Pu. The longer partial half-life determined up to now, of approximately 10^{26} s, is that of the ^{231}Pa nucleus against ^{23}F emission.

Shorter half-lives and larger branching ratios are expected in a new island of cluster emitters leading to daughter nuclei in the neighborhood of ^{100}Sn. Since 1984 we have predicted this island, with the best representative, ^{114}Ba, decaying by ^{12}C emission.

Usually only one kind of emitted cluster (the most probable) could be experimentally observed, the emission rate for the next one being smaller by some orders of magnitude. There are few exceptions: both F and Ne emitted by ^{231}Pa; Ne and Mg from ^{234}U, and Mg + Si from ^{238}Pu.

The main difficulty of such an experiment comes from the necessity to select a few rare events from an enormous background of α-particles. Either solid state track detectors (which are not sensitive to α-particles), or magnetic spectrometers (in which α-particles are deflected by a strong magnetic field), have been used to overcome this difficulty. The superconducting spectrometer SOLENO, at I. P. N. Orsay, has been employed since 1984 to detect and identify the ^{14}C clusters spontaneously emitted from ^{222,223,224,226}Ra parent nuclei. Moreover, its good energy resolution was exploited in 1989 to discover[24,25] a "*fine structure*" in the kinetic energy spectrum of ^{14}C emitted by ^{223}Ra. Cluster emission leading to excited states of the final fragments was considered for the first time[26] in 1986.

In the usual mechanism of fission, a significant part (about 25–35 MeV) of the released energy Q is used to deform and excite the fragments (which subsequently cool down by neutron and γ-ray emissions); hence the total kinetic energy of the fragments, TKE, is always smaller than Q. Since 1980 a new mechanism has been experimentally observed – *cold fission*[27,28,15] characterized by a very high TKE, practically exhausting the Q-value, and a compact scission configuration. Cold fission with compact shapes can be considered to be the inverse of a fusion process. Experimental data have been collected in two regions of nuclei: (a) thermal neutron induced fission on some targets like ^{233,235}U, ^{238}Np, ^{239,241}Pu, ^{245}Cm, and the spontaneous fission of ^{252}Cf; (b) the *bimodal* spontaneous mass-symmetrical fission of ^{258}Fm, ^{259,260}Md, ^{258,262}No, and ^{260}104. The yield of the cold fission mechanism is comparable to that of the usual fission events in the latter region, but it is much lower (about five–six orders of magnitude) in the former.

We have systematically studied the cold fission process viewed as cluster radioactivity within our ASAF model.[29,15] The cold fission properties of transuranium nuclei are dominated by the interplay between the magic number of neutrons, $N = 82$, and protons, $Z = 50$, in one or both fragments. The best conditions for symmetric cold fission are fulfilled by ^{264}Fm, leading to identical doubly magic fragments ^{132}Sn. A spectrum of the ^{234}U half-lives versus the mass and atomic numbers of the fragments, which is plotted in Fig. 7 below, illustrates the idea of a unified treatment of different decay modes over a wide range of mass asymmetry. Three distinct groups, α-decay, cluster radioactivities, and cold fission, can be seen, in good agreement with experimental results. For cluster decays our calculations were performed before the measurements.

II. METASTABILITY

A. Stable and radioactive nuclei

The deformation energy, E, of the system with constrained mass and charge asymmetry versus separation distance, R, between fragment centers can be used to study the stability of the parent nucleus relative to the division into two nuclei mentioned above. Typical examples of three different kinds of such curves are plotted in Fig. 1. By taking the energy of two final nuclei at infinite separation distance as the origin of potential energy, we get initially at $R = R_i$, before the beginning of the separation process, $E(R_i) = Q$.

The fission barrier height E_b is then given by the difference between the saddle-point energy E_{SP} and the Q-value:

$$E_b = E_{SP} - Q \tag{4}$$

The initial nucleus can be: *unstable* (curve 1 in Fig. 1) if $Q > 0$ and $E_b < 0$; metastable (curve 2 in Fig. 1) when both $Q > 0$ and $E_b > 0$, or stable if $Q < 0$ and $E_b > 0$.

In a metastable or radioactive state the two fragments are temporarily held together by the potential barrier. There is a finite probability for the penetration of the barrier by the quantum-mechanical tunneling effect. Gamow (and independently Condon and Gurney) gave the first explanation of the α-decay on this basis in 1928.

From the potential energy one can get a qualitative indication about the nuclear stability or metastability with respect to a given split. As can be seen in Fig. 2, from the energetical point of view the area of cluster (or heavy-ion

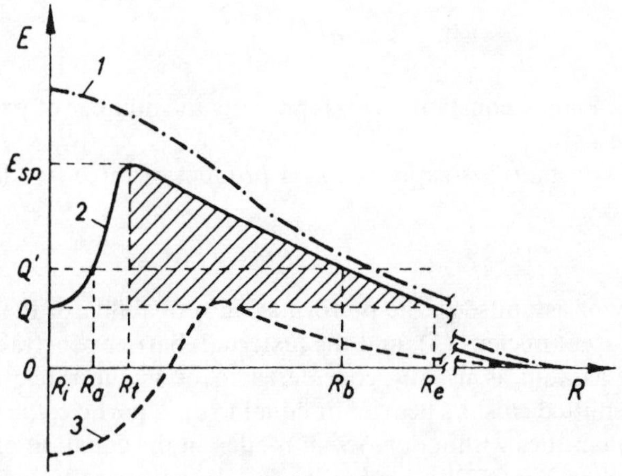

Fig. 1. Typical curves of deformation energy versus separation distance between fragments for a given mass and charge asymmetry. The parent nucleus can be unstable (1), metastable (2), or stable (3) relative to this fragmentation.

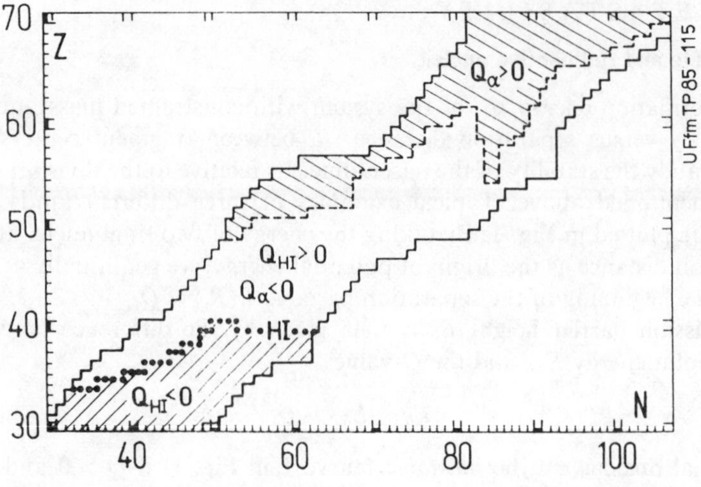

Fig. 2. Part of the nuclear chart showing that the region in which cluster (or heavy-ion (HI)) radioactivities are energetically allowed is extended beyond that of α-decay.

(HI)) radioactivities extends beyond that of α-decay. A dynamic theory has to be applied in order to estimate the emission rate, or the disintegration constant λ of the exponential decay law in time

$$\mathcal{N}(t) = \mathcal{N}(0)\exp(-\lambda t) \tag{5}$$

The partial decay half-life T and the corresponding disintegration width Γ are the related quantities:

$$\Gamma = \hbar\lambda = \frac{\hbar \ln 2}{T} \tag{6}$$

where $\hbar$ is the Planck constant. $\mathcal{N}(t)$ represents the number of parent nuclei at a certain time t.

The decay constant λ is expressed as a product of three (model-dependent) quantities:

$$\lambda = \nu SP \tag{7}$$

the frequency of assaults (ν), the preformation probability of the emitted cluster into the parent nucleus (S), and the (external) barrier penetrability (P). The frequency of assaults is usually considered to be in the range 10^{20}–10^{22} per second. For emitted clusters heavier or equal to an α-particle, the three (model-dependent) quantities within our ASAF model, in the equation of the half-life, $\log T(s) = \log[(\ln 2)/\nu(s^{-1})] - \log S - \log P$, take numerical values of the order -21, > 2, and > 12, respectively, leading to $T > 10^{-7}$ s.

Within the traditional theory of α-decay, the partial disintegration width is sometimes also expressed by $\Gamma = 2\gamma^2 P$, and hence the reduced width

$2\gamma^2 = \hbar\nu S = E_\nu S/\pi$, where $E_\nu = h\nu/2$ is the zero-point vibration energy, has the dimension of an energy.

Very frequently the penetrability is calculated by using the one-dimensional Wentzel–Kramers–Brillouin (WKB) approximation

$$P = \exp(-K);\ K = \frac{2}{\hbar}\int_{R_a}^{R_b} \sqrt{2B(R)\,E(R)}\,dR \tag{8}$$

where B is the nuclear inertia, equal to the reduced mass $\mu = mA_1A_2/A$ for separated fragments, m is the nucleon mass, E is the potential energy from which the Q-value has been subtracted out, R_a and R_b are the classical turning points.

B. Limits of stability. Proton and neutron drip lines

About 300 stable nuclides found in nature (the large black squares in Fig. 3) lie on (or very close to) the line of beta stability, which can be approximatted (following Green) by

$$A = (Z - 100)/0.6 + \sqrt{(Z-100)^2/0.36 + 200Z/0.3} \tag{9}$$

The majority of other nuclei[30] are β-radioactive,[31] and they also exhibit some other decay modes which could be even stronger (see the comprehensive table by Tuli in the present book). Beta-decay is an important source of nuclear structure data.

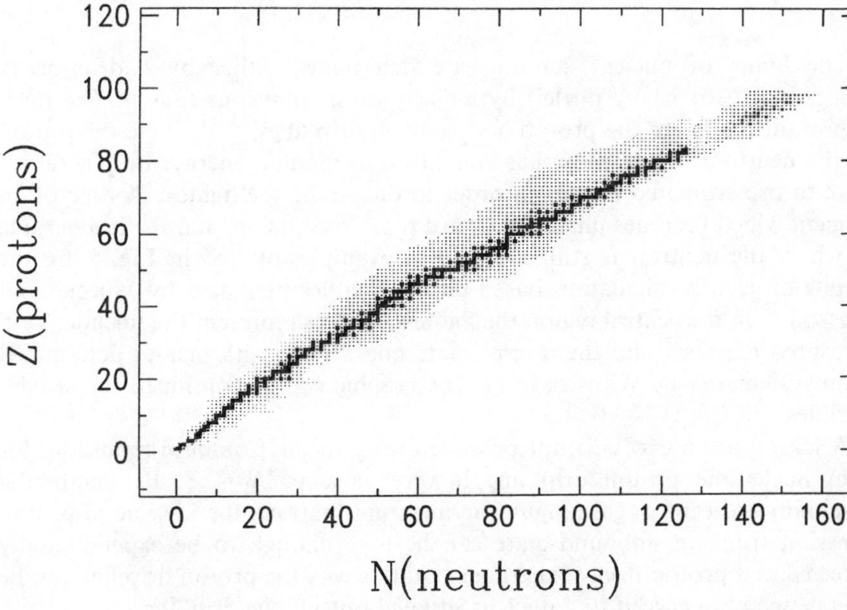

Fig. 3. Nuclear half-lives T_t. Light (small-area) squares express short log T_t, and dark (large-area) squares long log T_t.

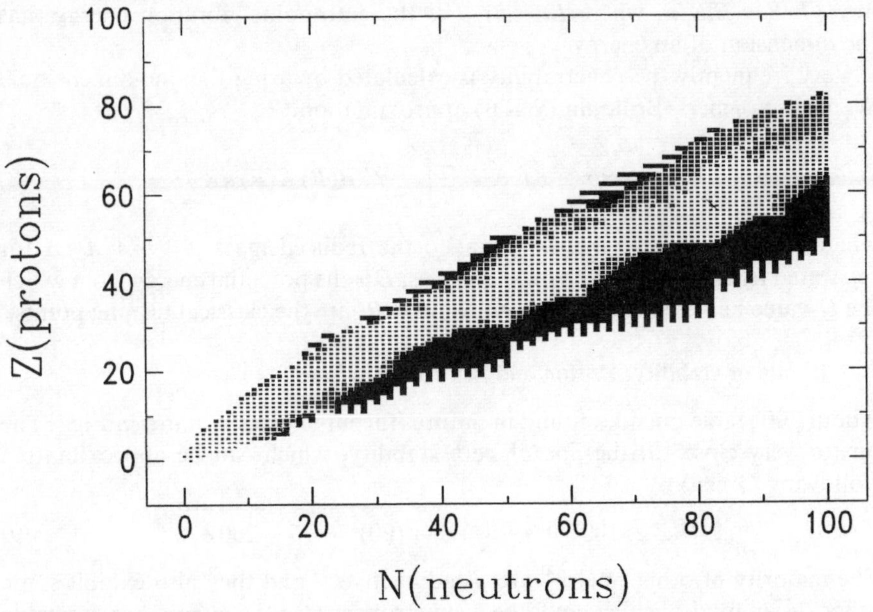

Fig. 4. Mass codes of light and intermediate-mass nuclides up to the proton and neutron drip lines. In the central regions the light points correspond to measured masses, $C = 0$, and the intermediate points to masses determined from systematics by Wapstra *et al.*, $C = 1$. The black areas denote estimated masses with $C = 3$.

The limits of nuclear stability are determined either by hadron-decay modes,[32] or (for heavy nuclei) by α-decay or spontaneous fission. The most important limits are the proton and neutron drip lines,[33–35] where the proton or the neutron, respectively, has roughly zero binding energy. In general we have to use estimated masses in order to determine the lightest isotope of an element which becomes unstable toward p or 2p emission, and the heavier one in which the neutron is still bound. For example, plotted in Fig. 4 are the results of such a calculation based on mass tables predicted by Jänecke and Masson.[36] In the central region the lightest squares represent the nuclides with measured masses[37] and the intermediate ones, those with masses determined from systematics by Wapstra *et al.* The graphic can be continued for heavier nuclides.

A large proton excess is not possible owing to the Coulomb repulsion; for light nuclei the proton drip line is very close to $N = Z$. By combining Coulomb and centrifugal (angular momentum) barriers the lifetime of proton emission from an unbound state can be long enough to be experimentally detected in a proton-decay experiment. In this way the proton drip line can be overstepped: we acceed to a nucleus situated outside the drip line.

The accessibility of nuclides very close to the drip lines has increased considerably. There have been few proton emitters observed experimentally until now, but many beta-delayed proton emitters. The proton drip line in the s–d shell ($Z = 8$–20) region has been clarified[34,35] by experiments using the LISE spectrometer at GANIL. Measurements on this line have been performed[35] up to $_{35}$Br. In 1994 a series of proton-rich In and Sn isotopes (including the doubly magic ^{100}Sn) were produced (by the nuclear fragmentation of ^{124}Xe projectiles at $A \cdot 1095$ MeV at the heavy-ion synchrotron SIS of GSI Darmstadt, and by ^{112}Sn + ^{nat}Ni reaction at GANIL).[38] In Darmstadt, the projectile fragments were separated in flight with the fragment separator SIS. Experiments on the ^{12}C radioactivity of proton-rich ^{114}Ba are in progress.[39,40] Neutron-deficient ^{228,229}Pu nuclei have been obtained in Dubna by using heavy-ion induced reactions on lead isotopes.[41]

On the other hand, a very large neutron excess may be encountered (for example, ^{176}Sn could be bound). As can be seen from Fig. 4 the neutron drip line has been reached only for the lightest nuclides. It is commonly believed that the neutron drip line has been mapped[42] up to $Z = 10$. By comparing Fig. 4 with Fig. 3 it is evident that, owing to the Coulomb forces, the proton drip line is much closer to the valley of stability than the neutron drip line.

The low-energy fission of neutron-deficient Ac, Th, Pa, and U isotopes has been investigated in inverse kinematics, using a new experimental technique: the projectile fragmentation of a relativistic ($950A$ MeV) ^{238}U primary beam.[43] About 50 new neutron-rich isotopes (of the elements $_{32}$Ge to $_{58}$Ce) have been observed by M. Bernas et al. at the FRS fragment separator. New neutron-rich Zr, Nb, Mo, Tc, and Ru isotopes have been discovered in fission.[44]

Close to the neutron drip line, nuclei (like ^{11}Li) might exhibit a halo structure[45–47] or (for example, ^{8}He and ^{181}Cs) neutron skin. In the former case collective vibration of the di-neutron halo against the core could give rise to a new kind of soft (at 1.2 MeV) electric dipole giant resonance.[48] The giant resonances studied by Danos and Greiner[1] in the early 1960s are still of great importance.

The halo of ^{11}Li (two-neutron separation energy $S_{2n} \simeq 315$ keV) is a result of the tunneling of two loosely bound neutrons away from the core ^{9}Li at a large distance. In a similar way one can have a "1n-halo" nucleus ^{11}Be (in which the odd neutron is bound by 502 keV). Other candidates could be ^{14}Be, ^{19}B, and ^{19,22}C. In these nuclei the mass and charge radii may differ by large amounts, and the molecular (or cluster) structure plays an important role. The density distributions[45] show an extended tail at low density. The two-neutron halo nuclei could have a similarity (with opposite sign because the nn force is attractive) with the two-electron atoms, where some correlations are present (the probability of finding the two electrons close to each other is smaller than in an independent-particle scheme). It was observed that pairing is weakened for large values of $N - Z$.

The existence of very heavy nuclei is prevented by the spontaneous fission process. When the number of protons becomes very high, the Coulomb repulsion is no longer counterbalanced by the strong interaction, no matter how many neutrons are present, and the fission barrier height may approach zero.

Between Th and the heavy Fm isotopes, the spontaneous fission half-life decreases by 30 orders of magnitude. Even at a finite positive (but small) barrier height the fission half-life could be very low in comparison with the partial half-life of any other disintegration mode. At GSI Darmstadt[13,49] it was shown that the shell-stabilizing property,[50] characteristic of the superheavy nuclei, is already present in the heaviest elements known to date: 107, 108, 109, up to 111. Owing to shell effects the fission barriers of these nuclei are so high that they decay mainly by α emission instead of spontaneous fission. In September 1992, it was suggested that IUPAC adopt the following names of the elements 107, 108, and 109: nielsbohrium (Ns), hassium* (Hs), and meitnerium (Mt), respectively. They were identified at SHIP in cold fusion–evaporation reactions, by means of α-decay.

In the attempt to come closer to the predicted doubly magic superheavy nucleus $^{298}_{184}114$, the element 110 has been recently produced[51] by using the (^{208}Pb, ^{62}Ni) reaction with the evaporation of one neutron. A production cross-section of about 3.3 picobarn was measured. The idea of a cold fusion reaction with magic nuclei partners[11,52] was again exploited. According to recently performed calculations[50,53] the superheavy nuclei would not form an island of stability surrounded by an instability sea. The shell effects of the predicted neutron deformed magic number $N = 162$ are acting to stabilize the neighbors, leading to a peninsula of superheavy nuclei. In this way one expects that ^{270}Hs would have a long-enough half-life to be measured.

At present nuclear physics is expanding in three directions: higher nuclear temperatures or densities toward a quark–gluon plasma;[54] higher spin (super- and hyper deformations, transitions from an ordered nuclear phase to a chaotic one); the production and spectroscopy of proton- and neutron-rich nuclei far from stability. Exotic ion (radioactive) beams from isotope separators and recoil spectrometers and large multi-detector arrays are among the most important technical achievements making this development possible. By using a Penning trap mass spectrometer (see the chapter by Bollen and Kluge), in which the mass is determined by measuring the cyclotron frequency in a stable homogeneous magnetic field, an extremely high mass resolution has been obtained (for example, to distinguish between the masses of the ground state and of an isomeric state of ^{84}Rb). The major heavy-ion facilities in the world are now producing radioactive nuclear beams and are using more than half of the total machine time for the investigation of radioactive nuclei.

Even at a relatively low excitation (about 3 MeV) new collective vibrations (scissors modes[55]) have been discovered and studied. At much higher excitation energy, multi-step excitations (for example, two-phonon giant dipole resonance produced in pion double-charge reactions) are expected to take place.

---

* From the Latin name of the region Hessen to which Darmstadt belongs.

A very interesting new field of atomic physics – atomic (or metallic) clusters[56] – is developing rapidly, with the benefit of a close analogy to nuclear cluster phenomena.[57]

About 40 nuclei with superdeformed bands[58] are known (not only Dy, Tb, Gd, Eu, Nd, Ce, La, Pb, Tl, Hg isotopes, but also lighter (^{16}O, ^{40,42}Ca) or heavier (spontaneously fissioning shape isomers[59]) ones). It is interesting to note a unifying point of view[60] according to which these phenomena may be viewed as nuclear molecular states.

Nuclear physics is linked to astrophysics, cosmology, and elementary particle physics, helping to unify our knowledge about the universe. The main site of nucleosynthesis for most of the elements is considered to be the stellar environment; a second one is the Big Bang (primordial nucleosynthesis). In both cases, the reactions of beta unstable nuclei play an important role. With a temperature as high as several hundred million Kelvin, a rapid proton burning (rp) process (such as the CNO cycle which contributes to the energy production in some stars like the sun) is possible: proton capture becomes more important than β decay in many proton-rich nuclei. The reaction ^{13}N(p,γ)^{14}O breaks the CNO cycle and initiates the hot CNO cycle. Also the excited states of ^{20}Na are important for determining the break-out rate. Similarly, for the rapid neutron capture processes, the (n,γ) reactions are relevant. In this case the inverse reactions (virtual γ, n) can be studied. One of the important reactions for the Big Bang nucleosynthesis is $^{8}\text{Li} + {}^{4}\text{He} \rightarrow {}^{11}\text{B} + \text{n}$.

Another subject of interest in astrophysics (element production in hot stars with a high degree of ionization) is the recently discovered β-decay of highly ionized $^{163}\text{Dy}^{66+}$ into bound electron states (the K and L shell of single ionized ^{163}Ho).[61] The process is equivalent to an inverse electron capture, and it shows that drastic changes in decay properties can be produced by ionization. In this way nuclear reactions involving short-lived nuclei in explosive processes in the universe could be simulated in the laboratory.[62] The nuclear structure effects reflected in the decay properties of the exotic $N \simeq 28$ S and Cl nuclei indicate[63] a rapid weakening of the $N = 28$ shell below ^{48}Ca. This result could be used to try to explain the unusual $^{48}\text{Ca}/^{46}\text{Ca}$ abundance ratio measured in the solar system.

The overlap between nuclear physics and elementary particle physics has become increasingly important. We admit that nucleons are made from quarks. There is a need to find some specific signatures of the quark structure in nuclear phenomena.

C. Particle and beta-delayed particle emission

Proton and two-proton radioactivity[64,65] determines the borderline of nuclear stability. The existing experimental information on one-proton emission[65,66] is summarized in Table 1. The odd–even staggering of the proton separation energy S_p, responsible for the non-regular structure of the drip line, leads to situations[67] where $S_p > 0$ for a given nucleus, while $S_{2p} < 0$. No successful

Table 1 Proton radioactivity, decay half-lives[65, 66]

Emitter	$\log T_t$ (s)	E_p (keV)	$\log T_p$ (s)	E_α (keV)	$\log T_\alpha$ (s)	$\log T_\beta$ (s)
^{53m}Co	-0.61 ± 0.02	1560	1.23			−0.70
^{109}I	-3.99 ± 0.02	813	−3.99	3757	−1.07	−0.60
^{112}Cs	-3.30 ± 0.08	807	−3.30			
^{113}Cs	-4.55 ± 0.09	959	−4.55	3473	1.83	−0.70
^{146}Tm	-1.14 ± 0.12	1189	−1.02	3482	10.48	−0.52
^{146m}Tm	-0.63 ± 0.05	1119	0.03			
^{147}Tm	-0.25 ± 0.03	1051	0.43	2821	16.85	−0.52
^{147m1}Tm	$-4 \div -2$	947	$-4 \div -2$			
^{147m2}Tm	-3.44 ± 0.04	1118	−3.44			
^{150}Lu	-1.46 ± 0.11	1263	−1.34	2978	16.60	−0.82
^{151}Lu	-1.07 ± 0.05	1233	−0.92	2823	18.30	−0.60
^{156}Ta	$-0.78^{+0.30}_{-0.18}$	1022	−0.70	4618	4.87	0.00
^{160}Re	-3.10 ± 0.08	1261	−2.88	6537	−2.88	−0.22

experiment on two-proton radioactivity has been reported so far. Possible candidates are: ^{39}Ti, ^{42}Cr, ^{45}Fe, ^{59}Ge, ^{152}Hf, ^{153}Ta, ^{154}W, ^{155}Re.

Alpha-decay[68] and spontaneous fission[69] are the main decay modes of some heavy nuclei. For this reason, the proton and neutron drip lines of some nuclides are not accessible.There are very short (under 1 s) half-lives relative to the α-decay of several proton-rich nuclides, isotopes of $_{70}$Yb, Lu, Hf, W, Pt, Po, At, Fr, Ra, Ac, Th, Pa, No, 106, Ns, Hs, Mt. Also the spontaneous fission half-lives[69] of both light ($A = 255$) and heavy ($A = 258$) Fm isotopes are of the order of 3.3 and 0.4 ms respectively. Even faster fission (see Table 2) takes place in the shape isomeric states.[70–72, 59] These are the first superdeformed states[73] that have been experimentally determined.

Cluster radioactivities are presented in section IV.G. Some of the cluster emitters are members of the natural radioactive families (see Fig. 5). Unlike the decay modes with mass numbers $A_e < 5$, producing a relatively slight change of the nuclear structure from the parent to the daughter, cluster radioactivities allow us to study transitions from a deformed parent nucleus, generally with a complex structure resulting from the mixing of several single-particle states, to a spherical daughter possessing a pure well-known state.

The particle-accompanied fission[74, 75] is a three-body process, and hence the energy spectra of the emitted particles (α, C, etc.) is bell shaped with a rather large width, unlike binary fragmentation (α-decay, cluster decay modes, etc.) where a monoenergetic line should be observed (of course we have to take into account straggling in the source and the detector resolution).

By studying the beta-delayed decay modes one can extract nuclear structure information. Beta strength functions at high excitation energies have been mainly obtained in such a way.

In the vicinity of the drip line the Q_β-values are rather high; when the

Table 2 Excitation energies and half-lives of spontaneous fission isomers

Nucleus		E_{II}, E^*_{II} (MeV)	Half-lives
Element	A		log T (ps)
Th	232	<4	
	233		4.23
U	236	2.75	5.06 ± 0.03
	238		≥ 3.0
		2.55	5.19 ± 0.08
Np	237	2.7 ± 0.3	4.60 ± 0.12
Pu	235	1.7 ± 2.6	4.48 ± 0.06
	236	4.0 ± 0.3	4.53 ± 0.09
		2.7 ± 0.3	1.54 ± 0.16
	237	$E^* = E + (0.3 \pm 0.15)$	6.05 ± 0.03
		$E = 2.3 \div 2.9$	5.06 ± 0.04
	238	$E^* = E + (1.3 \pm 0.3)$	3.81 ± 0.09
		$E = 2.4 \div 2.7$	2.85 ± 0.10
	239	$E^* = E + 0.2$	3.41 ± 0.40
		$E = 2.2 \div 2.6$	6.91 ± 0.04
	240	$E = 2.3 \div 2.6$	3.58 ± 0.03
	241		4.48 ± 0.06
		$E = 2.2 \div 2.6$	7.36 ± 0.02
	242		4.70 ± 0.20
			3.54 ± 0.07
	243	(1.8 ± 0.2)	4.78 ± 0.09
	244		2.58 ± 0.08
	245		4.59 ± 0.13
Am	237	$2.1 \div 2.4$	3.70 ± 0.15
	238	$2.3 \div 2.7$	7.54 ± 0.05
	239	$2.4 \div 2.5$	5.20 ± 0.10
	240	$2.6 \div 3.0$	8.95 ± 0.05
	241	$2.2 \div 2.3$	6.18 ± 0.14
	242	$2.3 \div 2.9$	10.15 ± 0.02
	243	$2.0 \div 2.3$	6.72 ± 0.04
	244	$1.6 \div 2.8$	9.04 ± 0.06
	245		5.81 ± 0.04
	246		7.86 ± 0.06
Cm	240		4.74 ± 0.09
			1.00 ± 0.11
	241	$2.0 \div 2.3$	4.18 ± 0.02
	242	$E^* = E + 1.3$	5.25 ± 0.15
		$E = 1.9 \pm 0.2$	1.60 ± 0.14
	243	$1.5 \div 2.3$	4.62 ± 0.06
	244		> 5.00
			≤ 0.70
	245	$1.7 \div 2.4$	4.11 ± 0.07
Bk	242		3.98 ± 0.08
			5.78 ± 0.07
	243	2.2 ± 0.2	3.70 ± 0.15
	244		5.91 ± 0.03
	245		3.30 ± 0.18

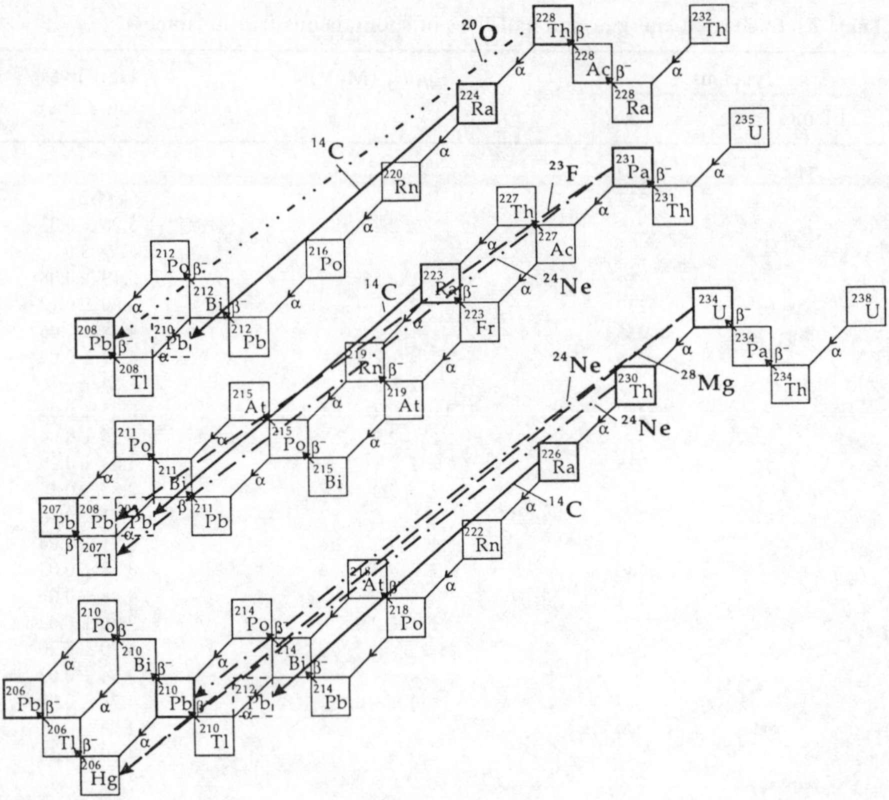

Fig. 5. ^{14}C, ^{20}O, ^{23}F, ^{24}Ne, and ^{28}Mg emission from some members of the three natural radioactive families.

corresponding daughters possess low separation energies for hadrons or clusters of nucleons, a beta-delayed particle emission takes place, usually from an excited state. In this way the particle emission is much faster; the half-life is determined by the preceding β-decay of the precursor.

In the following we give some examples of the beta-delayed decay modes observed so far. Near the proton drip line, the most frequently occurring delayed decay mode is beta-delayed proton radioactivity,[76,77] discovered by Karnaukhov in Dubna.

Examples of beta-delayed proton precursors are: ^{13}O, ^{17}Ne, ^{20}Mg, ^{23,25}Si, ^{27,28}P, ^{28}S, ^{31}Cl, ^{32,33}Ar, ^{37}Ca, ^{43}Cr, ^{46,48}Mn, ^{52}Ni, ^{65}Ge, ^{69}Se, ^{73}Kr, ^{73}Sr, ^{100,102}I, ^{103,105}Sn, ^{108,109}Te, ^{113,115}Xe, ^{114,118}Cs, ^{117,119}Ba, ^{123,125}Ce, ^{141}Gd, ^{148}Ho, ^{149}Er, ^{150}Tm, ^{152}Lu, ^{151,153}Yb, ^{154}Lu, ^{179,181}Hg, etc.

The isospin projections of beta-delayed two-proton precursors[78] identified so far (^{21,22}Al, ^{26}P, ^{27}S, ^{31}Ar, ^{35}Ca, ^{39}Ti, ^{43}Cr, ^{46}Mn) are $T_z = -2$ and $-5/2$. Very likely the two protons are emitted sequentially.

There is one example of a beta-delayed three-proton emitter.[79] The superallowed β-decay of ^{31}Ar feeds the isobaric analog state in ^{31}Cl at an excitation energy of 12.24 MeV. Several decay channels are opened: β–p, β–2p, β–3p (with an energy of 4.895 MeV), β–α, and β–αp.

The β^+-delayed α-decay (sometimes known as "long-range α") has been seen, for example, in ^{8}B, ^{12}N, ^{17}Ne, ^{20}Na, ^{24}Al, ^{28}P, ^{32}Cl, ^{36}K, ^{40}Sc, ^{44}V, ^{76}Rb, ^{110,112}I, ^{115}Xe, ^{114,118,120}Cs, ^{117}Ba, ^{181}Hg, ^{114,118,120}Cs. The β^--delayed α precursors are ^{8,9,11}Li, ^{11}Be, ^{12}B, ^{16}N, ^{30}Na, ^{212,214}Bi.

On the neutron-rich side[80,77] for intermediate-mass and heavy nuclei it is much more difficult to reach the drip line since a very large neutron excess is needed. Beta-delayed one-neutron emitters are, for example, the following nuclei: ^{9,11}Li, ^{15}B, ^{18,20}C, $^{18-22}$N, ^{23,24}O, $^{26-28}$Ne, ^{29,30}Na, ^{34,35}Al, $^{39-42}$P, ^{43,44}S, ^{49,50,52}K, ^{75,76,79}Cu, $^{79-81}$Zn, $^{79-84}$Ga, $^{84-85}$Ge, ^{85}As, $^{91-94}$Br, and many others. There are also such precursors in the fission fragment region (from Cu to La heavy isotopes).[81,82]

Besides ^{11}Li, which was the reference source of several new delayed particle decay modes, beta-delayed two-neutron decay has been observed in ^{14}Be, ^{17}B, and ^{19}C.

The ^{17}B nucleus is also very rich in different decay modes; not only the already-mentioned β–2n, but also β–3n, and β–4n have been identified.

Beta-delayed triton emission has been detected in ^{11}Li and ^{8}He, and beta-delayed deuteron radioactivity in ^{6}He. Such decay modes from the ground state of any parent nucleus would be forbidden from energetical considerations (negative Q-value). They are now allowed because the β-decay of the precursor feeds the excited states of its daughter.

The beta-delayed fission rates of neutron-rich nuclei have been investigated in ref. 83.

III. DECAY DYNAMICS

We have performed an extensive study of one-dimensional and two-dimensional fission dynamics over a wide range of mass asymmetry.[84,85] We have used the same Yukawa-plus-exponential (Y+EM) potential extended to fragments with different charge densities,[86] as in 1979 (when we showed that α-decay is a superasymmetric fission process). A smoothed neck parametrization of the nuclear shape[88] has been adopted in two-dimensional calculations.

A. Werner–Wheeler inertia tensor

The potential energy surface in a multi-dimensional hyperspace of deformation parameters $\beta_1, \beta_2, \ldots, \beta_n$ gives the generalized forces acting on the nucleus. Information concerning how the system reacts to these forces is contained in a tensor of inertial coefficients, or the effective mass parameters $\{B_{ij}\}$. The contribution of a shape change to the kinetic energy is expressed by

$$E_k = \frac{1}{2} \sum_{i,j=1}^{n} B_{ij}(\beta) \frac{d\beta_i}{dt} \frac{d\beta_j}{dt} \tag{10}$$

As an approximation to the incompressible irrotational flow, one can use the Werner–Wheeler assumption. For pure spheroidal deformation, the flow produced by using the Werner–Wheeler approximation is exactly irrotational. It allows analytical results to be obtained for two parametrizations of intersected spheres: the ASAFM "cluster-like" (R_2 = constant) and Pik-Pichak more compact (fragment volumes = constant) shapes.

The kinetic energy of a non-viscous fluid due to a shape change is written as

$$E_k = \frac{\sigma}{2} \left[\int_V \mathbf{v}^2 d^3r - \left(\int_V \dot{z} d^3r \right)^2 \Big/ \int_V d^3r \right] \tag{11}$$

if the system possesses a cylindrical symmetry relative to the z-axis. V is the volume assumed to be conserved, $\sigma = 3m/(4\pi r_o^3)$ is the mass density, $\mathbf{v}$ is the velocity; the nuclear radius constant $r_o = 1.16$ fm within Y+EM.

By assuming irrotational motion ($\nabla \times \mathbf{v} = rot\,\mathbf{v} = 0$), the velocity field may be derived from a scalar velocity potential φ, i.e. $\mathbf{v} = \nabla\varphi$. From the continuity equation of an incompressible fluid ($D\sigma/Dt = 0$) it follows that the Laplace equation, $\nabla^2\varphi = \Delta\varphi = 0$, should be satisfied with kinematical boundary conditions

$$\frac{DF}{\partial t} = \mathbf{v}\nabla F + \frac{\partial F}{\partial t} = 0 \tag{12}$$

where the surface equation for axially symmetric shapes in cylindrical coordinates (ρ, φ, z) is written as $F(r, t, \beta) = \rho - \rho_s(z, t, \beta) = 0$ in which ρ_s is the value of ρ on the surface. The velocity components, $\dot{z} = \partial\varphi/\partial z$ and $\dot{\rho} = \partial\varphi/\partial\rho$, are both functions of z and ρ.

In the Werner–Wheeler approximation the flow is considered to be a motion of circular layers of fluid, $\dot{z}$ is independent of ρ, and $\dot{\rho}$ is linear in ρ:

$$\dot{z} = \sum_i X_i(z, \beta)\dot{\beta}_i; \; \dot{\rho} = (\rho/\rho_s) \sum_i Y_i(z, \beta)\dot{\beta}_i \tag{13}$$

A vanishing total (convective) time derivative of the fluid volume to the right (or left) side of an arbitrary plane normal to the z-axis leads to

$$X_{il} = -\rho_s^{-2} \frac{\partial}{\partial\beta_i} \int_{z_{min}}^{z} \rho_s^2\, dz, \;\; X_{ir} = \rho_s^{-2} \frac{\partial}{\partial\beta_i} \int_{z}^{z_{max}} \rho_s^2\, dz \tag{14}$$

By requiring a vanishing normal component of the velocity at the surface, one has

$$Y_{ir(l)} = -\frac{\rho_s}{2} \frac{\partial}{\partial z} X_{ir(l)} \tag{15}$$

from which the functions X_i and Y_i are found as a sum of two terms for the left (l) and right (r) side of the shape.

After substitution in the relationship for the kinetic energy and comparison with the initial equation for E_k we find the following relationships for the components of the inertia tensor:

$$B_{ij} = \pi\sigma \int_{z_{min}}^{z_{max}} \rho_s^2 \left(X_i X_j + \frac{1}{2} Y_i Y_j \right) dz + B_{ij}^c \tag{16}$$

$$B_{ij}^c = -(\pi^2 \sigma / V) \int_{z_{min}}^{z_{max}} \rho_s^2 X_i dz \int_{z_{min}}^{z_{max}} \rho_s^2 X_j \, dz \tag{17}$$

where $\rho_s = \rho_s(z)$ is the nuclear surface equation in cylindrical coordinates, with z_{min}, z_{max} intercepts on the z-axis. The correction term for the center of mass motion B_{ij}^c is different from zero if the origin of z is not placed in the center of mass.

For another set of deformation parameters $\{\alpha\}$ describing the same shape,

$$B_{kl}(\alpha) = \sum B_{ij}(\beta) \frac{\partial \beta_i}{\partial \alpha_k} \frac{\partial \beta_j}{\partial \alpha_l} \tag{18}$$

Within the above-mentioned one-dimensional parametrizations, the contributions of the left (B_1) and right (B_2) sides to inertia[84] are given by

$$\begin{aligned} 4r_o^3 B_i(R)/3m = {} & (z_i')^2 V_i/\pi + (-1)^i 2z_i' R_i R_i' (R_i + D_i)^2 + \\ & (R_i R_i')^2 [2R_i^2/H_i - 4.5R_i - 3.5D_i + 6R_i \ln(2R_i/H_i)] \end{aligned} \tag{19}$$

and the correction added to the above terms

$$4r_o^3 B_c(R)/3m = -(3/4R_o^3)C^2; C = C_1 + C_2 \tag{20}$$

$$C_i = (z_i')^2 V_i/\pi + (-1)^i R_i R_i' (R_i + D_i)^2; (i = 1, 2) \tag{21}$$

where $D_1 = z_s - z_1$, $D_2 = z_2 - z_s$, $H_i = R_i - D_i$, z_s is the position of the intersection plane of the spheres, and z_i are the geometrical centers of the spheres. The superscript prime means the derivative with respect to R – the separation of centers. The above equations become very simple when the origin is at the center of the first sphere ($z_1' = 0$). When the origin is in the separation plane and at the center of mass, z_1' is replaced by $-D_1'$ and $(z_1 - z_c)'$, respectively. In all cases $z_2' = z_1' + R'$. Here $z_c - z_1$ is the distance of the center of mass relative to the center of the first sphere. By changing the shape coordinate from R to z_m (the distance between mass centers of the fragments), the inertia becomes

$$B(z_m) = B(R) \left(\frac{dR}{dz_m} \right)^2 \tag{22}$$

For cluster-like shapes, both $B(R)$ and $B(z_m)$ are increasing functions of the respective variable. On the contrary, $B(z_m)$ decreases but $B(R)$ increases for the more compact shapes. When the motion of the center of mass is not taken into account,[87] the inertia is much higher. For α-decay, the ratio of the wrong to the correct value of inertia may be as high as 30/4. One, arbitrarily chosen, independent shape coordinate (separation of mass centers z_m) cannot be better than the other (separation of geometrical centers, R) describing the same shapes, as it was sometimes argued, because the action integral is independent of this choice. This property follows immediately by substituting $B(R)$ with $B(z_m)$ in the equation for the action integral K. In fact the final result – the half-life – should not depend on the deformation coordinate.

B. Optimum fission trajectory

We have compared the decay dynamics for two parametrizations of nuclear shapes mentioned above (two intersected spheres with R_2 or V_2 = constant). As a result we have proved that the latter parametrization is not suitable for cluster radioactivities because a lower value of the action integral is obtained[84] for the former (cluster-like shapes) when the mass value of the light fragment is under 34. In this way we obtained a dynamical confirmation for the choice of the ASAFM two-center cluster-like shape.

As the next step, we introduce a second shape degree of freedom – the neck curvature. The asymmetric two-center shape[88] is obtained by smoothly joining two spheres of radii R_1 and R_2 with a third surface generated by rotating around the symmetry axis a circle of radius $R_3 = S$. Calculations have been performed[89] for a fixed mass asymmetry, and hence the (symmetric) inertia tensor has three components. Other multi-dimensional calculations are reported elsewhere.[90–92]

The multi-dimensional penetration problem is reduced to a one-dimensional one by defining an inertia scalar $B(s) = \sum B_{ij}(\beta)(d\beta_i/ds)(d\beta_j/ds)$ along a fission path given parametrically by the equations $\beta_i = \beta_i(s)$. In our case the trajectory equation is $S = S(R)$.[93]

The Euler–Lagrange equation associated with the minimization of the action integral may be written as

$$\frac{\partial F}{\partial S} = \frac{d}{dR}\left(\frac{\partial F}{\partial S'}\right) \tag{23}$$

in which

$$F = F(R, S, S') = [B(R, S, S')E(R, S)]^{1/2} \tag{24}$$

In such a manner, the following non-linear differential equation of second degree

$$DS'' + D_3 S'^3 + D_2 S'^2 + D_1 S' + D_0 = 0 \tag{25}$$

has been obtained, where the coefficients D are expressions containing partial derivatives with respect to S and R of the inertia tensor components and of the energy.[93]

We have performed calculations (also in a three-dimensional case, when R_2

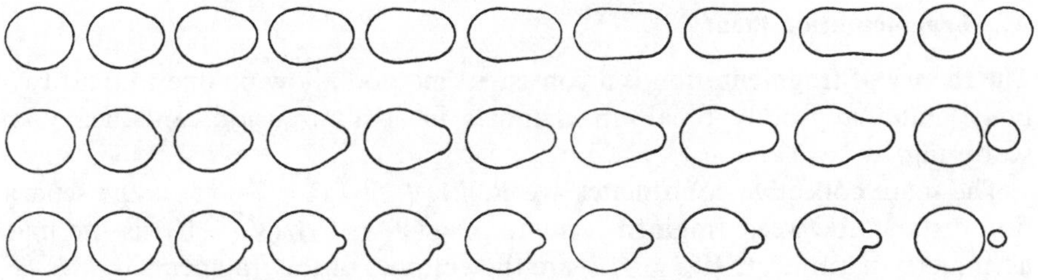

Fig. 6. Nuclear shapes along the optimum fission paths for cold fission with ^{100}Zr light fragment (at the top), ^{28}Mg radioactivity (middle), and α-decay (at the bottom) of ^{234}U, for $(R - R_i)/(R_t - R_i) = 0.1, 0.2, \ldots, 1.0$.

or η was allowed to vary[94]) for the α/-decay, ^{28}Mg radioactivity, and cold fission (with the light fragment ^{100}Zr) of ^{234}U – the first nucleus for which all three groups of decay modes have been experimentally detected. The nuclear shapes along the optimum fission paths are given in Fig. 6.

The neck influence is stronger when the mass asymmetry parameter is small; for a very large mass asymmetry the parametrization of two intersected spheres is a good approximation of the optimum fission path.

IV. CLUSTER DECAY MODELS

In order to predict cluster radioactivities and to arrive at a unified approach for cluster decay modes, alpha disintegration, and cold fission, before the first experimental confirmation in 1984, we developed and used a series of fission theories in a wide range of mass asymmetry, namely: fragmentation theory, numerical (NuSAF) and analytical (ASAF) superasymmetric fission models, and a semiempirical formula for α-decay.

Particularly useful has been ASAF, improved successively since 1980, allowing us to predict the half-lives for more than 150 different kinds of cluster radioactivities in 1984, including all cases experimentally determined so far. Several other models have been introduced since 1985. As normally expected, being intermediate phenomena between fission and α-decay cluster radioactivities have been treated either as extremely asymmetric cold fission phenomena or in a similar way to α-decay, but with heavier emitted particles.[95–107] since 1985 (see also the review papers[108, 109]), explaining existing measurements and (in some instances) making further predictions. We have discussed them elsewhere.[108] More recently, the emitted clusters have been interpreted as solitons on the nuclear surface.[110]

The main difference from model to model consists in the method used to calculate the preformation probability or the half-life. There are also different relationships for the nuclear radii, interaction potentials, as well as for the frequency of assaults (see eq. (7)).

A. Fragmentation theory

The theory of fragmentation is a consistent method allowing one to treat two-body and many-body break-up channels in fission, fusion, and heavy-ion scattering.[9,111]

The main collective coordinates are $R, \eta, \varepsilon, \beta_1, \beta_2$. $R = z_2 - z_1$ is the separation distance between fragment centers, $\eta = (V_1 - V_2)/(V_1 + V_2)$ is the mass asymmetry parameter, $V_i (i = 1, 2)$ are the volumes of the fragments, $\varepsilon = E'/E_o$ is the neck parameter, and $\beta_i = a_i/b_i$ are the fragment deformations. E_o is the actual barrier height and $E' = m\omega_{z_i}^2 z_i^2/2$ is the barrier of the two-center oscillator.

By assuming the adiabatic approximation, for each pair of (R, η) coordinates, the liquid drop energy $V_{LDM}(R, \eta, \varepsilon, \beta_1, \beta_2)$ can be minimized with respect to $\varepsilon, \beta_1, \beta_2$, leading to $E_{LDM}(R, \eta)$. Then the shell and pairing corrections $\delta U(R, \eta)$ are added. Also the collective mass tensor $B_{ik}(R, \eta)$ can be calculated by using the states of the two-center shell model (TCSM) supplemented with pairing forces.

The wave function $\Phi_k(R, \eta)$ is a solution of a Schroedinger equation in η:

$$\left(-\frac{\hbar^2}{2\sqrt{B_{\eta\eta}}} \frac{\partial}{\partial\eta} \frac{1}{\sqrt{B_{\eta\eta}}} \frac{\partial}{\partial\eta} + V(R, \eta) \right) \Phi_k(R, \eta) = E_k \Phi_k(R, \eta) \tag{26}$$

Here R is kept fixed and only the η degree of freedom is considered.

The fission fragment mass yield normalized to 200% is given by

$$Y(A_2) = |\Psi(R, \eta, \Theta)|^2 \sqrt{B_{\eta\eta}} (400/A) \tag{27}$$

where $B_{\eta\eta}$ is a diagonal component of the inertia (effective mass) tensor.

The example of the transition from asymmetric to symmetric fission in Fm isotopes[10] illustrates the dominant influence of shell effects on fission fragment mass asymmetry. This transition was confirmed by experiment.[69] Moreover, for bimodal fission, observed in this region of nuclei,[112,113] two fission paths (one for the usual "normal" mechanism and the other for cold fission) have been found[88] on the potential energy surface.

The nuclear levels and the wave functions, used to compute shell corrections and the mass tensor, are obtained within the TCSM. Alternatively, a folded-Yukawa potential has been used[92] for a nuclear shape parametrization obtained by smoothly joining two spheres with a third surface generated by rotating a circle of radius $R_3 = 1/c_3$ around the symmetry axis.

The side peaks in mass yields have also been obtained in many calculations[114] based on fragmentation theory, as for example in U and No. Clearly the mass yield in the above equation includes contributions for extreme break-ups of the fissioning nucleus, i.e. for $\eta \sim 0.8$–1.0. This means that nuclei emit light clusters. We are thus lead to the idea of cluster radioactivity and, furthermore, to the insight that there is no essential difference between cold fission and such a radioactive process. Obviously

$$P(\eta) = \int \Psi_o^*(R,\eta)\Psi_o(R,\eta)d^3R \tag{28}$$

can be interpreted as the cluster preformation probability.

In a similar manner, but with a proximity potential[115] and the mass parameters from ref. 116, the preformation probability[97,117] is calculated for a configuration of two touching spheres, and using a scaling factor. A rather questionable way of estimating the penetrability was adopted.

On the potential energy surface given in Fig. 7, the fission paths leading to various final asymmetries are shown by dashed lines. They are determined by minimizing[118] the WKB integral.

The inertia tensor has been computed by using the Werner–Wheeler approximation within a hydrodynamical model. A fixed neck radius has been assumed. The corresponding half-life obtained along each path in the upper part of Fig. 7 is plotted in the middle diagram, showing that, according to these calculations, the Ne radioactivity of ^{232}U has a larger probability than the spontaneous fission of the same nucleus. It is interesting to note the similarity with the time spectrum of the ^{234}U nucleus calculated within ASAFM plotted in the bottom diagram.

B. Numerical superasymmetric fission model

We used a first variant in 1979 to describe the α-decay as a fission process.[119,120] In a second approach, the reduced mass was replaced by the Werner–Wheeler inertia.[84] The one-dimensional two-center parametrization of the nuclear shape has been considered. Recently we developed[94] a new method to find the optimum fission trajectory in a three-dimensional deformation space by solving a non-linear system of differential equations with partial derivatives.

In this model the nuclear energy, replacing the Myers–Swiatecki liquid drop model surface energy, is given by the double folded-Yukawa-plus-exponential (Y+E) potential[121] energy. By taking into account the difference between the charge densities of the two fragments, this energy can be expressed[86] as a sum of two self-energies and one interaction energy between them:

$$\begin{aligned} E_Y = &-\frac{c_{s1}}{8\pi^2 r_0^2 a^4}\int_{V_1} d^3r_1 \int_{V_1}\left(\frac{r_{12}}{a}-2\right)\frac{e^{-r_{12}/a}}{r_{12}/a}\,d^3r_2 - \\ &\frac{2\sqrt{c_{s1}c_{s2}}}{8\pi^2 r_0^2 a^4}\int_{V_1} d^3r_1 \int_{V_2}\left(\frac{r_{12}}{a}-2\right)\frac{e^{-r_{12}/a}}{r_{12}/a}\,d^3r_2 - \\ &\frac{c_{s2}}{8\pi^2 r_0^2 a^4}\int_{V_2} d^3r_1 \int_{V_2}\left(\frac{r_{12}}{a}-2\right)\frac{e^{-r_{12}/a}}{r_{12}/a}\,d^3r_2 \end{aligned} \tag{29}$$

where $r_{12} = |\mathbf{r}_1 - \mathbf{r}_2|$, $a = 0.68$ fm is the diffusivity parameter, and $c_s = a_s(1 - \kappa I^2)$, $a_s = 21.13$ MeV is the surface energy constant, $\kappa = 2.3$ is the surface asymmetry constant, $I = (N - Z)/A$, and $r_0 = 1.16$ fm is the nuclear radius constant.

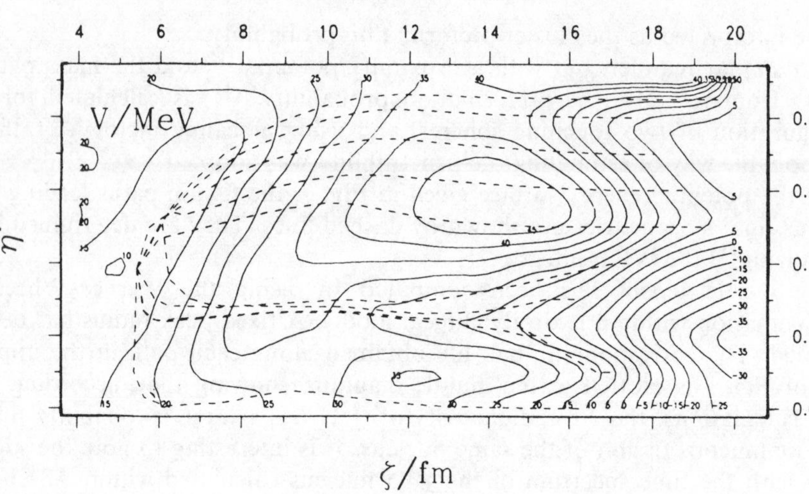
V/MeV
η^t
ζ/fm
4
6
8
10
12
14
16
18
20
0.8
0.6
0.4
0.2
0.0

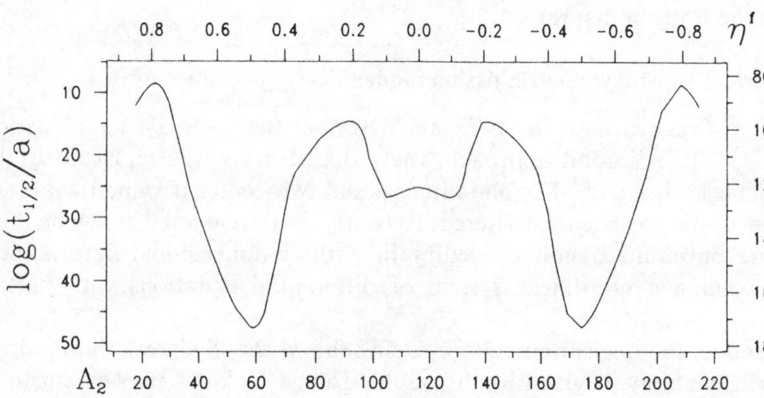
0.8 0.6 0.4 0.2 0.0 -0.2 -0.4 -0.6 -0.8 η^f
log($t_{1/2}$/a)
G
10
20
30
40
50
80
100
120
140
160
180
A_2 20 40 60 80 100 120 140 160 180 200 220

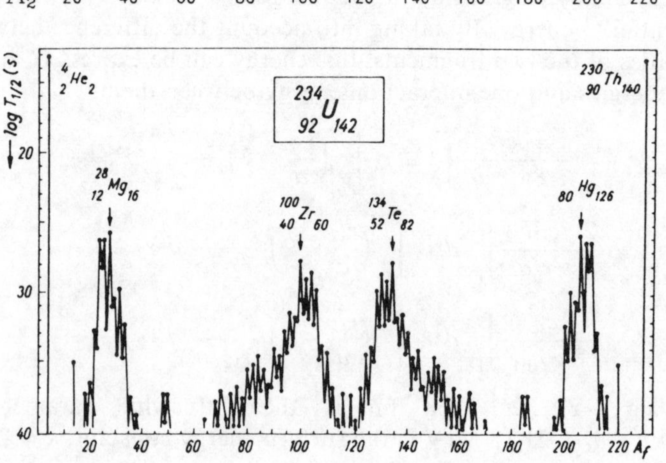
log $T_{1/2}$(s)
$^4_2He_2$
234
U
92 142
$^{230}_{90}Th_{140}$
$^{28}_{12}Mg_{16}$
$^{100}_{40}Zr_{60}$
$^{134}_{52}Te_{82}$
$_{80}Hg_{126}$
20
30
40
20 40 60 80 100 120 140 160 180 200 220 A_f

For overlapping fragments with axial symmetry, the six-fold integrals can be reduced to three-dimensional ones which are calculated numerically[86] by Gauss–Legendre quadrature. A spherical compound nucleus with a radius $R_0 = r_0 A^{1/3}$ gives

$$E_Y^0 = c_{s0} A^{2/3} \{1 - 3x^2 + (1 + 1/x)[2 + 3x(1 + x)]e^{-2/x}\} \tag{30}$$

in which $x = a/R_0$.

Similar to the (Y+EM) term, one has a Coulomb energy

$$E_C = \frac{\rho_{1e}^2}{2} \int_{V_1} d^3 r_1 \int_{V_1} \frac{d^3 r_2}{r_{12}} + \rho_{1e}\, \rho_{2e} \int_{V_1} d^3 r_1 \int_{V_2} \frac{d^3 r_2}{r_{12}} + \frac{\rho_{2e}^2}{2} \int_{V_2} d^3 r_1 \int_{V_2} \frac{d^3 r_2}{r_{12}} \tag{31}$$

The charge densities of the compound nucleus and of the two fragments are denoted by ρ_{0e}, ρ_{1e}, and ρ_{2e} respectively. The Coulomb energy of a spherical nucleus with mass number A is given by

$$E_C^0 = \frac{3e^2 Z^2}{5 r_0 A^{1/3}} \tag{32}$$

The total deformation energy is the sum of the Y+E, Coulomb, and the volume terms:

$$E = E_Y + E_C + E_V \tag{33}$$

By adding a correction term $E_{corr} = (Q_{exp} - Q_t)[1 - V_2(R)/V_{2f}]$ one can reproduce exactly the experimental Q-value. Here $V_{2f} = V_2(R_t)$ is the (final) volume of the emitted cluster. When the fragments are separated $(R > R_t = R_1 + R_2)$, analytical relationships are available:

$$E_C = \frac{Z_1 Z_2 e^2}{R} \tag{34}$$

$$E_Y = -4\left(\frac{a}{r_0}\right) \sqrt{c_{s1}\, c_{s2}} \left[g_1 g_2 \left(\frac{4+R}{a}\right) - g_2 f_1 - g_1 f_2 \right] \frac{e^{-R/a}}{R/a} \tag{35}$$

where

$$f_k = \left(\frac{R_k}{a}\right)^2 \sin\mathrm{h}\, \frac{R_k}{a} \tag{36}$$

Fig. 7. At the top: potential energy surfaces versus mass asymmetry η^f and separation distance between centers ζ for ^{232}U at a fixed neck coordinate $c_3 = 0.2$ fm^{-1}. The dashed lines are fission paths leading to a given final mass asymmetry. The corresponding half-life (in years) along these paths is plotted in the middle.[92] At the bottom is plotted the half-life spectrum for various clusters emitted from ^{234}U calculated within ASAF. A_f is the mass number of the emitted cluster.

$$g_k = \frac{R_k}{a} \cosh \frac{R_k}{a} - \sinh \frac{R_k}{a}, \quad (k = 1, 2) \tag{37}$$

and R is the distance between the centers of the two fragments.

The preformation, S, and the penetrability of the external barrier, P, are given by $S = \exp(-K_{ov})$, $P = \exp(-K_s)$. They are calculated within the WKB approximation according to eq. (8).

The zero-point vibration energy can be estimated from the stiffness of the potential barrier and the effective mass parameter:

$$E_v = \frac{\hbar}{R_t - R_i}\sqrt{\frac{E_b}{2\mu}} = \frac{4.5536}{R_t - R_i}\sqrt{\frac{AE_b}{A_e A_d}}; \; E_b = E(R_t) - Q \tag{38}$$

where R_t and $R_i = R_o - R_2$ are the separation distances at the touching and initial points, respectively. In this equation the energies are expressed in MeV and the distances in fm. For α-decay and ^{14}C radioactivity, one obtains values in the range 1.6–2.3 MeV and 1.0–1.35 MeV, respectively, and minima at the magic number of neutrons for the daughter nucleus.

For a three-dimensional parametrization of nuclear shape[93] the components of the nuclear inertia tensor were calculated numerically. We performed calculations for the α-decay, ^{28}Mg radioactivity, and cold fission (with the light fragment ^{100}Zr) of ^{234}U. In the absence of the smoothed neck, for two intersected spheres with a cusp, $K_{ov} = 7.1\%$, 35.2%, and 62.5% from the total value of $K = K_{ov} + K_s$. As a result of the minimization along the optimum dynamical path, K_{ov}/K has been reduced to $(K_{ov}/K)_{opt} = 5.9\%$, 32.9%, and 55%, respectively. A decrease of the potential barrier by the multiplication of $(E(R) - Q)$ by a factor $f^2 < 1$ (due to deformations or increased nuclear radii) will lead to an f-times lower value of the action integral $K = fK_0$.

In this way, the best fit (rms of log $T(s)$ values of 0.51, comparable to 0.49 of Blendowske–Fliessbach–Walliser (BFW)[101,122] and to 0.47 of ASAF) to 14 even–even cluster radioactivities measured so far[25,23] is obtained when $f = 0.8985 \simeq 0.9$. The experimental and theoretical results are plotted in Fig. 8 for 10 of these measured events versus the mass number of the parent nucleus. For ^{234}U and ^{238}Pu three different cluster emissions have been experimentally determined[20] we selected only one for each parent in Fig. 8 in order to simplify the presentation. A lower value of f for NuSAF than for the BFW model is understandable from the comparison of the corresponding potentials in section VI.

From Fig. 8 one can see that the variation of the pre-exponential factor $F = \ln 2/\nu$ can be neglected in comparison with that of penetrability P. Also the preformation probability S is very much dependent on the mass number of the emitted cluster, but it is less sensitive to the parent nucleus.

C. Analytical superasymmetric fission model

Since 1980, in order to make a systematic search for new cluster decay modes, we have had to take into account a large number of combinations of parent

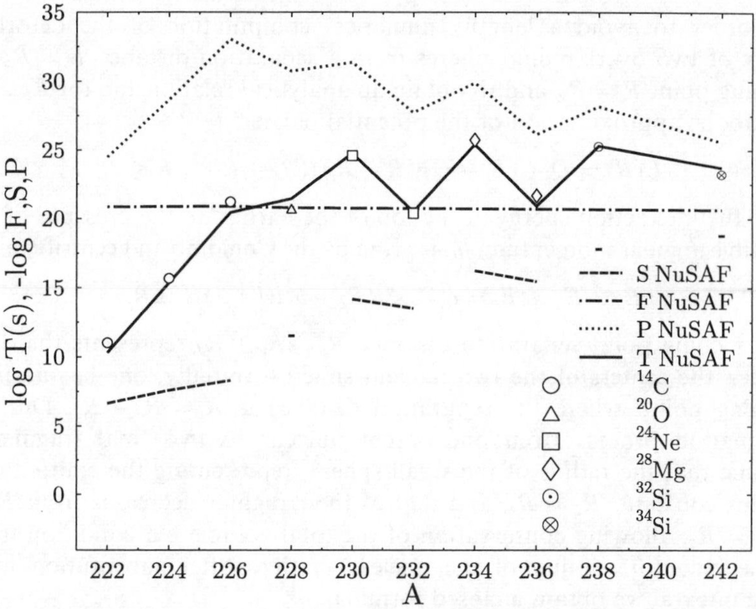

Fig. 8. Half-lives T of 10 even–even nuclei against cluster decays and the corresponding three model-dependent quantities (< 1): F (pre-exponential factor), S (preformation probability), and P (penetrability of the external barrier) calculated within the numerical superasymmetric fission model. The experimentally determined half-lives are also shown.

and emitted nuclei (at least $2000 \times 250 = 5 \cdot 10^5$). The corresponding large number of computations can be performed in a reasonable time by using an analytical relationship for the half-life. Such a formula has been obtained on the basis of the Myers–Swiatecki liquid drop model adjusted with a phenomenological shell correction in the spirit of the Strutinsky method.

Within ASAFM we can can take into account the allowed angular momenta, l, determined from the spin (J) and parity (π) conservation:

$$|J_f - J_i| \leq l \leq J_f + J_i;\ \pi_i \pi_f(-1)^l = 1 \tag{39}$$

The half-life (see eqs. (6)–(8)) of a parent nucleus AZ against the split into a cluster A_eZ_e and a daughter A_dZ_d is calculated by using the WKB approximation. We substitute in eq. (8) $B(R) = \mu$, and replace $E(R)$ by $[E(R) - E_{corr}] - Q$. In fact the action integral contains two terms $K = K_{ov} + K_s$, the former corresponding to the overlapping stage of the fragments (the limits of the integral R_a, R_t), and the latter to separated fragments (limits R_t, R_b). E_{corr} is a phenomenological correction energy similar to the Strutinsky shell correction, also taking into account the fact that the Myers–Swiatecki liquid drop model overestimates fission barrier heights, and the effective inertia in the overlapping region is different from the reduced mass.

In order to avoid a lengthy numerical computation of the deformation energy of two overlapping spheres from a separation distance $R = R_a$ to the touching point $R = R_t$, and to obtain an analytical relationship for K_{ov}, we use a parabolic approximation of the potential barrier:

$$E(R) = Q + (E_i - Q)[(R - R_i)/(R_t - R_i)]^2; \; R \leq R_t \tag{40}$$

where the interaction energy at the top of the barrier, in the presence of a non-negligible angular momentum, $l\hbar$, is given by the Coulomb and centrifugal terms:

$$E_i = E_C + E_l = e^2 Z_e Z_d / R_t + \hbar^2 l(l+1)/(2\mu R_t^2) \tag{41}$$

The touching point separation distance $R_t = R_e + R_d$ represents the distance between the centers of the two tangent spheres. Initially, one has an internal touching point, where the separation distance is $R_i = R_0 - R_e$. During the deformation process from one parent nucleus to two final fragments we presume that the radius of the small sphere, representing the emitted cluster, remains constant, $R_2 = R_e$, and that of the daughter decreases from $R_1 = R_0$ to $R_1 = R_d$, allowing conservation of the total volume – a condition imposed by the incompressibility of the nuclear matter. After substitution into the above integral we obtain a closed formula

$$K_{ov} = 0.2196(E_b^0 A_e A_d / A)^{1/2}(R_t - R_i)\left[\sqrt{1-b^2} - b^2 \ln \frac{1+\sqrt{1-b^2}}{b}\right] \tag{42}$$

where the parameter $b^2 = (E_{corr} + E^*)/E_b^0$ appears as a consequence of the following expression of the first turning point (defined by $E(R_a) = Q$):

$$R_a = R_i + (R_t - R_i)[(E_{corr} + E^*)/E_b^0]^{1/2} \tag{43}$$

in which E^* is the excitation energy concentrated in the separation degree of freedom, and $E_b^0 = E_i - Q$ is the barrier height before correction. This action integral allows us to estimate the preformation factor $S = \exp(-K_{ov})$.

For separated spherical fragments in the Coulomb-plus-centrifugal potential

$$E(R) = E_C R_t / R + E_l R_t^2 / R^2; \; R > R_t \tag{44}$$

it is not necessary to make any approximation; we get

$$K_s = 0.4392[(Q + E_{corr} + E^*) A_e A_d / A]^{1/2} R_b J_{rc} \tag{45}$$

with

$$J_{rc} = (c) \arccos\sqrt{(1-c+r)/(2-c)} - [(1-r)(1-c+r)]^{1/2} + \sqrt{1-c}\ln\left[\frac{2\sqrt{(1-c)(1-r)(1-c+r)} + 2 - 2c + cr}{r(2-c)}\right] \tag{46}$$

in which $r = R_t/R_b$ and $c = rE_C/(Q + E_{corr} + E^*)$. When the angular momentum $l = 0$, one has $c = 1$ and the above equations take a simpler form. The external turning point (defined as $E(R_b) = Q$) is given by

$$R_b = R_t E_C\{1/2 + [1/4 + (Q + E_{corr} + E^*)E_l/E_C^2]^{1/2}\}/(Q + E_{corr} + E^*) \quad (47)$$

The penetrability of the external part of the barrier is $P = \exp(-K_s)$.

In order to reduce the number of fitting parameters, we took $E_v = E_{corr}$ though it is evident that, owing to the exponential dependence, any small variation of E_{corr} induces a large change of T, and thus plays a more important role compared to the pre-exponential factor variation due to E_v (see eq. (8)). Both shell and pairing effects are included in $E_{corr} = a_i(A_e)Q$ ($i = 1, 2, 3, 4$ for even–even, odd–even, even–odd, and odd–odd parent nuclei). Pairing effects are clearly present: for a given cluster radioactivity we have four values of the coefficients a_i, the largest for the even–even parent and the smallest for the odd–odd one.[123] The shell effects for every cluster radioactivity are implicitly contained in the correction energy due to its proportionality with the Q-value, which is maximum when the daughter nucleus has a magic number of neutrons and protons. Another contribution of shell effects, which is always present, comes from the fact that we subtract from the deformation energy (with corrections) the true value of the released energy Q, obtained as a difference of experimentally determined atomic masses.

D. Cluster preformation in a fission model

We gave[124, 125] a new interpretation of the cluster preformation probability (more exactly of the touching point or scission configuration) within a fission model, as the penetrability of the prescission part of the barrier. This semiclassical method of calculating a spectroscopic factor could be useful in various applications. On this basis we have shown that preformation cluster models are equivalent to fission models and we introduced one universal curve for each kind of cluster radioactivity.

According to our semiclassical method, the preformation probability can be calculated as the penetrability of the prescission part of the fission barrier $S = \exp(-K_{ov})$, where K_{ov} is the corresponding action integral for overlapping fragments. We gave the ASAF relationship in the preceding section.

Unlike the approximation leading to the "universal curves" of even–even nuclei, where $\log S = -0.598(A_e - 1)$ depends only on the mass number of the emitted cluster, this time the nuclear structure effects could be taken into consideration. An example is given in Fig. 9, where (in the two upper curves) we have considered emitted clusters with $Z_e = N_e$. The heavier the cluster, the larger the difference between the probability to be formed in a nucleus leaving a daughter ^{208}Pb and the preformation in this doubly magic nucleus itself. Even higher cluster preformation probability is obtained for neutron-rich nuclei (^{14}C, ^{20}O, ^{24}Ne, ^{28,30}Mg, ^{34}Si) observed experimentally. An "experimental" value is defined as the ratio $S_{exp} = \lambda_{exp}/(\nu P_s)$.

The fact that a given cluster is preformed with a larger probability in a parent nucleus corresponding to a ^{208}Pb daughter than in the doubly magic nucleus ^{208}Pb is also illustrated in Fig. 10, in which we consider few clusters: α-particle; ^{14}C, and ^{24}Ne. Instead of only one value of S at a given mass number

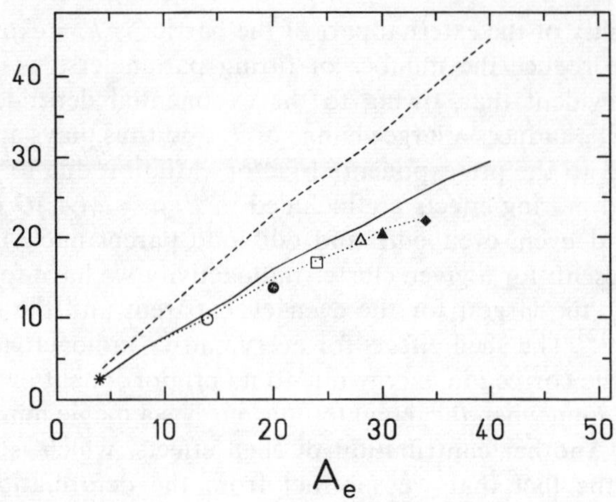

Fig. 9. Preformation probability of even–even emitted clusters with $Z_e = N_e$ from ^{208}Pb (dashed line), from various nuclei leading to ^{208}Pb daughter (solid line), compared to that of neutron-rich emitted nuclei ^{14}C, ^{20}O, ^{24}Ne, ^{28,30}Mg, ^{34}Si with ^{208}Pb heavy fragment (dotted line) and the experimental points.

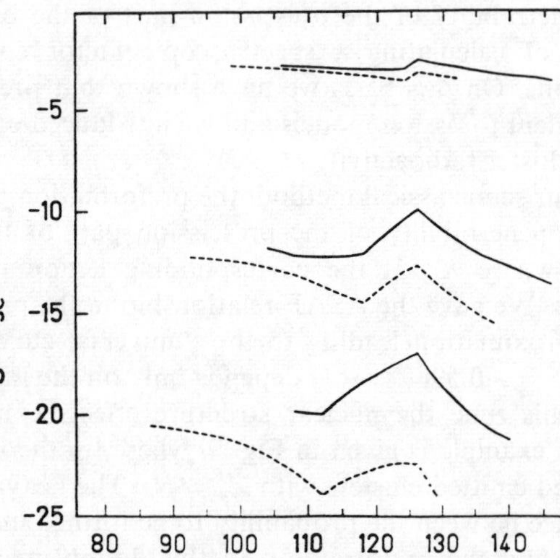

Fig. 10. In descending order from the top to the bottom we present the preformation probability of ^{4}He, ^{14}C, and ^{24}Ne versus neutron number of the daughter. For each cluster the parent nuclei are either Pb isotopes (dashed line) or heavier nuclides leading to heavy fragments which are Pb isotopes (solid lines).

A_e (a horizontal line) independent of Z_e, Z_d, A_d, one can see a whole range, increasing at a constant N_d from $Z = 82$ (dashed line) to $Z = 82 + Z_e$. The maximum value is reached at the magic neutron number of the daughter $N_d = 126$.

E. Universal curves

It is extremely difficult to find within a microscopic theory a convenient method to calculate the preformation probability S for clusters heavier than an α-particle. In order to study a broader range of emitted clusters[122] a convenient phenomenological law for the variation of S with A_e has been introduced:

$$S = S_\alpha^{(A_e - 1)/3} \tag{48}$$

where for even–even parent nuclei $S_\alpha = 0.0063$ was obtained by fitting calculated and experimental data on α-decay and cluster radioactivities with $A_e < 33$.

From eqs. (6) and (7) it follows that

$$\log T = \log \frac{\ln 2}{\nu} - \log S - \log P_s \tag{49}$$

If ν and S may be considered as constants for a given cluster radioactivity (given A_e), one can find a single universal curve, $\log T = f(\log P_s)$, for all even-even parent nuclei. Besides the above law of variation of S, a further assumption is needed, namely: the assault frequency ν should be independent of the emitted cluster and the daughter nucleus, $\nu(A_e, Z_e, A_d, Z_d) = \text{constant}$.

Starting from ASAF without correction E_{corr}, we got from a fit of experimental data: $S_\alpha = 0.0160694$ and $\nu = 10^{22.01}\ \text{s}^{-1}$, leading to

$$\log T = -\log P_s - 22.169 + 0.598(A_e - 1) \tag{50}$$

which is a straight line for a given A_e, with a slope equal to unity. The vertical distance between two universal curves corresponding to A_{e1} and A_{e2} is $0.598(A_{e2} - A_{e1})$.

For all cases of practical importance the daughter nucleus has a spherical shape (a doubly magic ^{208}Pb or one of its neighbors). When the cluster is preformed, by neglecting in the first approximation its deformation, we can assume a touching point configuration of two spherical nuclei with a separation distance $R = R_t$. For any combination of fragments A_eZ_e, A_dZ_d one can easily calculate the WKB penetrability of the external (Coulomb) part of the potential barrier $E(R) = Z_eZ_de^2/R$, between the classical turning points R_t and R_b. The electron charge $e = 1.43998$ MeV fm and R_b is defined by $E(R_b) = Q$. The nuclear inertia $B(R)$ at the touching point equals the reduced mass $\mu = m\mu_A = mA_dA_e/A$; hence

$$-\log P_s = 0.22873(\mu_A Z_d Z_e R_b)^{1/2}\left[\arccos\sqrt{r} - \sqrt{r(1-r)}\right] \tag{51}$$

where $r = R_t/R_b$, $R_t = 1.2249(A_d^{1/3} + A_e^{1/3})$, and $R_b = 1.43998 Z_d Z_e/Q$.

The advantage of such a relationship is evident. Up to now the 14 even–even half-life measurements are well reproduced within a ratio of 3.86, or rms = 0.587 orders of magnitude. A similar result is obtained with the "hand-pocket" formula.[126] Another argument of its universality is illustrated in Fig. 11 for α-decay and in Fig. 12 for ^{14}C radioactivity. Instead of having different lines for various parent nuclei, as in the "classical" (frequently called Geiger–Nuttal plot) systematics $\log T = f(Q^{-1/2})$, one gets practically only one line when $\log T = f(\log P_s)$ is plotted. Nevertheless one should be aware of its limitations. If we have a closer look, by plotting the deviations of log T values of 124 α emitters from the experimental ones versus the neutron number of the

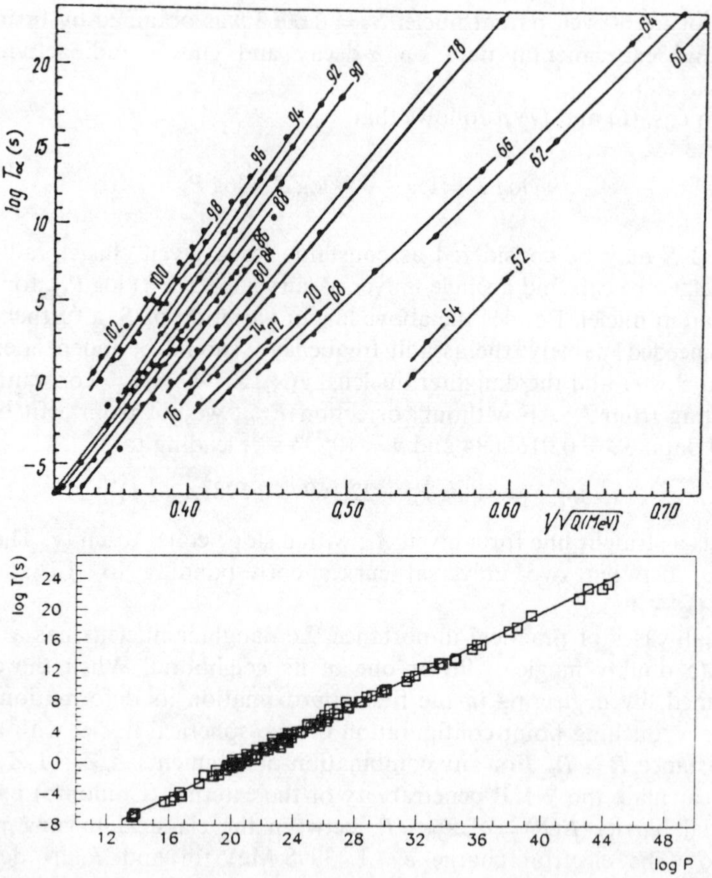

Fig. 11. Comparison of two systematics for α-decay of 124 even–even parent nuclei. In the upper part: variation of decimal logarithm of the partial half-lives with $Q^{-1/2}$. One has different lines for various elements (the atomic number Z is given on each curve). In the lower part: the "universal curve" (all points lie on a single line, when the variable is the decimal logarithm of the penetrability of external barrier P).

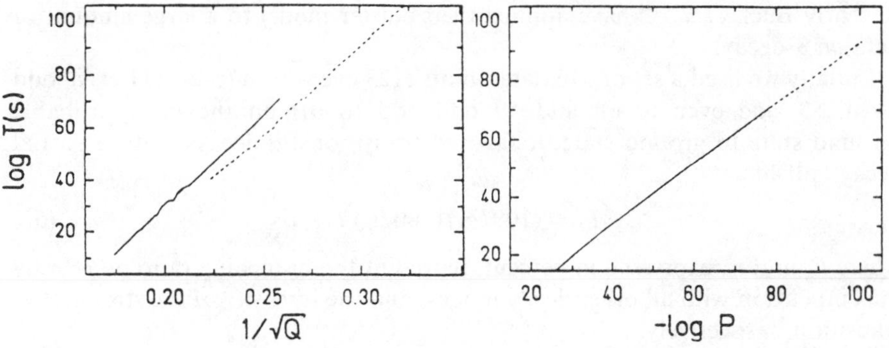

Fig. 12. Similar to Fig. 11: comparison of "classical" systematics (left hand side) with universal curve–like systematics (right hand side), calculated within ASAFM for ^{14}C radioactivity of even-even Ra isotopes (solid line) and of Pb isotopes (dashed line).

daughter nucleus, we can see the structure due to the shell effects. We have obtained a similar trend for the zero-point vibration energy, with a minimum at the closed shell number of neutrons $N_d = 126$, and absolute values decreasing with increasing cluster mass number, from α-decay to cold fission. If we introduce the shell effects in ν, including the variation with A_d, Z_d, A_e, Z_e, we can remove at least a part of these discrepancies.

We had to use calculated (ASAFM) values for the half-lives in both kinds of systematics presented in Fig. 12, but in principle a linear least-squares method through experimental points (if available) would give similar results. A similar result for α-decay has been obtained[127] by plotting $\log T$ values versus $Z_d^{0.6}/\sqrt{Q}$ for $Z \geq 72$. Large errors (up to five orders of magnitude) are obtained for lighter alpha emitters by using this[127] semiempirical formula.

F. Semiempirical formula for α-decay

As far back as 1911, Geiger and Nuttal found a simple dependence of the α-decay partial half-life on the α-particle range in air. Nowadays, very often a diagram of $\log T$ versus $Q^{-1/2}$ (see Fig. 11) is called a Geiger–Nuttal plot. There are many semiempirical relationships (see for example refs. 128–133, 127), allowing one to estimate the disintegration period if the kinetic energy of the emitted particle $E_\alpha = QA_1/A$ is known. The α-decay half-life of an even–even emitter can also be easily calculated by using NuSAF, ASAF models or the universal curve presented above.

In our semiempirical formula, based on the fission theory of α-decay,[134] we have improved the description of data in the neighborhood of the magic proton and neutron numbers, where the errors of the other relationships are large.

A computer program[135] allows the fitted parameters to be changed automatically every time a better set of experimental data is available. Besides the selected alpha-emitters[136] we have to mention two other large tables.[137, 138]

Recently Buck *et al.*[139] have applied their cluster model to a large amount of data on α-decay.

Initially we used a set of 376 data points (123 even–even (e–e), 111 even–odd (e–o), 83 odd–even (o–e), and 59 odd–odd (o–o)) on the most probable (ground state to ground state or favored transitions) α-decays, with a partial decay half-life

$$T_\alpha = (100/b_\alpha)(100/i_p)T_i \tag{52}$$

where b_α and i_p, expressed in percent, represent the branching ratio of α-decay in competition with all other decay modes, and the intensity of the strongest α-transition, respectively.

Our formula, based on the fission theory of α-decay, gives

$$\log T = 0.43429\, K_s\chi - 20.446 \tag{53}$$

where

$$K_s = 2.52956\, Z_{da}\,[A_{da}/(AQ\alpha)]^{1/2}\left[\arccos\sqrt{x} - \sqrt{x(1-x)}\right] \tag{54}$$

$x = 0.423Q_\alpha(1.5874 + A_{da}^{1/3})/Z_{da}$, and the numerical coefficient χ, close to unity, is a second-order polynomial

$$\chi = B_1 + B_2y + B_3z + B_4y^2 + B_5yz + B_6z^2 \tag{55}$$

in the reduced variables y and z, expressing the distance from the closest magic-plus-one neutron and proton numbers N_i and Z_i:

$$y \equiv (N - N_i)/(N_{i+1} - N_i);\ N_i < N \leq N_{i+1} \tag{56}$$

$$z \equiv (Z - Z_i)/(Z_{i+1} - Z_i);\ Z_i < Z \leq Z_{i+1} \tag{57}$$

with $N_i = \ldots, 51, 83, 127, 185, \ldots,$ $Z_i = \ldots, 29, 51, 83, 115, \ldots,$ and $Z_{da} = Z - 2$, $A_{da} = A - 4$. The coefficients B_i are given in Table 3.

Practically for even–even nuclei, the increased errors in the neighborhood of $N = 126$, present in all other cases, are smoothed out by our formula using the second-order polynomial approximation for χ. They are still present for the strongest α-decays of some even–odd and odd–odd parent nuclides. In fact for a non-even number of nucleons the structure effects become very important, and they should be carefully taken into account for every nucleus, not only globally.

Rurarz[140] has made predictions for nuclei far from stability with

Table 3 B_k parameter values

	B_1	B_2	B_3	B_4	B_5	B_6
e–e	0.985911	0.022841	0.024584	0.023279	−0.000716	−0.022562
o–e	1.000560	0.010783	0.050671	0.013919	0.043657	−0.079999
e–o	1.017560	−0.113054	0.019057	0.147320	0.230300	−0.101523
o–o	1.007740	−0.184136	0.260268	0.231900	0.326025	−0.407280

$62 < Z < 76$. In another variant[141] the shell effect on the formation factor is approximated by an empirical formula.

G. Experimental confirmation

The cluster emitters experimentally confirmed up to now (see the review papers[25, 23] and the references therein) are heavy transfrancium nuclei. The corresponding daughter nucleus is the doubly magic ^{208}Pb in eight of these cases.

In Table 4 we give our predictions of 1984 and of 1986 (after properly taking into account the even–odd effect). The predicted shell and pairing effects have been clearly confirmed. Generally the measured half-lives have confirmed, within one order of magnitude, the calculated values of ASAFM. This result does not look bad if we take into account that in spontaneous fission there are still disagreements of many orders of magnitude (three or even five orders). For the ^{14}C radioactivity of ^{223}Ra we have presented the global result. Details about the fine structure of different transitions observed in this case are discussed below.

Table 4 Comparison between measured and calculated half-lives

Parent		Emitted		Daughter		log T(s)			
Z	A	Z_e	A_e	Z_d	N_d	ASAFM			Experiment
						Early	Ref. 143	Ref. 123	
87	221	6	14	81	126	15.0	14.4	14.3	14.53 ± 0.10
88	221	6	14	82	125	13.8	14.3	14.2	13.39 ± 0.10
88	222	6	14	82	126	12.6	11.2	11.1	11.01 ± 0.06
88	223	6	14	82	127	14.8	15.2	15.1	15.15 ± 0.05
88	224	6	14	82	128	17.4	15.9	15.9	15.69 ± 0.12
88	226	6	14	82	130	22.4	21.0	20.9	21.22 ± 0.20
89	225	6	14	83	128	18.5	17.8	17.8	17.16 ± 0.10
90	228	8	20	82	126	22.4	21.9	21.9	20.72 ± 0.08
91	231	9	23	82	126	24.8	25.9	25.9	$26.02^{+0.76}_{-0.51}$
90	230	10	24	80	126	24.9	25.3	25.2	24.61 ± 0.09
91	231	10	24	81	126	22.0	23.4	23.3	22.89 ± 0.05
92	232	10	24	82	126	20.4	20.8	20.8	20.40 ± 0.05
92	233	10	24	82	127	23.1	24.8	25.2	24.84 ± 0.06
92	234	10	24	82	128	25.7	26.1	26.1	25.88 ± 0.30
92	233	10	25	82	126	23.3	25.0	25.7	24.84 ± 0.06
92	234	12	28	80	126	24.6	25.8	25.9	25.75 ± 0.06
94	236	12	28	82	126	19.8	21.0	21.1	21.68 ± 0.15
94	238	12	28	82	128	24.8	26.0	26.2	25.70 ± 0.25
94	238	12	30	82	126	24.4	25.7	26.2	25.70 ± 0.25
94	238	14	32	80	126	23.7	25.1	26.1	25.28 ± 0.16
96	242	14	34	82	126	20.9	22.3	23.5	23.24^2

[a] preliminary.

The unified approach of the three groups of decay modes (cold fission, cluster radioactivities, and α-decay) within ASAFM is best illustrated by the example of the ^{234}U nucleus (see Fig. 7 in section IV.A), for which all these processes have been measured. For α-decay, Ne, and Mg radioactivities the experimental half-lives are in good agreement with our calculations (in fact the predicted lifetimes for cluster radioactivities have been used as a guide to perform the experiment). Spontaneous cold fission of this nucleus was not measured, but the experiments on induced cold fission show that ^{100}Zr is indeed the most probable light fragment. It is interesting to note that in a plot of $\log T$ versus $(N - Z)/A$ with N = constant,[142] α-decay, cluster radioactivities, and spontaneous fission behave similarly.

Our predictions for the measured cluster decays during the period 1984–1994 are presented in Table 4. Many other interesting cases are tabulated elsewhere.[123] The particularly strong shell effects of the 126 neutron closure and of the 82 magic number of protons combine in this nucleus with a very favorable position from the liquid drop model point of view, namely almost the bottom of the valley of beta stability. None of the other doubly magic neighbors, ^{100}Sn and ^{132}Sn, are as privileged as ^{208}Pb. They are both far from stability, and the shell effects of $Z = 50$, $N = 50$, 82 magic numbers are weaker. Consequently the mass distributions in cold fission are very broad compared to cluster radioactivity and α-decay. Nevertheless, the cold and bimodal fission of heavy nuclei[15] prove that they are effective too. The ^{12}C emission from ^{114}Ba is discussed below.

We have reviewed elsewhere[15] the cold fission mechanism and our estimations of cold fission half-lives.[143]

V. ODD-MASS PARENT NUCLEI AND THE FINE STRUCTURE

Cluster transitions leading to excited states of the final fragments were considered for the first time by Martin Greiner and Werner Scheid.[26] When the fine structure of the ^{14}C radioactivity of ^{223}Ra was discovered,[24] it was shown that the transition toward the first excited state of the daughter nucleus is stronger than that to the ground state. In other words, as in the spontaneous fission of odd-mass nuclei, or in the fine structure of α-decay, one has a hindered and a favored transition, respectively. Spontaneous fission rates of odd-A nuclei are slower than those of e–e nuclei, indicating a higher fission barrier due to shell effects.

In both kinds of theories the physical explanation relies on the single-particle spectra of neutrons and/or protons. If the uncoupled nucleon is left in the same state both in the parent and the heavy fragment, the transition is favored. Otherwise the difference in structure leads to a large hindrance

$$H = T^{exp}/T_{e-e} \tag{58}$$

where T^{exp} is the measured partial half-life for a given transition, and T_{e-e} is the corresponding quantity for a hypothetical even–even equivalent, estimated either from systematics ($\log T$ versus $Q^{-1/2}$, for example)[25] or from a model. A transition is favored if $H \simeq 1$, and it is hindered if $H \gg 1$. Up to now theoretical considerations[144, 145] have only been partly successful in explaining the experimental result. New measurements performed by Hourany *et al.* are in favor of the interpretation given by Sheline and Ragnarsson.

Particularly interesting are the transitions in which at least one of the three partners possesses an odd number of nucleons, because of the spectroscopic information one can get. Unlike α-decay, where the initial and final states of the parent and daughter are not so far one from each other, in cluster radioactivities of odd-mass nuclides, one has a unique possibility to study a transition from a well-deformed parent nucleus with complex configuration mixing, to a spherical nucleus with a pure shell model wave function. In this way, one can get *direct spectroscopic information* on the spherical components of deformed states. In 1986, after properly taking into account the even–odd effect, we improved our earlier predictions. One can learn about some possible hindered transitions by analyzing the still existing discrepancies.

The sequence of spins and parities in the single-particle spectrum depends strongly on the values of the spin–orbit and l^2 correction coefficients κ and μ. Also, the presence of reflection asymmetry (the non-zero odd-multipole deformations $\varepsilon_3, \varepsilon_5)^2$ gives rise to parity mixing, like that experimentally observed in the region of Ra isotopes.

Within ASAFM we can easily simulate the even–even assumption. We can also take into account the angular momentum, l, determined from the spin and parity conservation. The penetrability P_s depends on l. The model dependent "experimental" spectroscopic factor can be defined as

$$S_{exp} = \lambda_{exp}/(\nu P_s) = S_{e-e}/H \tag{59}$$

In the following we shall present the results concerning cluster emission from the ground state of some odd mass number parent nuclei toward the ground state and first excited states of the corresponding daughters.[146] As can be seen from Table 5, the spin and parity of the (initial) ground state of ^{223}Ra is $J_i^{\pi_i} = 3/2^+$. In the final states of ^{209}Pb one has $J_f^{\pi_f} = 9/2^+$ (gs), $11/2^+$ (first excited state, at 0.779 MeV), and $15/2^-$ (second excited state, at 1.423 MeV). Consequently the allowed angular momenta are 4 and 6 units of $\hbar$ in the first two cases, and 7 and 9 in the last one.

The favored transitions are explained within the present version of ASAFM (one should allow an accuracy of ± 1 order of magnitude in any theoretical calculations), but the hindered one needs a larger action integral, K. Such an increase can be obtained with a larger potential barrier, due to the so-called *"specialization energy"*,[147] in a similar way to what has been done for the spontaneous fission of odd-mass nuclei.[148, 149] This energy arises from the

Table 5 Hindrance factors, $H = T^{exp}/T_{e-e}$ and $\log T_{e-e}$ for ^{14}C transitions from ^{223}Ra (gs $3/2^+$) to various excited states of ^{209}Pb calculated from the systematics (S) $\log T = f(Q^{-1/2})$ of three even–even neighbours (linear least squares) and within the analytical superasymmetric fission model (ASA). T is given in seconds

E^* (keV)	J_f^π	Q (MeV)	$\log T^{exp}$	l	$\log T_{e-e}^S$	$\log T_{e-e}^{ASA}$	H^S	H^{ASA}
0	$9/2^+$	31.852	15.79	0	13.21	13.36	380	269
				4		13.60		155
				6		13.86		85
778	$11/2^+$	31.073	15.13	0	14.79	14.90	2.2	1.7
				4		15.14		1.0
				6		15.40		0.5
1422	$15/2^-$	30.429	>18.08	0	16.13	16.22	>89	>72
				7		16.89		>15
				9		17.29		>6
1567	$5/2^+$	30.285	>18.09	0	16.44	16.52	>45	>37
				2		16.59		>32
				4		16.76		>21

conservation of spin and parity of the odd particle during the fission. With an increase of deformation (distance between centers within the single-particle two-center shell model) the odd nucleon may not be transferred on a lower energy at a level crossing, if in this way it cannot conserve the spin and parity. The corresponding barrier becomes higher and wider, compared to that of the even–even neighbor. The examples given in refs. 148, 149 show how the barrier height and width become larger due to this effect. If we introduce the correction, this would produce a larger area of the barrier $E(R) - E_{corr}$ integrated over the range of separation distances from the first to the second turning point, (R_a, R_b). We need a difference of about 15.76 MeV fm, in order to get a hindrance factor of 400.

We have performed similar calculations for other odd-mass parent nuclei.[146] The influence of the angular momentum is not very large, as can be seen by comparing the results obtained for $l = 0$ and for the allowed values of l. The T_{e-e}^S values obtained from the systematics of even-even neighbors (when the measurements are available, like in Table 5) are not essentially different from the calculated ones T_{e-e}^{ASA} within ASAFM.

We obtained a large hindrance factor ($H = 44$) for the ^{24}Ne decay of ^{233}U (gs $5/2^+$) to the gs ($9/2^+$) of ^{209}Pb; the best fit with $T^{exp} = 10^{24.84}$ s is obtained for the transition to the first excited state ($11/2^+$ at 778 keV) with $l = 4$, $T^{ASAFM} = 10^{24.89}$. Other hindered transitions could be: ^{23}F decay of ^{231}Pa (gs $3/2^-$) to the 0^+ gs of ^{208}Pb ($H = 12$); ^{14}C decay of ^{221}Fr (gs $5/2^-$) to the $1/2^+$ gs of ^{207}Tl ($H = 8.5$); ^{14}C decay of ^{221}Ra to the $1/2^-$ gs of ^{207}Pb ($H = 9$), and ^{24}Ne decay of ^{231}Pa to the gs of ^{207}Tl. The ^{14}C transition from ^{225}Ac (gs ($3/2^-$)) to the ($9/2^-$) gs of ^{211}Bi seems not to be hindered.

VI. CLUSTER DECAYS TO DAUGHTERS AROUND ^{100}SN AND NUCLEAR PROPERTIES

In 1984 we predicted the ^{12}C radioactivity of ^{114}Ba (see the references in refs. 16, 150), the ^{16}O radioactivity of neutron-deficient Ce isotopes, and a whole island of cluster emitters ending with daughters in the neighborhood of ^{100}Sn. At that time we stressed the importance of the magic numbers of protons and neutrons of the daughter, and the proton-richness of the parent nuclei. Also we mentioned the difficulty of performing reliable calculations owing to the lack of mass measurements in this region of nuclei very far from the line of beta stability. In 1988 we carried out new calculations with different mass estimations[151] and presented the results in comprehensive tables.[123]

In his review article[20] P. B. Price mentioned very clearly, among a few other experiments worth doing, the ^{12}C emission from ^{114}Ba. We have presented the new island of cluster emission, stressing the possible unusually high branching ratio with respect to α-decay, at many International Conferences (St. Petersburg 1989, Predeal Summer Schools in 1990 and 1992 (NATO Advanced Study Institute), Louvaine-la-Neuve, 1991, Bernkastel-Kues, 1992, Mainz, 1993). The possible detection of cluster radioactivity in this region of nuclei[152] has been invoked as one of the interesting experiments justifying the funding of a new accelerator.[153]

In spite of the good agreement with our predictions of the half-lives measured since 1984 in the transradium region (daughters around ^{208}Pb), and of the agreement of the calculations of other models (for example, the best representative BFW[101] of the preformed cluster models) with our calculations in this new region of nuclei, the preliminary results of the recently performed measurements[39, 40] seem to indicate an emission rate higher than expected, by assuming certain mass values estimated with different theoretical models.

This parent nucleus is very far from stability – almost on the proton drip line. It was not produced until very recently, and hence none of its properties, including the decay modes (β^+, α, etc.), has been experimentally determined. In Darmstadt[39] it was produced by a (^{58}Ni, ^{58}Ni) reaction followed by mass separation. The study of its decay properties is in progress. A deeper discussion of the physics we can learn from the experimental results will be possible when the masses, the decay energies, and the partial and total half-lives are determined. At present, we can investigate some of the nuclear properties able to accommodate an increased emission probability.

We have performed a large number of systematic calculations (25685 input masses times about 200 possible emitted clusters). From the results obtained in this way, we have usually selected those which have $T < 10^{25} y$ and a branching ratio relative to α-decay $b = T_\alpha / T > 10^{-18}$.

If no restriction is imposed on the half-lives T and branching ratios b, the region of cluster emitters (see Fig. 2) is broader than that of α-emitters.[154, 152, 150] As a general trend, light clusters are preferentially emitted from neutron-deficient nuclei and heavy ones from neutron-rich parents (see

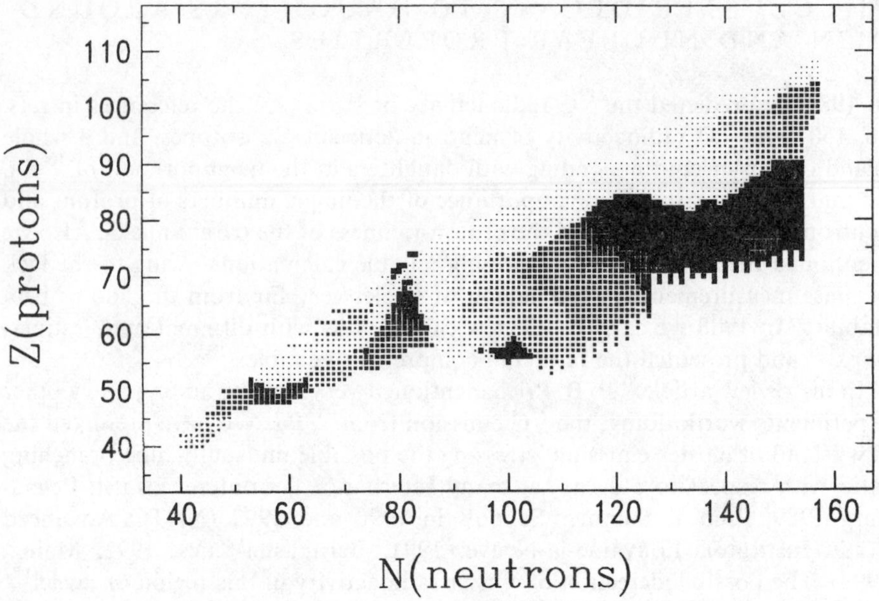

Fig. 13. Atomic numbers of the most probable cluster emitted from parent nuclides with Z protons and N neutrons. Light and heavy points correspond to low and high values, respectively, in the range 4–28 for Z_e.

Fig. 13). In Fig. 14 we have selected the regions of parent nuclei from which a cluster with a given atomic number (Be, C, O, Ne, Mg, and Si) is most probably spontaneously emitted. The cluster kinetic energy per nucleon, $E_k = QA_d/(AA_e)$, increases toward heavier proton-rich parents.

The half-life is more sensitive to the shell closure. Short half-lives appear not only in three regions of the proton-rich nuclei, due to the daughter neutron magic numbers 50, 82, and 126, where α-decay is in competition, but also for α-stable neutron-rich parents emitting neutron-rich clusters around ${}^{78}_{28}Ni_{50}$. The intensified regions at the top in the upper part of Fig. 15 are mainly due to the shell effects of the heavy fragment, while the island of cluster emitters lying in the neutron-rich α-stable part of the nuclear chart appears first of all as a consequence of the light fragment shell structure and, to a lesser extent, of the proton magic number of the daughter $Z_d = 50$.

In practice, with the experimental techniques presently available, the phenomena of cluster decays can only be detected when T is low enough and b is sufficiently high. By selecting $T < 10^{30}$ s, and $b > 10^{-18}$, the area of cluster emitter regions becomes considerably smaller (see the lower part of Fig. 15). Besides the main region of transradium parents, where experiments have been done, another one with $Z = 56$–64, $N = 58$–70 looks very promising. Much higher branching ratios than the highest (10^{-9}) measured so far, are expected to be found in this island. Its position is very close to the proton drip line.

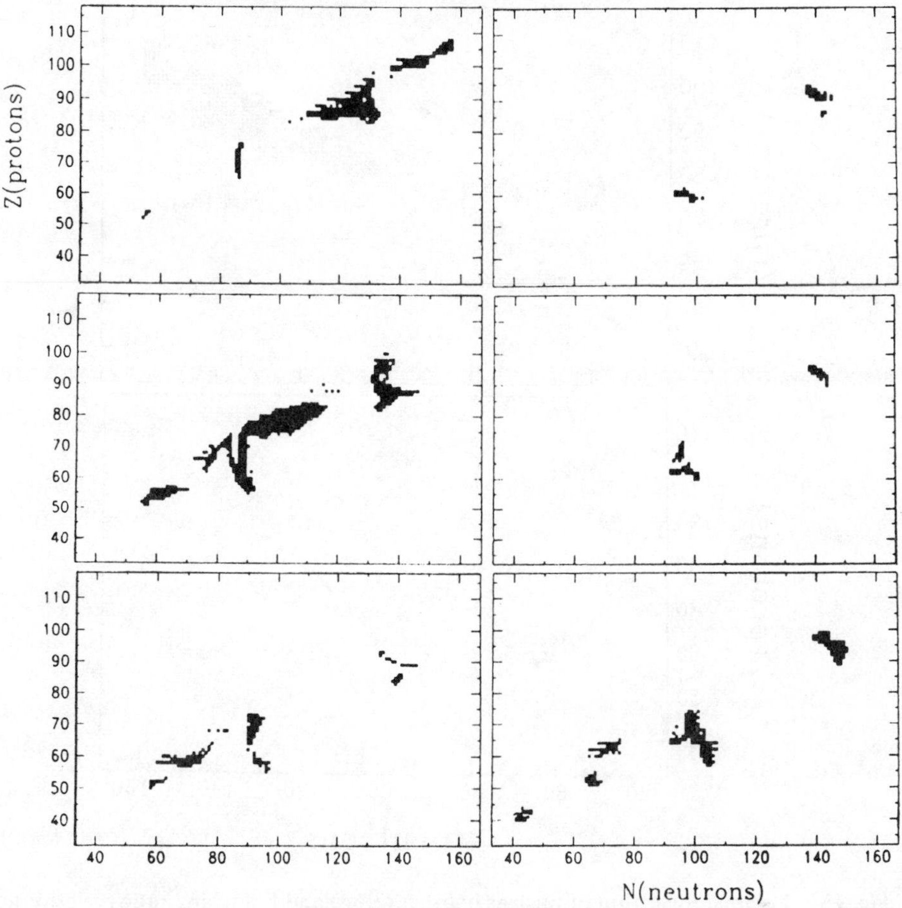

Fig. 14. The regions of the nuclear chart in which the most probable cluster emitted is Be (top left), C (middle left), O (bottom left), Ne (top right), Mg (middle right), and Si (bottom right).

Despite the large uncertainty in the estimated mass values of nuclei far from stability, we believe that there is a real chance to detect the ^{12}C, ^{16}O, ^{28}Si radioactivities (and maybe others) of some neutron-deficient nuclides with $Z = 56$–64 and $N = 58$–70. For example, both the ^{12}C radioactivity of ^{114}Ba and the ^{16}O radioactivity of ^{118}Ce from parent nuclei at the edge of the proton drip line lead to the same daughter ^{102}Sn. Even further from stability ^{112}Ba and ^{116}Ce, which would have a higher decay rate (the daughter would be ^{100}Sn), are unstable with respect to 1p and 2p emissions.

In a region of α-stable neutron-rich nuclei, the highest frequency of appearance of doubly magic ^{78}Ni and of ^{76}Fe$_{50}$ as the most probable emitted cluster is certainly a consequence of the shell and pairing effects acting around $N_e = 50$

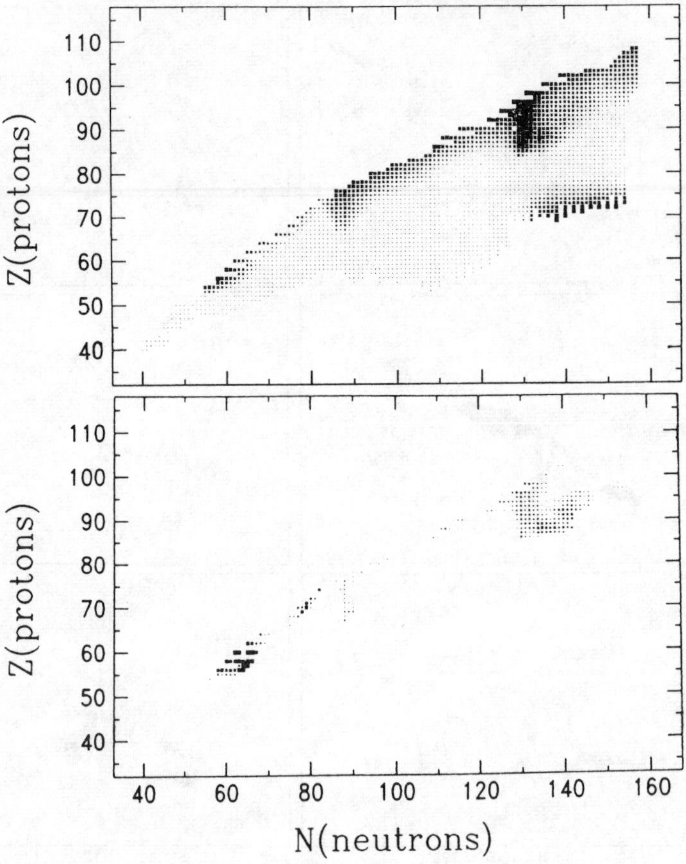

Fig. 15. Decimal logarithm of nuclear half-lives (top) and branching ratios relative to α-decay (bottom) for cluster radioactivities shown in Fig. 13. Light and heavy points correspond to long $\log T$ and low $\log b$, and short $\log T$ and high $\log b$, respectively. In the south-eastern island plotted above the parent nuclei are α-stable; hence $b \to \infty$. The shortest half-life is about 10^9 s and the selected range of branching ratio is 1–10^{-18}.

and $Z_e = 28$ magic neutron and proton numbers of the light fragment, and to a lesser extent of the magicity $Z_d = 50$ of the daughter number of protons. A comprehensive table has been published elsewhere.[123]

If we take into account the estimated[155] β-decay half-life of 0.3–0.5 s, the branching ratio with respect to this decay mode could be in the range 10^{-8}–10^{-12} s, which is measurable in an experiment on-line. There are difficulties connected with the high background produced by neutrons. In the experiment in progress at GSI Darmstadt,[39] one can considerably improve the experimental conditions by transporting the reaction products far from the region with a strong neutron background. The reaction cross-section, β-, α-, and C decays are planned to be determined. Radioactive beams could alternatively be used to produce nuclei far from stability which are cluster emitters. From the measured kinetic energy of the emitted cluster, and of the α- and β-particles we can get improved values of the implied masses, which in turn could be used to obtain more reliable estimates of half-lives and branching ratios.

We investigate the influence of nuclear masses, radii, and interaction potentials on the ^{12}C radioactivity of ^{114}Ba – the best representative of the new island of cluster emitters leading to daughter nuclei around the doubly magic ^{100}Sn. Three different models are considered: one derived by Blendowske, Fliessbach, and Walliser (BFW) from the many-body theory of α-decay, as well as our analytical (ASAF) and numerical (NuSAF) superasymmetric fission models. A Q-value larger by 1 MeV or an ASAF potential barrier reduced by 3% produce a half-life shorter by two orders of magnitude. A similar effect can be obtained within the BFW and NuSAF models by a decrease in the action integral of less than 10% and 5%, respectively. By increasing the radius constant within the ASAF or BFW models by 10% the half-life becomes shorter by three orders of magnitude.

By assuming that the same optimum value $K/K_0 = 0.9$ will also be obtained in the new island of cluster emitters the half-life will be shorter by two orders of magnitude than that estimated within the ASAF or BFW models. A similar reduction has been obtained by Buck *et al.*[156]

Our ASAF model is based on the liquid drop model developed in 1966–1967 by Myers and Swiatecki,[157] which is known to overestimate the fission barrier height E_b^0. This was the reason why we had introduced a correction energy E_{corr}, also playing the role of a phenomenological shell correction (in the spirit of Strutinsky's method). Up to now we have only considered a dependence of this quantity on the mass number of the emitted cluster and the Q-value. It seems that this quantity should also depend on the daughter nucleus, which would be reasonable in view of the important contribution of the Coulomb forces to the height of the barrier.

A higher value of E_{corr} produces a lower barrier height and an increased barrier penetrability. For example, we have initially at $Q = 20.75$ MeV, with a lowering of the potential barrier by 5.5% ($E_{corr} = 1.66$ MeV), a half-life of $T = 10^{6.57}$ s. By increasing E_{corr} by a factor of two (meaning that the barrier height is decreased by 11%), the half-life becomes shorter by about four orders of magnitude, $T = 10^{2.64}$ s.

An indication of the need for a lower potential barrier in the new region of nuclei can be deduced from Fig. 16. In this figure we compare the rms deviations from the experimental results

$$rms = \left\{ \left[\sum_{i=1}^{n} (\log T_i - \log T_{exp})^2 \right] \Big/ (n-1) \right\}^{1/2} \qquad (60)$$

in three groups of even–even α-emitters: three isotopes of Te, 13 isotopes of Po, and 124 parents (including Te and Po isotopes) with atomic numbers from $Z = 52$ to $Z = 102$ and mass numbers $A = 106$–256. The minimum value is

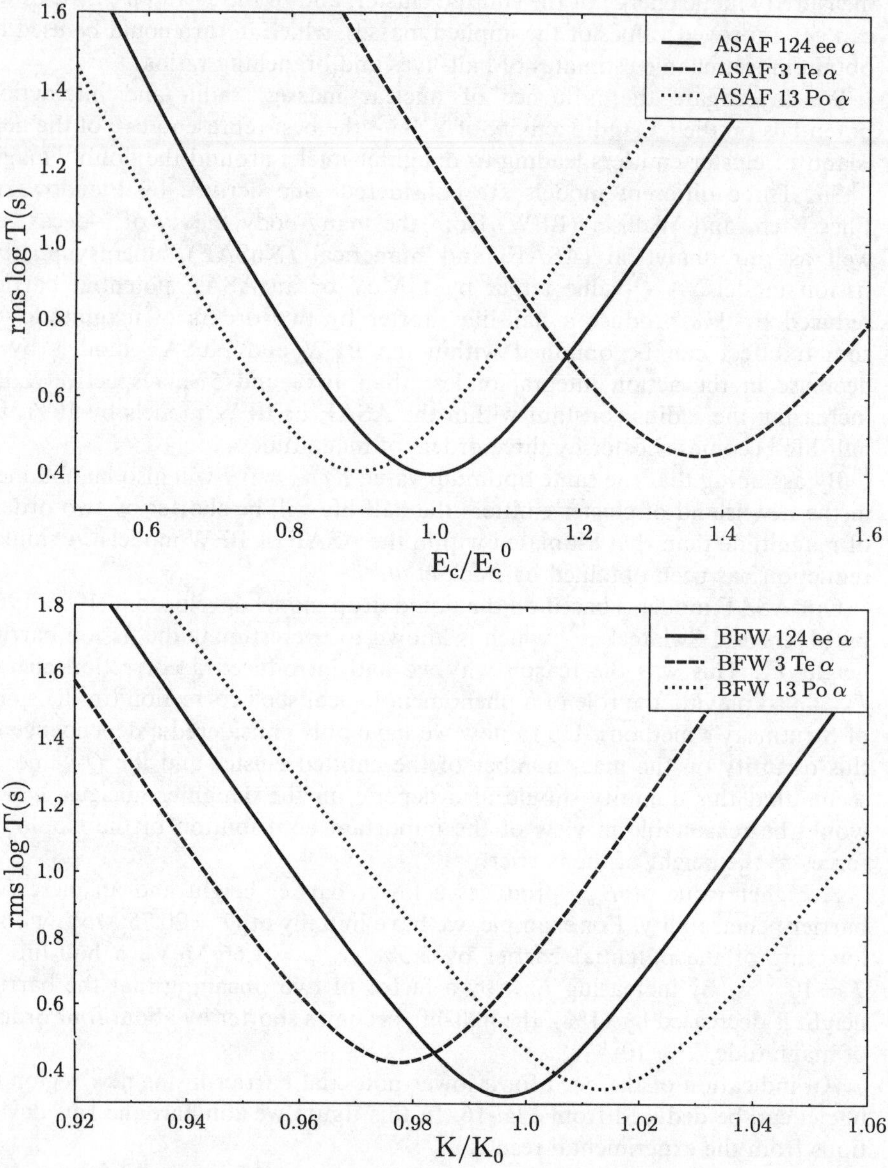

Fig. 16. Deviations of half-lives for α-decay calculated within ASAF (upper part) and BFW (lower part) from the experimental ones, versus the correction energy E_{corr}/E^0_{corr} and the action integral K/K_0, respectively. Three groups of even–even parent nuclei are considered: three Te isotopes, 13 Po isotopes, and 124 various nuclei with atomic numbers in the range 52 to 102.

obtained for $E_{corr}/E^0_{corr} = 1$ (or $f = 1$ within BFW) in the large collectivity of data, but at about $E_{corr}/E^0_{corr} = 1.4$ (or $f = 0.98$ within BFW) for the lightest parents leading to daughter nuclei not far from ^{100}Sn. A similar trend is not observed in the case of NuSAF, which gives almost the same optimum value $f = 0.94$ for all groups.

As we have discussed above, the larger f for NuSAF compared to BFW comes from the higher potential barrier of Y+E.[121] Both potentials Y+E and that used by BFW[158] have been successful for heavy-ion elastic scattering. A typical example may be seen in Fig. 17 for ^{12}C emission from ^{114}Ba. The touching point radii are similar, 8.04 fm in Y+E and 7.95 fm in BFW, but $E(R_t) = -50.1$ MeV and $E > Q$ beginning with $R = 8.6$ fm for BFW. In fact, from a similar figure[108] in which many other potentials used to study cluster emission have been compared, BFW presented the lowest and thinnest barrier.

When we get the experimental values of Q and T we can fix E_{corr}/E^0_{corr} or f (and consequently the potential barrier) into this new region of parent nuclei (see Fig. 18). The relative emission rate of various emitted clusters will not be essentially affected by this tuning; ^{12}C will remain the most probable emitted cluster from ^{114}Ba, and the Ba isotope with $A = 114$ will remain the parent with the highest emission rate (except for ^{112}Ba which is expected to decay by two-proton emission).

A lower potential barrier could also be an alternative explanation of the unexplained high cross-section of the low-energy (under-barrier) fusion. On

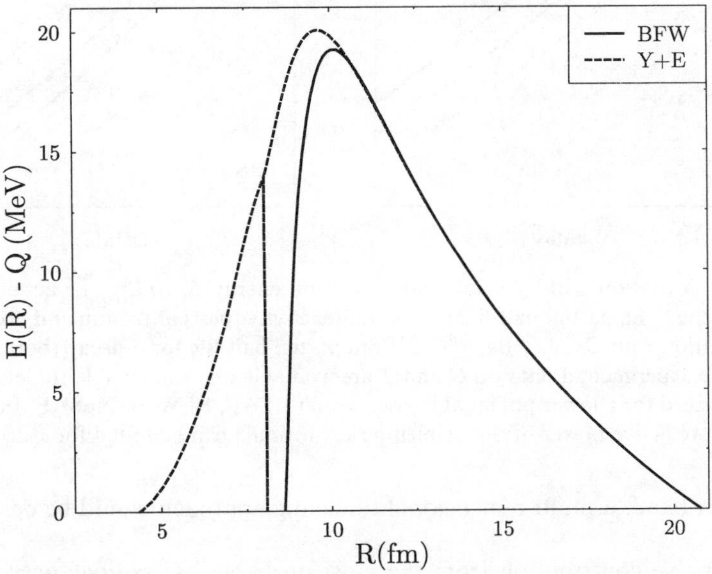

Fig. 17. Potential barrier $E(R) - Q$ for ^{12}C radioactivity of ^{114}Ba (assuming that $Q = 20.79$ MeV) within BFW and NuSAF models. The vertical line corresponds to the touching point separation distance.

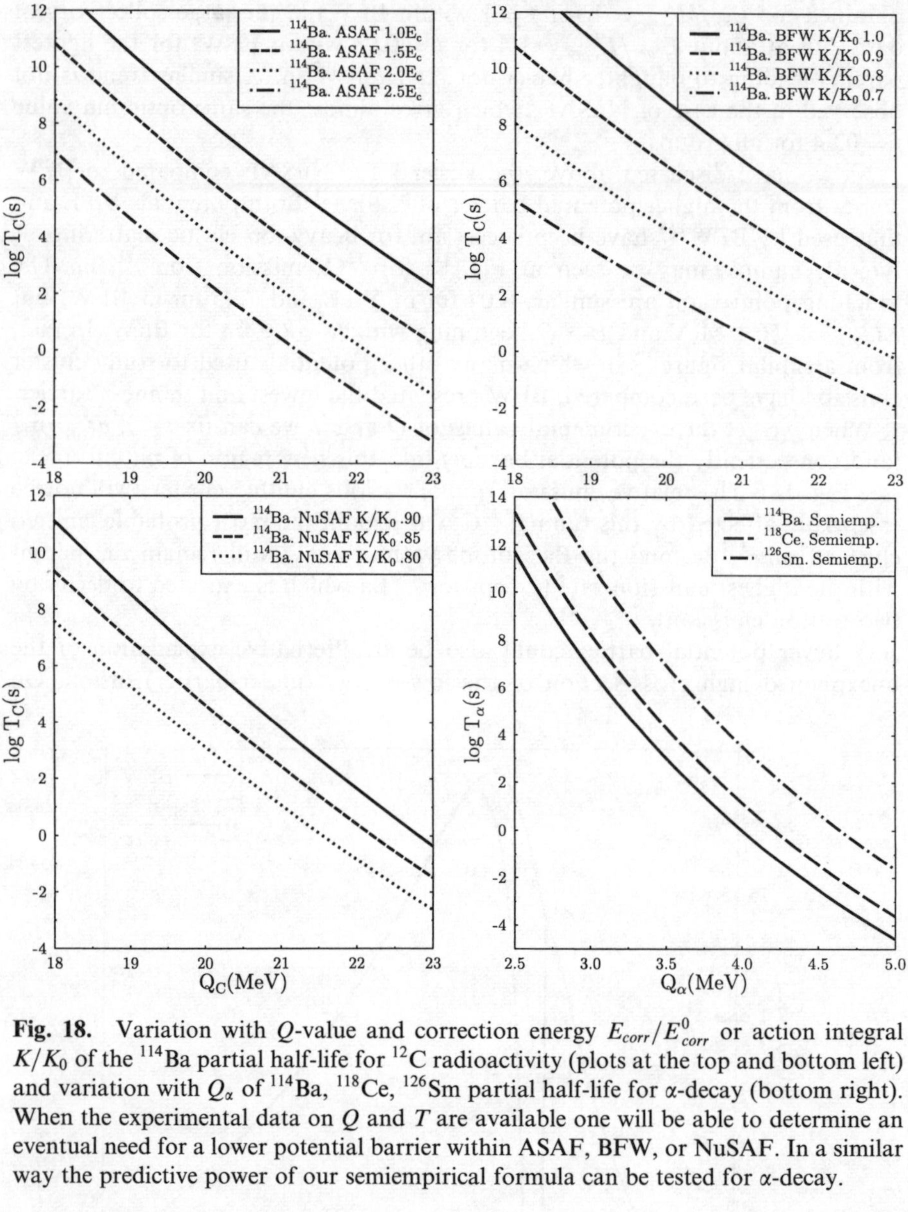

Fig. 18. Variation with Q-value and correction energy E_{corr}/E^0_{corr} or action integral K/K_0 of the ^{114}Ba partial half-life for ^{12}C radioactivity (plots at the top and bottom left) and variation with Q_α of ^{114}Ba, ^{118}Ce, ^{126}Sm partial half-life for α-decay (bottom right). When the experimental data on Q and T are available one will be able to determine an eventual need for a lower potential barrier within ASAF, BFW, or NuSAF. In a similar way the predictive power of our semiempirical formula can be tested for α-decay.

the other hand, a multi-dimensional tunneling approach would be desirable in our case.

A possible contribution from the most probable ^{12}C radioactivity of ^{112}Ba (producing the daughter ^{100}Sn) which would have an increased Q-value by about 2 MeV, and hence a shorter half-life by about four orders of magnitude), should be ruled out, at least for two reasons: mass (and on-line chemical

Table 6 Q-values for ^{12}C decay of ^{114}Ba according to different mass estimations

Mass code	11	9	8	12	3
Q(MeV)	20.75	20.20	19.97	18.34	18.16

separation) in the experiment,[39] and instability toward 2p emission according to all mass estimations.

Apart from the emitted particles, the nuclei implied in the β-, α-, and ^{12}C decays are ^{102}Sn, ^{110}Xe, ^{114}Cs, and ^{114}Ba. From these only Cs, Xe, and Sn are mentioned in the new mass table[159] as being measured (Cs) or having the mass determined from systematics (Xe, Sn). By taking only a few (more optimistic or with good reproduction of experimental data) of the (more than 15 available) mass values, both for the parent and daughter nuclei from the same table, one can obtain the Q-values given in Table 6 for the ^{12}C emission from ^{114}Ba, where the mass codes conventionally adopted by us are: 11 – Pearson *et al.*;[160] 9 – Möller and Nix;[161] 8 – Möller *et al.*;[162] 12 – Liran and Zeldes,[163] and 3 – Jänecke and Masson.[36]

For α-decay we have estimated with our semiempirical formula the half-lives, given in Fig. 18, bottom right. In this figure we show the results for other Q-values and parent nuclei too. Of course, similar figures can be drawn for the other ^{12}C, ^{16}O, and ^{28}Si emissions we mentioned long ago.

An increased effective radius (for example, due to the nuclear deformation) will also produce a similar effect of lowering the potential barrier. The best fit of Te isotope data is obtained if the nuclear radii are increased by 10% within ASAF (equivalent to $E_{corr}/E^0_{corr} = 1.4$) or by 5% within BFW (or $f = 0.98$).

The sensitivity of the half-life for the ^{12}C decay of ^{114}Ba to the variation of the nuclear radii and Q-value within the BFW and ASAF models can be estimated from Fig. 19. Around $Q = 20$ MeV a nuclear radius constant larger by 10% lowers the fission barrier and leads to a half-life shorter by three orders of magnitude.

In conclusion the measurement of ^{114}Ba partial half-lives and Q-values against ^{12}C and α-decay will allow us to determine the nuclear properties (masses, radii, interaction potentials) in this region of nuclei far from stability, by comparison with the corresponding calculations. We have analyzed the sensitivity of two fission models (ASAF, NuSAF) and of one model (BFW) derived from the traditional many-body theory of α-decay to the variation of these quantities. For light emitted clusters (He, C, O, etc.) the internal part of the barrier is much thinner than the external one. Consequently the main variation of the half-life is practically due to the (external) barrier penetrability (and to a lesser extent to the preformation probability).

Around $Q = 20$ MeV a Q-value larger by 1 MeV or an ASAF correction energy increased by 50% (reduction by 3% of the barrier height) produce a half-life shorter by two orders of magnitude. A similar effect can be obtained

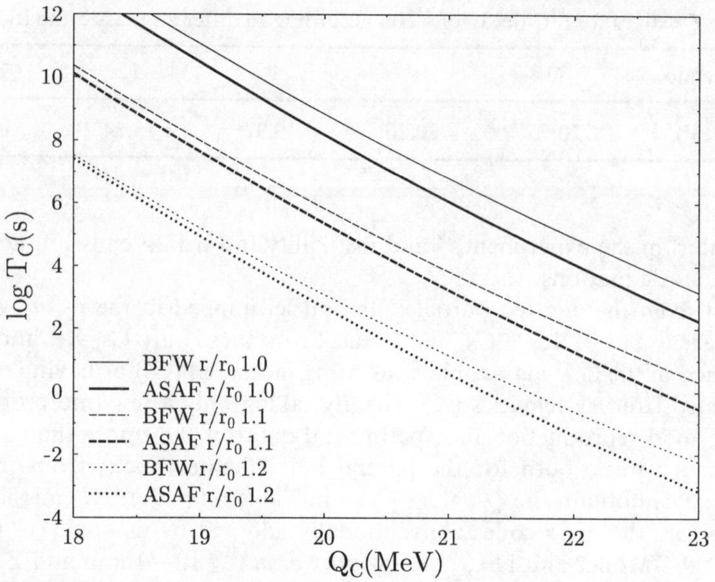

Fig. 19. The sensitivity of the half-life for ^{12}C decay of ^{114}Ba to the variation of nuclear radii and Q-value within the BFW and ASAF models.

within the BFW and NuSAF models by a decrease of the action integral (which may be due to a lower potential barrier) by less than 10% and 5%, respectively. By increasing the radius constant (which could simulate a deformation, for example) within the ASAF or BFW models by 10% the half-life becomes shorter by three orders of magnitude.

Both the ASAF and BFW models show similar results for α-decay and cluster radioactivities in various regions of the nuclear chart. The Y+E potential used by NuSAF leads to somewhat different behavior. If we adopt the same optimum value $K/K_0 = 0.9$ in the new island of cluster emitters, the half-life is two orders of magnitude shorter than that estimated within ASAF or BFW, pointing out the importance of the chosen nuclear potential.

We are sure that this interesting problem of nuclear physics we proposed in 1984 will continue to be a subject of future experimental and theoretical investigations.

ACKNOWLEDGEMENTS

This work was supported by the Bundesministerium für Forschung und Technologie, the Deutsche Forschungsgemeinschaft, the KfK Karlsruhe, and the Institute of Atomic Physics. One of us (DNP) has received a donation of computer equipment from the Soros Foundation for an Open Society.

REFERENCES

1. **Eisenberg, J. M. and Greiner, W.** *Nuclear Theory*, Vol. I: Nuclear Models. North-Holland, Amsterdam, 3rd edition (1987).
2. **Nazarewicz, W. and Ragnarsson, I.** Chapter in the present book.
3. *Nuclear Decay Modes*, Poenaru, D. N. and Greiner, W., editors, IOP, Bristol, to be published.
4. *Treatise on Heavy Ion Science, Vol. 8*, Bromley, D. A., editor, Plenum Press, New York (1989).
5. *Particle Emission from Nuclei, Vols. I–III,* Poenaru, D. N. and Ivaşcu, M., editors, CRC Press, Boca Raton, Florida (1989).
6. **Strutinsky, V. M.** *Nucl. Phys. A* **95**, 420 (1967).
7. **Brack, M., Damgaard, J., Jensen, A., Pauli, H. C., Strutinsky, V. M., and Wong, G. Y.** *Rev. Mod. Phys.* **44**, 320 (1972).
8. **Maruhn, J. A. and Greiner, W.** *Z. Phys.* **251**, 431 (1972).
9. **Maruhn, J. A., Greiner, W., and Scheid, W.** In *Heavy Ion Collisions, Bock, R., editor, Vol. 2, 399. North-Holland, Amsterdam (1980).*
10. **Lustig, H. J., Maruhn, J. A., and Greiner, W.** *J. Phys. G: Nucl. Phys.* **6**, L.25 (1980).
11. **Sandulescu, A., Gupta, R. K., Scheid, W., and Greiner, W.** *Phys. Lett. B* **60**, 225–228 (1976).
12. **Armbruster, P.** *Annu. Rev. Nucl. Part. Sci.* **35**, 135–194 (1985).
13. **Münzenberg, G.** *Rep. Prog. Phys.* **51**, 57–104 (1988).
14. **Săndulescu, A., Poenaru, D. N., and Greiner, W.** *Sov. J. Part. Nucl.* **11**, 528–541 (1980).
15. **Poenaru, D. N., Ivaşcu, M., and Greiner, W.** Chapter 7, pp. 203–235, in ref. 5., Vol. III.
16. **Greiner, W., Ivaşcu, M., Poenaru, D. N., and Săndulescu, A.** In ref. 4, p. 641–722.
17. **Rose, H. J. and Jones, G. A.** *Nature* **307**, 245–247 (1984).
18. **Poenaru, D. N., Ivaşcu, M., Săndulescu, A., and Greiner, W.** *J. Phys. G: Nucl. Phys.* **10**, L183–L189 (1984); *Phys. Rev. C* **32**, 572–581.
19. **Poenaru, D. N. and Ivaşcu, M.** In *Proc. International Summer School on Atomic and Nuclear Heavy-Ion Interactions, Poiana Brasov*, 277–331, Central Institute of Physics, Bucharest (1984).
20. **Price, P. B.** *Annu. Rev. Nucl. Part. Sci.* **39**, 19–42 (1989).
21. **Hourani, E., Hussonnois, M., and Poenaru, D. N.** *Ann. Phys. (Paris)* **14**, 311–345 (1989).
22. **Tretyakova, S. P., Ogloblin, A. A., Mikheev, V. L., and Zamyatnin, Y. S.** In *Clustering Phenomena in Atoms and Nuclei*, Brenner, M., Lönnroth, T., and Malik, F. B., editors, 283–292. Springer, Heidelberg (1992).
23. **Bonetti, R. and Guglielmetti, A.** Chapter in ref. 3, to be published.
24. **Brillard, L., Elayi, A. G., Hourani, E., Hussonnois, M., Le Du, J. F., Rosier, L. H., and Stab, L.** *C. R. Acad. Sci. Paris* **309**, 1105–1110 (1989).
25. **Hourani, E.** Chapter in ref. 3, to be published.
26. **Greiner, M. and Scheid, W.** *J. Phys. G: Nucl. Phys.* **12**, L229–L234 (1986).
27. **Gönnenwein, F.** In *The Nuclear Fission Process,* Wagemans, C., editor, 387. CRC Press, Boca Raton, Florida (1991).
28. **Hulet, E. K.** In *Proc. R. A. Welch Foundation Conference on Chemical Research: Fifty years with transuranium elements, Houston, 1990*, 279–310 (1990).

29. **Poenaru, D. N., Maruhn, J. A., Greiner, W., Ivascu, M., Mazilu, D., and Gherghescu, R.** *Z. Phys. A* **328**, 309–314 (1987).
30. **Hamilton, J. H.** In ref. 4, p. 3.
31. **Grotz, K. and Klapdor, H. V.** *The weak interaction in nuclear, particle and astrophysics*, Adam Hilger, Bristol (1990).
32. **Berlovich, E. Y. and Novikov, Y. N.** In *Modern Methods of Nuclear Spectroscopy*, 107–208. Nauka, Leningrad (1988).
33. **Hamilton, J. H., Hansen, P. G., and Zganjar, E. F.** *Rep. Prog. Phys.* **48**, 631–708 (1985).
34. **Détraz, C. and Vieira, D. J.** *Annu. Rev. Nucl. Part. Sci.* **39**, 407–465 (1989).
35. **Roeckl, E.** *Rep. Prog. Phys.* **55**, 1661 (1992).
36. **Jänecke, J. and Masson, P. J.** *At. Data Nucl. Data Tables* **39**, 265–271 (1988).
37. **Wapstra, A. H., Audi, G., and Hoekstra, R.** *At. Data Nucl. Data Tables* **39**, 281–287 (1988).
38. **Schneider, R.** *et al. Z. Phys. A* **348**, 241–242 (1994). **Lewitowicz, M.** *et al.* Phys. Lett. B **333**, 20–24 (1994).
39. **Price, P. B., Roeckl, E., Bonetti, R., and Guglielmetti, A.**, private communication.
40. **Oganessian, Y. T., Mikheev, V. L., and Tretyakova, S. P.**, private communication.
41. **Andreyev, A. N.** *et al. Z. Phys. A* **347**, 225–226 (1994).
42. **Guillemaud-Mueller, D.**et al. Phys. Rev. C **41**, 937–941 (1990).
43. **Schmidt, K. H.** *et al. Phys. Lett. B* **325**, 313–316 (1994).
44. **Äystö, J.** *et al. Phys. Rev. Lett.* **69**, 1167–1170 (1992).
45. **Tanihata, I.** *Nucl. Phys. A* **553**, 361c–372c (1993).
46. **Hansen, P. G. and Jonson, B.** *Europhys. Lett.* **4**, 409–414 (1987).
47. **Hansen, P. G.** *Nucl. Phys. A* **553**, 89c–106c (1993).
48. **Richter, A.** *Nucl. Phys. A* **553**, 417c–462c (1993).
49. **Armbruster, P.** *J. Phys. Soc. Japan, Suppl.* **58**, 232–248 (1989).
50. **Sobiczewski, A.** In *Nuclei Far From Stability/Atomic Masses and Fundamental Constants 1992, Inst. Phys. Conf. Ser. No 132*, Neugart, R. and Wöhr, A., editors, 403–412, IOP, Bristol (1993).
51. **Hofmann, S.** *et al. Z. Phys. A* **350**, 277–282 (1995).
52. **Oganessian, Y. T.** *Nucl. Phys. A* **239**, 353 (1975).
53. **Möller, P. and Nix, J. R.** *Nucl. Phys. A* **549**, 84–102 (1992); *J. Phys. G* **20**, 1681–1747 (1994).
54. **Stöcker, H. and Greiner, W.** *Phys. Rep.* **137**, 277–392 (1986).
55. **Troltenier, D., Maruhn, J. A., Greiner, W., and Rohozinski, S. G.** *Nucl. Phys. A* **494**, 235–243 (1989).
56. **Brack, M.** *Rev. Mod. Phys.* **65**, 677 (1993).
57. **Greiner, W.** *Z. Phys. A* **349**, 315–333 (1994).
58. **Twin, P., Nyako, B. M., Nelson, A. H., Simpson, J., Bentley, M. A., Cranmer-Gordon, H. W., Forsyth, P. D., Howe, D., Moktar, A. R., Morrison, J. D., Sharpey-Schafer, J. F., and Sletten, G.** *Phys. Rev. Lett.* **57**, 811–814 (1986).
59. **Poenaru, D. N., Ivascu, M., and Mazilu, D.** Chapter 2, Vol. III, pp. 41–61, in ref. 5.
60. **Greiner, W.** In *Clustering Phenomena in Atoms and Nuclei*, Brenner, M., Lönnroth, T., and Malik, F. B., editors, 213–234. Springer, Heidelberg (1992).
61. **Jung, M.** *et al. Nucl. Phys. A* **553**, 309c–312c (1993).
62. **Kubono, S.** *Comm. Astrophys.* **16**, 287 (1993).

63. **Sorlin, O.** *et al. Phys. Rev. C* **47**, 2941 (1993).
64. **Goldanski, V. I.** *Annu. Rev. Nucl. Sci.* **16**, 1 (1966).
65. **Hofmann, S.** Chapter in ref. 3, to be published.
66. **Page, R. D., Woods, P. J., Cunningham, R. A., Davinson, T., Davis, N. J., James, A. N., Livingston, K., Sellin, P. J., and Shotter, A. C.** *Phys. Rev. Lett.* **72**, 1798–1801 (1994).
67. **Äystö, J. and Cerny, J.** in ref. 4, p. 207.
68. **Roeckl, E.** Chapter in ref. 3, to be published.
69. **Hoffman, D. C. and Somerville, L. P.** Chapter 1, Vol. III, pp. 1–40 in ref. 5.
70. **Poenaru, D. N.** *Ann. Phys. (Paris)* **2**, 133–168 (1977).
71. **Bjørnholm, S. and Lynn, J. E.** *Rev. Mod. Phys.* **52**, 725 (1980).
72. **Metag, V., Habs, D., and Specht, H. J.** *Phys. Rep.* **65**, 1 (1980).
73. **Polikanov, S. M., Druin, V., Karnaukhov, V., Mikheev, V., Pleve, A., Skobelev, N., Subotin, V., Ter Akopian, G., and Fomichev, V.** *Sov. Phys. JETP* **15**, 1016 (1962).
74. **Wagemans, C.** Chapter 3, Vol. III, pp. 63–97, in ref. 5.
75. **Mutterer, M. and Theobald, J. P.** Chapter in ref. 3, to be published.
76. **Hardy, J. C. and Hagberg, E.** Chapter 4, Vol. III, pp. 99–131, in ref. 5.
77. **Jonson, B. and Nyman, G.** Chapter in ref. 3, to be published.
78. **Moltz, D. M. and Cerny, J.** Chapter 5, Vol. III, pp. 133–156, in ref. 5.
79. **Bazin, D.** *et al. Phys. Rev. C* **45**, 69–79 (1992).
80. **Hansen, P. G. and Jonson, B.** Chapter 6, Vol. III, pp. 157–201, in ref. 5.
81. **Rudstam, G., Aleklett, K., and Sihver, L.** *At. Data Nucl. Data Tables* **53**, 1–22 (1993).
82. **Hirsch, M., Staudt, A., and Klapdor-Kleingrothaus, H. V.** *At. Data Nucl. Data Tables* **51**, 243–271 (1992).
83. **Staudt, A. and Klapdor-Kleingrothaus, H. V.** *Nucl. Phys. A* **549**, 254–264 (1992).
84. **Poenaru, D. N., Maruhn, J. A., Greiner, W., Ivaşcu, M., Mazilu, D., and Ivaşcu, I.** *Z. Phys. A* **333**, 291–298 (1989).
85. **Depta, K., Maruhn, J. A., Wang, H. J., Săndulescu, A., Greiner, W., and Herrman, R.** *Int. J. Mod. Phys. A* **5**, 3901–3928 (1990).
86. **Poenaru, D. N., Ivascu, M., and Mazilu, D.** *Comput. Phys. Commun.* **19**, 205–214 (1980).
87. **Pik-Pichak, G. A.** *Sov. J. Nucl. Phys.* **44**, 923–929 (1987).
88. **Depta, K., Herrmann, R., Maruhn, J. A., and Greiner, W.** In *Dynamics of Collective Phenomena*, David, P., editor, 29–37. World Scientific, Singapore (1987).
89. **Poenaru, D. N., Mirea, M., Greiner, W., Căta, I., and Mazilu, D.** *Mod. Phys. Lett. A* **5**, 2101–2105 (1990).
90. **Kindo, T. and Iwamoto, A.** *Phys. Lett. B* **225**, 203–207 (1989).
91. **Klein, H.** PhD thesis, Universität Frankfurt am Main (1991).
92. **Schnabel, D. and Klein, H.**, private communication.
93. **Mirea, M., Poenaru, D. N., and Greiner, W.** *Nuovo Cimento* **105 A**, 571–580 (1992).
94. **Mirea, M., Poenaru, D. N., and Greiner, W.** *Z. Phys. A* **349**, 39–45 (1994).
95. **Shi, Y. J. and Swiatecki, W. J.** *Nucl. Phys. A* **464**, 205–222 (1987).
96. **Barranco, F., Broglia, R. A., and Bertsch, G. F.** *Phys. Rev. Lett.* **60**, 507–510 (1988).
97. **Malik, S. S. and Gupta, R. K.** *Phys. Rev. C* **39**, 1992–2000 (1989).
98. **Shanmugam, G. and Kamalaharan, B.** *Phys. Rev. C* **41**, 1184–1190 (1990).
99. **Landowne, S. and Dasso, C. H.** *Phys. Rev. C* **33**, 387–389 (1986).

100. **Iriondo, M., Jerrestam, D., and Liotta, R. J.** *Nucl. Phys. A* **454**, 252–266 (1986).
101. **Blendowske, R., Fliessbach, T., and Walliser, H.** *Nucl. Phys. A* **464**, 75–89 (1987).
102. **Rubchenia, V. A., Eismont, V. P., and Yavshits, S. G.** *Izv. Akad. Nauk SSSR* **50**, 1016–1020 (1986).
103. **Kadmenski, S. G., Kurgalin, S. D., Furman, V., and Chuvilski, Y. M.** *Sov. J. Nucl. Phys.* **51**, 32–38 (1990).
104. **Rotter, I.** *J. Phys. G: Nucl. Part. Phys.* **15**, 251–263 (1989).
105. **Ivaşcu, M. and Silişteanu, I.** *Nucl. Phys. A* **485**, 93–110 (1988).
106. **de Carvalho, H. G., Martins, J. B., and Tavares, O. A. P.** *Phys. Rev. C* **34**, 2261 (1986).
107. **Buck, B. and Merchant, A. C.** *Phys. Rev. C* **39**, 2097–2100 (1989).
108. **Poenaru, D. N. and Greiner, W.** In *Clustering Phenomena in Atoms and Nuclei*, Brenner, M., Lönnroth, T., and Malik, B., editors, 235–249, Springer, Heidelberg (1992).
109. **Săndulescu, A. and Greiner, W.** *Rep. Prog. Phys.* **55**, 1423–1481 (1992).
110. **Ludu, A., Săndulescu, A., and Greiner, W.** *Int. J. Mod. Phys. E* **1**, 169–200 (1992).
111. **Fink, H. J., Maruhn, J. A., Scheid, W., and Greiner, W.** *Z. Phys.* **268**, 321 (1974).
112. **Hulet, E. K., Wild, J. F., Dougan, R. J., Lougheed, R. W., Landrum, J. H., Dougan, A. D., Schädel, M., Hahn, R. L., Baisden, P. A., Hendersson, C. M., Dupzyk, R. L., Sümmerer, K., and Bethune, G. R.** *Phys. Rev. Lett.* **56**, 313–316 (1986).
113. **Hulet, E. K., Wild, J. F., Dougan, R. J., Lougheed, R. W., Landrum, J. H., Dougan, A. D., Baisden, P. A., Hendersson, C. M., Dupzyk, R. L., Hahn, R. L., Schädel, M., Sümmerer, K., and Bethune, G. R.** *Phys. Rev. C* **40**, 770–784 (1989).
114. **Săndulescu, A., Lustig, H. J., Hahn, J., and Greiner, W.** *J. Phys. G: Nucl. Phys.* **4**, L279 (1978).
115. **Shi, Y. J. and Swiatecki, W. J.** *Nucl. Phys. A* **438**, 450–460 (1985).
116. **Kröger, H. and Scheid, W.** *J. Phys. G: Nucl. Phys.* **6**, L85–L88 (1980).
117. **Gupta, R. K., Singh, S., Puri, R. K., and Scheid, W.** *Phys. Rev. C* **47** 561–566 (1993).
118. **Herrmann, R., Maruhn, J. A., and Greiner, W.** *J. Phys. G: Nucl. Phys.* **12**, L285–L289 (1986).
119. **Poenaru, D. N., Ivaşcu, M., and Săndulescu, A.** *J. Phys. G: Nucl. Phys.* **5**, L169–L173 (1979).
120. **Poenaru, D. N., Ivaşcu, M., and Săndulescu, A.** *J. Phys. Lett.* **40**, L465–L467 (1979).
121. **Scheid, W. and Greiner, W.** *Z. Phys.* **226**, 364 (1969). **Krappe, H. J., Nix, J. R. and Sierk, A. J.** *Phys. Rev. C* **20**, 992 (1979).
122. **Blendowske, R., Fliessbach, T., and Walliser, H.** Chapter in ref. 3, to be published.
123. **Poenaru, D. N., Schnabel, D., Greiner, W., Mazilu, D., and Gherghescu, R.** *At. Data Nucl. Data Tables* **48**, 231–327 (1991).
124. **Poenaru, D. N. and Greiner, W.** *J. Phys. G: Nucl. Part. Phys.* **17**, S443–S451 (1991).
125. **Poenaru, D. N. and Greiner, W.** *Phys. Scr.* **44**, 427–429 (1991).
126. **Blendowske, R. and Walliser, H.** *Phys. Rev. Lett.* **61**, 1930–1933 (1988).
127. **Brown, B. A.** *Phys. Rev. C* **46**, 811–814 (1992).

128. **Fröman, P. O.** *K. Dan. Vidensk. Selsk. Mat. Fys. Skr.* **1**, 3 (1957).
129. **Wapstra, A. H., Nijgh, G. J., and van Lieshout, R.** In *Nuclear Spectroscopy Tables*, 37. North-Holland, Amsterdam (1959).
130. **Taagepera, R. and Nurmia, M.** *Ann. Acad. Sci. Fenn.* **Ser. A**, no. 78 (1961).
131. **Viola Jr., V. E. and Seaborg, G. T.** *J. Inorg. Nucl. Chem.* **28**, 741 (1966).
132. **Keller, K. A. and Münzel, H.** *Z. Phys.* **255**, 419 (1972).
133. **Hornshøj, P., Hansen, P. G., Jonson, B., Ravn, H. L., Westgaard, L., and Nielsen, O. N.** *Nucl. Phys. A* **230**, 365 (1974).
134. **Poenaru, D. N. and Ivaşcu, M.** *J. Phys. (Paris)* **44**, 791–796 (1983).
135. **Poenaru, D. N., Ivaşcu, M., and Mazilu, D.** *Comput. Phys. Commun.* **25**, 297–309 (1982).
136. **Nichols, A. L.** Table in this book.
137. **Westmeier, W. and Merklin, A.** Report 29–1, Fachinformationszentrum, Karlsruhe (1985).
138. **Rytz, A.** *At. Data Nucl. Data Tables* **47**, 205–239 (1991).
139. **Buck, B., Merchant, A. C., and Perez, S. M.** *At. Data Nucl. Data Tables* **54**, 53–73 (1993).
140. **Rurarz, E.** Report INR 1950/IA/PL/A, Institute of Nuclear Research, Warsaw (1982).
141. **Hatsukawa, Y., Nakahara, H., and Hoffman, D. C.** *Phys. Rev. C* **42**, 674–682 (1990).
142. **Trofimov, Y. N.** *Sov. J. Nucl. Phys.* **52**, 961–963 (1990).
143. **Poenaru, D. N., Ivaşcu, M., Mazilu, D., Gherghescu, R., Depta, K., and Greiner, W.** Report NP-54, Central Institute of Physics, Bucharest (1986).
144. **Hussonnois, M., Le Du, J. F., Brillard, L., and Ardisson, G.** *Phys. Rev. C* **44**, 2884–2885 (1991).
145. **Sheline, R. K. and Ragnarsson, I.** *Phys. Rev. C* **44**, 2886–2887 (1991).
146. **Poenaru, D. N., Hourani, E., and Greiner, W.** *Ann. Phys. (Leipzig)* **3**, 107–117 (1994).
147. **Wheeler, J. A.** In *Niels Bohr and the Development of Physics*, Pauli, W., Rosenfeld, L., and Weisskopf, V., editors, 163–184. Pergamon Press, London (1955).
148. **Randrup, J., Larsson, S. E., Möller, P., Nilsson, S. G., Pomorski, K., and Sobiczewski, A.** *Phys. Rev. C* **13**, 229 (1976).
149. **Lojewski, Z. and Baran, A.** *Z. Phys., A* **322**, 695–700 (1985).
150. **Poenaru, D. N., Greiner, W., and Gherghescu, R.** *Phys. Rev. C* **47**, 2030–2037 (1993).
151. **Haustein, P. E.** *At. Data Nucl. Data Tables* **39**, 185–200 (1988).
152. **Poenaru, D. N. and Greiner, W.** In *Proc. 2nd International Conference on Radioactive Nuclear Beams, Louvain-la-Neuve, 1991*, Delbar, T., editor, 203–208, IOP, Bristol (1992).
153. **Dalpiaz, P.** Report LNL–INFN 58/92, Istituto Nationale di Fisica Nucleare, Legnaro (Padova) (1992).
154. **Poenaru, D. N., Greiner, W., Ivaşcu, M., and Săndulescu, A.** *Phys. Rev. C* **32**, 2198–2200 (1985).
155. **Hirsch, M., Staudt, A., Muto, K., and Klapdor-Kleingrothaus, H. V.** *At. Data Nucl. Data Tables* **53**, 165–194 (1993).
156. **Buck, B., Merchant, A. C., Perez, S. M., and Tripe, P.** *J. Phys. G: Nucl. Part. Phys.* **20**, 351–355 (1994).

157. **Myers, W. D. and Swiatecki, W. J.** *Nucl. Phys. A* **81**, 1 (1966).
158. **Christensen, P. R. and Winther, A.** *Phys. Lett. B* **65**, 19 (1976).
159. **Audi, G. and Wapstra, A. H.** *Nucl. Phys. A* **565**, 1–65 (1993).
160. **Aboussir, Y., Pearson, J. M., Dutta, A. K., and Tondeur, F.** *Nucl. Phys. A* **549**, 155–179 (1992).
161. **Möller, P. and Nix, J. R.** *At. Data Nucl. Data Tables* **39**, 213–223 (1988).
162. **Möller, P., Myers, W. D., Swiatecki, W. J., and Treiner, J.** *At. Data Nucl. Data Tables* **39**, 225–234 (1988).
163. **Liran, S. and Zeldes, N.** *At. Data Nucl. Data Tables* **17**, 431 (1976).

6

FUNDAMENTAL CONSTANTS AND ENERGY CONVERSION FACTORS

Barry N. Taylor and E. Richard Cohen

Table of contents

I. INTRODUCTION

This chapter gives the values of the basic fundamental physical constants and energy conversion factors of physics and chemistry currently recommended for international use by CODATA*.[1,2] These values, the most up-to-date, self-consistent set presently available, are a direct result of the 1986 CODATA least-squares adjustment of the fundamental constants carried out by the authors under the auspices of the CODATA Task Group on Fundamental Constants.[1,2] The 1986 CODATA set of recommended values replaced the 1973 set, its immediate predecessor, which resulted from the 1973 CODATA least-squares adjustment.[3,4] That adjustment was also carried out by the authors under the auspices of the Task Group.[3,4] The next adjustment of the constants, with its resulting set of recommended values, is scheduled for completion in late 1996 and will be implemented by the authors and Peter J. Mohr again under the sponsorship of the Task Group.

II. EXPLANATION OF TABLES

The 1986 CODATA set of recommended values is given in five tables. Table 1 is an abbreviated list containing the quantities that should be of greatest interest to most users. Table 2 is a much more complete compila-

---

*CODATA (Committee on Data for Science and Technology) was established in 1966 as an interdisciplinary committee of the International Council of Scientific Unions. It seeks to improve the compilation, critical evaluation, storage, and retrieval of data of importance to science and technology.

Table 1 Summary of the 1986 recommended values of the fundamental physical constants. An abbreviated list of the fundamental constants of physics and chemistry based on a least-squares adjustment with 17 degrees of freedom. The digits in parentheses are the one-standard-deviation uncertainty in the last digits of the given value. Since the uncertainties of many of these entries are correlated, the full covariance matrix must be used in evaluating the uncertainties of quantities computed from them

Quantity	Symbol	Value	Units	Relative uncertainty (ppm)
speed of light in vacuum	c	299 792 458	$\mathrm{m\,s^{-1}}$	(exact)
permeability of vacuum	μ_0	$4\pi\times10^{-7}$	$\mathrm{N\,A^{-2}}$	
		$=12.566\,370\,614\ldots$	$10^{-7}\ \mathrm{N\,A^{-2}}$	(exact)
permittivity of vacuum, $1/\mu_0c^2$	ε_0	$8.854\,187\,817\ldots$	$10^{-12}\ \mathrm{F\,m^{-1}}$	(exact)
Newtonian constant of gravitation	G	6.672 59(85)	$10^{-11}\ \mathrm{m^3\,kg^{-1}\,s^{-2}}$	128
Planck constant	h	6.626 075 5(40)	$10^{-34}\ \mathrm{J\,s}$	0.60
$h/2\pi$	$\hbar$	1.054 572 66(63)	$10^{-34}\ \mathrm{J\,s}$	0.60
elementary charge	e	1.602 177 33(49)	$10^{-19}\ \mathrm{C}$	0.30
magnetic flux quantum, $h/2e$	Φ_0	2.067 834 61(61)	$10^{-15}\ \mathrm{Wb}$	0.30
electron mass	m_e	9.109 389 7(54)	$10^{-31}\ \mathrm{kg}$	0.59
proton mass	m_p	1.672 623 1(10)	$10^{-27}\ \mathrm{kg}$	0.59
proton-electron mass ratio	m_p/m_e	1 836.152 701(37)		0.020
fine-structure constant, $\mu_0ce^2/2h$	α	7.297 353 08(33)	10^{-3}	0.045
inverse fine-structure constant	α^{-1}	137.035 989 5(61)		0.045
Rydberg constant, $m_ec\alpha^2/2h$	R_∞	10 973 731.534(13)	$\mathrm{m^{-1}}$	0.0012
Avogadro constant	N_A, L	6.022 136 7(36)	$10^{23}\ \mathrm{mol^{-1}}$	0.59
Faraday constant, N_Ae	F	96 485.309(29)	$\mathrm{C\,mol^{-1}}$	0.30
molar gas constant	R	8.314 510(70)	$\mathrm{J\,mol^{-1}\,K^{-1}}$	8.4
Boltzmann constant, R/N_A	k	1.380 658(12)	$10^{-23}\ \mathrm{J\,K^{-1}}$	8.5
Stefan-Boltzmann constant, $(\pi^2/60)k^4/\hbar^3c^2$	σ	5.670 51(19)	$10^{-8}\ \mathrm{W\,m^{-2}\,K^{-4}}$	34
		Non-SI units used with SI		
electron volt, $(e/\mathrm{C})\mathrm{J}=\{e\}\mathrm{J}$	eV	1.602 177 33(49)	$10^{-19}\ \mathrm{J}$	0.30
(unified) atomic mass unit, $1\ \mathrm{u}=m_u=\frac{1}{12}m(^{12}\mathrm{C})$	u	1.660 540 2(10)	$10^{-27}\ \mathrm{kg}$	0.59

tion. Table 3 is a list of related "maintained" units and "standard" values, while Table 4 contains a number of scientifically, technologically, and metrologically useful energy conversion factors. Finally, Table 5 is an extended covariance matrix containing the variances, covariances, and correlation coefficients of the variables of the adjustment and of a number of other constants included for convenience. Such a matrix is necessary, of course, because the variables in a least-squares adjustment are statistically correlated. Thus, with the exception of quantities that depend only on auxiliary constants, the uncertainty associated with a quantity calculated from these variables can be found only by using the full covariance matrix. (Auxiliary constants are either defined quantities with no uncertainty, or quantities such as the Rydberg constant R_∞ with assigned uncertainties sufficiently small that their values are not subject to adjustment. In the 1986 least-squares adjustment, the relative uncertainty of each auxiliary constant was no greater than 0.02 parts per million or ppm.)

In Table 5, K_V is the numerical value of the laboratory unit of voltage V_{76-BI} maintained at the Bureau International des Poids et Mesures (BIPM): $V_{76-BI} = K_V$ V. V_{76-BI} is based on the Josephson effect using a value of the

Table 2 1986 recommended values of the fundamental physical constants. The list of the fundamental constants of physics and chemistry based on a least-squares adjustment with 17 degrees of freedom. The digits in parentheses are the one-standard-deviation uncertainty in the last digits of the given value. Since the uncertainties of many of these entries are correlated, the full covariance matrix must be used in evaluating the uncertainties of quantities computed from them

Quantity	Symbol	Value	Units	Relative uncertainty (ppm)
GENERAL CONSTANTS				
Universal constants				
speed of light in vacuum	c	299 792 458	$\mathrm{m\,s^{-1}}$	(exact)
permeability of vacuum	μ_0	$4\pi \times 10^{-7}$	$\mathrm{N\,A^{-2}}$	
		$= 12.566\,370\,614\ldots$	$10^{-7}\ \mathrm{N\,A^{-2}}$	(exact)
permittivity of vacuum, $1/\mu_0 c^2$	ε_0	$8.854\,187\,817\ldots$	$10^{-12}\ \mathrm{F\,m^{-1}}$	(exact)
Newtonian constant of gravitation	G	6.672 59(85)	$10^{-11}\ \mathrm{m^3\,kg^{-1}\,s^{-2}}$	128
Planck constant	h	6.626 075 5(40)	$10^{-34}\ \mathrm{J\,s}$	0.60
in electron volts, $h/\{e\}$		4.135 669 2(12)	$10^{-15}\ \mathrm{eV\,s}$	0.30
$h/2\pi$	$\hbar$	1.054 572 66(63)	$10^{-34}\ \mathrm{J\,s}$	0.60
in electron volts, $\hbar/\{e\}$		6.582 122 0(20)	$10^{-16}\ \mathrm{eV\,s}$	0.30
Planck mass, $(\hbar c/G)^{1/2}$	m_P	2.176 71(14)	$10^{-8}\ \mathrm{kg}$	64
Planck length, $\hbar/m_\mathrm{P}c = (\hbar G/c^3)^{1/2}$	l_P	1.616 05(10)	$10^{-35}\ \mathrm{m}$	64
Planck time, $l_\mathrm{P}/c = (\hbar G/c^5)^{1/2}$	t_P	5.390 56(34)	$10^{-44}\ \mathrm{s}$	64
Electromagnetic constants				
elementary charge	e	1.602 177 33(49)	$10^{-19}\ \mathrm{C}$	0.30
	e/h	2.417 988 36(72)	$10^{14}\ \mathrm{A\,J^{-1}}$	0.30
magnetic flux quantum, $h/2e$	Φ_0	2.067 834 61(61)	$10^{-15}\ \mathrm{Wb}$	0.30
Josephson frequency-voltage quotient	$2e/h$	4.835 976 7(14)	$10^{14}\ \mathrm{Hz\,V^{-1}}$	0.30
quantized Hall conductance	e^2/h	3.874 046 14(17)	$10^{-5}\ \mathrm{S}$	0.045
quantized Hall resistance, $h/e^2 = \mu_0 c/2\alpha$	R_H	25 812.805 6(12)	Ω	0.045
Bohr magneton, $e\hbar/2m_e$	μ_B	9.274 015 4(31)	$10^{-24}\ \mathrm{J\,T^{-1}}$	0.34
in electron volts, $\mu_\mathrm{B}/\{e\}$		5.788 382 63(52)	$10^{-5}\ \mathrm{eV\,T^{-1}}$	0.089
in hertz, μ_B/h		1.399 624 18(42)	$10^{10}\ \mathrm{Hz\,T^{-1}}$	0.30
in wavenumbers, μ_B/hc		46.686 437(14)	$\mathrm{m^{-1}\,T^{-1}}$	0.30
in kelvins, μ_B/k		0.671 709 9(57)	$\mathrm{K\,T^{-1}}$	8.5
nuclear magneton, $e\hbar/2m_\mathrm{p}$	μ_N	5.050 786 6(17)	$10^{-27}\ \mathrm{J\,T^{-1}}$	0.34
in electron volts, $\mu_\mathrm{N}/\{e\}$		3.152 451 66(28)	$10^{-8}\ \mathrm{eV\,T^{-1}}$	0.089
in hertz, μ_N/h		7.622 591 4(23)	$\mathrm{MHz\,T^{-1}}$	0.30
in wavenumbers, μ_N/hc		2.542 622 81(77)	$10^{-2}\ \mathrm{m^{-1}\,T^{-1}}$	0.30
in kelvins, μ_N/k		3.658 246(31)	$10^{-4}\ \mathrm{K\,T^{-1}}$	8.5
ATOMIC CONSTANTS				
fine-structure constant, $\mu_0 ce^2/2h$	α	7.297 353 08(33)	10^{-3}	0.045
inverse fine-structure constant	α^{-1}	137.035 989 5(61)		0.045
Rydberg constant, $m_e c\alpha^2/2h$	R_∞	10 973 731.534(13)	$\mathrm{m^{-1}}$	0.0012
in hertz, $R_\infty c$		3.289 841 949 9(39)	$10^{15}\ \mathrm{Hz}$	0.0012
in joules, $R_\infty hc$		2.179 874 1(13)	$10^{-18}\ \mathrm{J}$	0.60
in eV, $R_\infty hc/\{e\}$		13.605 698 1(40)	eV	0.30
Bohr radius, $\alpha/4\pi R_\infty$	a_0	0.529 177 249(24)	$10^{-10}\ \mathrm{m}$	0.045
Hartree energy, $e^2/4\pi\varepsilon_0 a_0 = 2R_\infty hc$	E_h	4.359 748 2(26)	$10^{-18}\ \mathrm{J}$	0.60
in eV, $E_\mathrm{h}/\{e\}$		27.211 396 1(81)	eV	0.30
quantum of circulation	$h/2m_e$	3.636 948 07(33)	$10^{-4}\ \mathrm{m^2\,s^{-1}}$	0.089
	h/m_e	7.273 896 14(65)	$10^{-4}\ \mathrm{m^2\,s^{-1}}$	0.089
Electron				
electron mass	m_e	9.109 389 7(54)	$10^{-31}\ \mathrm{kg}$	0.59
		5.485 799 03(13)	$10^{-4}\ \mathrm{u}$	0.023
in electron volts, $m_e c^2/\{e\}$		0.510 999 06(15)	MeV	0.30
electron-muon mass ratio	m_e/m_μ	4.836 332 18(71)	10^{-3}	0.15
electron-proton mass ratio	m_e/m_p	5.446 170 13(11)	10^{-4}	0.020
electron-deuteron mass ratio	m_e/m_d	2.724 437 07(6)	10^{-4}	0.020
electron–α-particle mass ratio	m_e/m_α	1.370 933 54(3)	10^{-4}	0.021
electron specific charge	$-e/m_e$	−1.758 819 62(53)	$10^{11}\ \mathrm{C\,kg^{-1}}$	0.30
electron molar mass	$M(\mathrm{e}), M_e$	5.485 799 03(13)	$10^{-7}\ \mathrm{kg/mol}$	0.023
Compton wavelength, $h/m_e c$	λ_C	2.426 310 58(22)	$10^{-12}\ \mathrm{m}$	0.089
$\lambda_\mathrm{C}/2\pi = \alpha a_0 = \alpha^2/4\pi R_\infty$	$\bar{\lambda}_\mathrm{C}$	3.861 593 23(35)	$10^{-13}\ \mathrm{m}$	0.089

continued overleaf

Table 2 *continued*

Quantity	Symbol	Value	Units	Relative uncertainty (ppm)
		Electron *(Continued)*		
classical electron radius, $\alpha^2 a_0$	r_e	2.817 940 92(38)	10^{-15} m	0.13
Thomson cross section, $(8\pi/3)r_e^2$	σ_e	0.665 246 16(18)	10^{-28} m^2	0.27
electron magnetic moment	μ_e	928.477 01(31)	10^{-26} J T^{-1}	0.34
in Bohr magnetons	μ_e/μ_B	1.001 159 652 193(10)		1×10^{-5}
in nuclear magnetons	μ_e/μ_N	1 838.282 000(37)		0.020
electron magnetic moment anomaly, μ_e/μ_B-1	a_e	1.159 652 193(10)	10^{-3}	0.0086
electron g factor, $2(1+a_e)$	g_e	2.002 319 304 386(20)		1×10^{-5}
electron-muon magnetic moment ratio	μ_e/μ_μ	206.766 967(30)		0.15
electron-proton magnetic moment ratio	μ_e/μ_p	658.210 688 1(66)		0.010
		Muon		
muon mass	m_μ	1.883 532 7(11)	10^{-28} kg	0.61
		0.113 428 913(17)	u	0.15
in electron volts, $m_\mu c^2/\{e\}$		105.658 389(34)	MeV	0.32
muon-electron mass ratio	m_μ/m_e	206.768 262(30)		0.15
muon molar mass	$M(\mu), M_\mu$	1.134 289 13(17)	10^{-4} kg/mol	0.15
muon magnetic moment	μ_μ	4.490 451 4(15)	10^{-26} J T^{-1}	0.33
in Bohr magnetons,	μ_μ/μ_B	4.841 970 97(71)	10^{-3}	0.15
in nuclear magnetons,	μ_μ/μ_N	8.890 598 1(13)		0.15
muon magnetic moment anomaly, $[\mu_\mu/(e\hbar/2m_\mu)]-1$	a_μ	1.165 923 0(84)	10^{-3}	7.2
muon g factor, $2(1+a_\mu)$	g_μ	2.002 331 846(17)		0.0084
muon-proton magnetic moment ratio	μ_μ/μ_p	3.183 345 47(47)		0.15
		Proton		
proton mass	m_p	1.672 623 1(10)	10^{-27} kg	0.59
		1.007 276 470(12)	u	0.012
in electron volts, $m_p c^2/\{e\}$		938.272 31(28)	MeV	0.30
proton-electron mass ratio	m_p/m_e	1 836.152 701(37)		0.020
proton-muon mass ratio	m_p/m_μ	8.880 244 4(13)		0.15
proton specific charge	e/m_p	9.578 830 9(29)	10^7 C kg^{-1}	0.30
proton molar mass	$M(\mathrm{p}), M_p$	1.007 276 470(12)	10^{-3} kg/mol	0.012
proton Compton wavelength, $h/m_p c$	$\lambda_{C,p}$	1.321 410 02(12)	10^{-15} m	0.089
$\lambda_{C,p}/2\pi$	$\bar{\lambda}_{C,p}$	2.103 089 37(19)	10^{-16} m	0.089
proton magnetic moment	μ_p	1.410 607 61(47)	10^{-26} J T^{-1}	0.34
in Bohr magnetons	μ_p/μ_B	1.521 032 202(15)	10^{-3}	0.010
in nuclear magnetons	μ_p/μ_N	2.792 847 386(63)		0.023
diamagnetic shielding correction for protons in pure water, spherical sample, 25 °C, $1-\mu_p'/\mu_p$	σ_{H_2O}	25.689(15)	10^{-6}	
shielded proton moment (H_2O, sph., 25 °C)	μ_p'	1.410 571 38(47)	10^{-26} J T^{-1}	0.34
in Bohr magnetons	μ_p'/μ_B	1.520 993 129(17)	10^{-3}	0.011
in nuclear magnetons	μ_p'/μ_N	2.792 775 642(64)		0.023
proton gyromagnetic ratio	γ_p	26 752.212 8(81)	10^4 s^{-1}T^{-1}	0.30
	$\gamma_p/2\pi$	42.577 469(13)	MHz T^{-1}	0.30
uncorrected (H_2O, sph., 25 °C)	γ_p'	26 751.525 5(81)	10^4 s^{-1}T^{-1}	0.30
	$\gamma_p'/2\pi$	42.576 375(13)	MHz T^{-1}	0.30
		Neutron		
neutron mass	m_n	1.674 928 6(10)	10^{-27} kg	0.59
		1.008 664 904(14)	u	0.014
in electron volts, $m_n c^2/\{e\}$		939.565 63(28)	MeV	0.30
neutron-electron mass ratio	m_n/m_e	1 838.683 662(40)		0.022
neutron-proton mass ratio	m_n/m_p	1.001 378 404(9)		0.009
neutron molar mass	$M(\mathrm{n}), M_n$	1.008 664 904(14)	10^{-3} kg/mol	0.014
neutron Compton wavelength, $h/m_n c$	$\lambda_{C,n}$	1.319 591 10(12)	10^{-15} m	0.089
$\lambda_{C,n}/2\pi$	$\bar{\lambda}_{C,n}$	2.100 194 45(19)	10^{-16} m	0.089
neutron magnetic moment[a]	μ_n	0.966 237 07(40)	10^{-26} J T^{-1}	0.41
in Bohr magnetons	μ_n/μ_B	1.041 875 63(25)	10^{-3}	0.24
in nuclear magnetons	μ_n/μ_N	1.913 042 75(45)		0.24

continued overleaf

Table 2 *continued*

Quantity	Symbol	Value	Units	Relative uncertainty (ppm)
neutron-electron				
magnetic moment ratio	μ_n/μ_e	1.040 668 82(25)	10^{-3}	0.24
neutron-proton				
magnetic moment ratio	μ_n/μ_p	0.684 979 34(16)		0.24
		Deuteron		
deuteron mass	m_d	3.343 586 0(20)	10^{-27} kg	0.59
		2.013 553 214(24)	u	0.012
in electron volts, $m_dc^2/\{e\}$		1 875.613 39(57)	MeV	0.30
deuteron-electron mass ratio	m_d/m_e	3 670.483 014(75)		0.020
deuteron-proton mass ratio	m_d/m_p	1.999 007 496(6)		0.003
deuteron molar mass	$M(d), M_d$	2.013 553 214(24)	10^{-3} kg/mol	0.012
deuteron magnetic moment[a]	μ_d	0.433 073 75(15)	10^{-26} J T^{-1}	0.34
in Bohr magnetons,	μ_d/μ_B	0.466 975 447 9(91)	10^{-3}	0.019
in nuclear magnetons,	μ_d/μ_N	0.857 438 230(24)		0.028
deuteron-electron				
magnetic moment ratio	μ_d/μ_e	0.466 434 546 0(91)	10^{-3}	0.019
deuteron-proton				
magnetic moment ratio	μ_d/μ_p	0.307 012 203 5(51)		0.017
		PHYSICO-CHEMICAL CONSTANTS		
Avogadro constant	N_A, L	6.022 136 7(36)	10^{23} mol^{-1}	0.59
atomic mass constant				
$m_u=\frac{1}{12}m(^{12}C)$	m_u	1.660 540 2(10)	10^{-27} kg	0.59
in electron volts, $m_uc^2/\{e\}$		931.494 32(28)	MeV	0.30
Faraday constant, N_Ae	F	96 485.309(29)	C mol^{-1}	0.30
molar Planck constant	N_Ah	3.990 313 23(36)	10^{-10} J s mol^{-1}	0.089
	N_Ahc	0.119 626 58(11)	J m mol^{-1}	0.089
molar gas constant	R	8.314 510(70)	J mol^{-1} K^{-1}	8.4
Boltzmann constant, R/N_A	k	1.380 658(12)	10^{-23} J K^{-1}	8.5
in electron volts, $k/\{e\}$		8.617 385(73)	10^{-5} eV K^{-1}	8.4
in hertz, k/h		2.083 674(18)	10^{10} Hz K^{-1}	8.4
in wavenumbers, k/hc		69.503 87(59)	m^{-1} K^{-1}	8.4
molar volume (ideal gas), RT/p				
$T=273.15$ K, $p=101\,325$ Pa	V_m	0.022 414 10(19)	m^3 mol^{-1}	8.4
Loschmidt constant, N_A/V_m	n_0	2.686 763(23)	10^{25} m^{-3}	8.5
$T=273.15$ K, $p=100$ kPa	V_m	0.022 711 08(19)	m^3 mol^{-1}	8.4
Sackur-Tetrode constant				
(absolute entropy constant),[b]				
$\frac{5}{2}+\ln[(2\pi m_ukT_1/h^2)^{3/2}kT_1/p_0]$				
$T_1=1$ K, $p_0=100$ kPa	S_0/R	−1.151 693(21)		18
$p_0=101\,325$ Pa		−1.164 856(21)		18
Stefan-Boltzmann constant,				
$(\pi^2/60)k^4/\hbar^3c^2$	σ	5.670 51(19)	10^{-8} W m^{-2} K^{-4}	34
first radiation constant, $2\pi hc^2$	c_1	3.741 774 9(22)	10^{-16} W m^2	0.60
second radiation constant, hc/k	c_2	0.014 387 69(12)	m K	8.4
Wien displacement law constant,				
$b=\lambda_{max}T=c_2/4.965\,114\,23\ldots$	b	2.897 756(24)	10^{-3} m K	8.4

[a]The scalar magnitude of the neutron moment is listed here. The neutron magnetic dipole is directed oppositely to that of the proton, and corresponds to the dipole associated with a spinning negative charge distribution. The vector sum, $\mu_d=\mu_p+\mu_n$, is approximately satisfied.

[b]The entropy of an ideal monatomic gas of relative atomic weight A_r is given by

$$S=S_0+\tfrac{3}{2}R\ln A_r-R\ln(p/p_0)+\tfrac{5}{2}R\ln(T/\mathrm{K}).$$

Josephson frequency-to-voltage quotient $2e/h$ adopted in 1972 by the Comité Consultatif d'Électricité (CCE) of the Comité International des Poids et Mesures (CIPM), namely $E = 483\,594$ GHz V^{-1} exactly; thus $2e/h = E/K_V$. K_Ω is the numerical value of the BIPM as-maintained ohm as it existed on 1 January 1985, Ω_{BI85}, based on the mean resistance of a particular group of wire-wound standard resistors: $\Omega_{BI85} = K_\Omega\,\Omega$.

Table 3 Maintained units and standard values. A summary of "maintained" units and "standard" values and their relationship to SI units, based on a least-squares adjustment with 17 degrees of freedom. The digits in parentheses are the one-standard-deviation uncertainty in the last digits of the given value. Since the uncertainties of many of these entries are correlated, the full covariance matrix must be used in evaluating the uncertainties of quantities computed from them

Quantity	Symbol	Value	Units	Relative uncertainty (ppm)
electron volt, $(e/\mathrm{C})\mathrm{J} = \{e\}\mathrm{J}$	eV	1.602 177 33(49)	10^{-19} J	0.30
(unified) atomic mass unit, $1\ \mathrm{u} = m_\mathrm{u} = \frac{1}{12} m(^{12}\mathrm{C})$	u	1.660 540 2(10)	10^{-27} kg	0.59
standard atmosphere	atm	101 325	Pa	(exact)
standard acceleration of gravity	g_n	9.806 65	$\mathrm{m\,s^{-2}}$	(exact)
"As-maintained" electrical units				
BIPM maintained ohm, $\Omega_{69-\mathrm{BI}}$, $\Omega_{\mathrm{BI85}} \equiv \Omega_{69-\mathrm{BI}}$(January 1, 1985)	Ω_{BI85}	$1 - 1.563(50) \times 10^{-6} = 0.999\,998\,437(50)$	Ω	0.050
Drift rate of $\Omega_{69-\mathrm{BI}}$	$\frac{d\Omega_{69-\mathrm{BI}}}{dt}$	−0.056 6(15)	$\mu\Omega$/a	
BIPM maintained volt, $\mathrm{V}_{76-\mathrm{BI}} = 483\,594.0$ GHz$(h/2e)$	$\mathrm{V}_{76-\mathrm{BI}}$	$1 - 7.59(30) \times 10^{-6} = 0.999\,992\,41(30)$	V	0.30
BIPM maintained ampere, $\mathrm{A}_{\mathrm{BIPM}} = \mathrm{V}_{76-\mathrm{BI}}/\Omega_{69-\mathrm{BI}}$	$\mathrm{A}_{\mathrm{BI85}}$	$1 - 6.03(30) \times 10^{-6} = 0.999\,993\,97(30)$	A	0.30
X-ray standards				
Cu x unit: $\lambda(\mathrm{CuK}\alpha_1) \equiv 1537.400$ xu	xu$(\mathrm{CuK}\alpha_1)$	1.002 077 89(70)	10^{-13} m	0.70
Mo x unit: $\lambda(\mathrm{MoK}\alpha_1) \equiv 707.831$ xu	xu$(\mathrm{MoK}\alpha_1)$	1.002 099 38(45)	10^{-13} m	0.45
Å*: $\lambda(\mathrm{WK}\alpha_1) \equiv 0.2090100$ Å*	Å*	1.000 014 81(92)	10^{-10} m	0.92
lattice spacing of Si (in vacuum, 22.5 °C),[a]	a	0.543 101 96(11)	nm	0.21
$d_{220} = a/\sqrt{8}$	d_{220}	0.192 015 540(40)	nm	0.21
molar volume of Si, $M(\mathrm{Si})/\rho(\mathrm{Si}) = N_\mathrm{A} a^3/8$	V_m(Si)	12.058 817 9(89)	$\mathrm{cm^3/mol}$	0.74

[a]The lattice spacing of single-crystal Si can vary by parts in 10^7 depending on the preparation process. Measurements at PTB indicate also the possibility of distortions from exact cubic symmetry of the order of 0.2 ppm.

To use Table 5, note that the covariance between two quantities Q_k and Q_s that are functions of a common set of variables x_i $(i = 1, \ldots, N)$ is given by

$$v_{ks} = \sum_{i,j=1}^{N} \frac{\partial Q_k}{\partial x_i} \frac{\partial Q_s}{\partial x_j} v_{ij} \tag{1}$$

where v_{ij} is the covariance of x_i and x_j. In this general form, the units of v_{ij} are the product of the units of x_i and x_j and the units of v_{ks} are the product of the units of Q_k and Q_s. For most cases involving the fundamental constants, the variables x_i may be taken to be the fractional change in the physical quantity from some fiducial value, and the quantities Q can be expressed as powers of physical constants Z_j according to

$$Q_k = q_k \prod_{j=1}^{N} Z_i^{Y_{kj}} = q_k \prod_{j=1}^{N} Z_{0j}^{Y_{kj}} (1 + x_j)^{Y_{kj}} \tag{2}$$

where q_k is a constant. If the variances and covariances are then expressed in relative units eq. (1) becomes

$$v_{ks} = \sum_{i,j=1}^{N} Y_{ki} Y_{sj} v_{ij} \tag{3}$$

where the v_{ij} are to be expressed, for example, in (parts in 10^9).^{2} Equation (3) is the basis for the expansion of the covariance matrix to include e, h, m_e, N_A, and F in addition to the principal variables of the adjustment, α^{-1}, K_V, K_Ω, and μ_μ/μ_p. Since the standard deviation $\varepsilon_i = (v_{ii})^{1/2}$, setting $s = k$ in eq. (3) yields for the variance $\varepsilon_k^2 = v_{kk}$

$$\varepsilon_k^2 = \sum_{i=1}^{N} Y_{ki}^2 \varepsilon_i^2 + 2\sum_{j<1}^{N} Y_{ki} Y_{kj} r_{ij} \varepsilon_i \varepsilon_j \tag{4}$$

where the correlation coefficients r_{ij} are defined by $v_{ij} = r_{ij}(v_{ii}v_{jj})^{1/2} = r_{ij}\varepsilon_i\varepsilon_j$.

As an example of the use of eq. (4) and Table 5, consider the calculation of the relative one-standard-deviation uncertainty ε_k of the Bohr magneton $\mu_B = eh/4\pi m_e$. In terms of the variables of the 1986 adjustment this quantity is given by

$$\mu_B = [2\pi \mu_0 R_\infty E]^{-1} \cdot (\alpha^{-1})^{-3} K_V \tag{5}$$

where the quantities in the brackets to the left of the centered dot are taken to be exact. Then using eq. (4) with $i = 1$ corresponding to α^{-1} and $i = 2$ corresponding to K_V, and dropping the subscript k because there is only a single quantity, $Q = \mu_B$, one obtains

$$\varepsilon^2 = Y_1^2 v_{11} + 2Y_1 Y_2 v_{12} + Y_2^2 v_{22} \tag{6}$$

where $Y_1 = -3$ and $Y_2 = 1$. Substituting the appropriate entries from Table 5 into eq. (6) leads to

$$\varepsilon^2 = [9(1997) - 6(-1062) + 87\,988]\,(10^{-9})^2 \tag{7}$$

or $\varepsilon = 0.335$ ppm. Alternatively, one may evaluate the uncertainty of $\mu_B = eh/4\pi m_e$ directly from eq. (4) and Table 5: with $i = 5$ corresponding to e, $i = 6$ to h, and $i = 7$ to m_e, and with $Y_5 = 1$, $Y_6 = 1$, and $Y_7 = -1$, one finds

$$\varepsilon^2 = Y_5^2 v_{55} + 2Y_5 Y_6 v_{56} + 2Y_5 Y_7 v_{57} + Y_6^2 v_{66} + 2Y_6 Y_7 v_{67} + Y_7^2 v_{77} \tag{8}$$

$$= [92\,109 + 2(181\,159) - 2(175\,042) + 358\,197 - 2(349\,956) + 349\,702]\,(10^{-9})^2 \tag{9}$$

which also yields $\varepsilon = 0.335$ ppm.

The 1986 CODATA adjustment took into consideration all relevant data available up to 1 January 1986. In the intervening years, a number of new results have been reported that have important implications for the 1986 CODATA set of recommended values. (Many of these new results were obtained in response to the CCE's plan, first stated in 1986, to introduce new representations of the volt and ohm based on the Josephson and quantum Hall effects starting 1 January 1990.[5–8]) The new data and their impact are thoroughly discussed in ref. 9, where it is concluded that the new results lead to

Table 4 Energy conversion factors. To use this table note that all entries on the same line are equal; the unit at the top of a column applies to all of the values beneath it. Example: 1 eV = 806 544.10m^{-1} = 11 604.45 K

	J	kg	m^{-1}	Hz
1J =	1	$1/\{c^2\}$ 1.112 650 06 $\times 10^{-17}$	$1/\{hc\}$ 5.034 112 5(30) $\times 10^{24}$	$1/\{h\}$ 1.509 188 97(90) $\times 10^{33}$
1 kg =	$\{c^2\}$ 8.987 551 787 $\times 10^{16}$	1	$\{c/h\}$ 4.524 434 7(27) $\times 10^{41}$	$\{c^2/h\}$ 1.356 391 40(81) $\times 10^{50}$
1 m^{-1} =	$\{hc\}$ 1.986 447 5(12) $\times 10^{-25}$	$\{h/c\}$ 2.210 220 9(13) $\times 10^{-42}$	1	$\{c\}$ 299 792 458
1 Hz =	$\{h\}$ 6.626 075 5(40) $\times 10^{-34}$	$\{h/c^2\}$ 7.372 503 2(44) $\times 10^{-51}$	$1/\{c\}$ 3.335 640 952 $\times 10^{-9}$	1
1 K =	$\{k\}$ 1.380 658(12) $\times 10^{-23}$	$\{k/c^2\}$ 1.536 189(13) $\times 10^{-40}$	$\{k/hc\}$ 69.503 87(59)	$\{k/h\}$ 2.083 674(18) $\times 10^{10}$
1 eV =	$\{e\}$ 1.602 177 33(49) $\times 10^{-19}$	$\{e/c^2\}$ 1.782 662 70(54) $\times 10^{-36}$	$\{e/hc\}$ 806 554.10(24)	$\{e/h\}$ 2.417 988 36(72) $\times 10^{14}$
1 eV =	$\{m_u c^2\}$ 1.492 419 09(88) $\times 10^{-10}$	$\{m_u\}$ 1.660 540 2(10) $\times 10^{-27}$	$\{m_u c/h\}$ 7.153 005 63(67) $\times 10^{14}$	$\{m_u c^2/h\}$ 2.252 342 42(20) $\times 10^{23}$
1 hartree =	$\{2R_\infty hc\}$ 4.359 748 2(26) $\times 10^{-18}$	$\{2R_\infty h/c\}$ 4.850 874 1(29) $\times 10^{-35}$	$\{2R_\infty\}$ 21 974 463.067(26)	$\{2R_\infty c\}$ 6.579 683 899 9(78) $\times 10^{15}$

Table 4 continued

	K	eV	u	hartree
1J =	$1/\{k\}$ $7.242\,924(61) \times 10^{22}$	$1/\{e\}$ $6.241\,506\,4(19) \times 10^{18}$	$1/\{m_u c^2\}$ $6.700\,530\,8(40) \times 10^{9}$	$1/\{2R_\infty hc\}$ $2.293\,710\,4(14) \times 10^{17}$
1 kg =	$\{c^2/k\}$ $6.509\,616(55) \times 10^{39}$	$\{c^2/e\}$ $5.609\,586\,2(17) \times 10^{35}$	$1/\{m_u\}$ $6.022\,136\,7(36) \times 10^{26}$	$\{c/2R_\infty h\}$ $2.061\,484\,1(12) \times 10^{34}$
$1\ m^{-1}$ =	$\{hc/k\}$ $0.014\,387\,69(12)$	$\{hc/e\}$ $1.239\,842\,44(37) \times 10^{-6}$	$\{h/m_u c\}$ $1.331\,025\,22(12) \times 10^{-15}$	$1/\{2R_\infty\}$ $4.556\,335\,267\,2(54) \times 10^{-8}$
1 Hz =	$\{h/k\}$ $4.799\,216(41) \times 10^{-11}$	$\{h/e\}$ $4.135\,669\,2(12) \times 10^{-15}$	$\{h/m_u c^2\}$ $4.439\,822\,24(40) \times 10^{-24}$	$1/\{2R_\infty c\}$ $1.519\,829\,850\,8(18) \times 10^{-16}$
1 K =	1	$\{k/e\}$ $8.617\,385(73) \times 10^{-5}$	$\{k/m_u c^2\}$ $9.251\,140(78) \times 10^{-14}$	$\{k/2R_\infty hc\}$ $3.166\,829(27) \times 10^{-6}$
1 eV =	$\{e/k\}$ $11\,604.45(10)$	1	$\{e/m_u c^2\}$ $1.073\,543\,85(33) \times 10^{-9}$	$\{e/2R_\infty hc\}$ $0.036\,749\,309(11)$
1 u =	$\{m_u c^2/k\}$ $1.080\,947\,8(91) \times 10^{13}$	$\{m_u c^2/e\}$ $931.494\,32(28) \times 10^{6}$	1	$\{m_u c/2R_\infty h\}$ $3.423\,177\,25(31) \times 10^{7}$
1 hartree =	$\{2R_\infty hc/k\}$ $3.157\,733(27) \times 10^{5}$	$\{2R_\infty hc/e\}$ $27.211\,3961(81)$	$\{2R_\infty h/m_u c\}$ $2.921\,262\,69(26) \times 10^{-8}$	1

Table 5 Expanded matrix of variances, covariances, and correlation coefficients for the 1986 recommended set of fundamental physical constants. The elements of the covariance matrix appear on and above the major diagonal in (parts in $10^9)^2$; correlation coefficients appear in *italics* below the diagonal. The values are given to as many as six digits only as a matter of consistency. The correlation coefficient between m_e and N_A appears as −1.000 in this table because the auxiliary constants were considered to be exact in carrying out the least-squares adjustment. When the uncertainties of m_p/m_e and M_p are properly taken into account, the correlation coefficient is −0.999 and the variances of m_e and N_A are slightly increased

	α^{-1}	K_V	K_Ω	μ_μ/μ_p	e	h	m_e	N_A	F
α^{-1}	1 997	−1 062	925	3 267	−3 059	−4 121	−127	127	−2 932
K_V	*−0.080*	87 988	90	−1 737	89 050	177 038	174 914	−174 914	−85 864
K_Ω	*0.416*	*0.006*	2 477	1 513	−835	−744	1 105	−1 105	−1 939
μ_μ/μ_p	*0.498*	*−0.040*	*0.207*	21 523	−5 004	−6 742	−208	208	−4 796
e	*−0.226*	*0.989*	*−0.055*	*−0.112*	92 109	181 159	175 042	−175 042	−82 933
h	*−0.154*	*0.997*	*−0.025*	*−0.077*	*0.997*	358 197	349 956	−349 956	−168 797
m_e	*−0.005*	*0.997*	*0.038*	*−0.002*	*0.975*	*0.989*	349 702	−349 702	−174 660
N_A	*0.005*	*−0.997*	*−0.038*	*0.002*	*−0.975*	*−0.989*	*−1.000*	349 702	174 660
F	*−0.217*	*−0.956*	*−0.129*	*−0.108*	*−0.902*	*−0.931*	*−0.975*	*0.975*	91 727

values of the constants with uncertainties five to seven times smaller than the uncertainties assigned to the 1986 values. However, the changes in the values themselves are less than twice the 1986 assigned one-standard-deviation uncertainties and thus are not highly significant. Although much new data has become available since 1986, three new results dominate the analysis: a value of the Planck constant h obtained from a realization of the watt; a value of the fine-structure constant α obtained by equating the experimental value and theoretical expression for the magnetic moment anomaly of the electron a_e; and a value of the molar gas constant R obtained from a measurement of the speed of sound in argon. Because of their dominant role in determining the values and uncertainties of many of the constants, it is highly desirable that additional results of comparable uncertainty that corroborate these three data items be obtained before the 1996 least-squares adjustment is carried out.

REFERENCES

1. **Cohen, E. R. and Taylor, B. N.,** The 1986 Adjustment of the Fundamental Physical Constants, a Report of the CODATA Task Group on Fundamental Constants, CODATA Bulletin No. 63 (Pergamon, Fairview Park, Elmsford, NY 10523, USA, or Headington Hill Hall, Oxford OX3 OBW, UK, 1986).
2. **Cohen, E. R. and Taylor, B. N.,** The 1986 adjustment of the fundamental physical constants, *Rev. Mod. Phys.,* **59,** 1121, 1987.
3. Recommended Consistent Values of the Fundamental Physical Constants, 1973, a Report of the CODATA Task Group on Fundamental Constants, CODATA Bulletin No. 11 (CODATA Secretariat, 51 Blvd. de Monmorency, 75016 Paris, France, 1973).
4. **Cohen, E. R. and Taylor, B. N.,** The 1973 least-squares adjustment of the fundamental constants, *J. Phys. Chem. Ref. Data,* **2,** 663, 1973.

5. BIPM Com. Cons. Électricité, **17**, E6, E10, E88, and E92, 1986; BIPM Proc.-Verb. Com. Int. Poids et Mesures, **54**, 8, 1986; BIPM Comptes Rendus 18e Conf. Gén. Poids et Mesures, p. 100, 1987.
6. **Giacomo, P.,** News from the BIPM, *Metrologia,* **24**, 45, 1987; **25**, 113, 1988.
7. **Taylor, B. N. and Witt, T. J.,** New international reference standards based on the Josephson and quantum Hall effects, *Metrologia,* **26**, 47, 1989.
8. **Quinn, T. J.,** News from the BIPM, *Metrologia,* **26**, 69, 1989.
9. **Taylor, B. N. and Cohen, E. R.,** Recommended values of the fundamental physical constants: a status report, *J. Res. Natl Inst. Stand. Technol.,* **95**, 497, 1990.

7

SUMMARY TABLES OF PARTICLE PROPERTIES

Particle Data Group

Table of contents

SUMMARY TABLES OF PARTICLE PROPERTIES

July 1994

Particle Data Group

M. Aguilar-Benitez, R.M. Barnett, C. Caso, G. Conforto, R.L. Crawford, S. Eidelman, C. Grab, D.E. Groom, A. Gurtu, K.G. Hayes, J.J. Hernandez, K. Hikasa, G. Höhler, S. Kawabata, D.M. Manley, A. Manohar, L. Montanet, R.J. Morrison, H. Murayama, K. Olive, F.C. Porter, M. Roos, R.H. Schindler, R.E. Shrock, J. Stone, N.A. Törnqvist, T.G. Trippe, C.G. Wohl, and R.L. Workman

Technical Associates: B. Armstrong, K. Gieselmann, P. Lantero, G.S. Wagman

Reprinted from Physical Review **D50**, (August 1994)

(Approximate closing date for data: January 1, 1994)

GAUGE AND HIGGS BOSONS

γ

$I(J^{PC}) = 0,1(1^{--})$

Mass $m < 3 \times 10^{-27}$ eV
Charge $q < 2 \times 10^{-32}$ e
Mean life τ = Stable

g or gluon

$I(J^P) = 0(1^-)$

Mass $m = 0$ [a]
SU(3) color octet

W

$J = 1$

Charge = ± 1 e
Mass $m = 80.22 \pm 0.26$ GeV
$m_Z - m_W = 10.96 \pm 0.26$ GeV
$m_{W^+} - m_{W^-} = -0.2 \pm 0.6$ GeV
Full width $\Gamma = 2.08 \pm 0.07$ GeV

W^- modes are charge conjugates of the modes below.

W^+ DECAY MODES		Fraction (Γ_i/Γ)	Confidence level	p (MeV/c)
$e^+\nu$		(10.8 ± 0.4) %		40100
$\mu^+\nu$		(10.6 ± 0.7) %		40100
$\tau^+\nu$		(10.8 ± 1.0) %		40100
$\ell^+\nu$	[b]	(10.7 ± 0.5) %		40100
hadrons		(67.8 ± 1.5) %		–
$\pi^+\gamma$		$< 5 \times 10^{-4}$	95%	40110

Z

$J = 1$

Charge = 0
Mass $m = 91.187 \pm 0.007$ GeV [c]
Full width $\Gamma = 2.490 \pm 0.007$ GeV
$\Gamma(\ell^+\ell^-) = 83.84 \pm 0.27$ MeV [b]
$\Gamma(\text{invisible}) = 498.2 \pm 4.2$ MeV [d]
$\Gamma(\text{hadrons}) = 1740.7 \pm 5.9$ MeV
$\Gamma(\mu^+\mu^-)/\Gamma(e^+e^-) = 1.000 \pm 0.005$
$\Gamma(\tau^+\tau^-)/\Gamma(e^+e^-) = 0.998 \pm 0.005$ [e]
$g_V^\ell = -0.0377 \pm 0.0016$
$g_A^\ell = -0.5008 \pm 0.0008$

Asymmetry parameters

$A_e = 0.161 \pm 0.012$ [f] (S = 1.7)
$A_\tau = 0.141 \pm 0.021$ [f] (S = 1.2)

Charge asymmetry at Z pole

$A_{FB}^{(0,\ell)} = (1.59 \pm 0.18) \times 10^{-2}$
$A_{FB}^{(0,c)} = (5.8 \pm 2.2) \times 10^{-2}$
$A_{FB}^{(0,b)} = (10.7 \pm 1.3) \times 10^{-2}$

Z DECAY MODES			Fraction (Γ_i/Γ)	Confidence level	p (MeV/c)
e^+e^-			(3.366 ± 0.008) %		45600
$\mu^+\mu^-$			(3.367 ± 0.013) %		45600
$\tau^+\tau^-$			(3.360 ± 0.015) %		45600
$\ell^+\ell^-$		[b]	(3.367 ± 0.006) %		45600
invisible			(20.01 ± 0.16) %		45600
hadrons			(69.90 ± 0.15) %		–
$(u\overline{u}+c\overline{c})/2$			(9.7 ± 1.8) %		–
$(d\overline{d}+s\overline{s}+b\overline{b})/3$			(16.8 ± 1.2) %		–
$c\overline{c}$			(11.9 ± 1.4) %		–
$b\overline{b}$			(15.45 ± 0.21) %		–
$\pi^0\gamma$			$< 5.5 \times 10^{-5}$	95%	45600
$\eta\gamma$			$< 5.1 \times 10^{-5}$	95%	45600
$\omega\gamma$			$< 6.5 \times 10^{-4}$	95%	45600
$\eta'(958)\gamma$			$< 4.2 \times 10^{-5}$	95%	45600
$\gamma\gamma$			$< 5.5 \times 10^{-5}$	95%	45600
$\gamma\gamma\gamma$			$< 1.7 \times 10^{-5}$	95%	45600
$\pi^\pm W^\mp$		[g]	$< 7 \times 10^{-5}$	95%	10300
$\rho^\pm W^\mp$		[g]	$< 8.3 \times 10^{-5}$	95%	10300
$K^0 X$			(61.5 ± 0.6) %		–
$K^*(892)^+ X$			(51 ± 5) %		–
ΛX			(20.9 ± 0.6) %		–
$\Xi^- X$			(1.42 ± 0.14) %		–
$\Sigma(1385)^\pm X$			(2.6 ± 0.4) %		–
$\Xi(1530)^0 X$			$(4.4\pm1.0) \times 10^{-3}$		–
$\Omega^- X$			$(3.5\pm1.0) \times 10^{-3}$		–
$J/\psi(1S) X$			$(3.8\pm0.5) \times 10^{-3}$		–
$\chi_{c1}(1P) X$			$(7.5\pm3.0) \times 10^{-3}$		–
$(D^0/\overline{D}^0) X$			(28 ± 4) %		–
$D^\pm X$			(13.9 ± 2.1) %		–
$D^*(2010)^\pm X$		[g]	(12.5 ± 1.3) %		–
$B_s^0 X$			seen		–
anomalous γ+ hadrons		[h]	$< 3.2 \times 10^{-3}$	95%	–
$e^+e^-\gamma$		[h]	$< 5.2 \times 10^{-4}$	95%	45600
$\mu^+\mu^-\gamma$		[h]	$< 5.6 \times 10^{-4}$	95%	45600
$\tau^+\tau^-\gamma$		[h]	$< 7.3 \times 10^{-4}$	95%	45600
$\ell^+\ell^-\gamma\gamma$		[i]	$< 6.8 \times 10^{-6}$	95%	45600
$q\overline{q}\gamma\gamma$		[i]	$< 5.5 \times 10^{-6}$	95%	–
$\nu\overline{\nu}\gamma\gamma$		[i]	$< 3.1 \times 10^{-6}$	95%	45600
$e^\pm\mu^\mp$	LF	[g]	$< 6 \times 10^{-6}$	95%	45600
$e^\pm\tau^\mp$	LF	[g]	$< 1.3 \times 10^{-5}$	95%	45600
$\mu^\pm\tau^\mp$	LF	[g]	$< 1.9 \times 10^{-5}$	95%	45600

Searches for Higgs Bosons — H^0 and $H^\pm$

H^0 Mass $m >$ 58.4 GeV, CL = 95%

H_1^0 in Supersymmetric Models ($m_{H_1^0} < m_{H_2^0}$) [j]

Mass $m >$ 44 GeV, CL = 95% for $\tan\beta > 1$

A^0 Pseudoscalar Higgs Boson in Supersymmetric Models [j]

Mass $m >$ 22 GeV, CL = 95% for $50 > \tan\beta > 1$

$H^\pm$ Mass $m >$ 41.7 GeV, CL = 95%

See the Full Listings for a Note giving details of Higgs Bosons.

Searches for Heavy Bosons Other Than Higgs Bosons

Additional W Bosons

W_R — right-handed W
Mass $m >$ 406 GeV, CL = 90%
(assuming light right-handed neutrino)

W' with standard couplings decaying to $e\nu$, $\mu\nu$
Mass $m >$ 520 GeV, CL = 95%

Gauge & Higgs Boson Summary Table

Additional Z Bosons

Z'_{SM} with standard couplings
Mass $m >$ 412 GeV, CL = 95% ($p\overline{p}$ direct search)
Mass $m >$ 779 GeV, CL = 95% (electroweak fit)

Z_{LR} of $SU(2)_L \times SU(2)_R \times U(1)$
(with $g_L = g_R$)
Mass $m >$ 310 GeV, CL = 95% ($p\overline{p}$ direct search)
Mass $m >$ 389 GeV, CL = 95% (electroweak fit)

Z_χ of $SO(10) \rightarrow SU(5) \times U(1)_\chi$
(coupling constant derived from G.U.T.)
Mass $m >$ 340 GeV, CL = 95% ($p\overline{p}$ direct search)
Mass $m >$ 321 GeV, CL = 95% (electroweak fit)

Z_ψ of $E_6 \rightarrow SO(10) \times U(1)_\psi$
(coupling constant derived from G.U.T.)
Mass $m >$ 320 GeV, CL = 95% ($p\overline{p}$ direct search)
Mass $m >$ 160 GeV, CL = 95% (electroweak fit)

Z_η of $E_6 \rightarrow SU(3) \times SU(2) \times U(1) \times U(1)_\eta$
(coupling constant derived from G.U.T.;
charges are $Q_\eta = \sqrt{3/8}Q_\chi - \sqrt{5/8}Q_\psi$)
Mass $m >$ 340 GeV, CL = 95% ($p\overline{p}$ direct search)
Mass $m >$ 182 GeV, CL = 95% (electroweak fit)

Scalar Leptoquarks

Mass $m >$ 120 GeV, CL = 95% (1st generation, pair prod.)
Mass $m >$ 181 GeV, CL = 95% (1st gener., single prod.)
Mass $m >$ 44.5 GeV, CL = 95% (2nd gener., pair prod.)
Mass $m >$ 73 GeV, CL = 95% (2nd gener., single prod.)
Mass $m >$ 45 GeV, CL = 95% (3rd gener., pair prod.)
(last four limits are for charge $-1/3$, weak isoscalar)

Searches for Axions (A^0) and Other Very Light Bosons

The standard Peccei-Quinn axion is ruled out. Variants with reduced couplings or much smaller masses are constrained by various data. The Full Listings in the full *Review* contain a Note discussing axion searches.

The best limit for the half-life of neutrinoless double beta decay with Majoron emission is $> 7.2 \times 10^{24}$ years (CL = 90%).

NOTES

In this Summary Table:

When a quantity has "(S = ...)" to its right, the error on the quantity has been enlarged by the "scale factor" S, defined as $S = \sqrt{\chi^2/(N-1)}$, where N is the number of measurements used in calculating the quantity. We do this when $S > 1$, which often indicates that the measurements are inconsistent. When $S > 1.25$, we also show in the Full Listings an ideogram of the measurements. For more about S, see the Introduction.

A decay momentum p is given for each decay mode. For a 2-body decay, p is the momentum of each decay product in the rest frame of the decaying particle. For a 3-or-more-body decay, p is the largest momentum any of the products can have in this frame.

[*a*] Theoretical value. A mass as large as a few MeV may not be precluded.

[*b*] ℓ indicates each type of lepton (e, μ, and τ), not sum over them.

[*c*] The Z-boson mass listed here corresponds to a Breit-Wigner resonance parameter. It lies approximately 34 MeV above the real part of the position of the pole (in the energy plane) in the Z-boson propagator.

[*d*] This partial width takes into account Z decays into $\nu\overline{\nu}$ and any other possible undetected modes.

[*e*] This ratio has not been corrected for the τ mass.

[*f*] Here $A \equiv 2g_V g_A/(g_V^2+g_A^2)$.

[*g*] The value is for the sum of the charge states indicated.

[*h*] See the Z Full Listings for the γ energy range used in this measurement.

[*i*] For $m_{\gamma\gamma} = (60 \pm 5)$ GeV.

[*j*] The limits assume no invisible decays.

Lepton & Quark Summary Table

LEPTONS

Neutrinos

See the Full Listings for a Note giving details of neutrinos, masses, mixing, and the status of experimental searches.

ν_e

$J = \frac{1}{2}$

Mass m: The formal upper limit, as obtained from the m^2 average (see the Full Listings), is 5.1 eV at the 95% CL. Caution is urged in interpreting this result, since the m^2 average is positive with only a 3.5% probability. If the weighted average m^2 were forced to zero, the limit would increase to 7.0 eV.

Mean life/mass, $\tau/m_{\nu_e} > 300$ s/eV, CL = 90%

Magnetic moment $\mu < 1.08 \times 10^{-9}\ \mu_B$, CL = 90%

ν_μ

$J = \frac{1}{2}$

Mass $m < 0.27$ MeV, CL = 90%

Mean life/mass, $\tau/m_{\nu_\mu} > 15.4$ s/eV, CL = 90%

Magnetic moment $\mu < 7.4 \times 10^{-10}\ \mu_B$, CL = 90%

ν_τ

$J = \frac{1}{2}$

Mass $m < 31$ MeV, CL = 95%

Magnetic moment $\mu < 5.4 \times 10^{-7}\ \mu_B$, CL = 90%

e

$J = \frac{1}{2}$

Mass $m = 0.51099906 \pm 0.00000015$ MeV [a]

$= (5.48579903 \pm 0.00000013) \times 10^{-4}$ u

$(m_{e^+} - m_{e^-})/m < 4 \times 10^{-8}$, CL = 90%

$|q_{e^+} + q_{e^-}|/e < 4 \times 10^{-8}$

Magnetic moment $\mu = 1.001159652193 \pm 0.000000000010\ \mu_B$

$(g_{e^+} - g_{e^-})\ /\ g_{\text{average}} = (-0.5 \pm 2.1) \times 10^{-12}$

Electric dipole moment $d = (-0.3 \pm 0.8) \times 10^{-26}$ $e$cm

Mean life $\tau > 2.7 \times 10^{23}$ yr, CL = 68% [b]

μ

$J = \frac{1}{2}$

Mass $m = 105.658389 \pm 0.000034$ MeV [a]

$= 0.113428913 \pm 0.000000017$ u

Mean life $\tau = (2.19703 \pm 0.00004) \times 10^{-6}$ s

$\tau_{\mu^+}/\tau_{\mu^-} = 1.00002 \pm 0.00008$

$c\tau = 658.654$ m

Magnetic moment $\mu = 1.001165923 \pm 0.000000008\ e\hbar/2m_\mu$

$(g_{\mu^+} - g_{\mu^-})\ /\ g_{\text{average}} = (-2.6 \pm 1.6) \times 10^{-8}$

Electric dipole moment $d = (3.7 \pm 3.4) \times 10^{-19}$ $e$cm

Decay parameters [c]

$\rho = 0.7518 \pm 0.0026$

$\eta = -0.007 \pm 0.013$

$\delta = 0.749 \pm 0.004$

$\xi P_\mu = 1.003 \pm 0.008$ [d]

$\xi P_\mu \delta/\rho > 0.99682$, CL = 90% [d]

$\xi' = 1.00 \pm 0.04$

$\xi'' = 0.7 \pm 0.4$

$\alpha/A = (0 \pm 4) \times 10^{-3}$

$\alpha'/A = (0 \pm 4) \times 10^{-3}$

$\beta/A = (4 \pm 6) \times 10^{-3}$

$\beta'/A = (2 \pm 6) \times 10^{-3}$

$\overline{\eta} = 0.02 \pm 0.08$

μ^+ modes are charge conjugates of the modes below.

μ^- DECAY MODES		Fraction (Γ_i/Γ)	Confidence level	p (MeV/c)
$e^- \overline{\nu}_e \nu_\mu$		$\approx 100\%$		53
$e^- \overline{\nu}_e \nu_\mu \gamma$		[e] (1.4 ± 0.4) %		53
$e^- \overline{\nu}_e \nu_\mu e^+ e^-$		[f] $(3.4 \pm 0.4) \times 10^{-5}$		53
Lepton Family number (*LF*) violating modes				
$e^- \nu_e \overline{\nu}_\mu$	LF	[g] < 1.2 %	90%	53
$e^- \gamma$	LF	$< 4.9 \times 10^{-11}$	90%	53
$e^- e^+ e^-$	LF	$< 1.0 \times 10^{-12}$	90%	53
$e^- 2\gamma$	LF	$< 7.2 \times 10^{-11}$	90%	53

τ

$J = \frac{1}{2}$

Mass $m = 1777.1^{+0.4}_{-0.5}$ MeV

Mean life $\tau = (295.6 \pm 3.1) \times 10^{-15}$ s

$c\tau = 88.6\ \mu$m

Electric dipole moment $d < 5 \times 10^{-17}$ $e$cm, CL = 95%

Weak dipole moment $< 3.7 \times 10^{-17}$ $e$cm, CL = 95%

Decay parameters

See the τ Full Listings for a note concerning τ-decay parameters.

$\rho^\tau(e \text{ or } \mu) = 0.74 \pm 0.04$

$\rho^\tau(e) = 0.72 \pm 0.04$

$\rho^\tau(\mu) = 0.76 \pm 0.05$

$\xi^\tau(e \text{ or } \mu) = 0.90 \pm 0.18$

W-τ couplings $2g_A g_V/(g_A^2 + g_V^2) = 1.25^{+0.27}_{-0.24}$

τ^+ modes are charge conjugates of the modes below. "$h^\pm$" stands for $\pi^\pm$ or $K^\pm$. "ℓ" stands for e or μ. "Neutral" means neutral hadron whose decay products include γ's and/or π^0's.

τ^- DECAY MODES	Fraction (Γ_i/Γ)	Scale factor/ Confidence level	p (MeV/c)
Modes with one charged particle			
particle$^-$ ≥ 0 neutrals ν_τ ("1-prong")	(85.49 ± 0.24) %	S=1.5	–
$\mu^- \overline{\nu}_\mu \nu_\tau$	(17.65 ± 0.24) %	S=1.1	885
$\mu^- \overline{\nu}_\mu \nu_\tau \gamma$ ($E_\gamma > 37$ MeV)	$(2.3 \pm 1.1) \times 10^{-3}$		–
$e^- \overline{\nu}_e \nu_\tau$	(18.01 ± 0.18) %	S=1.1	889
$h^- \geq 0$ neutrals ν_τ	(49.83 ± 0.35) %	S=1.3	–
$h^- \nu_\tau$	(12.88 ± 0.34) %	S=1.2	–
$\pi^- \nu_\tau$	(11.7 ± 0.4) %	S=1.3	883
$K^- \geq 0$ neutrals ν_τ	(1.68 ± 0.24) %		–
$K^- \nu_\tau$	$(6.7 \pm 2.3) \times 10^{-3}$	S=1.3	820
$K^- \geq 1$ neutrals ν_τ	$(1.2^{+0.5}_{-0.6})$ %		–
$h^- \geq 1$ neutrals ν_τ	(36.9 ± 0.4) %	S=1.3	–
$h^- \pi^0 \nu_\tau$	(25.7 ± 0.4) %	S=1.7	–
$\pi^- \pi^0 \nu_\tau$	(25.2 ± 0.4) %	S=1.7	878
$h^- \geq 2\pi^0 \nu_\tau$	(11.2 ± 0.4) %	S=1.5	–
$h^- 2\pi^0 \nu_\tau$	(9.6 ± 0.4) %	S=1.5	–
$h^- \geq 3\pi^0 \nu_\tau$	(1.48 ± 0.26) %	S=1.7	–
$h^- 3\pi^0 \nu_\tau$	(1.28 ± 0.24) %	S=1.7	–
$h^- 4\pi^0 \nu_\tau$	$(1.9^{+1.1}_{-1.0}) \times 10^{-3}$	S=1.6	–
Modes with three charged particles			
$2h^- h^+ \geq 0$ neutrals ν_τ ("3-prong")	(14.38 ± 0.24) %	S=1.5	–
$h^- h^- h^+ \nu_\tau$	(8.42 ± 0.31) %	S=1.3	–
$h^- h^- h^+ \geq 1$ neutrals ν_τ	(5.63 ± 0.30) %	S=1.2	–
$h^- h^- h^+ 2\pi^0 \nu_\tau$	$(4.9 \pm 0.5) \times 10^{-3}$		–
$\omega \pi^- \geq 0$ neutrals ν_τ	(1.6 ± 0.4) %		–
$\omega \pi^- \nu_\tau$	(1.6 ± 0.5) %		708
$h^- \omega \pi^0 \nu_\tau$	$(4.0 \pm 0.6) \times 10^{-3}$		–
$K^- h^+ h^- \geq 0$ neutrals ν_τ	$< 6 \times 10^{-3}$	CL=90%	–
$K^- \pi^+ \pi^- \geq 0$ neutrals ν_τ	$(2.2^{+1.6}_{-1.3}) \times 10^{-3}$		–
$K^- K^+ \pi^- \nu_\tau$	$(2.2^{+1.7}_{-1.1}) \times 10^{-3}$		685
Modes with five charged particles			
$3h^- 2h^+ \geq 0$ neutrals ν_τ ("5-prong")	$(1.25 \pm 0.24) \times 10^{-3}$		–
$3h^- 2h^+ \nu_\tau$	$(5.6 \pm 1.6) \times 10^{-4}$		–
$3h^- 2h^+ \pi^0 \nu_\tau$	$(5.1 \pm 2.2) \times 10^{-4}$		–

Lepton & Quark Summary Table

Miscellaneous other allowed modes

$4h^- 3h^+ \geq 0$ neutrals ν_τ ("7-prong")	< 1.9	$\times 10^{-4}$	CL=90%	–
$K^*(892)^- \geq 0$ neutrals ν_τ	(1.43 ± 0.17)	%		–
$K^*(892)^- \nu_\tau$	(1.45 ± 0.18)	%		665
$K^*(892)^0 K^- \geq 0$ neutrals ν_τ	(3.2 ± 1.4)	$\times 10^{-3}$		–
$\overline{K}^*(892)^0 \pi^- \geq 0$ neutrals ν_τ	(3.8 ± 1.7)	$\times 10^{-3}$		–
$K^0 h^- \geq 0$ neutrals ν_τ	(1.30 ± 0.30)	%		–
$K^- K^0 \geq 0$ neutrals ν_τ	< 8	$\times 10^{-3}$	CL=90%	–
$K^0 K^- \nu_\tau$	< 2.6	$\times 10^{-3}$	CL=95%	737
$K^0 K^- \geq 1$ neutrals ν_τ	< 2.6	$\times 10^{-3}$	CL=95%	–
$K^0 h^+ h^- h^- \geq 0$ neutrals ν_τ	< 1.7	$\times 10^{-3}$	CL=95%	–
$K_2^*(1430)^- \nu_\tau$	< 3	$\times 10^{-3}$	CL=95%	314
$\eta \pi^- \geq 0$ neutrals ν_τ	< 1.3	%	CL=95%	–
$\eta \pi^- \nu_\tau$	< 3.4	$\times 10^{-4}$	CL=95%	798
$\eta \pi^- \pi^0 \nu_\tau$	(1.70 ± 0.28)	$\times 10^{-3}$		778
$\eta \pi^- \pi^0 \pi^0 \nu_\tau$	< 4.3	$\times 10^{-4}$	CL=95%	746
$\eta K^- \nu_\tau$	< 4.7	$\times 10^{-4}$	CL=95%	720
$\eta \pi^+ \pi^- \pi^- \geq 0$ neutrals ν_τ	< 3	$\times 10^{-3}$	CL=90%	–
$\eta \eta \pi^- \geq 0$ neutrals ν_τ	< 5	$\times 10^{-3}$	CL=90%	–
$\eta \eta \pi^- \nu_\tau$	< 1.1	$\times 10^{-4}$	CL=95%	637
$\eta \eta \pi^- \pi^0 \nu_\tau$	< 2.0	$\times 10^{-4}$	CL=95%	559

Lepton Family number (*LF*), Lepton number (*L*), or Baryon number (*B*) violating modes (In the modes below, ℓ means a sum over e and μ modes)

L means lepton number violation (*e.g.* $\tau^- \to e^+ \pi^- \pi^-$). Following common usage, *LF* means lepton family violation *and not* lepton number violation (*e.g.* $\tau^- \to e^- \pi^+ \pi^-$).

$e^- \gamma$	*LF*		< 1.2	$\times 10^{-4}$	CL=90%	889
$\mu^- \gamma$	*LF*		< 4.2	$\times 10^{-6}$	CL=90%	885
$e^- \pi^0$	*LF*		< 1.4	$\times 10^{-4}$	CL=90%	883
$\mu^- \pi^0$	*LF*		< 4.4	$\times 10^{-5}$	CL=90%	880
$e^- K^0$	*LF*		< 1.3	$\times 10^{-3}$	CL=90%	819
$\mu^- K^0$	*LF*		< 1.0	$\times 10^{-3}$	CL=90%	815
$e^- \eta$	*LF*		< 6.3	$\times 10^{-5}$	CL=90%	804
$\mu^- \eta$	*LF*		< 7.3	$\times 10^{-5}$	CL=90%	800
$e^- \rho^0$	*LF*		< 1.9	$\times 10^{-5}$	CL=90%	723
$\mu^- \rho^0$	*LF*		< 2.9	$\times 10^{-5}$	CL=90%	718
$e^- K^*(892)^0$	*LF*		< 3.8	$\times 10^{-5}$	CL=90%	665
$\mu^- K^*(892)^0$	*LF*		< 4.5	$\times 10^{-5}$	CL=90%	660
$\pi^- \gamma$	*L*		< 2.8	$\times 10^{-4}$	CL=90%	883
$\pi^- \pi^0$	*L*		< 3.7	$\times 10^{-4}$	CL=90%	878
$\ell^- \ell^- \ell^+$	*LF*	[*h*]	< 3.4	$\times 10^{-5}$	CL=90%	–
$e^- e^+ e^-$	*LF*		< 1.3	$\times 10^{-5}$	CL=90%	889
$(e \mu \mu)^-$	*LF*		< 2.7	$\times 10^{-5}$	CL=90%	882
$e^- \mu^+ \mu^-$	*LF*		< 1.9	$\times 10^{-5}$	CL=90%	882
$e^+ \mu^- \mu^-$	*LF*		< 1.6	$\times 10^{-5}$	CL=90%	882
$(\mu e e)^-$	*LF*		< 2.7	$\times 10^{-5}$	CL=90%	885
$\mu^- e^+ e^-$	*LF*		< 1.4	$\times 10^{-5}$	CL=90%	885
$\mu^+ e^- e^-$	*LF*		< 1.4	$\times 10^{-5}$	CL=90%	885
$\mu^- \mu^+ \mu^-$	*LF*		< 1.7	$\times 10^{-5}$	CL=90%	873
$\ell^\pm \pi^\mp \pi^-$	*LF,L*	[*h,i*]	< 6.3	$\times 10^{-5}$	CL=90%	–
$e^\mp \pi^\pm \pi^-$	*LF,L*	[*i*]	< 6.0	$\times 10^{-5}$	CL=90%	877
$e^- \pi^+ \pi^-$	*LF*		< 2.7	$\times 10^{-5}$	CL=90%	877
$e^+ \pi^- \pi^-$	*L*		< 1.7	$\times 10^{-5}$	CL=90%	877
$\mu^\mp \pi^\pm \pi^-$	*LF,L*	[*i*]	< 3.9	$\times 10^{-5}$	CL=90%	866
$\mu^- \pi^+ \pi^-$	*LF*		< 3.6	$\times 10^{-5}$	CL=90%	866
$\mu^+ \pi^- \pi^-$	*L*		< 3.9	$\times 10^{-5}$	CL=90%	866
$\ell^\pm \pi^\mp K^-$	*LF,L*	[*h,i*]	< 1.2	$\times 10^{-4}$	CL=90%	–
$(e \pi K)^-$, all charged	*LF,L*		< 7.7	$\times 10^{-5}$	CL=90%	814
$e^- \pi^\pm K^\mp$	*LF*	[*i*]	< 5.8	$\times 10^{-5}$	CL=90%	814
$e^- \pi^+ K^-$	*LF*		< 2.9	$\times 10^{-5}$	CL=90%	814
$e^- \pi^- K^+$	*LF*		< 5.8	$\times 10^{-5}$	CL=90%	814
$e^+ \pi^- K^-$	*L*		< 2.0	$\times 10^{-5}$	CL=90%	814
$(\mu \pi K)^-$, all charged	*LF,L*		< 7.7	$\times 10^{-5}$	CL=90%	800
$\mu^- \pi^\pm K^\mp$	*LF*	[*i*]	< 7.7	$\times 10^{-5}$	CL=90%	800
$\mu^- \pi^+ K^-$	*LF*		< 7.7	$\times 10^{-5}$	CL=90%	800
$\mu^- \pi^- K^+$	*LF*		< 7.7	$\times 10^{-5}$	CL=90%	800
$\mu^+ \pi^- K^-$	*L*		< 4.0	$\times 10^{-5}$	CL=90%	800
$\overline{p} \gamma$	*L,B*		< 2.9	$\times 10^{-4}$	CL=90%	641
$\overline{p} \pi^0$	*L,B*		< 6.6	$\times 10^{-4}$	CL=90%	632
$\overline{p} \eta$	*L,B*		< 1.30	$\times 10^{-3}$	CL=90%	476
e^- light spinless boson	*LF*		< 3.2	$\times 10^{-3}$	CL=95%	–
μ^- light spinless boson	*LF*		< 6	$\times 10^{-3}$	CL=95%	–

Number of Light Neutrino Types

(including ν_e, ν_μ, and ν_τ)

Number $N = 2.983 \pm 0.025$ (Standard Model fits to Z data)

Number $N = 2.97 \pm 0.17$ (Direct measurement of invisible Z width)

Heavy Lepton Searches

$L^\pm$ – charged lepton

Mass $m >$ 44.3 GeV, CL = 95% $m_\nu \approx 0$

$L^\pm$ – stable charged heavy lepton

Mass $m >$ 42.8 GeV, CL = 95%

L^0 – stable neutral heavy lepton

Mass $m >$ 45.0 GeV, CL = 95% (Dirac)

Mass $m >$ 39.5 GeV, CL = 95% (Majorana)

Neutral heavy lepton

Mass $m >$ 19.6 GeV, CL = 95% (all $|U_{\ell j}|^2$) (Dirac)

Mass $m >$ 45.7 GeV or $m < 25$, CL = 95% ($|U_{\ell j}|^2 > 10^{-13}$ (Dirac)

Searches for Massive Neutrinos and Lepton Mixing

For excited leptons, see Compositeness Limits below.

See the Full Listings for a Note giving details of neutrinos, masses, mixing, and the status of experimental searches.

No direct, uncontested evidence for massive neutrinos or lepton mixing has been obtained. Sample limits are:

ν oscillation: $\overline{\nu}_e \not\to \overline{\nu}_e$

$\Delta(m^2) < 0.0083$ eV^2, CL = 90% (if $\sin^2 2\theta = 1$)

$\sin^2 2\theta < 0.14$, CL = 68% (if $\Delta(m^2)$ is large)

ν oscillation: $\nu_\mu \to \nu_e$ (θ = mixing angle)

$\Delta(m^2) < 0.09$ eV^2, CL = 90% (if $\sin^2 2\theta = 1$)

$\sin^2 2\theta < 2.5 \times 10^{-3}$, CL = 90% (if $\Delta(m^2)$ is large)

Lepton & Quark Summary Table

QUARKS

The *u*-, *d*-, and *s*-quark masses are estimates of so-called "current-quark masses," in a mass-independent subtraction scheme such as $\overline{\text{MS}}$ at a scale $\mu \approx 1$ GeV. The *c*- and *b*-quark masses are estimated from charmonium, bottomonium, *D*, and *B* masses. They are the "running" masses in the $\overline{\text{MS}}$ scheme. These can be different from the heavy quark masses obtained in potential models.

u

$I(J^P) = \frac{1}{2}(\frac{1}{2}^+)$

Mass m = 2 to 8 MeV [j]   Charge = $\frac{2}{3}$ e $I_z = +\frac{1}{2}$
m_u/m_d = 0.25 to 0.70

d

$I(J^P) = \frac{1}{2}(\frac{1}{2}^+)$

Mass m = 5 to 15 MeV [j]   Charge = $-\frac{1}{3}$ e $I_z = -\frac{1}{2}$
m_s/m_d = 17 to 25

s

$I(J^P) = 0(\frac{1}{2}^+)$

Mass m = 100 to 300 MeV [j]   Charge = $-\frac{1}{3}$ e Strangeness = −1
$(m_s - (m_u + m_d)/2)/(m_d - m_u)$ = 34 to 51

c

$I(J^P) = 0(\frac{1}{2}^+)$

Mass m = 1.0 to 1.6 GeV   Charge = $\frac{2}{3}$ e Charm = +1

b

$I(J^P) = 0(\frac{1}{2}^+)$

Mass m = 4.1 to 4.5 GeV   Charge = $-\frac{1}{3}$ e Bottom = −1

Searches for *t* Quark

$I(J^P) = 0(\frac{1}{2}^+)$

Charge = $\frac{2}{3}$ e Top = +1

Mass $m >$ 62 GeV, CL = 95%   (all decays)
Mass $m >$ 131 GeV, CL = 95%   (assumes $t \to W b$ decay)
Mass $m = 174 \pm 10^{+13}_{-12}$ GeV   (top candidate events)
Mass $m = 169^{+16}_{-18}{}^{+17}_{-20}$ GeV   (Standard Model electroweak fit)

The first result is from a CDF $\Gamma(W)$ measurement; the second is from a DØ direct search; the third is from a CDF observation of top candidate events. CDF observes a 2.8σ effect which is not sufficient to firmly establish the existence of top but which, if interpreted as top, yields the third result. The fourth result is from a Standard Model electroweak fit to Z, W, and νN data not including direct m_t measurements. The central value assumes m_H = 300 GeV while the second upper (lower) error corresponds to m_H = 1000 (60) GeV.

Note Added in Proof

After the 1994 Edition of the Review of Particle Properties was published, the CDF and D0 experiments at Fermilab published clear observations of the top quark with high statistical significance. CDF (Abe, F., *et al.*, (1995). CDF Collaboration, *Phys. Rev. Lett.* **74,** 2626) finds a top mass of (176 ± 8 ± 10) GeV while D0 (Abachi, S., *et al.* (1995). D0 Collaboration, *Phys. Rev. Lett.* **74,** 2632) finds (199 + 19 − 21 ± 22) GeV, where the first errors are statistical and the second are systematic. The average of these two results, combining statistical and systematic errors in quadrature, gives a top mass of 180 ± 12 GeV.

Searches for *b′* (4th Generation) Quark

Mass $m >$ 85 GeV, CL = 95%   ($p\overline{p}$, charged current decays)
Mass $m >$ 46.0 GeV, CL = 95%   ($e^+ e^-$, all decays)

NOTES

In this Summary Table:

When a quantity has "(S = ...)" to its right, the error on the quantity has been enlarged by the "scale factor" S, defined as $S = \sqrt{\chi^2/(N-1)}$, where N is the number of measurements used in calculating the quantity. We do this when S > 1, which often indicates that the measurements are inconsistent. When S > 1.25, we also show in the Full Listings an ideogram of the measurements. For more about S, see the Introduction.

A decay momentum p is given for each decay mode. For a 2-body decay, p is the momentum of each decay product in the rest frame of the decaying particle. For a 3-or-more-body decay, p is the largest momentum any of the products can have in this frame.

[*a*] The masses of the e and μ are most precisely known in u (unified atomic mass units). The conversion factor to MeV, 1 u = 931.49432(28) MeV, is less well known than are the masses in u.

[*b*] This is the best "electron disappearance" limit. The best limit for the mode $e^- \to \nu\gamma$ is > 2.35×10^{25} yr (CL=68%).

[*c*] See the "Note on Muon Decay Parameters" in the μ Full Listings for definitions and details.

[*d*] P_μ is the longitudinal polarization of the muon from pion decay. In standard V−A theory, $P_\mu = 1$ and $\rho = \delta = 3/4$.

[*e*] This only includes events with the γ energy > 10 MeV. Since the $e^-\overline{\nu}_e\nu_\mu$ and $e^-\overline{\nu}_e\nu_\mu\gamma$ modes cannot be clearly separated, we regard the latter mode as a subset of the former.

[*f*] See the μ Full Listings for the energy limits used in this measurement.

[*g*] A test of additive vs. multiplicative lepton family number conservation.

[*h*] ℓ means a sum over e and μ modes.

[*i*] The value is for the sum of the charge states indicated.

[*j*] The ratios m_u/m_d and m_s/m_d are extracted from pion and kaon masses using chiral symmetry. The estimates of u and d masses are not without controversy and remain under active investigation. Within the literature there are even suggestions that the u quark could be essentially massless. The s-quark mass is estimated from SU(3) splittings in hadron masses.

Meson Summary Table

LIGHT UNFLAVORED MESONS
$(S = C = B = 0)$

For $I = 1$ (π, b, ρ, a): $u\overline{d}$, $(u\overline{u}-d\overline{d})/\sqrt{2}$, $d\overline{u}$;
for $I = 0$ $(\eta, \eta', h, h', \omega, \phi, f, f')$: $c_1(u\overline{u} + d\overline{d}) + c_2(s\overline{s})$

$\pi^\pm$

$I^G(J^P) = 1^-(0^-)$

Mass $m = 139.56995 \pm 0.00035$ MeV [a]
Mean life $\tau = (2.6030 \pm 0.0024) \times 10^{-8}$ s
$c\tau = 7.804$ m

$\pi^\pm \to \ell^\pm \nu\gamma$ form factors [b]

$F_V = 0.017 \pm 0.008$
$F_A = 0.0116 \pm 0.0016$ (S = 1.3)
$R = 0.059^{+0.009}_{-0.008}$

π^- modes are charge conjugates of the modes below.

π^+ DECAY MODES		Fraction (Γ_i/Γ)	Confidence level	p (MeV/c)
$\mu^+\nu_\mu$	[c]	(99.98770 ± 0.00004) %		30
$\mu^+\nu_\mu\gamma$	[d]	$(1.24 \pm 0.25) \times 10^{-4}$		30
$e^+\nu_e$	[c]	$(1.230 \pm 0.004) \times 10^{-4}$		70
$e^+\nu_e\gamma$	[d]	$(1.61 \pm 0.23) \times 10^{-7}$		70
$e^+\nu_e\pi^0$		$(1.025 \pm 0.034) \times 10^{-8}$		4
$e^+\nu_e e^+ e^-$		$(3.2 \pm 0.5) \times 10^{-9}$		70
$e^+\nu_e\nu\overline{\nu}$		$< 5 \times 10^{-6}$	90%	70
Lepton Family number (*LF*) or Lepton number (*L*) violating modes				
$\mu^+\overline{\nu}_e$	*L* [e]	$< 1.5 \times 10^{-3}$	90%	30
$\mu^+\nu_e$	*LF* [e]	$< 8.0 \times 10^{-3}$	90%	30
$\mu^- e^+ e^+ \nu$	*LF*	$< 1.6 \times 10^{-6}$	90%	30

π^0

$I^G(J^{PC}) = 1^-(0^{-+})$

Mass $m = 134.9764 \pm 0.0006$ MeV [a]
$m_{\pi^\pm} - m_{\pi^0} = 4.5936 \pm 0.0005$ MeV
Mean life $\tau = (8.4 \pm 0.6) \times 10^{-17}$ s (S = 3.0)
$c\tau = 25.1$ nm

π^0 DECAY MODES		Fraction (Γ_i/Γ)	Scale factor/ Confidence level	p (MeV/c)
2γ		(98.798 ± 0.032) %	S=1.1	67
$e^+e^-\gamma$		(1.198 ± 0.032) %	S=1.1	67
γ positronium		$(1.82 \pm 0.29) \times 10^{-9}$		67
$e^+e^+e^-e^-$		$(3.14 \pm 0.30) \times 10^{-5}$		67
e^+e^-		$(7.5 \pm 2.0) \times 10^{-8}$		67
4γ		$< 2 \times 10^{-8}$	CL=90%	67
$\nu\overline{\nu}$	[f]	$< 8.3 \times 10^{-7}$	CL=90%	67
$\nu_e\overline{\nu}_e$		$< 1.7 \times 10^{-6}$	CL=90%	67
$\nu_\mu\overline{\nu}_\mu$		$< 3.1 \times 10^{-6}$	CL=90%	67
$\nu_\tau\overline{\nu}_\tau$		$< 2.1 \times 10^{-6}$	CL=90%	67
Charge conjugation (*C*) or Lepton Family number (*LF*) violating modes				
3γ	*C*	$< 3.1 \times 10^{-8}$	CL=90%	67
$\mu^+e^- + e^-\mu^+$	*LF*	$< 1.72 \times 10^{-8}$	CL=90%	26

η

$I^G(J^{PC}) = 0^+(0^{-+})$

Mass $m = 547.45 \pm 0.19$ MeV (S = 1.6)
Full width $\Gamma = 1.20 \pm 0.11$ keV [g] (S = 1.8)

C-nonconserving decay parameters [h]

$\pi^+\pi^-\pi^0$	Left-right asymmetry	$= (0.09 \pm 0.17) \times 10^{-2}$
$\pi^+\pi^-\pi^0$	Sextant asymmetry	$= (0.18 \pm 0.16) \times 10^{-2}$
$\pi^+\pi^-\pi^0$	Quadrant asymmetry	$= (-0.17 \pm 0.17) \times 10^{-2}$
$\pi^+\pi^-\gamma$	Left-right asymmetry	$= (0.9 \pm 0.4) \times 10^{-2}$
$\pi^+\pi^-\gamma$	β (*D*-wave)	$= 0.05 \pm 0.06$ (S = 1.5)

η DECAY MODES		Fraction (Γ_i/Γ)	Scale factor/ Confidence level	p (MeV/c)
neutral modes		(70.8 ± 0.8) %	S=1.2	–
2γ	[g]	(38.8 ± 0.5) %	S=1.2	274
$3\pi^0$		(31.9 ± 0.4) %	S=1.2	180
$\pi^0 2\gamma$		$(7.1 \pm 1.4) \times 10^{-4}$		258
charged modes		(29.2 ± 0.8) %	S=1.2	–
$\pi^+\pi^-\pi^0$		(23.6 ± 0.6) %	S=1.2	175
$\pi^+\pi^-\gamma$		(4.88 ± 0.15) %	S=1.2	236
$e^+e^-\gamma$		$(5.0 \pm 1.2) \times 10^{-3}$		274
$\mu^+\mu^-\gamma$		$(3.1 \pm 0.4) \times 10^{-4}$		253
e^+e^-		$< 3 \times 10^{-4}$	CL=90%	274
$\mu^+\mu^-$		$(5.7 \pm 0.8) \times 10^{-6}$		253
$\pi^+\pi^-e^+e^-$		$(1.3^{+1.3}_{-0.8}) \times 10^{-3}$		236
$\pi^+\pi^-2\gamma$		$< 2.1 \times 10^{-3}$		236
$\pi^+\pi^-\pi^0\gamma$		$< 6 \times 10^{-4}$	CL=90%	175
$\pi^0\mu^+\mu^-\gamma$		$< 3 \times 10^{-6}$	CL=90%	211
Charge conjugation (*C*), Parity (*P*), or Charge conjugation × Parity (*CP*) violating modes				
$\pi^+\pi^-$	*P,CP*	$< 1.5 \times 10^{-3}$		236
3γ	*C*	$< 5 \times 10^{-4}$	CL=95%	274
$\pi^0e^+e^-$	*C* [i]	$< 4 \times 10^{-5}$	CL=90%	258
$\pi^0\mu^+\mu^-$	*C* [i]	$< 5 \times 10^{-6}$	CL=90%	211

$\rho(770)$

$I^G(J^{PC}) = 1^+(1^{--})$

Mass $m = 769.9 \pm 0.8$ MeV (S = 1.8)
Full width $\Gamma = 151.2 \pm 1.2$ MeV
$\Gamma_{ee} = 6.77 \pm 0.32$ keV

$\rho(770)$ DECAY MODES		Fraction (Γ_i/Γ)	Scale factor/ Confidence level	p (MeV/c)
$\pi\pi$		~ 100 %		359
$\rho(770)^\pm$ decays				
$\pi^\pm\gamma$		$(4.5 \pm 0.5) \times 10^{-4}$	S=2.2	372
$\pi^\pm\eta$		$< 6 \times 10^{-3}$	CL=84%	147
$\pi^\pm\pi^+\pi^-\pi^0$		$< 2.0 \times 10^{-3}$	CL=84%	250
$\rho(770)^0$ decays				
$\pi^+\pi^-\gamma$		$(9.9 \pm 1.6) \times 10^{-3}$		359
$\pi^0\gamma$		$(7.9 \pm 2.0) \times 10^{-4}$		373
$\eta\gamma$		$(3.8 \pm 0.7) \times 10^{-4}$		190
$\mu^+\mu^-$	[j]	$(4.60 \pm 0.28) \times 10^{-5}$		370
e^+e^-	[j]	$(4.46 \pm 0.21) \times 10^{-5}$		385
$\pi^+\pi^-\pi^0$		$< 1.2 \times 10^{-4}$	CL=90%	320
$\pi^+\pi^-\pi^+\pi^-$		$< 2 \times 10^{-4}$	CL=90%	247
$\pi^+\pi^-\pi^0\pi^0$		$< 4 \times 10^{-5}$	CL=90%	253

$\omega(782)$

$I^G(J^{PC}) = 0^-(1^{--})$

Mass $m = 781.94 \pm 0.12$ MeV (S = 1.5)
Full width $\Gamma = 8.43 \pm 0.10$ MeV
$\Gamma_{ee} = 0.60 \pm 0.02$ keV

$\omega(782)$ DECAY MODES	Fraction (Γ_i/Γ)	Confidence level	p (MeV/c)
$\pi^+\pi^-\pi^0$	(88.8 ± 0.7) %		327
$\pi^0\gamma$	(8.5 ± 0.5) %		379
$\pi^+\pi^-$	(2.21 ± 0.30) %		365
neutrals (excluding $\pi^0\gamma$)	$(5.3^{+8.7}_{-3.5}) \times 10^{-3}$		–
$\eta\gamma$	$(8.3 \pm 2.1) \times 10^{-4}$		199
$\pi^0e^+e^-$	$(5.9 \pm 1.9) \times 10^{-4}$		379
$\pi^0\mu^+\mu^-$	$(9.6 \pm 2.3) \times 10^{-5}$		349
e^+e^-	$(7.15 \pm 0.19) \times 10^{-5}$		391
$\pi^+\pi^-\pi^0\pi^0$	< 2 %	90%	261
$\pi^+\pi^-\gamma$	$< 3.6 \times 10^{-3}$	95%	365
$\pi^+\pi^-\pi^+\pi^-$	$< 1 \times 10^{-3}$	90%	256
$\pi^0\pi^0\gamma$	$< 4 \times 10^{-4}$	90%	367
$\mu^+\mu^-$	$< 1.8 \times 10^{-4}$	90%	376

$\eta'(958)$ $\qquad I^G(J^{PC}) = 0^+(0^{-+})$

Mass $m = 957.77 \pm 0.14$ MeV
Full width $\Gamma = 0.201 \pm 0.016$ MeV (S = 1.3)

$\eta'(958)$ DECAY MODES	Fraction (Γ_i/Γ)	Scale factor/ Confidence level	p (MeV/c)
$\pi^+\pi^-\eta$	(43.7 ±1.5) %	S=1.1	232
$\rho^0\gamma$	(30.2 ±1.3) %	S=1.1	169
$\pi^0\pi^0\eta$	(20.8 ±1.3) %	S=1.2	239
$\omega\gamma$	(3.02±0.30) %		160
$\gamma\gamma$	(2.12±0.13) %	S=1.2	479
$3\pi^0$	(1.55±0.26) × 10^{-3}		430
$\mu^+\mu^-\gamma$	(1.04±0.26) × 10^{-4}		467
$\pi^+\pi^-\pi^0$	< 5 %	CL=90%	427
$\pi^0\rho^0$	< 4 %	CL=90%	118
$\pi^+\pi^-$	< 2 %	CL=90%	458
$\pi^0 e^+ e^-$	< 1.3 %	CL=90%	469
$\eta e^+ e^-$	< 1.1 %	CL=90%	322
$\pi^+\pi^+\pi^-\pi^-$	< 1 %	CL=90%	372
$\pi^+\pi^+\pi^-\pi^-$ neutrals	< 1 %	CL=95%	–
$\pi^+\pi^+\pi^-\pi^-\pi^0$	< 1 %	CL=90%	298
6π	< 1 %	CL=90%	189
$\pi^+\pi^- e^+ e^-$	< 6 × 10^{-3}	CL=90%	458
$\pi^0\pi^0$	< 9 × 10^{-4}	CL=90%	459
$\pi^0\gamma\gamma$	< 8 × 10^{-4}	CL=90%	469
$4\pi^0$	< 5 × 10^{-4}	CL=90%	379
3γ	< 1.0 × 10^{-4}	CL=90%	479
$\mu^+\mu^-\pi^0$	< 6.0 × 10^{-5}	CL=90%	445
$\mu^+\mu^-\eta$	< 1.5 × 10^{-5}	CL=90%	274
$\pi^+\pi^-\gamma$ (including $\rho^0\gamma$)	(27.9 ±2.3) %		–
e^+e^-	< 2.1 × 10^{-7}	CL=90%	479

$f_0(980)$ was $S(975)$ $\qquad I^G(J^{PC}) = 0^+(0^{++})$

Mass $m = 980 \pm 10$ MeV
Full width Γ = 40 to 400 MeV

$f_0(980)$ DECAY MODES	Fraction (Γ_i/Γ)	Confidence level	p (MeV/c)
$\pi\pi$	(78.1 ±2.4) %		470
$K\overline{K}$	(21.9 ±2.4) %		–
$\gamma\gamma$	(1.19±0.33) × 10^{-5}		490
e^+e^-	< 3 × 10^{-7}	90%	490

$a_0(980)$ was $\delta(980)$ $\qquad I^G(J^{PC}) = 1^-(0^{++})$

Mass $m = 982.4 \pm 1.4$ MeV
Full width Γ = 50 to 300 MeV

$a_0(980)$ DECAY MODES	Fraction (Γ_i/Γ)	p (MeV/c)
$\eta\pi$	dominant	319
$K\overline{K}$	seen	–
$\gamma\gamma$	seen	491

$\phi(1020)$ $\qquad I^G(J^{PC}) = 0^-(1^{--})$

Mass $m = 1019.413 \pm 0.008$ MeV
Full width $\Gamma = 4.43 \pm 0.06$ MeV
$\Gamma_{ee} = 1.37 \pm 0.05$ keV

$\phi(1020)$ DECAY MODES	Fraction (Γ_i/Γ)	Scale factor/ Confidence level	p (MeV/c)
K^+K^-	(49.1 ±0.9) %	S=1.3	127
$K^0_L K^0_S$	(34.3 ±0.7) %	S=1.2	110
$\rho\pi$	(12.9 ±0.7) %		181
$\pi^+\pi^-\pi^0$	(2.5 ±0.9) %	S=1.1	462
$\eta\gamma$	(1.28±0.06) %	S=1.2	363
$\pi^0\gamma$	(1.31±0.13) × 10^{-3}		501
e^+e^-	(3.09±0.07) × 10^{-4}		510
$\mu^+\mu^-$	(2.48±0.34) × 10^{-4}		499
$\eta e^+ e^-$	(1.3 $^{+0.8}_{-0.6}$) × 10^{-4}		363
$\pi^+\pi^-$	(8 $^{+5}_{-4}$) × 10^{-5}	S=1.5	490
$\omega\gamma$	< 5 %	CL=84%	210
$\rho\gamma$	< 2 %	CL=84%	219
$\pi^+\pi^-\gamma$	< 7 × 10^{-3}	CL=90%	490
$f_0(980)\gamma$	< 2 × 10^{-3}	CL=90%	39
$\pi^0\pi^0\gamma$	< 1 × 10^{-3}	CL=90%	492
$\pi^+\pi^-\pi^+\pi^-$	< 8.7 × 10^{-4}	CL=90%	410
$\eta'(958)\gamma$	< 4.1 × 10^{-4}	CL=90%	60
$\pi^+\pi^+\pi^-\pi^-\pi^0$	< 1.5 × 10^{-4}	CL=95%	341
$\pi^0 e^+ e^-$	< 1.2 × 10^{-4}	CL=90%	501
$\pi^0\eta\gamma$	< 2.5 × 10^{-3}	CL=90%	346
$a_0(980)\gamma$	< 5 × 10^{-3}	CL=90%	36

$h_1(1170)$ $\qquad I^G(J^{PC}) = 0^-(1^{+-})$

Mass $m = 1170 \pm 20$ MeV
Full width $\Gamma = 360 \pm 40$ MeV

$h_1(1170)$ DECAY MODES	Fraction (Γ_i/Γ)	p (MeV/c)
$\rho\pi$	seen	310

$b_1(1235)$ $\qquad I^G(J^{PC}) = 1^+(1^{+-})$

Mass $m = 1231 \pm 10$ MeV [k]
Full width $\Gamma = 142 \pm 8$ MeV (S = 1.1)

$b_1(1235)$ DECAY MODES	Fraction (Γ_i/Γ)	Confidence level	p (MeV/c)
$\omega\pi$	dominant		348
[D/S amplitude ratio = 0.26 ± 0.04]			
$\pi^\pm\gamma$	(1.6±0.4) × 10^{-3}		608
$\eta\rho$	seen		–
$\pi^+\pi^+\pi^-\pi^0$	< 50 %	84%	536
$(K\overline{K})^\pm\pi^0$	< 8 %	90%	248
$K^0_S K^0_L \pi^\pm$	< 6 %	90%	238
$K^0_S K^0_S \pi^\pm$	< 2 %	90%	238
$\pi\phi$	< 1.5 %	84%	146

$a_1(1260)$ $\qquad I^G(J^{PC}) = 1^-(1^{++})$

Mass $m = 1230 \pm 40$ MeV [k]
Full width $\Gamma \sim 400$ MeV

$a_1(1260)$ DECAY MODES	Fraction (Γ_i/Γ)	Confidence level	p (MeV/c)
$\rho\pi$	dominant		356
$\pi\gamma$	seen		607
$\pi(\pi\pi)_{S\text{-wave}}$	[k] <0.7 %	90%	575

$f_2(1270)$ $\qquad I^G(J^{PC}) = 0^+(2^{++})$

Mass $m = 1275 \pm 5$ MeV [k]
Full width $\Gamma = 185 \pm 20$ MeV [k]

$f_2(1270)$ DECAY MODES	Fraction (Γ_i/Γ)	Scale factor/ Confidence level	p (MeV/c)
$\pi\pi$	(84.9 $^{+2.5}_{-1.3}$) %	S=1.3	622
$\pi^+\pi^-2\pi^0$	(6.9 $^{+1.5}_{-2.7}$) %	S=1.4	562
$K\overline{K}$	(4.6 ±0.5) %	S=2.8	403
$2\pi^+2\pi^-$	(2.8 ±0.4) %	S=1.2	559
$\eta\eta$	(4.5 ±1.0) × 10^{-3}	S=2.4	327
$4\pi^0$	(3.0 ±1.0) × 10^{-3}		564
$\gamma\gamma$	(1.32$^{+0.18}_{-0.16}$) × 10^{-5}	S=1.1	637
$\eta\pi\pi$	< 8 × 10^{-3}	CL=95%	475
$K^0K^-\pi^+$ + c.c.	< 3.4 × 10^{-3}	CL=95%	293
e^+e^-	< 9 × 10^{-9}	CL=90%	637

Meson Summary Table

$f_1(1285)$

$I^G(J^{PC}) = 0^+(1^{++})$

Mass $m = 1282 \pm 5$ MeV [k]
Full width $\Gamma = 24 \pm 3$ MeV [k]

$f_1(1285)$ DECAY MODES	Fraction (Γ_i/Γ)	Scale factor/ Confidence level	p (MeV/c)
4π	(29 ± 6) %		563
$\pi^0\pi^0\pi^+\pi^-$	$(15 {}^{+9}_{-8})$ %	S=1.1	–
$2\pi^+2\pi^-$	(15 ± 6) %		563
$\rho^0\pi^+\pi^-$	dominates $2\pi^+ 2\pi^-$		340
$4\pi^0$	$< 7 \times 10^{-4}$	CL=90%	568
$\eta\pi\pi$	(54 ± 15) %		479
$a_0(980)\pi$ [ignoring $a_0(980) \to K\overline{K}$]	(44 ± 7) %	S=1.1	234
$\eta\pi\pi$ [excluding $a_0(980)\pi$]	$(10 {}^{+7}_{-6})$ %	S=1.1	–
$K\overline{K}\pi$	(9.7 ± 1.6) %	S=1.2	308
$K\overline{K}^*(892)$	not seen		–
$\gamma\rho^0$	(6.6 ± 1.3) %	S=1.5	410
$\phi\gamma$	$(8.0 \pm 3.1) \times 10^{-4}$		236

$\eta(1295)$

$I^G(J^{PC}) = 0^+(0^{-+})$

Mass $m = 1295 \pm 4$ MeV
Full width $\Gamma = 53 \pm 6$ MeV

$\eta(1295)$ DECAY MODES	Fraction (Γ_i/Γ)	p (MeV/c)
$\eta\pi^+\pi^-$	seen	488
$a_0(980)\pi$	seen	245

$f_0(1300)$
was $f_0(1400)$
was $\epsilon(1200)$

$I^G(J^{PC}) = 0^+(0^{++})$

Mass $m = 1000$-1500 MeV
Full width $\Gamma = 150$ to 400 MeV
$\Gamma_{\gamma\gamma} = 5.4 \pm 2.3$ keV
$\Gamma_{ee} < 20$ eV, CL = 90%

$f_0(1300)$ DECAY MODES	Fraction (Γ_i/Γ)	p (MeV/c)
$\pi\pi$	$(93.6 {}^{+1.9}_{-1.5})$ %	–
$K\overline{K}$	(7.5 ± 0.9) %	–
$\eta\eta$	seen	–
$\gamma\gamma$	seen	–
e^+e^-	not seen	–

$\pi(1300)$

$I^G(J^{PC}) = 1^-(0^{-+})$

Mass $m = 1300 \pm 100$ MeV [k]
Full width $\Gamma = 200$ to 600 MeV

$\pi(1300)$ DECAY MODES	Fraction (Γ_i/Γ)	p (MeV/c)
$\rho\pi$	seen	406
$\pi(\pi\pi)_{S\text{-wave}}$	seen	612

$a_2(1320)$

$I^G(J^{PC}) = 1^-(2^{++})$

Mass $m = 1318.4 \pm 0.6$ MeV (S = 1.1) (3π and $K^\pm K^0_S$ modes)
Full width $\Gamma = 107 \pm 5$ MeV [k] ($K^\pm K^0_S$ and $\eta\pi$ modes)

$a_2(1320)$ DECAY MODES	Fraction (Γ_i/Γ)	Scale factor/ Confidence level	p (MeV/c)
$\rho\pi$	(70.1 ± 2.7) %	S=1.2	419
$\eta\pi$	(14.5 ± 1.2) %		535
$\omega\pi\pi$	(10.6 ± 3.2) %	S=1.3	362
$K\overline{K}$	(4.9 ± 0.8) %		437
$\eta'(958)\pi$	$(5.7 \pm 1.1) \times 10^{-3}$		287
$\pi^\pm\gamma$	$(2.8 \pm 0.6) \times 10^{-3}$		652
$\gamma\gamma$	$(9.7 \pm 1.0) \times 10^{-6}$		659
$\pi^+\pi^-\pi^-$	< 8 %	CL=90%	621
e^+e^-	$< 2.3 \times 10^{-7}$	CL=90%	659

$f_1(1420)$ [l]

$I^G(J^{PC}) = 0^+(1^{++})$

Mass $m = 1426.8 \pm 2.3$ MeV (S = 1.3)
Full width $\Gamma = 52 \pm 4$ MeV

$f_1(1420)$ DECAY MODES	Fraction (Γ_i/Γ)	p (MeV/c)
$K\overline{K}\pi$	dominant	439
$\eta\pi\pi$	possibly seen	571

$\omega(1420)$ [m]

$I^G(J^{PC}) = 0^-(1^{--})$

Mass $m = 1419 \pm 31$ MeV
Full width $\Gamma = 174 \pm 60$ MeV

$\omega(1420)$ DECAY MODES	Fraction (Γ_i/Γ)	p (MeV/c)
$\rho\pi$	dominant	488

$\eta(1440)$ [n]
was $\iota(1440)$

$I^G(J^{PC}) = 0^+(0^{-+})$

Mass $m = 1420 \pm 20$ MeV [k]
Full width $\Gamma = 60 \pm 30$ MeV [k]

$\eta(1440)$ DECAY MODES	Fraction (Γ_i/Γ)	p (MeV/c)
$K\overline{K}\pi$	seen	433
$\eta\pi\pi$	seen	567
$a_0(980)\pi$	seen	350
4π	seen	640

$\rho(1450)$ [o]

$I^G(J^{PC}) = 1^+(1^{--})$

Mass $m = 1465 \pm 25$ MeV [k]
Full width $\Gamma = 310 \pm 60$ MeV [k]

$\rho(1450)$ DECAY MODES	Fraction (Γ_i/Γ)	Confidence level	p (MeV/c)
$\pi\pi$	seen		719
4π	seen		665
e^+e^-	seen		732
$\eta\rho$	<4 %		317
$\omega\pi$	<2.0 %	95%	512
$\phi\pi$	<1 %		358
$K\overline{K}$	$<1.6 \times 10^{-3}$	95%	541

$f_1(1510)$

$I^G(J^{PC}) = 0^+(1^{++})$

Mass $m = 1512 \pm 4$ MeV
Full width $\Gamma = 35 \pm 15$ MeV

$f_1(1510)$ DECAY MODES	Fraction (Γ_i/Γ)	p (MeV/c)
$K\overline{K}^*(892)$ + c.c.	seen	292

$f'_2(1525)$

$I^G(J^{PC}) = 0^+(2^{++})$

Mass $m = 1525 \pm 5$ MeV [k]
Full width $\Gamma = 76 \pm 10$ MeV [k]

$f'_2(1525)$ DECAY MODES	Fraction (Γ_i/Γ)	p (MeV/c)
$K\overline{K}$	$(71.2 {}^{+2.0}_{-2.5})$ %	581
$\eta\eta$	$(27.9 {}^{+2.5}_{-2.0})$ %	531
$\pi\pi$	$(8.2 \pm 1.6) \times 10^{-3}$	750
$\gamma\gamma$	$(1.23 \pm 0.22) \times 10^{-6}$	763

$f_0(1590)$ $I^G(J^{PC}) = 0^+(0^{++})$

Seen by one group only.
Mass $m = 1581 \pm 10$ MeV
Full width $\Gamma = 180 \pm 17$ MeV (S = 1.2)

$f_0(1590)$ DECAY MODES	Fraction (Γ_i/Γ)	p (MeV/c)
$\eta\eta'(958)$	dominant	234
$\eta\eta$	large	570
$4\pi^0$	large	732

$\omega(1600)$ [p] $I^G(J^{PC}) = 0^-(1^{--})$

Mass $m = 1662 \pm 13$ MeV
Full width $\Gamma = 280 \pm 24$ MeV

$\omega(1600)$ DECAY MODES	Fraction (Γ_i/Γ)	p (MeV/c)
$\rho\pi$	seen	644
$\omega\pi\pi$	seen	610
e^+e^-	seen	831

$\omega_3(1670)$ $I^G(J^{PC}) = 0^-(3^{--})$

Mass $m = 1668 \pm 5$ MeV
Full width $\Gamma = 173 \pm 11$ MeV [k]

$\omega_3(1670)$ DECAY MODES	Fraction (Γ_i/Γ)	p (MeV/c)
$\rho\pi$	seen	647
$\omega\pi\pi$	seen	614
$b_1(1235)\pi$	possibly seen	359

$\pi_2(1670)$ $I^G(J^{PC}) = 1^-(2^{-+})$

Mass $m = 1670 \pm 20$ MeV [k]
Full width $\Gamma = 240 \pm 15$ MeV [k] (S = 1.1)
$\Gamma_{ee} = 1.35 \pm 0.26$ keV

$\pi_2(1670)$ DECAY MODES	Fraction (Γ_i/Γ)	p (MeV/c)
$f_2(1270)\pi$	(56.2±3.2) %	325
$\pi^\pm\pi^+\pi^-$	(53 ±4) %	–
$\rho\pi$	(31 ±4) %	649
$f_0(1300)\pi$	(8.7±3.4) %	–
$K\overline{K}^*(892)$+ c.c.	(4.2±1.4) %	453
$\gamma\gamma$	(5.6±1.1) × 10^{-6}	835
$\eta\pi$	< 5 %	738
$\pi^\pm 2\pi^+ 2\pi^-$	< 5 %	734

$\phi(1680)$ $I^G(J^{PC}) = 0^-(1^{--})$

Not a well-established resonance.
Mass $m = 1680 \pm 50$ MeV [k]
Full width $\Gamma = 150 \pm 50$ MeV [k]

$\phi(1680)$ DECAY MODES	Fraction (Γ_i/Γ)	p (MeV/c)
$K\overline{K}^*(892)$+ c.c.	dominant	462
$K^0_S K\pi$	seen	619
$K\overline{K}$	seen	680
e^+e^-	seen	840
$\omega\pi\pi$	not seen	621

$\rho_3(1690)$ $I^G(J^{PC}) = 1^+(3^{--})$

J^P from the 2π and $K\overline{K}$ modes.
Mass $m = 1691 \pm 5$ MeV [k] (2π, $K\overline{K}$, and $K\overline{K}\pi$ modes)
Full width $\Gamma = 215 \pm 20$ MeV [k] (2π, $K\overline{K}$, and $K\overline{K}\pi$ modes)

$\rho_3(1690)$ DECAY MODES	Fraction (Γ_i/Γ)	Scale factor	p (MeV/c)
4π	(71.1 ± 1.9) %		788
$\pi^\pm\pi^+\pi^-\pi^0$	(67 ±22) %		788
$\pi\pi$	(23.6 ± 1.3) %		834
$\omega\pi$	(16 ± 6) %		656
$K\overline{K}\pi$	(3.8 ± 1.2) %		628
$K\overline{K}$	(1.58± 0.26) %	1.2	686
$\eta\pi^+\pi^-$	seen		728

$\rho(1700)$ [o] $I^G(J^{PC}) = 1^+(1^{--})$

Mass $m = 1700 \pm 20$ MeV [k] ($\eta\rho^0$ and mixed modes)
Full width $\Gamma = 235 \pm 50$ MeV [k] ($\eta\rho^0$, $\pi^+\pi^-$, and mixed modes)

$\rho(1700)$ DECAY MODES	Fraction (Γ_i/Γ)	p (MeV/c)
$\rho\pi\pi$	dominant	640
$\rho^0\pi^+\pi^-$	large	640
$\rho^\pm\pi^\mp\pi^0$	[q] large	642
$2(\pi^+\pi^-)$	large	792
$\pi^+\pi^-$	seen	838
$K\overline{K}^*(892)$+ c.c.	seen	479
$\eta\rho$	seen	533
$K\overline{K}$	seen	692
e^+e^-	seen	850

$f_J(1710)$ was $\theta(1690)$ $I^G(J^{PC}) = 0^+(\text{even}^{++})$

Mass $m = 1709 \pm 5$ MeV
Full width $\Gamma = 140 \pm 12$ MeV

$f_J(1710)$ DECAY MODES	Fraction (Γ_i/Γ)	p (MeV/c)
$K\overline{K}$	seen	697
$\pi\pi$	seen	843

$\phi_3(1850)$ $I^G(J^{PC}) = 0^-(3^{--})$

Mass $m = 1854 \pm 7$ MeV
Full width $\Gamma = 87^{+28}_{-23}$ MeV (S = 1.2)

$\phi_3(1850)$ DECAY MODES	Fraction (Γ_i/Γ)	p (MeV/c)
$K\overline{K}$	seen	785
$K\overline{K}^*(892)$+ c.c.	seen	602

$f_2(2010)$ $I^G(J^{PC}) = 0^+(2^{++})$

Seen by one group only.
Mass $m = 2011^{+60}_{-80}$ MeV
Full width $\Gamma = 202 \pm 60$ MeV

$f_2(2010)$ DECAY MODES	Fraction (Γ_i/Γ)	p (MeV/c)
$\phi\phi$	seen	–

Meson Summary Table

$f_4(2050)$

$I^G(J^{PC}) = 0^+(4^{++})$

Mass $m = 2044 \pm 11$ MeV (S = 1.4)
Full width $\Gamma = 208 \pm 13$ MeV (S = 1.2)

$f_4(2050)$ DECAY MODES	Fraction (Γ_i/Γ)	p (MeV/c)
$\omega\omega$	(26 ±6) %	658
$\pi\pi$	(17.0±1.5) %	1012
$K\overline{K}$	($6.8^{+3.4}_{-1.8}$) × 10^{-3}	895
$\eta\eta$	(2.1±0.8) × 10^{-3}	863
$4\pi^0$	< 1.2 %	977

$f_2(2300)$

$I^G(J^{PC}) = 0^+(2^{++})$

Mass $m = 2297 \pm 28$ MeV
Full width $\Gamma = 149 \pm 40$ MeV

$f_2(2300)$ DECAY MODES	Fraction (Γ_i/Γ)	p (MeV/c)
$\phi\phi$	seen	529

$f_2(2340)$

$I^G(J^{PC}) = 0^+(2^{++})$

Mass $m = 2339 \pm 60$ MeV
Full width $\Gamma = 319^{+80}_{-70}$ MeV

$f_2(2340)$ DECAY MODES	Fraction (Γ_i/Γ)	p (MeV/c)
$\phi\phi$	seen	573

STRANGE MESONS
($S = \pm1, C = B = 0$)

$K^+ = u\overline{s}$, $K^0 = d\overline{s}$, $\overline{K}^0 = \overline{d}s$, $K^- = \overline{u}s$, similarly for K^*'s

$K^\pm$

$I(J^P) = \frac{1}{2}(0^-)$

Mass $m = 493.677 \pm 0.016$ MeV (S = 2.8)
Mean life $\tau = (1.2371 \pm 0.0029) \times 10^{-8}$ s (S = 2.2)
$c\tau = 3.709$ m

Slope parameter g [r]

(See Full Listings for quadratic coefficients)

$K^+ \to \pi^+\pi^+\pi^- = -0.2154 \pm 0.0035$ (S = 1.4)
$K^- \to \pi^-\pi^-\pi^+ = -0.217 \pm 0.007$ (S = 2.5)
$K^\pm \to \pi^\pm\pi^0\pi^0 = 0.594 \pm 0.019$ (S = 1.3)

$K^\pm$ decay form factors [b,s]

K^+_{e3} $\lambda_+ = 0.0286 \pm 0.0022$
$K^+_{\mu3}$ $\lambda_+ = 0.033 \pm 0.008$ (S = 1.6)
$K^+_{\mu3}$ $\lambda_0 = 0.004 \pm 0.007$ (S = 1.6)
K^+_{e3} $|f_S/f_+| = 0.084 \pm 0.023$ (S = 1.2)
K^+_{e3} $|f_T/f_+| = 0.38 \pm 0.11$ (S = 1.1)
$K^+_{\mu3}$ $|f_T/f_+| = 0.02 \pm 0.12$
$K^+ \to e^+\nu_e\gamma$ $|F_A + F_V| = 0.148 \pm 0.010$
$K^+ \to \mu^+\nu_\mu\gamma$ $|F_A + F_V| < 0.23$, CL = 90%
$K^+ \to e^+\nu_e\gamma$ $|F_A - F_V| < 0.49$
$K^+ \to \mu^+\nu_\mu\gamma$ $|F_A - F_V| = -2.2$ to 0.3

K^- modes are charge conjugates of the modes below.

K^+ DECAY MODES		Fraction (Γ_i/Γ)	Scale factor/ Confidence level	p (MeV/c)
$\mu^+\nu_\mu$		(63.51±0.18) %	S=1.3	236
$e^+\nu_e$		(1.55±0.07) × 10^{-5}		247
$\pi^+\pi^0$		(21.16±0.14) %	S=1.1	205
$\pi^+\pi^+\pi^-$		(5.59±0.05) %	S=1.9	125
$\pi^+\pi^0\pi^0$		(1.73±0.04) %	S=1.2	133
$\pi^0\mu^+\nu_\mu$		(3.18±0.08) %	S=1.5	215
Called $K^+_{\mu3}$.				
$\pi^0 e^+\nu_e$		(4.82±0.06) %	S=1.3	228
Called K^+_{e3}.				
$\pi^0\pi^0 e^+\nu_e$		(2.1 ±0.4) × 10^{-5}		206
$\pi^+\pi^- e^+\nu_e$		(3.91±0.17) × 10^{-5}		203
$\pi^+\pi^-\mu^+\nu_\mu$		(1.4 ±0.9) × 10^{-5}		151
$\pi^0\pi^0\pi^0 e^+\nu_e$		< 3.5 × 10^{-6}	CL=90%	135
$\pi^+\gamma\gamma$		[t] < 1 × 10^{-6}	CL=90%	227
$\pi^+3\gamma$		[t] < 1.0 × 10^{-4}	CL=90%	227
$e^+\nu_e\nu\overline{\nu}$		< 6 × 10^{-5}	CL=90%	247
$\mu^+\nu_\mu\nu\overline{\nu}$		< 6.0 × 10^{-6}	CL=90%	236
$\mu^+\nu_\mu e^+e^-$		(1.06±0.32) × 10^{-6}		236
$e^+\nu_e e^+e^-$		($2.1^{+2.1}_{-1.1}$) × 10^{-7}		247
$\mu^+\nu_\mu\mu^+\mu^-$		< 4.1 × 10^{-7}	CL=90%	185
$\mu^+\nu_\mu\gamma$		[t,u] (5.50±0.28) × 10^{-3}		236
$\pi^+\pi^0\gamma$		[t,u] (2.75±0.15) × 10^{-4}		205
$\pi^+\pi^0\gamma$(DE)		[t,v] (1.8 ±0.4) × 10^{-5}		205
$\pi^+\pi^+\pi^-\gamma$		[t,u] (1.04±0.31) × 10^{-4}		125
$\pi^+\pi^0\pi^0\gamma$		[t,u] ($7.4^{+5.5}_{-2.9}$) × 10^{-6}		133
$\pi^0\mu^+\nu_\mu\gamma$		[t,u] < 6.1 × 10^{-5}	CL=90%	215
$\pi^0 e^+\nu_e\gamma$		[t,u] (2.62±0.20) × 10^{-4}		228
$\pi^0 e^+\nu_e\gamma$(SD)		[w] < 5.3 × 10^{-5}	CL=90%	228
$\pi^0\pi^0 e^+\nu_e\gamma$		< 5 × 10^{-6}	CL=90%	206
Lepton Family number (*LF*), Lepton number (*L*), $\Delta S = \Delta Q$ (*SQ*) violating modes, or $\Delta S = 1$ weak neutral current (*S1*) modes				
$\pi^+\pi^+ e^-\overline{\nu}_e$	*SQ*	< 1.2 × 10^{-8}	CL=90%	203
$\pi^+\pi^+\mu^-\overline{\nu}_\mu$	*SQ*	< 3.0 × 10^{-6}	CL=95%	151
$\pi^+ e^+e^-$	*S1*	(2.74±0.23) × 10^{-7}		227
$\pi^+\mu^+\mu^-$	*S1*	< 2.3 × 10^{-7}	CL=90%	172
$\pi^+\nu\overline{\nu}$	*S1*	< 5.2 × 10^{-9}	CL=90%	227
$\mu^-\nu e^+e^+$	*LF*	< 2.0 × 10^{-8}	CL=90%	236
$\mu^+\nu_e$	*LF*	[e] < 4 × 10^{-3}	CL=90%	236
$\pi^+\mu^+e^-$	*LF*	< 2.1 × 10^{-10}	CL=90%	214
$\pi^+\mu^-e^+$	*LF*	< 7 × 10^{-9}	CL=90%	214
$\pi^-\mu^+e^+$	*L*	< 7 × 10^{-9}	CL=90%	214
$\pi^- e^+e^+$	*L*	< 1.0 × 10^{-8}	CL=90%	227
$\pi^-\mu^+\mu^+$	*L*	< 1.5 × 10^{-4}	CL=90%	172
$\mu^+\overline{\nu}_e$	*L*	[e] < 3.3 × 10^{-3}	CL=90%	236
$\pi^0 e^+\overline{\nu}_e$	*L*	[e] < 3 × 10^{-3}	CL=90%	228

K^0

$I(J^P) = \frac{1}{2}(0^-)$

50% K_S, 50% K_L
Mass $m = 497.672 \pm 0.031$ MeV
$m_{K^0} - m_{K^\pm} = 3.995 \pm 0.034$ MeV (S = 1.1)

K_S^0

$I(J^P) = \frac{1}{2}(0^-)$

Mean life $\tau = (0.8926 \pm 0.0012) \times 10^{-10}$ s

$c\tau = 2.676$ cm

CP-violation parameters [x]

$\mathrm{Im}(\eta_{+-0})^2 < 0.12$, CL = 90%

$\mathrm{Im}(\eta_{000})^2 < 0.1$, CL = 90%

K_S^0 DECAY MODES		Fraction (Γ_i/Γ)	Scale factor/ Confidence level	p (MeV/c)
$\pi^+\pi^-$		(68.61 ± 0.28) %	S=1.2	206
$\pi^0\pi^0$		(31.39 ± 0.28) %	S=1.2	209
$\pi^+\pi^-\gamma$	[u,y]	$(1.78 \pm 0.05) \times 10^{-3}$		206
$\gamma\gamma$		$(2.4 \pm 1.2) \times 10^{-6}$		249
$\pi^+\pi^-\pi^0$		$< 8.5 \times 10^{-5}$	CL=90%	133
$3\pi^0$		$< 3.7 \times 10^{-5}$	CL=90%	139
$\pi^\pm e^\mp \nu$	[z]	$(6.68 \pm 0.10) \times 10^{-4}$	S=1.3	229
$\pi^\pm \mu^\mp \nu$	[z]	$(4.66 \pm 0.07) \times 10^{-4}$	S=1.2	216
$\Delta S = 1$ weak neutral current ($S1$) modes				
$\mu^+\mu^-$	$S1$	$< 3.2 \times 10^{-7}$	CL=90%	225
e^+e^-	$S1$	$< 1.0 \times 10^{-5}$	CL=90%	249
$\pi^0 e^+ e^-$	$S1$	$< 1.1 \times 10^{-6}$	CL=90%	231

K_L^0

$I(J^P) = \frac{1}{2}(0^-)$

$m_{K_L} - m_{K_S} = (0.5333 \pm 0.0027) \times 10^{10}\ \hbar\ \mathrm{s}^{-1}$ (S = 1.2)

$= (3.510 \pm 0.018) \times 10^{-12}$ MeV

Mean life $\tau = (5.17 \pm 0.04) \times 10^{-8}$ s

$c\tau = 15.49$ m

Slope parameter g [r]

(See Full Listings for quadratic coefficients)

$K_L^0 \to \pi^+\pi^-\pi^0 = 0.670 \pm 0.014$ (S = 1.6)

K_L decay form factors [s]

K_{e3}^0 $\lambda_+ = 0.0300 \pm 0.0016$ (S = 1.2)

$K_{\mu 3}^0$ $\lambda_+ = 0.034 \pm 0.005$ (S = 2.3)

$K_{\mu 3}^0$ $\lambda_0 = 0.025 \pm 0.006$ (S = 2.3)

K_{e3}^0 $|f_S/f_+| < 0.04$, CL = 68%

K_{e3}^0 $|f_T/f_+| < 0.23$, CL = 68%

$K_{\mu 3}^0$ $|f_T/f_+| = 0.12 \pm 0.12$

$K_L \to e^+e^-\gamma$: $\alpha_{K^*} = -0.28 \pm 0.08$

CP-violation parameters [x]

$\delta = (0.327 \pm 0.012)$%

$|\eta_{00}| = (2.259 \pm 0.023) \times 10^{-3}$ (S = 1.1)

$|\eta_{+-}| = (2.269 \pm 0.023) \times 10^{-3}$ (S = 1.1)

$|\eta_{00}/\eta_{+-}| = 0.9955 \pm 0.0023$ [aa] (S = 1.8)

$\epsilon'/\epsilon = (1.5 \pm 0.8) \times 10^{-3}$ [aa] (S = 1.8)

$\phi_{+-} = (44.3 \pm 0.8)°$

$\phi_{00} = (43.3 \pm 1.3)°$

$\Delta S = -\Delta Q$ in $K_{\ell 3}^0$ decay

Re $x = 0.006 \pm 0.018$ (S = 1.3)

Im $x = -0.003 \pm 0.026$ (S = 1.2)

K_L^0 DECAY MODES		Fraction (Γ_i/Γ)	Scale factor/ Confidence level	p (MeV/c)
$3\pi^0$		(21.6 ± 0.8) %	S=1.5	139
$\pi^+\pi^-\pi^0$		(12.38 ± 0.21) %	S=1.5	133
$\pi^\pm\mu^\mp\nu$	[q]	(27.0 ± 0.4) %	S=1.3	216
Called $K_{\mu 3}^0$.				
$\pi^\pm e^\mp \nu$	[q]	(38.7 ± 0.5) %	S=1.4	229
Called K_{e3}^0.				
2γ		$(5.73 \pm 0.27) \times 10^{-4}$	S=2.0	249
$\pi^0 2\gamma$	[bb]	$(1.70 \pm 0.28) \times 10^{-6}$		231
$\pi^0\pi^\pm e^\mp \nu$	[q]	$(5.18 \pm 0.29) \times 10^{-5}$		207
$(\pi\mu\,\mathrm{atom})\nu$		$(1.05 \pm 0.11) \times 10^{-7}$		216
$\pi^\pm e^\mp \nu_e \gamma$	[q,u,bb]	(1.3 ± 0.8) %		229
$\pi^+\pi^-\gamma$	[u,bb]	$(4.61 \pm 0.14) \times 10^{-5}$		206
$\pi^0\pi^0\gamma$		$< 5.6 \times 10^{-6}$		–
Charge conjugation × Parity (CP) or Lepton Family number (LF) violating modes, or $\Delta S = 1$ weak neutral current ($S1$) modes				
$\pi^+\pi^-$	CPV	$(2.03 \pm 0.04) \times 10^{-3}$	S=1.2	206
$\pi^0\pi^0$	CPV	$(9.14 \pm 0.34) \times 10^{-4}$	S=1.8	209
$\mu^+\mu^-$	$S1$	$(7.4 \pm 0.4) \times 10^{-9}$		225
$\mu^+\mu^-\gamma$	$S1$	$(2.8 \pm 2.8) \times 10^{-7}$		225
e^+e^-	$S1$	$< 4.1 \times 10^{-11}$	CL=90%	249
$e^+e^-\gamma$	$S1$	$(9.1 \pm 0.5) \times 10^{-6}$		249
$e^+e^-\gamma\gamma$	$S1$ [bb]	$(6.6 \pm 3.2) \times 10^{-7}$		249
$\pi^+\pi^-e^+e^-$	$S1$	$< 2.5 \times 10^{-6}$	CL=90%	206
$\mu^+\mu^-e^+e^-$	$S1$	$< 4.9 \times 10^{-6}$	CL=90%	225
$e^+e^-e^+e^-$	$S1$ [cc]	$(3.9 \pm 0.7) \times 10^{-8}$		249
$\pi^0\mu^+\mu^-$	$CP,S1$ [dd]	$< 5.1 \times 10^{-9}$	CL=90%	177
$\pi^0 e^+e^-$	$CP,S1$ [dd]	$< 4.3 \times 10^{-9}$	CL=90%	231
$\pi^0\nu\bar{\nu}$	$CP,S1$ [ee]	$< 2.2 \times 10^{-4}$	CL=90%	231
$e^\pm\mu^\mp$	LF [q]	$< 3.3 \times 10^{-11}$	CL=90%	238

$K^*(892)$

$I(J^P) = \frac{1}{2}(1^-)$

$K^*(892)^\pm$ mass $m = 891.59 \pm 0.24$ MeV (S = 1.1)

$K^*(892)^0$ mass $m = 896.10 \pm 0.28$ MeV (S = 1.4)

$K^*(892)^\pm$ full width $\Gamma = 49.8 \pm 0.8$ MeV

$K^*(892)^0$ full width $\Gamma = 50.5 \pm 0.6$ MeV (S = 1.1)

$K^*(892)$ DECAY MODES	Fraction (Γ_i/Γ)	Confidence level	p (MeV/c)
$K\pi$	~ 100 %		291
$K^0\gamma$	$(2.30 \pm 0.20) \times 10^{-3}$		310
$K^\pm\gamma$	$(1.01 \pm 0.09) \times 10^{-3}$		309
$K\pi\pi$	$< 7 \times 10^{-4}$	95%	224

$K_1(1270)$

$I(J^P) = \frac{1}{2}(1^+)$

Mass $m = 1273 \pm 7$ MeV [k]

Full width $\Gamma = 90 \pm 20$ MeV [k]

$K_1(1270)$ DECAY MODES	Fraction (Γ_i/Γ)	p (MeV/c)
$K\rho$	(42 ± 6) %	76
$K_0^*(1430)\pi$	(28 ± 4) %	–
$K^*(892)\pi$	(16 ± 5) %	301
$K\omega$	(11.0 ± 2.0) %	–
$K f_0(1300)$	(3.0 ± 2.0) %	–

$K_1(1400)$

$I(J^P) = \frac{1}{2}(1^+)$

Mass $m = 1402 \pm 7$ MeV

Full width $\Gamma = 174 \pm 13$ MeV (S = 1.6)

$K_1(1400)$ DECAY MODES	Fraction (Γ_i/Γ)	p (MeV/c)
$K^*(892)\pi$	(94 ± 6) %	401
$K\rho$	(3.0 ± 3.0) %	298
$K f_0(1300)$	(2.0 ± 2.0) %	–
$K\omega$	(1.0 ± 1.0) %	285

Meson Summary Table

$K^*(1410)$

$I(J^P) = \frac{1}{2}(1^-)$

Mass $m = 1412 \pm 12$ MeV (S = 1.1)
Full width $\Gamma = 227 \pm 22$ MeV (S = 1.1)

$K^*(1410)$ DECAY MODES	Fraction (Γ_i/Γ)	Confidence level	p (MeV/c)
$K^*(892)\pi$	> 40 %	95%	408
$K\pi$	(6.6±1.3) %		611
$K\rho$	< 7 %	95%	309

$K_0^*(1430)$

$I(J^P) = \frac{1}{2}(0^+)$

Mass $m = 1429 \pm 6$ MeV
Full width $\Gamma = 287 \pm 23$ MeV

$K_0^*(1430)$ DECAY MODES	Fraction (Γ_i/Γ)	p (MeV/c)
$K\pi$	(93±10) %	621

$K_2^*(1430)$

$I(J^P) = \frac{1}{2}(2^+)$

$K_2^*(1430)^\pm$ mass $m = 1425.4 \pm 1.3$ MeV (S = 1.1)
$K_2^*(1430)^0$ mass $m = 1432.4 \pm 1.3$ MeV
$K_2^*(1430)^\pm$ full width $\Gamma = 98.4 \pm 2.3$ MeV
$K_2^*(1430)^0$ full width $\Gamma = 109 \pm 5$ MeV (S = 1.9)

$K_2^*(1430)$ DECAY MODES	Fraction (Γ_i/Γ)	Scale factor/ Confidence level	p (MeV/c)
$K\pi$	(49.7±1.2) %		622
$K^*(892)\pi$	(25.2±1.7) %		423
$K^*(892)\pi\pi$	(13.0±2.3) %		375
$K\rho$	(8.8±0.8) %	S=1.2	331
$K\omega$	(2.9±0.8) %		319
$K^+\gamma$	(2.4±0.5) $\times 10^{-3}$		627
$K\eta$	($1.4^{+2.8}_{-0.9}$) $\times 10^{-3}$	S=1.1	492
$K\omega\pi$	< 7.2 $\times 10^{-4}$	CL=95%	110
$K^0\gamma$	< 9 $\times 10^{-4}$	CL=90%	631

$K^*(1680)$

$I(J^P) = \frac{1}{2}(1^-)$

Mass $m = 1714 \pm 20$ MeV (S = 1.1)
Full width $\Gamma = 323 \pm 110$ MeV (S = 4.2)

$K^*(1680)$ DECAY MODES	Fraction (Γ_i/Γ)	p (MeV/c)
$K\pi$	(38.7±2.5) %	779
$K\rho$	($31.4^{+4.7}_{-2.1}$) %	571
$K^*(892)\pi$	($29.9^{+2.2}_{-4.7}$) %	615

$K_2(1770)$ [ff]
was $L(1770)$

$I(J^P) = \frac{1}{2}(2^-)$

Mass $m = 1773 \pm 8$ MeV
Full width $\Gamma = 186 \pm 14$ MeV

$K_2(1770)$ DECAY MODES	Fraction (Γ_i/Γ)	p (MeV/c)
$K\pi\pi$		–
$K_2^*(1430)\pi$	dominant	287
$K^*(892)\pi$	seen	653
$K f_2(1270)$	seen	–
$K\phi$	seen	441
$K\omega$	seen	608

$K_3^*(1780)$

$I(J^P) = \frac{1}{2}(3^-)$

Mass $m = 1770 \pm 10$ MeV (S = 1.7)
Full width $\Gamma = 164 \pm 17$ MeV (S = 1.1)

$K_3^*(1780)$ DECAY MODES	Fraction (Γ_i/Γ)	Scale factor/ Confidence level	p (MeV/c)
$K\rho$	(45 ±4) %	S=1.4	612
$K^*(892)\pi$	(27.3±3.2) %	S=1.5	651
$K\pi$	(19.3±1.0) %		810
$K\eta$	(8.0±1.5) %	S=1.4	715
$K_2^*(1430)\pi$	< 21 %	CL=95%	284

$K_2(1820)$

$I(J^P) = \frac{1}{2}(2^-)$

Mass $m = 1816 \pm 13$ MeV
Full width $\Gamma = 276 \pm 35$ MeV

$K_2(1820)$ DECAY MODES	Fraction (Γ_i/Γ)	p (MeV/c)
$K\phi$	possibly seen	481
$K_2^*(1430)\pi$	seen	325
$K^*(892)\pi$	seen	680
$K f_2(1270)$	seen	186
$K\omega$	seen	638

$K_4^*(2045)$

$I(J^P) = \frac{1}{2}(4^+)$

Mass $m = 2045 \pm 9$ MeV (S = 1.1)
Full width $\Gamma = 198 \pm 30$ MeV

$K_4^*(2045)$ DECAY MODES	Fraction (Γ_i/Γ)	p (MeV/c)
$K\pi$	(9.9±1.2) %	958
$K^*(892)\pi\pi$	(9 ±5) %	800
$K^*(892)\pi\pi\pi$	(7 ±5) %	764
$\rho K\pi$	(5.7±3.2) %	742
$\omega K\pi$	(5.0±3.0) %	736
$\phi K\pi$	(2.8±1.4) %	591
$\phi K^*(892)$	(1.4±0.7) %	363

CHARMED MESONS ($C = \pm1$)

$D^+ = c\overline{d}$, $D^0 = c\overline{u}$, $\overline{D}^0 = \overline{c}u$, $D^- = \overline{c}d$, similarly for D^*'s

$D^\pm$

$I(J^P) = \frac{1}{2}(0^-)$

Mass $m = 1869.4 \pm 0.4$ MeV

Mean life $\tau = (1.057 \pm 0.015) \times 10^{-12}$ s

$c\tau = 317\ \mu$m

D^- modes are charge conjugates of the modes below.

D^+ DECAY MODES		Fraction (Γ_i/Γ)	Scale factor/ Confidence level	p (MeV/c)
Inclusive modes				
e^+ anything		(17.2 ±1.9) %		–
K^- anything		(24.2 ±2.8) %	S=1.4	–
$\overline{K}^0$ anything + K^0 anything		(59 ±7) %		–
K^+ anything		(5.8 ±1.4) %		–
η anything	[gg]	< 13 %	CL=90%	–
Leptonic and semileptonic modes				
$\mu^+\nu_\mu$		< 7.2 × 10^{-4}	CL=90%	932
$\overline{K}^0$ "e^+" ν_e	[hh]	(6.7 ±0.8) %		868
$\overline{K}^0 e^+\nu_e$		(6.6 ±0.9) %		868
$\overline{K}^0\mu^+\nu_\mu$		(7.0 $^{+3.0}_{-2.0}$) %		865
$\overline{K}^0\ell^+\nu_\ell$		(6.7 ±3.5) %		–
$K^-\pi^+e^+\nu_e$		(4.2 $^{+0.9}_{-0.7}$) %		863
$\overline{K}^*(892)^0 e^+\nu_e$ × B($\overline{K}^{*0} \to K^-\pi^+$)		(3.2 ±0.33) %		720
$K^-\pi^+e^+\nu_e$ nonresonant		< 7 × 10^{-3}	CL=90%	863
$K^-\pi^+\mu^+\nu_\mu$		(3.2 ±1.7) %		851
$\overline{K}^*(892)^0\mu^+\nu_\mu$ × B($\overline{K}^{*0} \to K^-\pi^+$)		(3.0 ±0.4) %		715
$K^-\pi^+\mu^+\nu_\mu$ nonresonant		(2.7 ±1.1) × 10^{-3}		851
$(\overline{K}^*(892)\pi)^0 e^+\nu_e$		< 1.2 %	CL=90%	714
$(\overline{K}\pi\pi)^0 e^+\nu_e$ non-$\overline{K}^*(892)$		< 9 × 10^{-3}	CL=90%	846
$K^-\pi^+\pi^0\mu^+\nu_\mu$		< 1.4 × 10^{-3}	CL=90%	825
$\pi^0\ell^+\nu_\ell$	[ii]	(5.7 ±2.2) × 10^{-3}		–

Fractions of some of the following modes with resonances have already appeared above as submodes of particular charged-particle modes.

$\overline{K}^*(892)^0$ "e^+" ν_e	[hh]	(4.8 ±0.4) %		720
$\overline{K}^*(892)^0 e^+\nu_e$		(4.8 ±0.5) %		720
$\overline{K}^*(892)^0\mu^+\nu_\mu$		(4.5 ±0.6) %	S=1.1	715
$\rho^0 e^+\nu_e$		< 3.7 × 10^{-3}	CL=90%	776
$\rho^0\mu^+\nu_\mu$		(2.0 $^{+1.5}_{-1.3}$) × 10^{-3}		772
$\phi e^+\nu_e$		< 2.09 %	CL=90%	657
$\phi\mu^+\nu_\mu$		< 3.72 %	CL=90%	651
$\eta'(958)\mu^+\nu_\mu$		< 9 × 10^{-3}	CL=90%	684
Hadronic modes with one or three K's				
$\overline{K}^0\pi^+$		(2.74±0.29) %		862
$K^-\pi^+\pi^+$	[jj]	(9.1 ±0.6) %		845
$\overline{K}^*(892)^0\pi^+$ × B($\overline{K}^{*0} \to K^-\pi^+$)		(1.5 ±0.3) %		712
$\overline{K}_0(1430)^0\pi^+$ × B($\overline{K}^*(1430)^0 \to K^-\pi^+$)		(2.3 ±0.3) %		368
$\overline{K}^*(1680)^0\pi^+$ × B($\overline{K}^*(1680)^0 \to K^-\pi^+$)		(2.6 ±1.3) × 10^{-3}		65
$K^-\pi^+\pi^+$ nonresonant		(7.3 ±1.4) %		845
$\overline{K}^0\pi^+\pi^0$	[jj]	(9.7 ±3.0) %	S=1.1	845
$\overline{K}^0\rho^+$		(6.6 ±2.5) %		680
$\overline{K}^*(892)^0\pi^+$ × B($\overline{K}^{*0} \to \overline{K}^0\pi^0$)		(0.7 ±0.2) %		712
$\overline{K}^0\pi^+\pi^0$ nonresonant		(1.3 ±1.1) %		845
$K^-\pi^+\pi^+\pi^0$	[jj]	(6.4 ±1.1) %		816
$\overline{K}^*(892)^0\rho^+$ total × B($\overline{K}^{*0} \to K^-\pi^+$)		(1.4 ±0.9) %		423
$\overline{K}_1(1400)^0\pi^+$ × B($\overline{K}_1(1400)^0 \to K^-\pi^+\pi^0$)		(2.2 ±0.6) %		390
$K^-\rho^+\pi^+$ total		(3.1 ±1.1) %		616
$\overline{K}^*(892)^0\pi^+\pi^0$ total × B($\overline{K}^{*0} \to K^-\pi^+$)		(4.5 ±0.9) %		687
$\overline{K}^*(892)^0\pi^+\pi^0$ 3-body × B($\overline{K}^{*0} \to K^-\pi^+$)		(2.8 ±0.9) %		687
$K^*(892)^-\pi^+\pi^+$ 3-body × B($K^{*-} \to K^-\pi^0$)		(1.4 ±0.6) %		688
$K^-\pi^+\pi^+\pi^0$ nonresonant	[kk]	(1.2 ±0.6) %		816
$\overline{K}^0\pi^+\pi^+\pi^-$	[jj]	(7.0 ±1.0) %		814
$\overline{K}^0 a_1(1260)^+$ × B($a_1(1260)^+ \to \pi^+\pi^+\pi^-$)		(4.0 ±0.8) %		328
$\overline{K}_1(1400)^0\pi^+$ × B($\overline{K}_1(1400)^0 \to \overline{K}^0\pi^+\pi^-$)		(2.2 ±0.6) %		390
$K^*(892)^-\pi^+\pi^+$ 3-body × B($K^{*-} \to \overline{K}^0\pi^-$)		(1.4 ±0.6) %		688
$\overline{K}^0\rho^0\pi^+$ total		(4.2 ±0.9) %		614
$\overline{K}^0\pi^+\pi^+\pi^-$ nonresonant		(8 ±4) × 10^{-3}		814
$K^-\pi^+\pi^+\pi^+\pi^-$		(8.2 ±1.4) × 10^{-3}		772
$\overline{K}^*(892)^0\pi^+\pi^+\pi^-$ × B($\overline{K}^{*0} \to K^-\pi^+$)		(6.8 ±1.8) × 10^{-3}		642
$\overline{K}^*(892)^0\rho^0\pi^+$ × B($\overline{K}^{*0} \to K^-\pi^+$)		(5.1 ±2.2) × 10^{-3}		242
$K^-\pi^+\pi^+\pi^0\pi^0$		(2.2 $^{+5.0}_{-0.9}$) %		775
$\overline{K}^0\pi^+\pi^+\pi^-\pi^0$		(5.4 $^{+3.0}_{-1.4}$) %		773
$\overline{K}^0\pi^+\pi^+\pi^+\pi^-\pi^-$		(8 ±7) × 10^{-4}		714
$K^-\pi^+\pi^+\pi^+\pi^-\pi^0$		(2.0 ±1.8) × 10^{-3}		718
$\overline{K}^0\overline{K}^0K^+$		(3.1 ±0.7) %		545

Fractions of some of the following modes with resonances have already appeared above as submodes of particular charged-particle modes.

$\overline{K}^0\rho^+$		(6.6 ±2.5) %		680
$\overline{K}^0 a_1(1260)^+$		(8.1 ±1.7) %		328
$\overline{K}^0 a_2(1320)^+$		< 3 × 10^{-3}	CL=90%	199
$\overline{K}^*(892)^0\pi^+$		(2.2 ±0.4) %		712
$\overline{K}^*(892)^0\rho^+$ total		(2.1 ±1.4) %		423
$\overline{K}^*(892)^0\rho^+$ S-wave	[kk]	(1.7 ±1.6) %		423
$\overline{K}^*(892)^0\rho^+$ P-wave		< 1 × 10^{-3}	CL=90%	423
$\overline{K}^*(892)^0\rho^+$ D-wave		(10 ±7) × 10^{-3}		423
$\overline{K}^*(892)^0\rho^+$ D-wave longitudinal		< 7 × 10^{-3}	CL=90%	423
$\overline{K}_1(1270)^0\pi^+$		< 7 × 10^{-3}	CL=90%	487
$\overline{K}_1(1400)^0\pi^+$		(5.0 ±1.3) %		390
$\overline{K}^*(1410)^0\pi^+$		< 7 × 10^{-3}	CL=90%	382
$\overline{K}_0^*(1430)^0\pi^+$		(3.4 ±0.4) %		368
$\overline{K}^*(1680)^0\pi^+$		(1.0 ±0.5) %		65
$\overline{K}^*(892)^0\pi^+\pi^0$ total		(6.7 ±1.4) %		687
$\overline{K}^*(892)^0\pi^+\pi^0$ 3-body		(4.2 ±1.4) %		687
$K^*(892)^-\pi^+\pi^+$ 3-body		(2.1 ±0.9) %		688
$K^-\rho^+\pi^+$ total		(3.1 ±1.1) %		616
$K^-\rho^+\pi^+$ 3-body		(1.1 ±0.4) %		616
$\overline{K}^0\rho^0\pi^+$ total		(4.2 ±0.9) %	CL=90%	614
$\overline{K}^0\rho^0\pi^+$ 3-body		(5 ±5) × 10^{-3}		614
$\overline{K}^0 f_0(980)\pi^+$		< 5 × 10^{-3}	CL=90%	461
$\overline{K}^*(892)^0\pi^+\pi^+\pi^-$		(1.02±0.27) %		642
$\overline{K}^*(892)^0\rho^0\pi^+$		(7.7 ±3.3) × 10^{-3}		242
Pionic modes				
$\pi^+\pi^0$		(2.5 ±0.7) × 10^{-3}		925
$\pi^+\pi^+\pi^-$		(3.2 ±0.6) × 10^{-3}		908
$\rho^0\pi^+$		< 1.4 × 10^{-3}	CL=90%	769
$\pi^+\pi^+\pi^-$ nonresonant		(2.5 ±0.7) × 10^{-3}		908
$\pi^+\pi^+\pi^-\pi^0$		(1.9 $^{+1.5}_{-1.2}$) %		883
$\eta\pi^+$ × B($\eta \to \pi^+\pi^-\pi^0$)		(1.8 ±0.6) × 10^{-3}		848
$\omega\pi^+$ × B($\omega \to \pi^+\pi^-\pi^0$)		< 6 × 10^{-3}	CL=90%	764
$\pi^+\pi^+\pi^+\pi^-\pi^-$		(1.0 $^{+0.8}_{-0.7}$) × 10^{-3}		845
$\pi^+\pi^+\pi^+\pi^-\pi^-\pi^0$		(2.9 $^{+2.9}_{-2.0}$) × 10^{-3}		799

Meson Summary Table

Fractions of some of the following modes with resonances have already appeared above as submodes of particular charged-particle modes.

$\rho^0 \pi^+$		$< 1.4 \times 10^{-3}$	CL=90%	769
$\eta \pi^+$		$(7.5 \pm 2.5) \times 10^{-3}$		848
$\omega \pi^+$		$< 7 \times 10^{-3}$	CL=90%	764
$\eta \rho^+$		< 1.2 %	CL=90%	658
$\eta'(958) \pi^+$		$< 9 \times 10^{-3}$	CL=90%	680
$\eta'(958) \rho^+$		< 1.5 %	CL=90%	355
Hadronic modes with two K's				
$\overline{K}^0 K^+$		$(7.8 \pm 1.7) \times 10^{-3}$		792
$K^+ K^- \pi^+$		(1.13 ± 0.13) %		744
$\phi \pi^+ \times B(\phi \to K^+ K^-)$		$(3.3 \pm 0.4) \times 10^{-3}$		647
$\overline{K}^*(892)^0 K^+ \times B(\overline{K}^{*0} \to K^- \pi^+)$		$(3.4 \pm 0.7) \times 10^{-3}$		610
$K^+ K^- \pi^+$ nonresonant		$(4.6 \pm 0.9) \times 10^{-3}$		744
$K^+ K^- \pi^+ \pi^0$				682
$\phi \pi^+ \pi^0 \times B(\phi \to K^+ K^-)$		(1.2 ± 0.5) %		619
$\phi \rho^+ \times B(\phi \to K^+ K^-)$		$< 7 \times 10^{-3}$	CL=90%	268
$K^+ K^- \pi^+ \pi^0$ non-ϕ		$(1.5 {}^{+0.7}_{-0.6})$ %		682
$K^+ \overline{K}^0 \pi^+ \pi^-$		< 2 %	CL=90%	678
$K^0 K^- \pi^+ \pi^+$		(1.0 ± 0.6) %		678
$K^*(892)^+ \overline{K}^*(892)^0 \times B^2(K^* \to K\pi^+)$		(1.2 ± 0.5) %		273
$K^0 K^- \pi^+ \pi^+$ non-$K^{*+} \overline{K}^{*0}$		$< 7.9 \times 10^{-3}$	CL=90%	678
$K^+ K^- \pi^+ \pi^+ \pi^-$				600
$\phi \pi^+ \pi^+ \pi^- \times B(\phi \to K^+ K^-)$		$< 1 \times 10^{-3}$	CL=90%	566
$K^+ K^- \pi^+ \pi^+ \pi^-$ nonresonant		< 3 %	CL=90%	600

Fractions of the following modes with resonances have already appeared above as submodes of particular charged-particle modes.

$\phi \pi^+$		$(6.7 \pm 0.8) \times 10^{-3}$		647
$\overline{K}^*(892)^0 K^+$		$(5.1 \pm 1.0) \times 10^{-3}$		610
$\phi \pi^+ \pi^0$		(2.3 ± 1.0) %		619
$\phi \rho^+$		< 1.5 %	CL=90%	268
$K^*(892)^+ \overline{K}^*(892)^0$		(2.6 ± 1.1) %		273
$\phi \pi^+ \pi^+ \pi^-$		$< 2 \times 10^{-3}$	CL=90%	566

Doubly Cabibbo suppressed (DC) modes, $\Delta C = 1$ weak neutral current ($C1$) modes, or Lepton Family number (LF) or Lepton number (L) violating modes

$K^+ \pi^+ \pi^-$	DC		$< 5 \times 10^{-3}$	CL=90%	845
$K^+ K^+ K^-$	DC		$(5.2 \pm 2.0) \times 10^{-3}$		550
ϕK^+	DC		$(3.9 {}^{+2.2}_{-1.9}) \times 10^{-4}$		527
$\pi^+ e^+ e^-$	$C1$		$< 2.5 \times 10^{-3}$	CL=90%	929
$\pi^+ \mu^+ \mu^-$	$C1$		$< 2.9 \times 10^{-3}$	CL=90%	917
$K^+ e^+ e^-$		[ll]	$< 4.8 \times 10^{-3}$	CL=90%	870
$K^+ \mu^+ \mu^-$		[ll]	$< 9.2 \times 10^{-3}$	CL=90%	856
$\pi^+ e^{\pm} \mu^{\mp}$	LF	[q]	$< 3.8 \times 10^{-3}$	CL=90%	926
$\pi^+ e^+ \mu^-$	LF		$< 3.3 \times 10^{-3}$	CL=90%	926
$\pi^+ e^- \mu^+$	LF		$< 3.3 \times 10^{-3}$	CL=90%	926
$K^+ e^+ \mu^-$	LF		$< 3.4 \times 10^{-3}$	CL=90%	866
$K^+ e^- \mu^+$	LF		$< 3.4 \times 10^{-3}$	CL=90%	866
$\pi^- e^+ e^+$	L		$< 4.8 \times 10^{-3}$	CL=90%	929
$\pi^- \mu^+ \mu^+$	L		$< 6.8 \times 10^{-3}$	CL=90%	917
$\pi^- e^+ \mu^+$	L		$< 3.7 \times 10^{-3}$	CL=90%	926
$K^- e^+ e^+$	L		$< 9.1 \times 10^{-3}$	CL=90%	870
$K^- \mu^+ \mu^+$	L		$< 4.3 \times 10^{-3}$	CL=90%	856
$K^- e^+ \mu^+$	L		$< 4.0 \times 10^{-3}$	CL=90%	866

D^0

$I(J^P) = \frac{1}{2}(0^-)$

Mass $m = 1864.6 \pm 0.5$ MeV

$|m_{D^0_1} - m_{D^0_2}| < 20 \times 10^{10}\ \hbar\ s^{-1}$, CL = 90% [mm]

$m_{D^\pm} - m_{D^0} = 4.78 \pm 0.10$ MeV

Mean life $\tau = (0.415 \pm 0.004) \times 10^{-12}$ s

$c\tau = 124.4\ \mu$m

$|\tau_{D^0_1} - \tau_{D^0_2}|/\tau_{D^0} < 0.17$, CL = 90% [mm]

$\Gamma(K^+ \pi^- (\text{via } \overline{D}^0))/\Gamma(K^- \pi^+) < 0.0037$, CL = 90%

$\Gamma(\mu^- X (\text{via } \overline{D}^0))/\Gamma(\mu^+ X) < 0.0056$, CL = 90%

$[\Gamma(D^0 \to K^+ K^-) - \Gamma(\overline{D}^0 \to K^+ K^-)]/\text{sum} < 0.45$, CL = 90%

$\overline{D}^0$ modes are charge conjugates of the modes below.

D^0 DECAY MODES		Fraction (Γ_i/Γ)	Scale factor/ Confidence level	p (MeV/c)
Inclusive modes				
e^+ anything		(7.7 ± 1.2) %	S=1.1	–
μ^+ anything		(10.0 ± 2.6) %		–
K^- anything		(53 ± 4) %	S=1.3	–
$\overline{K}^0$ anything + K^0 anything		(42 ± 5) %		–
K^+ anything		$(3.4 {}^{+0.6}_{-0.4})$ %		–
η anything	[gg]	< 13 %	CL=90%	–
Semileptonic modes				
K^- "e^+" ν_e	[hh]	(3.68 ± 0.21) %	S=1.1	867
$K^- e^+ \nu_e$		(3.80 ± 0.22) %	S=1.1	867
$K^- \mu^+ \nu_\mu$		(3.2 ± 0.4) %		864
$K^- \pi^0 e^+ \nu_e$	[nn]	$(1.6 {}^{+1.3}_{-0.5})$ %		861
$\overline{K}^0 \pi^- e^+ \nu_e$	[nn]	$(2.8 {}^{+1.7}_{-0.9})$ %		860
$\overline{K}^*(892)^- e^+ \nu_e \times B(K^{*-} \to \overline{K}^0 \pi^+)$		(1.3 ± 0.3) %		719
$\overline{K}^*(892)^0 \pi^- e^+ \nu_e$	[oo]	< 1.3 %	CL=90%	709
$K^- \pi^+ \pi^- \mu^+ \nu_\mu$		$< 1.2 \times 10^{-3}$	CL=90%	821
$(\overline{K}^*(892)\pi)^- \mu^+ \nu_\mu$		$< 1.4 \times 10^{-3}$	CL=90%	694
$\pi^- e^+ \nu_e$		$(3.9 {}^{+2.3}_{-1.2}) \times 10^{-3}$		927

A fraction of the following resonance mode has already appeared above as a submode of a particular charged-particle mode.

$K^*(892)^- e^+ \nu_e$		(2.0 ± 0.4) %		719
Hadronic modes with one or three K's				
$K^- \pi^+$		(4.01 ± 0.14) %		861
$\overline{K}^0 \pi^0$		(2.05 ± 0.26) %	S=1.1	860
$\overline{K}^0 \pi^+ \pi^-$	[jj]	(5.3 ± 0.6) %	S=1.2	842
$\overline{K}^0 \rho^0$		(1.10 ± 0.18) %		676
$\overline{K}^0 f_0(980) \times B(f_0 \to \pi^+ \pi^-)$		$(2.4 \pm 1.0) \times 10^{-3}$		549
$\overline{K}^0 f_2(1270) \times B(f_2 \to \pi^+ \pi^-)$		$(2.6 \pm 1.2) \times 10^{-3}$		263
$\overline{K}^0 f_0(1300) \times B(f_0 \to \pi^+ \pi^-)$		$(4.3 \pm 1.7) \times 10^{-3}$		223
$K^*(892)^- \pi^+ \times B(K^{*-} \to \overline{K}^0 \pi^-)$		(3.3 ± 0.4) %		711
$K_0^*(1430)^- \pi^+ \times B(K_0^*(1430)^- \to \overline{K}^0 \pi^-)$		$(7 \pm 3) \times 10^{-3}$		364
$\overline{K}^0 \pi^+ \pi^-$ nonresonant		(1.43 ± 0.26) %		842
$K^- \pi^+ \pi^0$	[jj]	(13.8 ± 1.0) %	S=1.1	844
$K^- \rho^+$		(10.4 ± 1.3) %		678
$K^*(892)^- \pi^+ \times B(K^{*-} \to K^- \pi^0)$		(1.6 ± 0.2) %		711
$\overline{K}^*(892)^0 \pi^0 \times B(\overline{K}^{*0} \to K^- \pi^+)$		(2.0 ± 0.3) %		709
$K^- \pi^+ \pi^0$ nonresonant		$(6.0 \pm 2.7) \times 10^{-3}$		844
$\overline{K}^0 \pi^0 \pi^0$				843
$\overline{K}^*(892)^0 \pi^0 \times B(\overline{K}^{*0} \to \overline{K}^0 \pi^0)$		(1.0 ± 0.2) %		709
$\overline{K}^0 \pi^0 \pi^0$ nonresonant		$(7.6 \pm 2.1) \times 10^{-3}$		843

$K^-\pi^+\pi^+\pi^-$	[jj]	$(8.1 \pm 0.5)\,\%$		812
$K^-\pi^+\rho^0$ total		$(6.8 \pm 0.5)\,\%$		612
$K^-\pi^+\rho^0$ 3-body		$(5.1 \pm 2.3) \times 10^{-3}$		612
$\overline{K}^*(892)^0\rho^0$		$(1.1 \pm 0.3)\,\%$		418
$\times\ B(\overline{K}^{*0} \to K^-\pi^+)$				
$K^- a_1(1260)^+$		$(3.9 \pm 0.6)\,\%$		327
$\times\ B(a_1(1260)^+ \to \pi^+\pi^+\pi^-)$				
$\overline{K}^*(892)^0\pi^+\pi^-$ total		$(1.6 \pm 0.4)\,\%$		683
$\times\ B(\overline{K}^{*0} \to K^-\pi^+)$				
$\overline{K}^*(892)^0\pi^+\pi^-$ 3-body		$(1.01 \pm 0.22)\,\%$		683
$\times\ B(\overline{K}^{*0} \to K^-\pi^+)$				
$K_1(1270)^-\pi^+$		$(3.5 \pm 1.1) \times 10^{-3}$		483
$\times\ B(K_1(1270)^- \to K^-\pi^+\pi^-)$				
$K^-\pi^+\pi^+\pi^-$ nonresonant		$(1.89 \pm 0.28)\,\%$		812
$\overline{K}^0\pi^+\pi^-\pi^0$	[jj]	$(9.8 \pm 1.4)\,\%$	S=1.1	812
$\overline{K}^0\eta \times B(\eta \to \pi^+\pi^-\pi^0)$		$(1.61 \pm 0.26) \times 10^{-3}$		772
$\overline{K}^0\omega \times B(\omega \to \pi^+\pi^-\pi^0)$		$(1.8 \pm 0.4)\,\%$		670
$K^*(892)^-\rho^+$		$(3.9 \pm 1.6)\,\%$		422
$\times\ B(K^{*-} \to \overline{K}^0\pi^-)$				
$\overline{K}^*(892)^0\rho^0$		$(5.3 \pm 1.4) \times 10^{-3}$		418
$\times\ B(\overline{K}^{*0} \to \overline{K}^0\pi^0)$				
$K_1(1270)^-\pi^+$	[kk]	$(5.0 \pm 1.5) \times 10^{-3}$		483
$\times\ B(K_1(1270)^- \to \overline{K}^0\pi^-\pi^0)$				
$\overline{K}^*(892)^0\pi^+\pi^-$ 3-body		$(5.1 \pm 1.1) \times 10^{-3}$		683
$\times\ B(\overline{K}^{*0} \to \overline{K}^0\pi^0)$				
$\overline{K}^0\pi^+\pi^-\pi^0$ nonresonant		$(2.1 \pm 2.1)\,\%$		812
$K^-\pi^+\pi^0\pi^0$		$(15 \pm 5)\,\%$		815
$K^-\pi^+\pi^+\pi^-\pi^0$		$(4.3 \pm 0.4)\,\%$		771
$\overline{K}^*(892)^0\pi^+\pi^-\pi^0$		$(1.3 \pm 0.6)\,\%$		641
$\times\ B(\overline{K}^{*0} \to K^-\pi^+)$				
$\overline{K}^*(892)^0\eta$		$(3.0 \pm 0.8) \times 10^{-3}$		580
$\times\ B(\overline{K}^{*0} \to K^-\pi^+)$				
$\times\ B(\eta \to \pi^+\pi^-\pi^0)$				
$K^-\pi^+\omega \times B(\omega \to \pi^+\pi^-\pi^0)$		$(2.8 \pm 0.5)\,\%$		605
$\overline{K}^*(892)^0\omega$		$(7 \pm 3) \times 10^{-3}$		406
$\times\ B(\overline{K}^{*0} \to K^-\pi^+)$				
$\times\ B(\omega \to \pi^+\pi^-\pi^0)$				
$\overline{K}^0\pi^+\pi^+\pi^-\pi^-$		$(5.6 \pm 1.7) \times 10^{-3}$		768
$\overline{K}^0\pi^+\pi^-\pi^0\pi^0(\pi^0)$		$(10.6\ ^{+7.3}_{-3.0})\,\%$		771
$\overline{K}^0K^+K^-$		$(9.1 \pm 1.2) \times 10^{-3}$		544
$\overline{K}^0\phi \times B(\phi \to K^+K^-)$		$(4.2 \pm 0.6) \times 10^{-3}$		520
$\overline{K}^0K^+K^-$ non-ϕ		$(4.9 \pm 0.9) \times 10^{-3}$		544
$K^0_S K^0_S K^0_S$		$(8.6 \pm 2.5) \times 10^{-4}$		538
$K^+K^-\overline{K}^0\pi^0$		$(7.2\ ^{+4.8}_{-3.5}) \times 10^{-3}$		435

Fractions of many of the following modes with resonances have already appeared above as submodes of particular charged-particle modes. (Modes for which there are only upper limits and $\overline{K}^*(892)\rho$ submodes only appear below.)

$\overline{K}^0\eta$		$(6.8 \pm 1.1) \times 10^{-3}$		772
$\overline{K}^0\rho^0$		$(1.10 \pm 0.18)\,\%$		676
$K^-\rho^+$		$(10.4 \pm 1.3)\,\%$	S=1.2	679
$\overline{K}^0\omega$		$(2.0 \pm 0.4)\,\%$		670
$\overline{K}^0\eta'(958)$		$(1.66 \pm 0.29)\,\%$		565
$\overline{K}^0 f_0(980)$		$(4.6 \pm 2.0) \times 10^{-3}$		549
$\overline{K}^0\phi$		$(8.3 \pm 1.2) \times 10^{-3}$	S=1.1	520
$K^- a_1(1260)^+$		$(7.9 \pm 1.2)\,\%$		327
$\overline{K}^0 a_1(1260)^0$		$< 1.9\ \%$	CL=90%	322
$\overline{K}^0 f_2(1270)$		$(4.6 \pm 2.1) \times 10^{-3}$		263
$\overline{K}^0 f_0(1300)$		$(6.9 \pm 2.7) \times 10^{-3}$		223
$K^- a_2(1320)^+$		$< 2 \times 10^{-3}$	CL=90%	197
$K^*(892)^-\pi^+$		$(4.9 \pm 0.6)\,\%$	S=1.3	711
$\overline{K}^*(892)^0\pi^0$		$(3.0 \pm 0.4)\,\%$		709
$\overline{K}^*(892)^0\pi^+\pi^-$ total		$(2.4 \pm 0.6)\,\%$		683
$\overline{K}^*(892)^0\pi^+\pi^-$ 3-body		$(1.52 \pm 0.33)\,\%$		683
$K^-\pi^+\rho^0$ total		$(6.8 \pm 0.5)\,\%$		612
$K^-\pi^+\rho^0$ 3-body		$(5.1 \pm 2.3) \times 10^{-3}$		612
$\overline{K}^*(892)^0\rho^0$		$(1.6 \pm 0.4)\,\%$		418
$\overline{K}^*(892)^0\rho^0$ transverse		$(1.6 \pm 0.5)\,\%$		418
$\overline{K}^*(892)^0\rho^0$ *S*-wave		$(3.0 \pm 0.6)\,\%$		418
$\overline{K}^*(892)^0\rho^0$ *S*-wave long.		$< 3 \times 10^{-3}$	CL=90%	418
$\overline{K}^*(892)^0\rho^0$ *P*-wave		$< 3 \times 10^{-3}$	CL=90%	418
$\overline{K}^*(892)^0\rho^0$ *D*-wave		$(2.1 \pm 0.6)\,\%$		418
$K^*(892)^-\rho^+$		$(5.9 \pm 2.4)\,\%$		422
$K^*(892)^-\rho^+$ longitudinal		$(2.8 \pm 1.2)\,\%$		422
$K^*(892)^-\rho^+$ transverse		$(3.1 \pm 1.8)\,\%$		422
$K^*(892)^-\rho^+$ *P*-wave		$< 1.5\ \%$	CL=90%	422
$K^-\pi^+ f_0(980)$		$< 1.1\ \%$	CL=90%	459
$\overline{K}^*(892)^0 f_0(980)$		$< 7 \times 10^{-3}$	CL=90%	–
$K_1(1270)^-\pi^+$	[kk]	$(1.04 \pm 0.31)\,\%$		483
$K_1(1400)^-\pi^+$		$< 1.2\ \%$	CL=90%	386
$\overline{K}_1(1400)^0\pi^0$		$< 3.7\ \%$	CL=90%	387
$K^*(1410)^-\pi^+$		$< 1.2\ \%$	CL=90%	378
$K^*_0(1430)^-\pi^+$		$(1.1 \pm 0.4)\,\%$		364
$K^*_2(1430)^-\pi^+$		$< 8 \times 10^{-3}$	CL=90%	367
$\overline{K}^*_2(1430)^0\pi^0$		$< 4 \times 10^{-3}$	CL=90%	363
$\overline{K}^*(892)^0\pi^+\pi^-\pi^0$		$(1.9 \pm 0.9)\,\%$		641
$\overline{K}^*(892)^0\eta$		$(1.9 \pm 0.5)\,\%$		580
$K^-\pi^+\omega$		$(3.1 \pm 0.6)\,\%$		605
$\overline{K}^*(892)^0\omega$		$(1.1 \pm 0.5)\,\%$		406
$K^-\pi^+\eta'(958)$		$(7.5 \pm 2.0) \times 10^{-3}$		479
$\overline{K}^*(892)^0\eta'(958)$		$< 1.1 \times 10^{-3}$	CL=90%	100

Pionic modes

$\pi^+\pi^-$	$(1.59 \pm 0.12) \times 10^{-3}$		922
$\pi^0\pi^0$	$(8.8 \pm 2.3) \times 10^{-4}$		922
$\pi^+\pi^-\pi^0$	$(1.6 \pm 1.1)\,\%$	S=2.7	907
$\pi^+\pi^+\pi^-\pi^-$	$(8.3 \pm 0.9) \times 10^{-3}$		880
$\pi^+\pi^+\pi^-\pi^-\pi^0$	$(1.9 \pm 0.4)\,\%$		844
$\pi^+\pi^+\pi^+\pi^-\pi^-\pi^-$	$(4.0 \pm 3.0) \times 10^{-4}$		795

Hadronic modes with two *K*'s

K^+K^-	$(4.54 \pm 0.29) \times 10^{-3}$		791
$K^0\overline{K}^0$	$(1.1 \pm 0.4) \times 10^{-3}$		788
$K^0K^-\pi^+$	$(6.3 \pm 1.1) \times 10^{-3}$	S=1.2	739
$\overline{K}^*(892)^0K^0$	$< 1{,}0 \times 10^{-3}$	CL=90%	605
$\times\ B(\overline{K}^{*0} \to K^-\pi^+)$			
$K^*(892)^+K^-$	$(2.3 \pm 0.5) \times 10^{-3}$		610
$\times\ B(K^{*+} \to K^0\pi^+)$			
$K^0K^-\pi^+$ nonresonant	$(2.4 \pm 2.4) \times 10^{-3}$		739
$\overline{K}^0K^+\pi^-$	$(4.9 \pm 1.0) \times 10^{-3}$		739
$K^*(892)^0\overline{K}^0$	$< 5 \times 10^{-4}$	CL=90%	605
$\times\ B(K^{*0} \to K^+\pi^-)$			
$K^*(892)^-K^+$	$(1.2 \pm 0.7) \times 10^{-3}$		610
$\times\ B(K^{*-} \to \overline{K}^0\pi^-)$			
$\overline{K}^0K^+\pi^-$ nonresonant	$(4.0\ ^{+2.4}_{-2.0}) \times 10^{-3}$		739
$K^+K^-\pi^+\pi^-$	$(2.4 \pm 0.5) \times 10^{-3}$		677
$\phi\pi^+\pi^- \times B(\phi \to K^+K^-)$	$(1.3 \pm 0.4) \times 10^{-3}$		614
$\phi\rho^0 \times B(\phi \to K^+K^-)$	$(1.0 \pm 0.25) \times 10^{-3}$		260
$K^*(892)^0K^-\pi^+$ + c.c. $\times$ $B(K^{*0} \to K^+\pi^-)$	$(5\ ^{+9}_{-5}) \times 10^{-4}$		528
$K^*(892)^0\overline{K}^*(892)^0$ $\times\ B^2(K^{*0} \to K^+\pi^-)$	$(1.3\ ^{+0.7}_{-0.6}) \times 10^{-3}$		257
$K^+K^-\pi^+\pi^-$ non-ϕ	$(1.7 \pm 0.5) \times 10^{-3}$		677
$K^+K^-\pi^+\pi^-$ nonresonant	$(8\ ^{+90}_{-8}) \times 10^{-5}$		677
$K^+K^-\pi^+\pi^-\pi^0$	$(3.1 \pm 2.0) \times 10^{-3}$		600

Fractions of the following modes with resonances have already appeared above as submodes of particular charged-particle modes.

$\overline{K}^*(892)^0K^0$	$< 1.5 \times 10^{-3}$	CL=90%	605
$K^*(892)^+K^-$	$(3.4 \pm 0.8) \times 10^{-3}$		610
$K^*(892)^0\overline{K}^0$	$< 8 \times 10^{-4}$	CL=90%	605
$K^*(892)^-K^+$	$(1.8 \pm 1.0) \times 10^{-3}$		610
$\phi\pi^+\pi^-$	$(2.6 \pm 0.7) \times 10^{-3}$		614
$\phi\rho^0$	$(1.9 \pm 0.5) \times 10^{-3}$		260
$K^*(892)^0K^-\pi^+$ + c.c.	$(8\ ^{+13}_{-8}) \times 10^{-4}$		528
$K^*(892)^0\overline{K}^*(892)^0$	$(2.9\ ^{+1.6}_{-1.3}) \times 10^{-3}$		257

Doubly Cabibbo suppressed (*DC*) modes,
$\Delta C = 2$ forbidden via mixing (*C2M*) modes,
$\Delta C = 1$ weak neutral current (*C1*) modes, or
Lepton Family number (*LF*) violating modes

$K^+\pi^-$	*DC*		$(3.1 \pm 1.4) \times 10^{-4}$		861
$K^+\pi^-$ (via $\overline{D}^0$)	*C2M*		$< 1.5 \times 10^{-4}$	CL=90%	861
$K^+\pi^+\pi^-\pi^-$	*DC*		$< 1.5 \times 10^{-3}$	CL=90%	812
μ^- anything (via $\overline{D}^0$)	*C2M*		$< 6 \times 10^{-4}$	CL=90%	–
e^+e^-	*C1*		$< 1.3 \times 10^{-4}$	CL=90%	932
$\mu^+\mu^-$	*C1*		$< 1.1 \times 10^{-5}$	CL=90%	926
$\overline{K}^0e^+e^-$			$< 1.7 \times 10^{-3}$	CL=90%	866
$\rho^0e^+e^-$	*C1*		$< 4.5 \times 10^{-4}$	CL=90%	773
$\rho^0\mu^+\mu^-$	*C1*		$< 8.1 \times 10^{-4}$	CL=90%	756
$\mu^\pm e^\mp$	*LF*	[q]	$< 1.0 \times 10^{-4}$	CL=90%	929

Meson Summary Table

$D^*(2007)^0$

$I(J^P) = \frac{1}{2}(1^-)$
I, J, P need confirmation.

Mass $m = 2006.7 \pm 0.5$ MeV
$m_{D^{*0}} - m_{D^0} = 142.12 \pm 0.07$ MeV
Full width $\Gamma < 2.1$ MeV, CL = 90%

$\overline{D}^*(2007)^0$ modes are charge conjugates of modes below.

$D^*(2007)^0$ DECAY MODES	Fraction (Γ_i/Γ)	p (MeV/c)
$D^0\pi^0$	(63.6 ± 2.8) %	43
$D^0\gamma$	(36.4 ± 2.8) %	137

$D^*(2010)^\pm$

$I(J^P) = \frac{1}{2}(1^-)$
I, J, P need confirmation.

Mass $m = 2010.0 \pm 0.5$ MeV
$m_{D^*(2010)^+} - m_{D^+} = 140.64 \pm 0.09$ MeV
$m_{D^*(2010)^+} - m_{D^0} = 145.42 \pm 0.05$ MeV
Full width $\Gamma < 0.131$ MeV, CL = 90%

$D^*(2010)^-$ modes are charge conjugates of the modes below.

$D^*(2010)^\pm$ DECAY MODES	Fraction (Γ_i/Γ)	p (MeV/c)
$D^0\pi^+$	(68.1 ± 1.3) %	39
$D^+\pi^0$	(30.8 ± 0.8) %	38
$D^+\gamma$	$(1.1^{+1.4}_{-0.7})$ %	136

$D_1(2420)^0$

$I(J^P) = \frac{1}{2}(1^+)$
I, J, P need confirmation.

Mass $m = 2422.8 \pm 3.2$ MeV (S = 1.6)
Full width $\Gamma = 18^{+6}_{-4}$ MeV

$\overline{D}_1(2420)^0$ modes are charge conjugates of modes below.

$D_1(2420)^0$ DECAY MODES	Fraction (Γ_i/Γ)	p (MeV/c)
$D^*(2010)^+\pi^-$	seen	355
$D^+\pi^-$	not seen	474

$D_2^*(2460)$

$I(J^P) = \frac{1}{2}(2^+)$

$J^P = 2^+$ assignment strongly favored (ALBRECHT 89B).

Mass $m_{D_2^*(2460)^0} = 2457.7 \pm 1.9$ MeV
Mass $m_{D_2^*(2460)^\pm} = 2456 \pm 6$ MeV (S = 2.0)
$m_{D_2^*(2460)^\pm} - m_{D_2^*(2460)^0} = 2 \pm 5$ MeV (S = 1.4)
Full width $\Gamma_{D_2^*(2460)^0} = 21 \pm 5$ MeV
Full width $\Gamma_{D_2^*(2460)^\pm} = 23 \pm 10$ MeV

$\overline{D}_2^*(2460)$ modes are charge conjugates of modes below.

$D_2^*(2460)$ DECAY MODES	Fraction (Γ_i/Γ)	p (MeV/c)
$D_2^*(2460)^0 \to D^+\pi^-$	seen	503
$D_2^*(2460)^0 \to D^*(2010)^+\pi^-$	seen	387
$D_2^*(2460)^\pm \to D^0\pi^+$	seen	505

CHARMED, STRANGE MESONS ($C = S = \pm1$)

$D_s^+ = c\overline{s}$, $D_s^- = \overline{c}s$, similarly for D_s^*'s

$D_s^\pm$ was $F^\pm$

$I(J^P) = 0(0^-)$

Mass $m = 1968.5 \pm 0.7$ MeV (S = 1.2)
$m_{D_s^\pm} - m_{D^\pm} = 99.1 \pm 0.6$ MeV (S = 1.1)
Mean life $\tau = (0.467 \pm 0.017) \times 10^{-12}$ s
$c\tau = 140$ μm

Branching fractions for modes below with a resonance in the final state include all the decay modes of the resonance. D_s^- modes are charge conjugates of the modes below.

Nearly all other modes are measured relative to the $\phi\pi^+$ mode. However, none of the determinations of the $\phi\pi^+$ branching fraction are direct measurements: all rely on calculated relations between D^+ and D_s^+ decay widths, on estimates of D_s^+ cross sections, or on other model-dependent assumptions. Thus a better determination of the $\phi\pi^+$ branching fraction could cause the other branching fractions to slide up or down, all together.

D_s^+ DECAY MODES		Fraction (Γ_i/Γ)	Scale factor/ Confidence level	p (MeV/c)
Inclusive modes				
K^- anything		$(13\ ^{+14}_{-12})$ %		–
$\overline{K}^0$ anything + K^0 anything		(39 ± 28) %		–
K^+ anything		$(20\ ^{+18}_{-14})$ %		–
non-$K\overline{K}$ anything		(64 ± 17) %		–
e^+ anything		< 20 %	CL=90%	–
Leptonic and semileptonic modes				
$\mu^+\nu_\mu$		$(5.9 \pm 2.2) \times 10^{-3}$	S=1.1	981
$\phi\ell^+\nu_\ell$	[pp]	(1.88 ± 0.29) %		–
$\eta\mu^+\nu_\mu + \eta'(958)\mu^+\nu_\mu$		(7.4 ± 3.2) %		–
$\eta\mu^+\nu_\mu$				905
$\eta'(958)\mu^+\nu_\mu$		< 3.0 %	CL=90%	747
Hadronic modes with two K's (including from ϕ's)				
$K^+\overline{K}^0$		(3.5 ± 0.7) %		850
$K^+K^-\pi^+$	[qq]	(4.8 ± 0.7) %		805
$\phi\pi^+$		(3.5 ± 0.4) %		712
$K^+\overline{K}^*(892)^0$		(3.3 ± 0.5) %		682
$K^+K^-\pi^+$ nonresonant		$(8.7 \pm 3.2) \times 10^{-3}$		805
$K^0\overline{K}^0\pi^+$				802
$K^*(892)^+\overline{K}^0$		(4.2 ± 1.0) %		683
$K^+K^-\pi^+\pi^0$				748
$\phi\pi^+\pi^0$		(8 ± 4) %		687
$\phi\rho^+$		$(6.5\ ^{+1.6}_{-1.8})$ %		407
$\phi\pi^+\pi^0$ 3-body		< 2.5 %	CL=90%	687
$K^+K^-\pi^+\pi^0$ non-ϕ		< 8 %	CL=90%	748
$K^+\overline{K}^0\pi^+\pi^-$		< 2.7 %	CL=90%	744
$K^0K^-\pi^+\pi^+$		(4.2 ± 1.1) %		744
$K^*(892)^+\overline{K}^*(892)^0$		(5.6 ± 2.1) %		412
$K^0K^-\pi^+\pi^+$ non-$K^{*+}\overline{K}^{*0}$		< 2.8 %	CL=90%	744
$K^+K^-\pi^+\pi^+\pi^-$				673
$\phi\pi^+\pi^+\pi^-$		(1.8 ± 0.5) %		640
$K^+K^-\pi^+\pi^+\pi^-$ non-ϕ		$(3.0\ ^{+3.0}_{-2.0}) \times 10^{-3}$		673

Other hadronic modes

$\pi^+\pi^+\pi^-$	(1.35 ± 0.31) %		959
$\quad\rho^0\pi^+$	< 2.8 $\times 10^{-3}$	CL=90%	827
$\quad f_0(980)\pi^+$	$(10 \pm 4) \times 10^{-3}$		732
$\quad\pi^+\pi^+\pi^-$ nonresonant	(1.01 ± 0.35) %		959
$\pi^+\pi^+\pi^-\pi^0$	< 12 %	CL=90%	935
$\quad\eta\pi^+$	(1.9 ± 0.4) %		902
$\quad\omega\pi^+$	< 1.7 %	CL=90%	822
$\pi^+\pi^+\pi^+\pi^-\pi^-$	$(3.0 {}^{+4.0}_{-3.0}) \times 10^{-3}$		899
$\pi^+\pi^+\pi^-\pi^0\pi^0$			902
$\quad\eta\rho^+$	(10.0 ± 2.2) %		727
$\quad\eta\pi^+\pi^0$ 3-body	< 2.9 %	CL=90%	787
$\pi^+\pi^+\pi^+\pi^-\pi^-\pi^0$	(4.9 ± 3.2) %		856
$\quad\eta'(958)\pi^+$	(4.7 ± 1.4) %		743
$\pi^+\pi^+\pi^+\pi^-\pi^-\pi^0\pi^0$			803
$\quad\eta'(958)\rho^+$	(12.0 ± 3.0) %		470
$\quad\eta'(958)\pi^+\pi^0$ 3-body	< 3.0 %	CL=90%	720
$K^0\pi^+$	< 7 $\times 10^{-3}$	CL=90%	916
$K^+\pi^+\pi^-$	$(3.0 {}^{+4.0}_{-3.0}) \times 10^{-3}$		900
$K^+K^-K^+$			628
$\quad\phi K^+$	< 2.5 $\times 10^{-3}$	CL=90%	607

$D_s^{*\pm}$

$I(J^P) = ?(?^?)$

Mass $m = 2110.0 \pm 1.9$ MeV (S = 1.2)

$m_{D_s^{*\pm}} - m_{D_s^\pm} = 141.6 \pm 1.8$ MeV (S = 1.2)

Full width $\Gamma < 4.5$ MeV, CL = 90%

D_s^{*-} modes are charge conjugates of the modes below.

D_s^{*+} DECAY MODES	Fraction (Γ_i/Γ)	p (MeV/c)
$D_s^+\gamma$	dominant	137

$D_{s1}(2536)^\pm$

$I(J^P) = 0(1^+)$

I, J, P need confirmation.

Mass $m = 2535.35 \pm 0.34$ MeV

Full width $\Gamma < 2.3$ MeV, CL = 90%

$D_{s1}(2536)^-$ modes are charge conjugates of the modes below.

$D_{s1}(2536)^+$ DECAY MODES	Fraction (Γ_i/Γ)	p (MeV/c)
$D^*(2010)^+K^0$	seen	150
$D^*(2007)^0K^+$	seen	169
D^+K^0	not seen	382
D^0K^+	not seen	392
$D_s^{*+}\gamma$	possibly seen	389

BOTTOM MESONS
$(B = \pm 1)$

$B^+ = u\bar{b}$, $B^0 = d\bar{b}$, $\overline{B}^0 = \bar{d}b$, $B^- = \bar{u}b$, similarly for B^*'s

$B^\pm$

$I(J^P) = \frac{1}{2}(0^-)$

I, J, P need confirmation. Quantum numbers shown are quark-model predictions. Measurements which do not identify the charge state of B also appear here.

Mass $m_{B^\pm} = 5278.7 \pm 2.0$ MeV

Mean life $\tau = (1.54 \pm 0.11) \times 10^{-12}$ s

Mean life τ (avg over B hadrons) $= (1.537 \pm 0.021) \times 10^{-12}$ s [a]

$c\tau = 462\ \mu$m

B^- modes are charge conjugates of the modes below.

Only data from $\Upsilon(4S)$ decays are used for branching fractions, with rare exceptions. The branching fractions listed below assume a 50:50 $B^0\overline{B}^0$:B^+B^- production ratio at the $\Upsilon(4S)$. We have attempted to bring older measurements up to date by rescaling their assumed $\Upsilon(4S)$ production ratio to 50:50 and their assumed D, D_s, D^*, and ψ branching ratios to current values whenever this would effect our averages and best limits significantly.

Indentation is used to indicate a subchannel of a previous reaction. All resonant subchannels have been corrected for resonance branching fractions to the final state so the sum of the subchannel branching fractions can exceed that of the final state.

B^+ DECAY MODES	Fraction (Γ_i/Γ)	Scale factor/ Confidence level	p (MeV/c)
Semileptonic modes			
$B^+ \to \overline{D}^0\ell^+\nu$	[ii] (1.6 ± 0.7) %		–
$B^+ \to \overline{D}^*(2007)^0\ell^+\nu$	[ii] (6.6 ± 2.2) %		–
$B^+ \to \pi^0e^+\nu_e$	< 2.2 $\times 10^{-3}$	CL=90%	2638
$B^+ \to \omega\ell^+\nu_\ell$	[ii] < 2.1 $\times 10^{-4}$	CL=90%	–
$\quad B^+ \to \omega\mu^+\nu_\mu$	seen		2580
$B^+ \to \rho^0\ell^+\nu_\ell$	[ii] < 2.1 $\times 10^{-4}$	CL=90%	–
D, D^*, or D_s modes			
$B^+ \to \overline{D}^0\pi^+$	$(5.3 \pm 0.5) \times 10^{-3}$		2308
$B^+ \to \overline{D}^0\rho^+$	(1.34 ± 0.18) %		2237
$B^+ \to \overline{D}^0\pi^+\pi^+\pi^-$	(1.1 ± 0.4) %		2289
$\quad B^+ \to \overline{D}^0\pi^+\pi^+\pi^-$ nonresonant	$(5 \pm 4) \times 10^{-3}$		2289
$\quad B^+ \to \overline{D}^0\pi^+\rho^0$	$(4.2 \pm 3.0) \times 10^{-3}$		2208
$\qquad B^+ \to \overline{D}^0 a_1(1260)^+$	$(5 \pm 4) \times 10^{-3}$		2123
$\quad B^+ \to D^*(2010)^-\pi^+\pi^+$	$(2.1 \pm 0.6) \times 10^{-3}$		2247
$B^+ \to D^-\pi^+\pi^+$	< 1.4 $\times 10^{-3}$	CL=90%	2299
$B^+ \to \overline{D}^*(2007)^0\pi^+$	$(5.2 \pm 0.8) \times 10^{-3}$		2255
$B^+ \to \overline{D}^*(2007)^0\rho^+$	(1.55 ± 0.31) %		2182
$B^+ \to \overline{D}^*(2007)^0\pi^+\pi^+\pi^-$	$(9.4 \pm 2.6) \times 10^{-3}$		2236
$B^+ \to D^*(2010)^-\pi^+\pi^+\pi^0$	(1.5 ± 0.7) %		2235
$B^+ \to D^*(2010)^-\pi^+\pi^+\pi^+\pi^-$	< 1 %	CL=90%	2217
$B^+ \to \overline{D}_1^*(2420)^0\pi^+$	$(1.1 \pm 0.5) \times 10^{-3}$		2081
$B^+ \to \overline{D}_1^*(2420)^0\rho^+$	< 1.4 $\times 10^{-3}$	CL=90%	1996
$B^+ \to \overline{D}_2^*(2460)^0\pi^+$	< 1.3 $\times 10^{-3}$	CL=90%	2064
$B^+ \to \overline{D}_2^*(2460)^0\rho^+$	< 4.7 $\times 10^{-3}$	CL=90%	1979
$B^+ \to \overline{D}^0D_s^+$	(1.7 ± 0.6) %		1814
$B^+ \to \overline{D}^0D_s^{*+}$	(1.2 ± 1.0) %		1735
$B^+ \to \overline{D}^*(2007)^0D_s^+$	(1.0 ± 0.7) %		1737
$B^+ \to \overline{D}^*(2007)^0D_s^{*+}$	(2.4 ± 1.3) %		1652
$B^+ \to D_s^+\pi^0$	< 2.1 $\times 10^{-4}$	CL=90%	2270
$B^+ \to D_s^{*+}\pi^0$	< 3.4 $\times 10^{-4}$	CL=90%	2215
$B^+ \to D_s^+\eta$	< 5 $\times 10^{-4}$	CL=90%	2235
$B^+ \to D_s^{*+}\eta$	< 8 $\times 10^{-4}$	CL=90%	2178
$B^+ \to D_s^+\rho^0$	< 4 $\times 10^{-4}$	CL=90%	2197
$B^+ \to D_s^{*+}\rho^0$	< 5 $\times 10^{-4}$	CL=90%	2139
$B^+ \to D_s^+\omega$	< 5 $\times 10^{-4}$	CL=90%	2195
$B^+ \to D_s^{*+}\omega$	< 7 $\times 10^{-4}$	CL=90%	2137
$B^+ \to D_s^+a_1(1260)^0$	< 2.3 $\times 10^{-3}$	CL=90%	2079

Meson Summary Table

Mode	Fraction	Scale factor / Confidence level	p (MeV/c)
$B^+ \to D_s^{*+} a_1(1260)^0$	$< 1.7 \times 10^{-3}$	CL=90%	2015
$B^+ \to D_s^+ \phi$	$< 3.3 \times 10^{-4}$	CL=90%	2140
$B^+ \to D_s^{*+} \phi$	$< 4 \times 10^{-4}$	CL=90%	2080
$B^+ \to D_s^+ \overline{K}^0$	$< 1.1 \times 10^{-3}$	CL=90%	2241
$B^+ \to D_s^{*+} \overline{K}^0$	$< 1.2 \times 10^{-3}$	CL=90%	2185
$B^+ \to D_s^+ \overline{K}^*(892)^0$	$< 5 \times 10^{-4}$	CL=90%	2171
$B^+ \to D_s^{*+} \overline{K}^*(892)^0$	$< 5 \times 10^{-4}$	CL=90%	2111
$B^+ \to D_s^- \pi^+ K^+$	$< 9 \times 10^{-4}$	CL=90%	2222
$B^+ \to D_s^{*-} \pi^+ K^+$	$< 1.2 \times 10^{-3}$	CL=90%	2165
$B^+ \to D_s^- \pi^+ K^*(892)^+$	$< 7 \times 10^{-3}$	CL=90%	2137
$B^+ \to D_s^{*-} \pi^+ K^*(892)^+$	$< 9 \times 10^{-3}$	CL=90%	2076
Charmonium modes			
$B^+ \to J/\psi(1S) K^+$	$(1.02 \pm 0.14) \times 10^{-3}$		1683
$B^+ \to J/\psi(1S) K^+ \pi^+ \pi^-$	$(1.4 \pm 0.6) \times 10^{-3}$		1612
$B^+ \to J/\psi(1S) K^*(892)^+$	$(1.7 \pm 0.5) \times 10^{-3}$		1571
$B^+ \to \psi(2S) K^+$	$(6.9 \pm 3.1) \times 10^{-4}$	S=1.3	1284
$B^+ \to \psi(2S) K^*(892)^+$	$< 3.0 \times 10^{-3}$	CL=90%	1115
$B^+ \to \psi(2S) K^*(892)^+ \pi^+ \pi^-$	$(1.9 \pm 1.2) \times 10^{-3}$		909
$B^+ \to \chi_{c1}(1P) K^+$	$(1.0 \pm 0.4) \times 10^{-3}$		1411
$B^+ \to \chi_{c1}(1P) K^*(892)^+$	$< 2.1 \times 10^{-3}$	CL=90%	1265
K or K^* modes			
$B^+ \to K^0 \pi^+$	$< 1.0 \times 10^{-4}$	CL=90%	2614
$B^+ \to K^*(892)^0 \pi^+$	$< 1.5 \times 10^{-4}$	CL=90%	2561
$B^+ \to K^+ \pi^- \pi^+$ (no charm)	$< 1.9 \times 10^{-4}$	CL=90%	2609
$B^+ \to K_1(1400)^0 \pi^+$	$< 2.6 \times 10^{-3}$	CL=90%	2451
$B^+ \to K_2^*(1430)^0 \pi^+$	$< 6.8 \times 10^{-4}$	CL=90%	2443
$B^+ \to K^+ \rho^0$	$< 8 \times 10^{-5}$	CL=90%	2559
$B^+ \to K^*(892)^+ \pi^+ \pi^-$	$< 1.1 \times 10^{-3}$	CL=90%	2556
$B^+ \to K^*(892)^+ \rho^0$	$< 9.0 \times 10^{-4}$	CL=90%	2505
$B^+ \to K_1(1400)^+ \rho^0$	$< 7.8 \times 10^{-4}$	CL=90%	2388
$B^+ \to K_2^*(1430)^+ \rho^0$	$< 1.5 \times 10^{-3}$	CL=90%	2382
$B^+ \to K^+ K^- K^+$	$< 3.5 \times 10^{-4}$	CL=90%	2522
$B^+ \to K^+ \phi$	$< 9 \times 10^{-5}$	CL=90%	2516
$B^+ \to K^*(892)^+ K^+ K^-$	$< 1.6 \times 10^{-3}$	CL=90%	2466
$B^+ \to K^*(892)^+ \phi$	$< 1.3 \times 10^{-3}$	CL=90%	2460
$B^+ \to K_1(1400)^+ \phi$	$< 1.1 \times 10^{-3}$	CL=90%	2339
$B^+ \to K_2^*(1430)^+ \phi$	$< 3.4 \times 10^{-3}$	CL=90%	2332
$B^+ \to K^+ f_0(980)$	$< 8 \times 10^{-5}$	CL=90%	2524
$B^+ \to K^*(892)^+ \gamma$	$(5.7 \pm 3.3) \times 10^{-5}$		2564
$B^+ \to K_1(1270)^+ \gamma$	$< 7.3 \times 10^{-3}$	CL=90%	2486
$B^+ \to K_1(1400)^+ \gamma$	$< 2.2 \times 10^{-3}$	CL=90%	2453
$B^+ \to K_2^*(1430)^+ \gamma$	$< 1.4 \times 10^{-3}$	CL=90%	2447
$B^+ \to K^*(1680)^+ \gamma$	$< 1.9 \times 10^{-3}$	CL=90%	2361
$B^+ \to K_3^*(1780)^+ \gamma$	$< 5.5 \times 10^{-3}$	CL=90%	2343
$B^+ \to K_4^*(2045)^+ \gamma$	$< 9.9 \times 10^{-3}$	CL=90%	2243
Light unflavored meson modes			
$B^+ \to \pi^+ \pi^0$	$< 2.4 \times 10^{-4}$	CL=90%	2636
$B^+ \to \pi^+ \pi^+ \pi^-$	$< 1.9 \times 10^{-4}$	CL=90%	2630
$B^+ \to \rho^0 \pi^+$	$< 1.5 \times 10^{-4}$	CL=90%	2581
$B^+ \to \pi^+ f_0(980)$	$< 1.4 \times 10^{-4}$	CL=90%	2546
$B^+ \to \pi^+ f_2(1270)$	$< 2.4 \times 10^{-4}$	CL=90%	2483
$B^+ \to \pi^+ \pi^0 \pi^0$	$< 8.9 \times 10^{-4}$	CL=90%	2631
$B^+ \to \rho^+ \pi^0$	$< 5.5 \times 10^{-4}$	CL=90%	2581
$B^+ \to \pi^+ \pi^- \pi^+ \pi^0$	$< 4.0 \times 10^{-3}$	CL=90%	2621
$B^+ \to \rho^+ \rho^0$	$< 1.0 \times 10^{-3}$	CL=90%	2525
$B^+ \to a_1(1260)^+ \pi^0$	$< 1.7 \times 10^{-3}$	CL=90%	2494
$B^+ \to a_1(1260)^0 \pi^+$	$< 9.0 \times 10^{-4}$	CL=90%	2494
$B^+ \to \omega \pi^+$	$< 4.0 \times 10^{-4}$	CL=90%	2580
$B^+ \to \eta \pi^+$	$< 7.0 \times 10^{-4}$	CL=90%	2609
$B^+ \to \pi^+ \pi^+ \pi^+ \pi^- \pi^-$	$< 8.6 \times 10^{-4}$	CL=90%	2608
$B^+ \to \rho^0 a_1(1260)^+$	$< 6.2 \times 10^{-4}$	CL=90%	2433
$B^+ \to \rho^0 a_2(1320)^+$	$< 7.2 \times 10^{-4}$	CL=90%	2411
$B^+ \to \pi^+ \pi^+ \pi^+ \pi^- \pi^- \pi^0$	$< 6.3 \times 10^{-3}$	CL=90%	2592
$B^+ \to a_1(1260)^+ a_1(1260)^0$	< 1.3 %	CL=90%	2335
Baryon modes			
$B^+ \to p \overline{p} \pi^+$	$< 1.6 \times 10^{-4}$	CL=90%	2438
$B^+ \to p \overline{p} \pi^+ \pi^+ \pi^-$	$< 5.2 \times 10^{-4}$	CL=90%	2369
$B^+ \to p \overline{\Lambda}$	$< 6 \times 10^{-5}$	CL=90%	2430
$B^+ \to p \overline{\Lambda} \pi^+ \pi^-$	$< 2.0 \times 10^{-4}$	CL=90%	2367
$B^+ \to \overline{\Delta}^0 p$	$< 3.8 \times 10^{-4}$	CL=90%	2402
$B^+ \to \Delta^{++} \overline{p}$	$< 1.5 \times 10^{-4}$	CL=90%	2402

Lepton Family number (LF) or Lepton number (L) violating modes, or $\Delta B = 1$ weak neutral current ($B1$) modes

Mode		Fraction	Confidence level	p (MeV/c)
$B^+ \to \pi^+ e^+ e^-$	$B1$	$< 3.9 \times 10^{-3}$	CL=90%	2638
$B^+ \to \pi^+ \mu^+ \mu^-$	$B1$	$< 9.1 \times 10^{-3}$	CL=90%	2633
$B^+ \to K^+ e^+ e^-$	$B1$	$< 6 \times 10^{-5}$	CL=90%	2616
$B^+ \to K^+ \mu^+ \mu^-$	$B1$	$< 1.7 \times 10^{-4}$	CL=90%	2612
$B^+ \to K^*(892)^+ e^+ e^-$	$B1$	$< 6.9 \times 10^{-4}$	CL=90%	2564
$B^+ \to K^*(892)^+ \mu^+ \mu^-$	$B1$	$< 1.2 \times 10^{-3}$	CL=90%	2560
$B^+ \to \pi^+ e^+ \mu^-$	LF	$< 6.4 \times 10^{-3}$	CL=90%	2636
$B^+ \to \pi^+ e^- \mu^+$	LF	$< 6.4 \times 10^{-3}$	CL=90%	2636
$B^+ \to K^+ e^+ \mu^-$	LF	$< 6.4 \times 10^{-3}$	CL=90%	2615
$B^+ \to K^+ e^- \mu^+$	LF	$< 6.4 \times 10^{-3}$	CL=90%	2615
$B^+ \to \pi^- e^+ e^+$	L	$< 3.9 \times 10^{-3}$	CL=90%	2638
$B^+ \to \pi^- \mu^+ \mu^+$	L	$< 9.1 \times 10^{-3}$	CL=90%	2633
$B^+ \to \pi^- e^+ \mu^+$	L	$< 6.4 \times 10^{-3}$	CL=90%	2636
$B^+ \to K^- e^+ e^+$	L	$< 3.9 \times 10^{-3}$	CL=90%	2616
$B^+ \to K^- \mu^+ \mu^+$	L	$< 9.1 \times 10^{-3}$	CL=90%	2612
$B^+ \to K^- e^+ \mu^+$	L	$< 6.4 \times 10^{-3}$	CL=90%	2615

B DECAY MODES

$\overline{B}$ modes are charge conjugates of the modes below.

For the following modes, the charge of B was not determined. The measurements are for an admixture of B mesons at the $\Upsilon(4S)$ unless otherwise indicated by a footnote and a "$\overline{b}$" instead of "B" in the initial state.

Mode		Fraction	Scale factor / Confidence level	p (MeV/c)
Semileptonic and leptonic modes				
$B \to e^+ \nu_e$ anything	[ss]	(10.4 ± 0.4) %	S=1.3	–
$B \to \overline{D}^*(2010) e^+ \nu_e$		(7.0 ± 2.3) %		–
$B \to \overline{p} e^+ \nu_e$ anything		$< 1.6 \times 10^{-3}$	CL=90%	–
$B \to \mu^+ \nu_\mu$ anything	[ss]	(10.3 ± 0.5) %		–
$B \to \ell^+ \nu_\ell$ anything	[ii, ss]	(10.43 ± 0.24) %		–
$B \to D^- \ell^+ \nu_\ell$ anything	[ii]	(2.7 ± 0.8) %		–
$B \to \overline{D}^0 \ell^+ \nu_\ell$ anything	[ii]	(7.0 ± 1.4) %		–
$B \to D^{**} \ell^+ \nu_\ell$	[ii, tt]	(2.7 ± 0.7) %		–
$B \to D_s^- \ell^+ \nu_\ell$ anything	[ii]	$< 9 \times 10^{-3}$	CL=90%	–
$B \to D_s^- \ell^+ \nu_\ell K^+$ anything	[ii]	$< 6 \times 10^{-3}$	CL=90%	–
$B \to D_s^- \ell^+ \nu_\ell K^0$ anything	[ii]	$< 9 \times 10^{-3}$	CL=90%	–
$B \to K^+ \ell^+ \nu_\ell$ anything	[ii]	(5.6 ± 1.0) %		–
$B \to K^- \ell^+ \nu_\ell$ anything	[ii]	(1.0 ± 0.6) %		–
$B \to K^0 / \overline{K}^0 \ell^+ \nu_\ell$ anything	[ii]	(4.1 ± 0.8) %		–
$\overline{b} \to \tau^+ \nu_\tau$ anything	[uu]	(4.1 ± 1.0) %		–
D, D^*, or D_s modes				
$B \to D^-$ anything		(26 ± 4) %		–
$B \to \overline{D}^0$ anything		(54 ± 6) %		–
$B \to D^*(2010)^-$ anything		(23 ± 4) %	S=1.4	–
$B \to D_s^\pm$ anything	[q]	(8.9 ± 1.1) %		–
$B \to D_s D$, $D_s^* D$, $D_s D^*$, or $D_s^* D^*$	[q]	(5.0 ± 0.9) %		–
$B \to D^*(2010) \gamma$		$< 1.1 \times 10^{-3}$	CL=90%	–
$B \to D_s^+ \pi^-$, $D_s^{*+} \pi^-$, $D_s^+ \rho^-$, $D_s^{*+} \rho^-$, $D_s^+ \pi^0$, $D_s^{*+} \pi^0$, $D_s^+ \eta$, $D_s^{*+} \eta$, $D_s^+ \rho^0$, $D_s^{*+} \rho^0$, $D_s^+ \omega$, $D_s^{*+} \omega$		$< 5 \times 10^{-4}$	CL=90%	–
Charmonium modes				
$B \to J/\psi(1S)$ anything		(1.30 ± 0.17) %		–
$B \to \psi(2S)$ anything		$(4.6 \pm 2.0) \times 10^{-3}$		–
$B \to \chi_{c1}(1P)$ anything		(1.1 ± 0.4) %		–
K or K^* modes				
$B \to K^\pm$ anything	[q]	(85 ± 11) %		–
$B \to K^0 / \overline{K}^0$ anything		(63 ± 8) %		–
$b \to s \gamma$	[vv]	$< 1.2 \times 10^{-3}$	CL=90%	–
$B \to K^*(892) \gamma$		$< 2.4 \times 10^{-4}$	CL=90%	–
$B \to K_1(1400) \gamma$		$< 4.1 \times 10^{-4}$	CL=90%	–
$B \to K_2^*(1430) \gamma$		$< 8.3 \times 10^{-4}$	CL=90%	–
$B \to K_2(1770) \gamma$		$< 1.2 \times 10^{-3}$	CL=90%	–
$B \to K_3^*(1780) \gamma$		$< 3.0 \times 10^{-3}$	CL=90%	–
$B \to K_4^*(2045) \gamma$		$< 1.0 \times 10^{-3}$	CL=90%	–
Light unflavored meson modes				
$B \to \phi$ anything		(2.3 ± 0.8) %		–

Baryon modes			
$B \to$ charmed-baryon anything	(6.4 ± 1.1) %		–
$B \to \overline{\Sigma}_c^{--}$ anything	$(4.8 \pm 2.5) \times 10^{-3}$		–
$B \to \overline{\Sigma}_c^{-}$ anything	< 1.1 %	CL=90%	–
$B \to \overline{\Sigma}_c^{0}$ anything	$(5.3 \pm 2.5) \times 10^{-3}$		–
$B \to \overline{\Sigma}_c^{0} N (N = p \text{ or } n)$	$< 1.7 \times 10^{-3}$	CL=90%	–
$B \to p$anything + $\overline{p}$anything	(8.0 ± 0.5) %		–
$B \to p$(direct) anything + $\overline{p}$(direct) anything	(5.6 ± 0.7) %		–
$B \to \Lambda$anything + $\overline{\Lambda}$anything	(4.0 ± 0.5) %		–
$B \to \Xi^-$ anything + $\overline{\Xi}^+$ anything	$(2.7 \pm 0.6) \times 10^{-3}$		–
$B \to$ baryons anything	(6.8 ± 0.6) %		–
$B \to p\overline{p}$anything	(2.47 ± 0.23) %		–
$B \to \Lambda\overline{p}$anything + $\overline{\Lambda}p$anything	(2.5 ± 0.4) %		–
$B \to \Lambda\overline{\Lambda}$anything	$< 5 \times 10^{-3}$	CL=90%	–
$\Delta B = 1$ weak neutral current ($B1$) modes			
$\overline{b} \to e^+ e^-$ anything	$B1$ [vv] $< 2.4 \times 10^{-3}$		–
$\overline{b} \to \mu^+ \mu^-$ anything	$B1$ [vv] $< 5.0 \times 10^{-5}$	CL=90%	–

B^0

$I(J^P) = \frac{1}{2}(0^-)$

I, J, P need confirmation. Quantum numbers shown are quark-model predictions.

Mass $m_{B^0} = 5279.0 \pm 2.0$ MeV

$m_{B^0} - m_{B^\pm} = 0.34 \pm 0.29$ MeV (S = 1.1)

Mean life $\tau = (1.50 \pm 0.11) \times 10^{-12}$ s

$c\tau = 449\ \mu$m

$\tau_{B^+}/\tau_{B^0} = 0.98 \pm 0.09$

B^0-$\overline{B}^0$ mixing parameters

$\chi_d = 0.156 \pm 0.024$

$\Delta m_{B^0} = m_{B^0_H} - m_{B^0_L} = (0.51 \pm 0.06) \times 10^{12}\ \hbar\ s^{-1}$

$x_d = \Delta m_{B^0}/\Gamma_{B^0} = 0.71 \pm 0.06$ [ww]

$\overline{B}^0$ modes are charge conjugates of the modes below. Reactions indicate the weak decay vertex and do not include mixing. Decays in which the charge of the B is not determined are in the $B^\pm$ section.

Only data from $\Upsilon(4S)$ decays are used for branching fractions, with rare exceptions. The branching fractions listed below assume a 50:50 $B^0\overline{B}^0$:B^+B^- production ratio at the $\Upsilon(4S)$. We have attempted to bring older measurements up to date by rescaling their assumed $\Upsilon(4S)$ production ratio to 50:50 and their assumed D, D_s, D^*, and ψ branching ratios to current values whenever this would effect our averages and best limits significantly.

Indentation is used to indicate a subchannel of a previous reaction. All resonant subchannels have been corrected for resonance branching fractions to the final state so the sum of the subchannel branching fractions can exceed that of the final state.

B^0 DECAY MODES	Fraction (Γ_i/Γ)	Confidence level	p (MeV/c)
Semileptonic and leptonic modes			
$\ell^+ \nu_\ell$ anything	[ii] (9.5 ± 1.6) %		–
$D^- \ell^+ \nu_\ell$	[ii] (1.9 ± 0.5) %		–
$D^*(2010)^- \ell^+ \nu_\ell$	[ii] (4.4 ± 0.4) %		–
$\rho^- \ell^+ \nu_\ell$	[ii] $< 4.1 \times 10^{-4}$	90%	–
$\pi^- \mu^+ \nu_\mu$	seen		2636
D, D^*, or D_s modes			
$D^- \pi^+$	$(3.0 \pm 0.4) \times 10^{-3}$		2306
$D^- \rho^+$	$(7.8 \pm 1.4) \times 10^{-3}$		2236
$\overline{D}^0 \pi^+ \pi^-$	$< 1.6 \times 10^{-3}$	90%	2301
$D^*(2010)^- \pi^+$	$(2.6 \pm 0.4) \times 10^{-3}$		2254
$D^- \pi^+ \pi^+ \pi^-$	$(8.0 \pm 2.5) \times 10^{-3}$		2287
$(D^- \pi^+ \pi^+ \pi^-)$ nonresonant	$(3.9 \pm 1.9) \times 10^{-3}$		2287
$D^- \pi^+ \rho^0$	$(1.1 \pm 1.0) \times 10^{-3}$		2207
$D^- a_1(1260)^+$	$(6.0 \pm 3.3) \times 10^{-3}$		2121
$D^*(2010)^- \pi^+ \pi^0$	(1.5 ± 0.5) %		2247
$D^*(2010)^- \rho^+$	$(7.3 \pm 1.5) \times 10^{-3}$		2181
$D^*(2010)^- \pi^+ \pi^+ \pi^-$	(1.19 ± 0.27) %		2235
$(D^*(2010)^- \pi^+ \pi^+ \pi^-)$ nonresonant	$(0.0 \pm 2.5) \times 10^{-3}$		2235
$D^*(2010)^- \pi^+ \rho^0$	$(5.7 \pm 3.1) \times 10^{-3}$		2151
$D^*(2010)^- a_1(1260)^+$	(1.5 ± 0.7) %		2061
$D^*(2010)^- \pi^+ \pi^+ \pi^- \pi^0$	(3.4 ± 1.8) %		2218
$\overline{D}_2^*(2460)^- \pi^+$	$< 2.2 \times 10^{-3}$	90%	2065
$\overline{D}_2^*(2460)^- \rho^+$	$< 4.9 \times 10^{-3}$	90%	1980
$D^- D_s^+$	$(8 \pm 4) \times 10^{-3}$		1812
$D^*(2010)^- D_s^+$	(1.2 ± 0.6) %		1735
$D^- D_s^{*+}$	(2.1 ± 1.5) %		1733
$D^*(2010)^- D_s^{*+}$	(2.0 ± 1.2) %		1650
$D_s^+ \pi^-$	$< 2.9 \times 10^{-4}$	90%	2270
$D_s^{*+} \pi^-$	$< 5 \times 10^{-4}$	90%	2215
$D_s^+ \rho^-$	$< 7 \times 10^{-4}$	90%	2198
$D_s^{*+} \rho^-$	$< 8 \times 10^{-4}$	90%	2140
$D_s^+ a_1(1260)^-$	$< 2.7 \times 10^{-3}$	90%	2079
$D_s^{*+} a_1(1260)^-$	$< 2.2 \times 10^{-3}$	90%	2015
$D_s^- K^+$	$< 2.4 \times 10^{-4}$	90%	2242
$D_s^{*-} K^+$	$< 1.8 \times 10^{-4}$	90%	2186
$D_s^- K^*(892)^+$	$< 1.0 \times 10^{-3}$	90%	2172
$D_s^{*-} K^*(892)^+$	$< 1.2 \times 10^{-3}$	90%	2113
$D_s^- \pi^+ K^0$	$< 6 \times 10^{-3}$	90%	2221
$D_s^{*-} \pi^+ K^0$	$< 3.2 \times 10^{-3}$	90%	2164
$D_s^- \pi^+ K^*(892)^0$	$< 4 \times 10^{-3}$	90%	2136
$D_s^{*-} \pi^+ K^*(892)^0$	$< 2.1 \times 10^{-3}$	90%	2075
$\overline{D}^0 \pi^0$	$< 4.8 \times 10^{-4}$	90%	2308
$\overline{D}^0 \rho^0$	$< 5.5 \times 10^{-4}$	90%	2238
$\overline{D}^0 \eta$	$< 6.8 \times 10^{-4}$	90%	2274
$\overline{D}^0 \eta'$	$< 8.6 \times 10^{-4}$	90%	2197
$\overline{D}^0 \omega$	$< 6.3 \times 10^{-4}$	90%	2235
$\overline{D}^*(2007)^0 \pi^0$	$< 9.7 \times 10^{-4}$	90%	2256
$\overline{D}^*(2007)^0 \rho^0$	$< 1.17 \times 10^{-3}$	90%	2182
$\overline{D}^*(2007)^0 \eta$	$< 6.9 \times 10^{-4}$	90%	2220
$\overline{D}^*(2007)^0 \eta'$	$< 2.7 \times 10^{-3}$	90%	2140
$\overline{D}^*(2007)^0 \omega$	$< 2.1 \times 10^{-3}$	90%	2180
Charmonium modes			
$J/\psi(1S) K^0$	$(7.5 \pm 2.1) \times 10^{-4}$		1682
$J/\psi(1S) K^+ \pi^-$	$(1.2 \pm 0.6) \times 10^{-3}$		1652
$J/\psi(1S) K^*(892)^0$	$(1.58 \pm 0.28) \times 10^{-3}$		1569
$\psi(2S) K^0$	$< 8 \times 10^{-4}$	90%	1283
$\psi(2S) K^+ \pi^-$	$< 1 \times 10^{-3}$	90%	1238
$\psi(2S) K^*(892)^0$	$(1.4 \pm 0.9) \times 10^{-3}$		1113
$\chi_{c1}(1P) K^0$	$< 2.7 \times 10^{-3}$	90%	1410
$\chi_{c1}(1P) K^*(892)^0$	$< 2.1 \times 10^{-3}$	90%	1263
K or K^* modes			
$K^+ \pi^-$	$< 2.6 \times 10^{-5}$	90%	2615
$K^+ K^-$	$< 7 \times 10^{-6}$	90%	2593
$K^0 \pi^+ \pi^-$	$< 4.4 \times 10^{-4}$	90%	2609
$K^0 \rho^0$	$< 3.2 \times 10^{-4}$	90%	2559
$K^0 f_0(980)$	$< 3.6 \times 10^{-4}$	90%	2523
$K^*(892)^+ \pi^-$	$< 3.8 \times 10^{-4}$	90%	2562
$K_2^*(1430)^+ \pi^-$	$< 2.6 \times 10^{-3}$	90%	2445
$K^0 K^+ K^-$	$< 1.3 \times 10^{-3}$	90%	2522
$K^0 \phi$	$< 4.2 \times 10^{-4}$	90%	2516
$K^*(892)^0 \pi^+ \pi^-$	$< 1.4 \times 10^{-3}$	90%	2556
$K^*(892)^0 \rho^0$	$< 4.6 \times 10^{-4}$	90%	2504
$K^*(892)^0 f_0(980)$	$< 1.7 \times 10^{-4}$	90%	2467
$K_1(1400)^+ \pi^-$	$< 1.1 \times 10^{-3}$	90%	2451
$K^*(892)^0 K^+ K^-$	$< 6.1 \times 10^{-4}$	90%	2465
$K^*(892)^0 \phi$	$< 3.2 \times 10^{-4}$	90%	2459
$K_1(1400)^0 \rho^0$	$< 3.0 \times 10^{-3}$	90%	2388
$K_1(1400)^0 \phi$	$< 5.0 \times 10^{-3}$	90%	2339
$K_2^*(1430)^0 \rho^0$	$< 1.1 \times 10^{-3}$	90%	2380
$K_2^*(1430)^0 \phi$	$< 1.4 \times 10^{-3}$	90%	2330
$K^*(892)^0 \gamma$	$(4.0 \pm 1.9) \times 10^{-5}$		2563
$K_1(1270)^0 \gamma$	$< 7.0 \times 10^{-3}$	90%	2486
$K_1(1400)^0 \gamma$	$< 4.3 \times 10^{-3}$	90%	2453
$K_2^*(1430)^0 \gamma$	$< 4.0 \times 10^{-4}$	90%	2445
$K^*(1680)^0 \gamma$	$< 2.0 \times 10^{-3}$	90%	2361
$K_3^*(1780)^0 \gamma$	< 1.0 %	90%	2343
$K_4^*(2045)^0 \gamma$	$< 4.3 \times 10^{-3}$	90%	2243

Meson Summary Table

		Light unflavored meson modes			
$\pi^+\pi^-$		< 2.9	$\times 10^{-5}$	90%	2636
$\pi^+\pi^-\pi^0$		< 7.2	$\times 10^{-4}$	90%	2631
$\rho^0\pi^0$		< 4.0	$\times 10^{-4}$	90%	2582
$\rho^\mp\pi^\pm$		[q] < 5.2	$\times 10^{-4}$	90%	2581
$\pi^+\pi^-\pi^+\pi^-$		< 6.7	$\times 10^{-4}$	90%	2621
$\rho^0\rho^0$		< 2.8	$\times 10^{-4}$	90%	2525
$a_1(1260)^\mp\pi^\pm$		[q] < 4.9	$\times 10^{-4}$	90%	2494
$a_2(1320)^\mp\pi^\pm$		[q] < 3.0	$\times 10^{-4}$	90%	2473
$\pi^+\pi^-\pi^0\pi^0$		< 3.1	$\times 10^{-3}$	90%	2622
$\rho^+\rho^-$		< 2.2	$\times 10^{-3}$	90%	2525
$a_1(1260)^0\pi^0$		< 1.1	$\times 10^{-3}$	90%	2494
$\omega\pi^0$		< 4.6	$\times 10^{-4}$	90%	2580
$\eta\pi^0$		< 1.8	$\times 10^{-3}$	90%	2609
$\pi^+\pi^+\pi^-\pi^-\pi^0$		< 9.0	$\times 10^{-3}$	90%	2609
$a_1(1260)^+\rho^-$		< 3.4	$\times 10^{-3}$	90%	2433
$a_1(1260)^0\rho^0$		< 2.4	$\times 10^{-3}$	90%	2433
$\pi^+\pi^+\pi^+\pi^-\pi^-\pi^-$		< 3.0	$\times 10^{-3}$	90%	2591
$a_1(1260)^+a_1(1260)^-$		< 2.8	$\times 10^{-3}$	90%	2335
$\pi^+\pi^+\pi^+\pi^-\pi^-\pi^-\pi^0$		< 1.1	%	90%	2572
		Baryon modes			
$p\overline{p}$		< 3.4	$\times 10^{-5}$	90%	2467
$p\overline{p}\pi^+\pi^-$		< 2.5	$\times 10^{-4}$	90%	2406
$p\overline{\Lambda}\pi^-$		< 1.8	$\times 10^{-4}$	90%	2401
$\Delta^0\overline{\Delta}^0$		< 1.5	$\times 10^{-3}$	90%	2334
$\Delta^{++}\Delta^{--}$		< 1.1	$\times 10^{-4}$	90%	2334
$\overline{\Sigma}_c^{--}\Delta^{++}$		< 1.2	$\times 10^{-3}$	90%	1839
		Lepton Family number (*LF*) violating modes, $\Delta B = 2$ forbidden decay via mixing (*B2M*) modes, or $\Delta B = 1$ weak neutral current (*B1*) modes			
e^+e^-	*B1*	< 5.9	$\times 10^{-6}$	90%	2639
$\mu^+\mu^-$	*B1*	< 5.9	$\times 10^{-6}$	90%	2637
$K^0e^+e^-$	*B1*	< 3.0	$\times 10^{-4}$	90%	2616
$K^0\mu^+\mu^-$	*B1*	< 3.6	$\times 10^{-4}$	90%	2612
$K^*(892)^0e^+e^-$	*B1*	< 2.9	$\times 10^{-4}$	90%	2563
$K^*(892)^0\mu^+\mu^-$	*B1*	< 2.3	$\times 10^{-5}$	90%	2559
$e^\pm\mu^\mp$	*LF*	[q] < 5.9	$\times 10^{-6}$	90%	2638
$e^\pm\tau^\mp$	*LF*	[q] < 5.3	$\times 10^{-4}$	90%	2340
$\mu^\pm\tau^\mp$	*LF*	[q] < 8.3	$\times 10^{-4}$	90%	2339

B^*

$I(J^P) = \frac{1}{2}(1^-)$

I, *J*, *P* need confirmation. Quantum numbers shown are quark-model predictions.

Mass $m_{B^*} = 5324.8 \pm 2.1$ MeV

$m_{B^*} - m_B = 46.0 \pm 0.6$ MeV

BOTTOM, STRANGE MESONS $(B = \pm 1, S = \mp 1)$

$B_s^0 = s\overline{b}$, $\overline{B}_s^0 = \overline{s}b$, similarly for B_s^*'s

B_s^0

$I(J^P) = \frac{1}{2}(0^-)$

I, *J*, *P* need confirmation. Quantum numbers shown are quark-model predictions.

Mass $m_{B_s^0} = 5375 \pm 6$ MeV (S = 1.3)

Mean life $\tau = (1.34^{+0.32}_{-0.27}) \times 10^{-12}$ s (S = 1.4)

B_s^0-$\overline{B}_s^0$ mixing parameters

$\chi_s = 0.62 \pm 0.13$

$\Delta m_{B_s^0} = m_{B_{sH}^0} - m_{B_{sL}^0} > 1.8 \times 10^{12}\ \hbar\ s^{-1}$, CL = 95%

$x_s = \Delta m_{B_s^0}/\Gamma_{B_s^0} > 2.0$, CL = 95%

B_s^0 DECAY MODES	Fraction (Γ_i/Γ)	p (MeV/c)
D_s^- anything	seen	–
$D_s^-\ell^+\nu_\ell$ anything (ℓ means sum of e and μ)	seen	–
$D_s^-\pi^+$	seen	2325
$J/\psi(1S)\phi$	seen	1594
$\psi(2S)\phi$	seen	1128

HEAVY QUARK SEARCHES

Searches for Top and Fourth Generation Hadrons

See the sections "Searches for t Quark" and "Searches for b' (4th Generation) Quark" at the end of the QUARKS section.

$c\overline{c}$ MESONS

$\eta_c(1S)$ or $\eta_c(2980)$

$I^G(J^{PC}) = 0^+(0^{-+})$

Mass $m = 2978.8 \pm 1.9$ MeV (S = 1.8)

Full width $\Gamma = 10.3^{+3.8}_{-3.4}$ MeV

$\eta_c(1S)$ DECAY MODES	Fraction (Γ_i/Γ)	Confidence level	p (MeV/c)
Decays involving hadronic resonances			
$\eta'(958)\pi\pi$	(4.1 ± 1.7) %		1319
$\rho\rho$	(2.6 ± 0.9) %		1275
$K^*(892)^0K^-\pi^+$ + c.c.	(2.0 ± 0.7) %		1273
$K^*(892)\overline{K}^*(892)$	$(8.5 \pm 3.1) \times 10^{-3}$		1193
$\phi\phi$	$(7.1 \pm 2.8) \times 10^{-3}$		1086
$a_0(980)\pi$	< 2 %	90%	1323
$a_2(1320)\pi$	< 2 %	90%	1193
$K^*(892)\overline{K}$ + c.c.	< 1.28 %	90%	1307
$f_2(1270)\eta$	< 1.1 %	90%	1142
$\omega\omega$	$< 3.1 \times 10^{-3}$	90%	1268
Decays into stable hadrons			
$K\overline{K}\pi$	(6.6 ± 1.8) %		1378
$\eta\pi\pi$	(4.9 ± 1.8) %		1425
$\pi^+\pi^-K^+K^-$	$(2.0\ ^{+0.7}_{-0.6})$ %		1342
$2(\pi^+\pi^-)$	(1.2 ± 0.4) %		1457
$p\overline{p}$	$(1.2 \pm 0.4) \times 10^{-3}$		1157
$K\overline{K}\eta$	< 3.1 %	90%	1262
$\pi^+\pi^-p\overline{p}$	< 1.2 %	90%	1023
$\Lambda\overline{\Lambda}$	$< 2 \times 10^{-3}$	90%	987
Radiative decays			
$\gamma\gamma$	$(6\ ^{+6}_{-5}) \times 10^{-4}$		1489

$J/\psi(1S)$
or $J/\psi(3097)$

$I^G(J^{PC}) = 0^-(1^{--})$

Mass $m = 3096.88 \pm 0.04$ MeV
Full width $\Gamma = 88 \pm 5$ keV
$\Gamma_{ee} = 5.26 \pm 0.37$ keV (Assuming $\Gamma_{ee} = \Gamma_{\mu\mu}$)

$J/\psi(1S)$ DECAY MODES	Fraction (Γ_i/Γ)	Scale factor/ Confidence level	p (MeV/c)
hadrons	(86.0 ±2.0) %		–
virtual $\gamma \to$ hadrons	(17.0 ±2.0) %		–
e^+e^-	(5.99±0.25) %		1548
$\mu^+\mu^-$	(5.97±0.25) %	S=1.1	1545
Decays involving hadronic resonances			
$\rho\pi$	(1.28±0.10) %		1449
$\rho^0\pi^0$	(4.2 ±0.5) × 10^{-3}		1449
$a_2(1320)\rho$	(1.09±0.22) %		1125
$\omega\pi^+\pi^+\pi^-\pi^-$	(8.5 ±3.4) × 10^{-3}		1392
$\omega\pi^+\pi^-$	(7.2 ±1.0) × 10^{-3}		1435
$K^*(892)^0\overline{K}{}^*_2(1430)^0$ + c.c.	(6.7 ±2.6) × 10^{-3}		1005
$\omega K^*(892)\overline{K}$ + c.c.	(5.3 ±2.0) × 10^{-3}		1098
$\omega f_2(1270)$	(4.3 ±0.6) × 10^{-3}		1143
$K^+\overline{K}{}^*(892)^-$ + c.c.	(5.0 ±0.4) × 10^{-3}		1373
$K^0\overline{K}{}^*(892)^0$ + c.c.	(4.2 ±0.4) × 10^{-3}		1371
$\omega\pi^0\pi^0$	(3.4 ±0.8) × 10^{-3}		1436
$b_1(1235)^\pm\pi^\mp$	[q] (3.0 ±0.5) × 10^{-3}		1299
$\omega K^\pm K^0_S\pi^\mp$	[q] (3.0 ±0.7) × 10^{-3}		1210
$b_1(1235)^0\pi^0$	(2.3 ±0.6) × 10^{-3}		1299
$\phi K^*(892)\overline{K}$ + c.c.	(2.04±0.28) × 10^{-3}		969
$\omega K\overline{K}$	(1.9 ±0.4) × 10^{-3}		1268
$\omega f_J(1710) \to \omega K\overline{K}$	(4.8 ±1.1) × 10^{-4}		878
$\phi 2(\pi^+\pi^-)$	(1.60±0.32) × 10^{-3}		1318
$\Delta(1232)^{++}\overline{p}\pi^-$	(1.6 ±0.5) × 10^{-3}		1030
$\omega\eta$	(1.58±0.16) × 10^{-3}		1394
$\phi K\overline{K}$	(1.48±0.22) × 10^{-3}		1179
$\phi f_J(1710) \to \phi K\overline{K}$	(3.6 ±0.6) × 10^{-4}		875
$p\overline{p}\omega$	(1.30±0.25) × 10^{-3}	S=1.3	769
$\Delta(1232)^{++}\overline{\Delta}(1232)^{--}$	(1.10±0.29) × 10^{-3}		938
$\Sigma(1385)^-\overline{\Sigma}(1385)^+$ (or c.c.)	[q] (1.03±0.13) × 10^{-3}		692
$p\overline{p}\eta'(958)$	(9 ±4) × 10^{-4}	S=1.7	596
$\phi f'_2(1525)$	(8 ±4) × 10^{-4}	S=2.7	871
$\phi\pi^+\pi^-$	(8.0 ±1.2) × 10^{-4}		1365
$\phi K^\pm K^0_S\pi^\mp$	[q] (7.2 ±0.9) × 10^{-4}		1114
$\omega f_1(1420)$	(6.8 ±2.4) × 10^{-4}		1062
$\phi\eta$	(6.5 ±0.7) × 10^{-4}		1320
$\Xi(1530)^-\overline{\Xi}{}^+$	(5.9 ±1.5) × 10^{-4}		597
$pK^-\overline{\Sigma}(1385)^0$	(5.1 ±3.2) × 10^{-4}		645
$\omega\pi^0$	(4.2 ±0.6) × 10^{-4}	S=1.4	1447
$\phi\eta'(958)$	(3.3 ±0.4) × 10^{-4}		1192
$\phi f_0(980)$	(3.2 ±0.9) × 10^{-4}	S=1.9	1182
$\Xi(1530)^0\overline{\Xi}{}^0$	(3.2 ±1.4) × 10^{-4}		608
$\Sigma(1385)^-\overline{\Sigma}{}^+$ (or c.c.)	[q] (3.1 ±0.5) × 10^{-4}		857
$\phi f_1(1285)$	(2.6 ±0.5) × 10^{-4}	S=1.1	1032
$\rho\eta$	(1.93±0.23) × 10^{-4}		1398
$\omega\eta'(958)$	(1.67±0.25) × 10^{-4}		1279
$\omega f_0(980)$	(1.4 ±0.5) × 10^{-4}		1271
$\rho\eta'(958)$	(1.05±0.18) × 10^{-4}		1283
$p\overline{p}\phi$	(4.5 ±1.5) × 10^{-5}		527
$a_2(1320)^\pm\pi^\mp$	[q] < 4.3 × 10^{-3}	CL=90%	1263
$K\overline{K}{}^*_2(1430)$ + c.c.	< 4.0 × 10^{-3}	CL=90%	1159
$K^*_2(1430)^0\overline{K}{}^*_2(1430)^0$	< 2.9 × 10^{-3}	CL=90%	588
$K^*(892)^0\overline{K}{}^*(892)^0$	< 5 × 10^{-4}	CL=90%	1263
$\phi f_2(1270)$	< 3.7 × 10^{-4}	CL=90%	1036
$p\overline{p}\rho$	< 3.1 × 10^{-4}	CL=90%	779
$\phi\eta(1440) \to \phi\eta\pi\pi$	< 2.5 × 10^{-4}	CL=90%	946
$\omega f'_2(1525)$	< 2.2 × 10^{-4}	CL=90%	1003
$\Sigma(1385)^0\overline{\Lambda}$	< 2 × 10^{-4}	CL=90%	911
$\Delta(1232)^+\overline{p}$	< 1 × 10^{-4}	CL=90%	1100
$\Sigma^0\overline{\Lambda}$	< 9 × 10^{-5}	CL=90%	1032
$\phi\pi^0$	< 6.8 × 10^{-6}	CL=90%	1377
Decays into stable hadrons			
$2(\pi^+\pi^-)\pi^0$	(3.37±0.26) %		1496
$3(\pi^+\pi^-)\pi^0$	(2.9 ±0.6) %		1433
$\pi^+\pi^-\pi^0$	(1.50±0.20) %		1533
$\pi^+\pi^-\pi^0K^+K^-$	(1.20±0.30) %		1368
$4(\pi^+\pi^-)\pi^0$	(9.0 ±3.0) × 10^{-3}		1345
$\pi^+\pi^-K^+K^-$	(7.2 ±2.3) × 10^{-3}		1407
$K\overline{K}\pi$	(6.1 ±1.0) × 10^{-3}		1440
$p\overline{p}\pi^+\pi^-$	(6.0 ±0.5) × 10^{-3}	S=1.3	1107
$2(\pi^+\pi^-)$	(4.0 ±1.0) × 10^{-3}		1517
$3(\pi^+\pi^-)$	(4.0 ±2.0) × 10^{-3}		1466
$n\overline{n}\pi^+\pi^-$	(4 ±4) × 10^{-3}		1106
$\Sigma\overline{\Sigma}$	(3.8 ±0.5) × 10^{-3}		992
$2(\pi^+\pi^-)K^+K^-$	(3.1 ±1.3) × 10^{-3}		1320
$p\overline{p}\pi^+\pi^-\pi^0$	[xx] (2.3 ±0.9) × 10^{-3}	S=1.9	1033
$p\overline{p}$	(2.14±0.10) × 10^{-3}		1232
$p\overline{p}\eta$	(2.09±0.18) × 10^{-3}		948
$p\overline{n}\pi^-$	(2.00±0.10) × 10^{-3}		1174
$n\overline{n}$	(1.9 ±0.5) × 10^{-3}		1231
$\Xi\overline{\Xi}$	(1.8 ±0.4) × 10^{-3}	S=1.8	818
$\Lambda\overline{\Lambda}$	(1.35±0.14) × 10^{-3}	S=1.2	1074
$p\overline{p}\pi^0$	(1.09±0.09) × 10^{-3}		1176
$\Lambda\overline{\Sigma}{}^-\pi^+$ (or c.c.)	[q] (1.06±0.12) × 10^{-3}		945
$pK^-\overline{\Lambda}$	(8.9 ±1.6) × 10^{-4}		876
$2(K^+K^-)$	(7.0 ±3.0) × 10^{-4}		1131
$pK^-\overline{\Sigma}{}^0$	(2.9 ±0.8) × 10^{-4}		820
K^+K^-	(2.37±0.31) × 10^{-4}		1468
$\Lambda\overline{\Lambda}\pi^0$	(2.2 ±0.7) × 10^{-4}		998
$\pi^+\pi^-$	(1.47±0.23) × 10^{-4}		1542
$K^0_SK^0_L$	(1.08±0.14) × 10^{-4}		1466
$\Lambda\overline{\Sigma}$ + c.c.	< 1.5 × 10^{-4}	CL=90%	1032
$K^0_SK^0_S$	< 5.2 × 10^{-6}	CL=90%	1466
Radiative decays			
$\gamma\eta_c(1S)$	(1.3 ±0.4) %		116
$\gamma\pi^+\pi^-2\pi^0$	(8.3 ±3.1) × 10^{-3}		1518
$\gamma\eta\pi\pi$	(6.1 ±1.0) × 10^{-3}		1487
$\gamma\eta(1440) \to \gamma K\overline{K}\pi$	[n] (9.1 ±1.8) × 10^{-4}		1223
$\gamma\eta(1440) \to \gamma\gamma\rho^0$	(6.4 ±1.4) × 10^{-5}		1223
$\gamma\rho\rho$	(4.5 ±0.8) × 10^{-3}		1343
$\gamma\eta'(958)$	(4.31±0.30) × 10^{-3}		1400
$\gamma 2\pi^+2\pi^-$	(2.8 ±0.5) × 10^{-3}	S=1.9	1517
$\gamma f_4(2050)$	(2.7 ±0.7) × 10^{-3}		874
$\gamma\omega\omega$	(1.59±0.33) × 10^{-3}		1337
$\gamma\eta(1440) \to \gamma\rho^0\rho^0$	(1.4 ±0.4) × 10^{-3}		1223
$\gamma f_2(1270)$	(1.38±0.14) × 10^{-3}		1286
$\gamma f_J(1710) \to \gamma K\overline{K}$	(9.7 ±1.2) × 10^{-4}		1075
$\gamma\eta$	(8.6 ±0.8) × 10^{-4}		1500
$\gamma f_1(1420) \to \gamma K\overline{K}\pi$	(8.3 ±1.5) × 10^{-4}		1220
$\gamma f_1(1285)$	(6.5 ±1.0) × 10^{-4}		1283
$\gamma f'_2(1525)$	(6.3 ±1.0) × 10^{-4}		1173
$\gamma\phi\phi$	(4.0 ±1.2) × 10^{-4}	S=2.1	1166
$\gamma p\overline{p}$	(3.8 ±1.0) × 10^{-4}		1232
$\gamma\eta(2225)$	(2.9 ±0.6) × 10^{-4}		834
$\gamma\eta(1760) \to \gamma\rho^0\rho^0$	(1.3 ±0.9) × 10^{-4}		1048
$\gamma\pi^0$	(3.9 ±1.3) × 10^{-5}		1546
$\gamma p\overline{p}\pi^+\pi^-$	< 7.9 × 10^{-4}	CL=90%	1107
$\gamma\gamma$	< 5 × 10^{-4}	CL=90%	1548
$\gamma\Lambda\overline{\Lambda}$	< 1.3 × 10^{-4}	CL=90%	1074
3γ	< 5.5 × 10^{-5}	CL=90%	1548

Meson Summary Table

$\chi_{c0}(1P)$ or $\chi_{c0}(3415)$

$I^G(J^{PC}) = 0^+(0^{++})$

Mass $m = 3415.1 \pm 1.0$ MeV
Full width $\Gamma = 14 \pm 5$ MeV

$\chi_{c0}(1P)$ DECAY MODES	Fraction (Γ_i/Γ)	Confidence level	p (MeV/c)
Hadronic decays			
$2(\pi^+\pi^-)$	(3.7 ± 0.7) %		1679
$\pi^+\pi^-K^+K^-$	(3.0 ± 0.7) %		1580
$\rho^0\pi^+\pi^-$	(1.6 ± 0.5) %		1608
$3(\pi^+\pi^-)$	(1.5 ± 0.5) %		1633
$K^+\overline{K}^*(892)^0\pi^-$ + c.c.	(1.2 ± 0.4) %		1522
$\pi^+\pi^-$	$(7.5\pm2.1)\times10^{-3}$		1702
K^+K^-	$(7.1\pm2.4)\times10^{-3}$		1635
$\pi^+\pi^-p\overline{p}$	$(5.0\pm2.0)\times10^{-3}$		1320
$\pi^0\pi^0$	$(3.1\pm0.6)\times10^{-3}$		1702
$\eta\eta$	$(2.5\pm1.1)\times10^{-3}$		1617
$p\overline{p}$	$<9.0 \times10^{-4}$	90%	1427
Radiative decays			
$\gamma J/\psi(1S)$	$(6.6\pm1.8)\times10^{-3}$		303
$\gamma\gamma$	$(4.0\pm2.3)\times10^{-4}$		1708

$\chi_{c1}(1P)$ or $\chi_{c1}(3510)$

$I^G(J^{PC}) = 0^+(1^{++})$

Mass $m = 3510.53 \pm 0.12$ MeV
Full width $\Gamma = 0.88 \pm 0.14$ MeV

$\chi_{c1}(1P)$ DECAY MODES	Fraction (Γ_i/Γ)	p (MeV/c)
Hadronic decays		
$3(\pi^+\pi^-)$	(2.2 ± 0.8) %	1683
$2(\pi^+\pi^-)$	(1.6 ± 0.5) %	1727
$\pi^+\pi^-K^+K^-$	$(9\ \pm4\ )\times10^{-3}$	1632
$\rho^0\pi^+\pi^-$	$(3.9\pm3.5)\times10^{-3}$	1659
$K^+\overline{K}^*(892)^0\pi^-$ + c.c.	$(3.2\pm2.1)\times10^{-3}$	1576
$\pi^+\pi^-p\overline{p}$	$(1.4\pm0.9)\times10^{-3}$	1381
$p\overline{p}$	$(8.6\pm1.2)\times10^{-5}$	1483
$\pi^+\pi^-+K^+K^-$	$<2.1 \times10^{-3}$	–
Radiative decays		
$\gamma J/\psi(1S)$	(27.3 ± 1.6) %	389

$\chi_{c2}(1P)$ or $\chi_{c2}(3555)$

$I^G(J^{PC}) = 0^+(2^{++})$

Mass $m = 3556.17 \pm 0.13$ MeV
Full width $\Gamma = 2.00 \pm 0.18$ MeV

$\chi_{c2}(1P)$ DECAY MODES	Fraction (Γ_i/Γ)	Confidence level	p (MeV/c)
Hadronic decays			
$2(\pi^+\pi^-)$	$(2.2\ \pm0.5\ )$ %		1751
$\pi^+\pi^-K^+K^-$	$(1.9\ \pm0.5\ )$ %		1656
$3(\pi^+\pi^-)$	$(1.2\ \pm0.8\ )$ %		1707
$\rho^0\pi^+\pi^-$	$(7\ \pm4\ )\times10^{-3}$		1683
$K^+\overline{K}^*(892)^0\pi^-$ + c.c.	$(4.8\ \pm2.8\ )\times10^{-3}$		1601
$\pi^+\pi^-p\overline{p}$	$(3.3\ \pm1.3\ )\times10^{-3}$		1410
$\pi^+\pi^-$	$(1.9\ \pm1.0\ )\times10^{-3}$		1773
K^+K^-	$(1.5\ \pm1.1\ )\times10^{-3}$		1708
$p\overline{p}$	$(10.0\ \pm1.0\ )\times10^{-5}$		1510
$\pi^0\pi^0$	$(1.10\pm0.28)\times10^{-3}$		1773
$\eta\eta$	$(8\ \pm5\ )\times10^{-4}$		1692
$J/\psi(1S)\pi^+\pi^-\pi^0$	<1.5 %	90%	185
Radiative decays			
$\gamma J/\psi(1S)$	$(13.5\ \pm1.1\ )$ %		430
$\gamma\gamma$	$(1.6\ \pm0.5\ )\times10^{-4}$		1778

$\psi(2S)$ or $\psi(3685)$

$I^G(J^{PC}) = 0^-(1^{--})$

Mass $m = 3686.00 \pm 0.09$ MeV
Full width $\Gamma = 277 \pm 31$ keV (S = 1.1)
$\Gamma_{ee} = 2.14 \pm 0.21$ keV (Assuming $\Gamma_{ee} = \Gamma_{\mu\mu}$)

$\psi(2S)$ DECAY MODES	Fraction (Γ_i/Γ)	Scale factor/ Confidence level	p (MeV/c)
hadrons	(98.10 ± 0.30) %		–
virtual$\gamma\rightarrow$ hadrons	$(2.9\ \pm0.4\ )$ %		–
e^+e^-	$(8.8\ \pm1.3\ )\times10^{-3}$		1843
$\mu^+\mu^-$	$(7.7\ \pm1.7\ )\times10^{-3}$		1840
Decays into $J/\psi(1S)$ and anything			
$J/\psi(1S)$ anything	$(57\ \pm4\ )$ %		–
$J/\psi(1S)$ neutrals	$(23.2\ \pm2.6\ )$ %		–
$J/\psi(1S)\pi^+\pi^-$	$(32.4\ \pm2.6\ )$ %		477
$J/\psi(1S)\pi^0\pi^0$	$(18.4\ \pm2.7\ )$ %		481
$J/\psi(1S)\eta$	$(2.7\ \pm0.4\ )$ %	S=1.7	200
$J/\psi(1S)\pi^0$	$(9.7\ \pm2.1\ )\times10^{-4}$		527
Hadronic decays			
$3(\pi^+\pi^-)\pi^0$	$(3.5\ \pm1.6\ )\times10^{-3}$		1746
$2(\pi^+\pi^-)\pi^0$	$(3.1\ \pm0.7\ )\times10^{-3}$		1799
$\pi^+\pi^-K^+K^-$	$(1.6\ \pm0.4\ )\times10^{-3}$		1726
$\pi^+\pi^-p\overline{p}$	$(8.0\ \pm2.0\ )\times10^{-4}$		1491
$K^+\overline{K}^*(892)^0\pi^-$ + c.c.	$(6.7\ \pm2.5\ )\times10^{-4}$		1673
$2(\pi^+\pi^-)$	$(4.5\ \pm1.0\ )\times10^{-4}$		1817
$\rho^0\pi^+\pi^-$	$(4.2\ \pm1.5\ )\times10^{-4}$		1751
$\overline{p}p$	$(1.9\ \pm0.5\ )\times10^{-4}$		1586
$3(\pi^+\pi^-)$	$(1.5\ \pm1.0\ )\times10^{-4}$		1774
$\overline{p}p\pi^0$	$(1.4\ \pm0.5\ )\times10^{-4}$		1543
K^+K^-	$(1.0\ \pm0.7\ )\times10^{-4}$		1776
$\pi^+\pi^-\pi^0$	$(9\ \pm5\ )\times10^{-5}$		1830
$\pi^+\pi^-$	$(8\ \pm5\ )\times10^{-5}$		1838
$\Lambda\overline{\Lambda}$	$<4 \times10^{-4}$	CL=90%	1467
$\Xi^-\overline{\Xi}^+$	$<2 \times10^{-4}$	CL=90%	1285
$\rho\pi$	$<8.3 \times10^{-5}$	CL=90%	1760
$K^+K^-\pi^0$	$<2.96 \times10^{-5}$	CL=90%	1754
$K^+\overline{K}^*(892)^-$ + c.c.	$<1.79 \times10^{-5}$	CL=90%	1698
Radiative decays			
$\gamma\chi_{c0}(1P)$	$(9.3\ \pm0.8\ )$ %		261
$\gamma\chi_{c1}(1P)$	$(8.7\ \pm0.8\ )$ %		171
$\gamma\chi_{c2}(1P)$	$(7.8\ \pm0.8\ )$ %		127
$\gamma\eta_c(1S)$	$(2.8\ \pm0.6\ )\times10^{-3}$		639
$\gamma\pi^0$	$<5.4 \times10^{-3}$	CL=95%	1841
$\gamma\eta'(958)$	$<1.1 \times10^{-3}$	CL=90%	1719
$\gamma\gamma$	$<1.6 \times10^{-4}$	CL=90%	1843
$\gamma\eta(1440)\rightarrow\gamma K\overline{K}\pi$	[n] $<1.2 \times10^{-4}$	CL=90%	1569

$\psi(3770)$

$I^G(J^{PC}) = ?^?(1^{--})$

Mass $m = 3769.9 \pm 2.5$ MeV (S = 1.8)
Full width $\Gamma = 23.6 \pm 2.7$ MeV (S = 1.1)
$\Gamma_{ee} = 0.26 \pm 0.04$ keV (S = 1.2)

$\psi(3770)$ DECAY MODES	Fraction (Γ_i/Γ)	Scale factor	p (MeV/c)
$D\overline{D}$	dominant		242
e^+e^-	$(1.12\pm0.17)\times10^{-5}$	1.2	1885

$\psi(4040)$ [yy]

$I^G(J^{PC}) = ?^?(1^{--})$

Mass $m = 4040 \pm 10$ MeV
Full width $\Gamma = 52 \pm 10$ MeV
$\Gamma_{ee} = 0.75 \pm 0.15$ keV

$\psi(4040)$ DECAY MODES	Fraction (Γ_i/Γ)	p (MeV/c)
e^+e^-	$(1.4\pm0.4)\times10^{-5}$	2020
$D^0\overline{D}^0$	seen	777
$D^*(2007)^0\overline{D}^0$ + c.c.	seen	578
$D^*(2007)^0\overline{D}^*(2007)^0$	seen	232

$\psi(4160)$ [yy]

$I^G(J^{PC}) = ?^?(1^{--})$

Mass $m = 4159 \pm 20$ MeV
Full width $\Gamma = 78 \pm 20$ MeV
$\Gamma_{ee} = 0.77 \pm 0.23$ keV

$\psi(4160)$ DECAY MODES	Fraction (Γ_i/Γ)	p (MeV/c)
e^+e^-	$(10\pm4) \times 10^{-6}$	2079

$\psi(4415)$ [yy]

$I^G(J^{PC}) = ?^?(1^{--})$

Mass $m = 4415 \pm 6$ MeV
Full width $\Gamma = 43 \pm 15$ MeV (S = 1.8)
$\Gamma_{ee} = 0.47 \pm 0.10$ keV

$\psi(4415)$ DECAY MODES	Fraction (Γ_i/Γ)	p (MeV/c)
hadrons	dominant	–
e^+e^-	$(1.1\pm0.4) \times 10^{-5}$	2207

$b\overline{b}$ MESONS

$\Upsilon(1S)$ or $\Upsilon(9460)$

$I^G(J^{PC}) = ?^?(1^{--})$

Mass $m = 9460.37 \pm 0.21$ MeV (S = 2.7)
Full width $\Gamma = 52.5 \pm 1.8$ keV
$\Gamma_{ee} = 1.32 \pm 0.03$ keV

$\Upsilon(1S)$ DECAY MODES	Fraction (Γ_i/Γ)	Scale factor/ Confidence level	p (MeV/c)
$\tau^+\tau^-$	(2.97 ± 0.35) %		4384
e^+e^-	(2.52 ± 0.17) %		4730
$\mu^+\mu^-$	(2.48 ± 0.07) %	S=1.1	4729
Hadronic decays			
$J/\psi(1S)$ anything	$(1.1 \pm0.4) \times 10^{-3}$		4223
$\rho\pi$	$< 2 \times 10^{-4}$	CL=90%	4698
$\pi^+\pi^-$	$< 5 \times 10^{-4}$	CL=90%	4728
K^+K^-	$< 5 \times 10^{-4}$	CL=90%	4704
$p\overline{p}$	$< 9 \times 10^{-4}$	CL=90%	4636
Radiative decays			
$\gamma 2h^+2h^-$	$(7.0 \pm1.5) \times 10^{-4}$		4720
$\gamma 3h^+3h^-$	$(5.4 \pm2.0) \times 10^{-4}$		4703
$\gamma 4h^+4h^-$	$(7.4 \pm3.5) \times 10^{-4}$		4679
$\gamma\pi^+\pi^-K^+K^-$	$(2.9 \pm0.9) \times 10^{-4}$		4686
$\gamma 2\pi^+2\pi^-$	$(2.5 \pm0.9) \times 10^{-4}$		4720
$\gamma 3\pi^+3\pi^-$	$(2.5 \pm1.2) \times 10^{-4}$		4703
$\gamma 2\pi^+2\pi^-K^+K^-$	$(2.4 \pm1.2) \times 10^{-4}$		4658
$\gamma\pi^+\pi^-p\overline{p}$	$(1.5 \pm0.6) \times 10^{-4}$		4604
$\gamma 2\pi^+2\pi^-p\overline{p}$	$(4 \pm6) \times 10^{-5}$		4563
$\gamma 2K^+2K^-$	$(2.0 \pm2.0) \times 10^{-5}$		4601
$\gamma\eta'(958)$	$< 1.3 \times 10^{-3}$	CL=90%	4682
$\gamma\eta$	$< 3.5 \times 10^{-4}$	CL=90%	4714
$\gamma f_2'(1525)$	$< 1.4 \times 10^{-4}$	CL=90%	4607
$\gamma f_2(1270)$	$< 1.3 \times 10^{-4}$	CL=90%	4644
$\gamma\eta(1440)$	$< 8.2 \times 10^{-5}$	CL=90%	4624
$\gamma f_J(1710) \rightarrow \gamma K\overline{K}$	$< 2.6 \times 10^{-4}$	CL=90%	4576
$\gamma f_4(2220) \rightarrow \gamma K^+K^-$	$< 1.5 \times 10^{-5}$	CL=90%	4469

$\chi_{b0}(1P)$ [zz] **or $\chi_{b0}(9860)$**

$I^G(J^{PC}) = ?^?(0$ preferred $^{++})$
J needs confirmation.

Mass $m = 9859.8 \pm 1.3$ MeV

$\chi_{b0}(1P)$ DECAY MODES	Fraction (Γ_i/Γ)	Confidence level	p (MeV/c)
$\gamma\Upsilon(1S)$	<6 %	90%	391

$\chi_{b1}(1P)$ [zz] **or $\chi_{b1}(9890)$**

$I^G(J^{PC}) = ?^?(1^{++})$
J needs confirmation.

Mass $m = 9891.9 \pm 0.7$ MeV

$\chi_{b1}(1P)$ DECAY MODES	Fraction (Γ_i/Γ)	p (MeV/c)
$\gamma\Upsilon(1S)$	(35 ± 8) %	422

$\chi_{b2}(1P)$ [zz] **or $\chi_{b2}(9915)$**

$I^G(J^{PC}) = ?^?(2^{++})$
J needs confirmation.

Mass $m = 9913.2 \pm 0.6$ MeV

$\chi_{b2}(1P)$ DECAY MODES	Fraction (Γ_i/Γ)	p (MeV/c)
$\gamma\Upsilon(1S)$	(22 ± 4) %	443

$\Upsilon(2S)$ or $\Upsilon(10023)$

$I^G(J^{PC}) = ?^?(1^{--})$

Mass $m = 10.02330 \pm 0.00031$ GeV
Full width $\Gamma = 44 \pm 7$ keV

$\Upsilon(2S)$ DECAY MODES	Fraction (Γ_i/Γ)	Confidence level	p (MeV/c)
$\Upsilon(1S)\pi^+\pi^-$	(18.5 ± 0.8) %		475
$\Upsilon(1S)\pi^0\pi^0$	(8.8 ± 1.1) %		480
$\tau^+\tau^-$	(1.7 ± 1.6) %		4686
$\mu^+\mu^-$	(1.31 ± 0.21) %		5011
e^+e^-	seen		5012
$\Upsilon(1S)\pi^0$	$< 8 \times 10^{-3}$	90%	531
$\Upsilon(1S)\eta$	$< 2 \times 10^{-3}$	90%	127
$J/\psi(1S)$ anything	$< 6 \times 10^{-3}$	90%	4533
Radiative decays			
$\gamma\chi_{b1}(1P)$	(6.7 ± 0.9) %		131
$\gamma\chi_{b2}(1P)$	(6.6 ± 0.9) %		110
$\gamma\chi_{b0}(1P)$	(4.3 ± 1.0) %		162
$\gamma f_J(1710)$	$< 5.9 \times 10^{-4}$	90%	4866
$\gamma f_2'(1525)$	$< 5.3 \times 10^{-4}$	90%	4896
$\gamma f_2(1270)$	$< 2.41 \times 10^{-4}$	90%	4931

$\chi_{b0}(2P)$ [zz] **or $\chi_{b0}(10235)$**

$I^G(J^{PC}) = ?^?(0$ preferred $^{++})$
J needs confirmation.

Mass $m = 10.2321 \pm 0.0006$ GeV

$\chi_{b0}(2P)$ DECAY MODES	Fraction (Γ_i/Γ)	p (MeV/c)
$\gamma\Upsilon(2S)$	(4.6 ± 2.1) %	210
$\gamma\Upsilon(1S)$	$(9 \pm6) \times 10^{-3}$	746

$\chi_{b1}(2P)$ [zz] **or $\chi_{b1}(10255)$**

$I^G(J^{PC}) = ?^?(1$ preferred $^{++})$
J needs confirmation.

Mass $m = 10.2552 \pm 0.0005$ GeV
$m_{\chi_{b1}(2P)} - m_{\chi_{b0}(2P)} = 23.5 \pm 1.0$ MeV

$\chi_{b1}(2P)$ DECAY MODES	Fraction (Γ_i/Γ)	Scale factor	p (MeV/c)
$\gamma\Upsilon(2S)$	(21 ± 4) %	1.5	229
$\gamma\Upsilon(1S)$	(8.5 ± 1.3) %	1.3	764

$\chi_{b2}(2P)$ [zz] **or $\chi_{b2}(10270)$**

$I^G(J^{PC}) = ?^?(2$ preferred $^{++})$
J needs confirmation.

Mass $m = 10.2685 \pm 0.0004$ GeV
$m_{\chi_{b2}(2P)} - m_{\chi_{b1}(2P)} = 13.5 \pm 0.6$ MeV

$\chi_{b2}(2P)$ DECAY MODES	Fraction (Γ_i/Γ)	p (MeV/c)
$\gamma\Upsilon(2S)$	(16.2 ± 2.4) %	242
$\gamma\Upsilon(1S)$	(7.1 ± 1.0) %	776

Meson Summary Table

$\Upsilon(3S)$ or $\Upsilon(10355)$

$I^G(J^{PC}) = ?^?(1^{--})$

Mass $m = 10.3553 \pm 0.0005$ GeV
Full width $\Gamma = 26.3 \pm 3.5$ keV

$\Upsilon(3S)$ DECAY MODES	Fraction (Γ_i/Γ)	Scale factor	p (MeV/c)
$\Upsilon(2S)$anything	(10.6 ±0.8) %		296
$\Upsilon(2S)\pi^+\pi^-$	(2.8 ±0.6) %	2.2	177
$\Upsilon(2S)\pi^0\pi^0$	(2.00±0.32) %		190
$\Upsilon(2S)\gamma\gamma$	(5.0 ±0.7) %		–
$\Upsilon(1S)\pi^+\pi^-$	(4.48±0.21) %		814
$\Upsilon(1S)\pi^0\pi^0$	(2.06±0.28) %		816
$\mu^+\mu^-$	(1.81±0.17) %		5177
e^+e^-	seen		5177
Radiative decays			
$\gamma\chi_{b2}(2P)$	(11.4 ±0.8) %	1.3	87
$\gamma\chi_{b1}(2P)$	(11.3 ±0.6) %		100
$\gamma\chi_{b0}(2P)$	(5.4 ±0.6) %	1.1	123

$\Upsilon(4S)$ or $\Upsilon(10580)$

$I^G(J^{PC}) = ?^?(1^{--})$

Mass $m = 10.5800 \pm 0.0035$ GeV
Full width $\Gamma = 23.8 \pm 2.2$ MeV
$\Gamma_{ee} = 0.24 \pm 0.05$ keV (S = 1.7)

$\Upsilon(4S)$ DECAY MODES	Fraction (Γ_i/Γ)	Confidence level	p (MeV/c)
e^+e^-	$(1.01\pm0.21) \times 10^{-5}$		5290
D^{*+}anything + c.c.	< 7.4 %	90%	5099
ϕanything	$< 2.3 \times 10^{-3}$	90%	5240
$\Upsilon(1S)$anything	$< 4 \times 10^{-3}$	90%	1053

$\Upsilon(10860)$

$I^G(J^{PC}) = ?^?(1^{--})$

Mass $m = 10.865 \pm 0.008$ GeV (S = 1.1)
Full width $\Gamma = 110 \pm 13$ MeV
$\Gamma_{ee} = 0.31 \pm 0.07$ keV (S = 1.3)

$\Upsilon(10860)$ DECAY MODES	Fraction (Γ_i/Γ)	p (MeV/c)
e^+e^-	$(2.8\pm0.7) \times 10^{-6}$	5432

$\Upsilon(11020)$

$I^G(J^{PC}) = ?^?(1^{--})$

Mass $m = 11.019 \pm 0.008$ GeV
Full width $\Gamma = 79 \pm 16$ MeV
$\Gamma_{ee} = 0.130 \pm 0.030$ keV

$\Upsilon(11020)$ DECAY MODES	Fraction (Γ_i/Γ)	p (MeV/c)
e^+e^-	$(1.6\pm0.5) \times 10^{-6}$	5509

Searches for Top and Fourth Generation Hadrons

See the sections "Searches for t Quark" and "Searches for b' (4th Generation) Quark" at the end of the QUARKS section.

NOTES

In this Summary Table:

When a quantity has "(S = ...)" to its right, the error on the quantity has been enlarged by the "scale factor" S, defined as $S = \sqrt{\chi^2/(N-1)}$, where N is the number of measurements used in calculating the quantity. We do this when $S > 1$, which often indicates that the measurements are inconsistent. When $S > 1.25$, we also show in the Full Listings an ideogram of the measurements. For more about S, see the Introduction.

A decay momentum p is given for each decay mode. For a 2-body decay, p is the momentum of each decay product in the rest frame of the decaying particle. For a 3-or-more-body decay, p is the largest momentum any of the products can have in this frame.

[a] The $\pi^\pm$ mass has increased by three (old) standard deviations since our 1992 edition, and the π^0 mass, which is determined using the mass difference ($m_{\pi^\pm} - m_{\pi^0}$), has increased accordingly. See the "Note on the Charged Pion Mass" in the $\pi^\pm$ Full Listings for a discussion.

[b] See the "Note on $\pi^\pm \to \ell^\pm\nu\gamma$ and $K^\pm \to \ell^\pm\nu\gamma$ Form Factors" in the $\pi^\pm$ Full Listings for definitions and details.

[c] Measurements of $\Gamma(e^+\nu_e)/\Gamma(\mu^+\nu_\mu)$ always include decays with γ's, and measurements of $\Gamma(e^+\nu_e\gamma)$ and $\Gamma(\mu^+\nu_\mu\gamma)$ never include low-energy γ's. Therefore, since no clean separation is possible, we consider the modes with γ's to be subreactions of the modes without them, and let $[\Gamma(e^+\nu_e) + \Gamma(\mu^+\nu_\mu)]/\Gamma_{total} = 100\%$.

[d] See the $\pi^\pm$ Full Listings for the energy limits used in this measurement; low-energy γ's are not included.

[e] Derived from an analysis of neutrino-oscillation experiments.

[f] Astrophysical and cosmological arguments give limits of order 10^{-13}; see the π^0 Full Listings.

[g] See the "Note on the Decay Width $\Gamma(\eta \to \gamma\gamma)$" in the η Full Listings.

[h] See the "Note on η Decay Parameters" in the η Full Listings.

[i] C parity forbids this to occur as a single-photon process.

[j] The e^+e^- branching fraction is from $e^+e^- \to \pi^+\pi^-$ experiments only. The $\omega\rho$ interference is then due to $\omega\rho$ mixing only, and is expected to be small. If $e\mu$ universality holds, $\Gamma(\rho^0 \to \mu^+\mu^-) = \Gamma(\rho^0 \to e^+e^-) \times 0.99785$.

[k] This is only an educated guess; the error given is larger than the error on the average of the published values. See the Full Listings for details.

[l] See the "Note on the $f_1(1420)$" in the $f_1(1420)$ Full Listings.

[m] See also the $\omega(1600)$ Full Listings.

[n] See the "Note on the $\eta(1440)$" in the $\eta(1440)$ Full Listings.

[o] See the "Note on the $\rho(1450)$ and the $\rho(1700)$" in the $\rho(1700)$ Full Listings.

[p] See also the $\omega(1420)$ Full Listings.

[q] The value is for the sum of the charge states indicated.

[r] The definition of the slope parameter g of the $K \to 3\pi$ Dalitz plot is as follows (see also "Note on Dalitz Plot Parameters for $K \to 3\pi$ Decays" in the $K^\pm$ Full Listings):

$$|M|^2 = 1 + g(s_3 - s_0)/m_{\pi^+}^2 + \cdots .$$

[s] For more details and definitions of parameters see the Full Listings.

[t] See the $K^\pm$ Full Listings for the energy limits used in this measurement.

[u] Most of this radiative mode, the low-momentum γ part, is also included in the parent mode listed without γ's.

[v] Direct-emission branching fraction.

[w] Structure-dependent part.

[x] The CP-violation parameters are defined as follows (see also "Note on CP Violation in $K_S \to 3\pi$" and "Note on CP Violation in K^0_L Decay" in the Full Listings):

$$\eta_{+-} = |\eta_{+-}|e^{i\phi_{+-}} = \frac{A(K^0_L \to \pi^+\pi^-)}{A(K^0_S \to \pi^+\pi^-)} = \epsilon + \epsilon'$$

$$\eta_{00} = |\eta_{00}|e^{i\phi_{00}} = \frac{A(K^0_L \to \pi^0\pi^0)}{A(K^0_S \to \pi^0\pi^0)} = \epsilon - 2\epsilon'$$

$$\delta = \frac{\Gamma(K^0_L \to \pi^-\ell^+\nu) - \Gamma(K^0_L \to \pi^+\ell^-\nu)}{\Gamma(K^0_L \to \pi^-\ell^+\nu) + \Gamma(K^0_L \to \pi^+\ell^-\nu)},$$

$$\mathrm{Im}(\eta_{+-0})^2 = \frac{\Gamma(K^0_S \to \pi^+\pi^-\pi^0)^{CP\ \mathrm{viol.}}}{\Gamma(K^0_L \to \pi^+\pi^-\pi^0)},$$

$$\mathrm{Im}(\eta_{000})^2 = \frac{\Gamma(K^0_S \to \pi^0\pi^0\pi^0)}{\Gamma(K^0_L \to \pi^0\pi^0\pi^0)}.$$

where for the last two relations CPT is assumed valid, i.e., $\mathrm{Re}(\eta_{+-0}) \simeq 0$ and $\mathrm{Re}(\eta_{000}) \simeq 0$.

[y] See the K^0_S Full Listings for the energy limits used in this measurement.

[z] Calculated from K^0_L semileptonic rates and the K^0_S lifetime assuming $\Delta S = \Delta Q$.

[aa] ϵ'/ϵ is derived from $|\eta_{00}/\eta_{+-}|$ measurements using theoretical input on phases.

[bb] See the K^0_L Full Listings for the energy limits used in this measurement.

[cc] $m_{e^+e^-}$ >470 MeV

[dd] Allowed by higher-order electroweak interactions.

[ee] Violates CP in leading order. Test of direct CP violation since the indirect CP-violating and CP-conserving contributions are expected to be suppressed.

[ff] See the note in the $L(1770)$ Full Listings in Reviews of Modern Physics **56** No. 2 Pt. II (1984), p. S200.

[gg] This is a weighted average of $D^\pm$ (44%) and D^0 (56%) branching fractions. See "D^+ and $D^0 \to$ (ηanything) / (total D^+ and D^0)" under "D^+ Branching Ratios" in the Full Listings.

[hh] This value combines the e^+ and μ^+ branching fractions, making a small phase-space adjustment to the μ^+ fraction to be able to use it as an e^+ fraction; hence the "e^+." In fact, some of the e^+ measurements already use μ^+ events in this way.

[ii] ℓ indicates e or μ mode, not sum over modes.

[jj] The branching fractions for this mode may differ from the sum of the submodes that contribute to it, due to interference effects. See the relevant papers in the Full Listings.

[kk] The two experiments determining this ratio are in serious disagreement. See the Full Listings.

[ll] This mode is not a useful test for a ΔC=1 weak neutral current because both quarks must change flavor in this decay.

[mm] The D^0_1-D^0_2 limits are inferred from the limit on $D^0 \to \overline{D}^0 \to K^+\pi^-$.

[nn] See the "Note on Semileptonic Decays of D and B Mesons" in the D^+ Full Listings for a comparison of inclusive and summed-inclusive branching fractions.

[oo] The limit on $(\overline{K}^*(892)\pi)^- \mu^+\nu_\mu$ just below is much stronger.

[pp] For now, we average together measurements of the $\phi e^+\nu_e$ and $\phi\mu^+\nu_\mu$ branching fractions.

[qq] This branching fraction is calculated from appropriate fractions of the next three branching fractions.

[rr] For admixture of B hadrons at LEP and Tevatron energies.

[ss] These values are model dependent. See note on "Semileptonic Decays" in the B^+ Full Listings.

[tt] D^{**} stands for the sum of the $D(1\,^1P_1)$, $D(1\,^3P_0)$, $D(1\,^3P_1)$, $D(1\,^3P_2)$, $D(2\,^1S_0)$, and $D(2\,^1S_1)$ resonances.

[uu] B^0, B^+, B^0_s, and B baryon states not separated.

[vv] B^0, B^+, and B^0_s not separated.

[ww] Derived from measurements of χ_d and of Δm_{B^0} times B^0 mean life.

[xx] Includes $p\overline{p}\pi^+\pi^-\gamma$ and excludes $p\overline{p}\eta$, $p\overline{p}\omega$, $p\overline{p}\eta'$.

[yy] J^{PC} known by production in e^+e^- via single photon annihilation. I^G is not known; interpretation of this state as a single resonance is unclear because of the expectation of substantial threshold effects in this energy region.

[zz] Spectroscopic labeling for these states is theoretical, pending experimental information.

Meson Summary Table

See also the table of suggested $q\overline{q}$ quark-model assignments in the Quark Model section.

- • Indicates particles that appear in the preceding Meson Summary Table. We do not regard the other entries as being established.
- † Indicates that the value of J given is preferred, but needs confirmation.

LIGHT UNFLAVORED ($S = C = B = 0$)

	$I^G(J^{PC})$		$I^G(J^{PC})$
• $\pi^\pm$	$1^-(0^-)$	• $\omega_3(1670)$	$0^-(3^{--})$
• π^0	$1^-(0^{-+})$	• $\pi_2(1670)$	$1^-(2^{-+})$
• η	$0^+(0^{-+})$	• $\phi(1680)$	$0^-(1^{--})$
• $\rho(770)$	$1^+(1^{--})$	• $\rho_3(1690)$	$1^+(3^{--})$
• $\omega(782)$	$0^-(1^{--})$	• $\rho(1700)$	$1^+(1^{--})$
• $\eta'(958)$	$0^+(0^{-+})$	$X(1700)$	$\text{even}^+(?^{?+})$
• $f_0(980)$	$0^+(0^{++})$	• $f_J(1710)$	$0^+(\text{even}^{++})$
• $a_0(980)$	$1^-(0^{++})$	$X(1740)$	$0^+(\text{even}^{++})$
• $\phi(1020)$	$0^-(1^{--})$	$\eta(1760)$	$0^+(0^{-+})$
• $h_1(1170)$	$0^-(1^{+-})$	$\pi(1770)$	$1^-(0^{-+})$
• $b_1(1235)$	$1^+(1^{+-})$	$X(1775)$	$1^-(?^{-+})$
• $a_1(1260)$	$1^-(1^{++})$	$f_2(1810)$	$0^+(2^{++})$
• $f_2(1270)$	$0^+(2^{++})$	$X(1830)$	$1^-(?^{?+})$
• $f_1(1285)$	$0^+(1^{++})$	• $\phi_3(1850)$	$0^-(3^{--})$
• $\eta(1295)$	$0^+(0^{-+})$	$\eta_2(1870)$	$0^+(2^{-+})$
• $f_0(1300)$	$0^+(0^{++})$	$X(1910)$	$0^+(?^{?+})$
• $\pi(1300)$	$1^-(0^{-+})$	$X(1950)$	$0^+(\text{even}^{++})$
• $a_2(1320)$	$1^-(2^{++})$	• $f_2(2010)$	$0^+(2^{++})$
$f_0(1370)$	$0^+(0^{++})$	$a_4(2040)$	$1^-(4^{++})$
$h_1(1380)$	$?^-(1^{+?})$	$a_3(2050)$	$1^-(3^{++})$
$\hat{\rho}(1405)$	$1^-(1^{-+})$	• $f_4(2050)$	$0^+(4^{++})$
• $f_1(1420)$	$0^+(1^{++})$	$\pi_2(2100)$	$1^-(2^{-+})$
• $\omega(1420)$	$0^-(1^{--})$	$f_2(2150)$	$0^+(2^{++})$
$f_2(1430)$	$0^+(2^{++})$	$\rho(2150)$	$1^+(1^{--})$
• $\eta(1440)$	$0^+(0^{-+})$	$X(2200)$	$?^?(\text{even}^{++})$
• $\rho(1450)$	$1^+(1^{--})$	$\rho(2210)$	$1^+(1^{--})$
• $f_1(1510)$	$0^+(1^{++})$	$f_4(2220)$	$0^+(4^{++})$
$f_2(1520)$	$0^+(2^{++})$	$\eta(2225)$	$0^+(0^{-+})$
• $f'_2(1525)$	$0^+(2^{++})$	$\rho_3(2250)$	$1^+(3^{--})$
$f_0(1525)$	$0^+(0^{++})$	• $f_2(2300)$	$0^+(2^{++})$
• $f_0(1590)$	$0^+(0^{++})$	$f_4(2300)$	$0^+(4^{++})$
• $\omega(1600)$	$0^-(1^{--})$	• $f_2(2340)$	$0^+(2^{++})$
$X(1600)$	$2^+(2^{++})$	$\rho_5(2350)$	$1^+(5^{--})$
$f_2(1640)$	$0^+(2^{++})$	$a_6(2450)$	$1^-(6^{++})$
		$f_6(2510)$	$0^+(6^{++})$
		$X(3250)$	$?^?(?^{??})$

OTHER LIGHT UNFLAVORED ($S = C = B = 0$)

$e^+e^-(1100\text{–}2200)$	$?^?(1^{--})$
$\overline{N}N(1100\text{–}3600)$	
$X(1900\text{–}3600)$	

STRANGE ($S = \pm1$, $C = B = 0$)

	$I(J^P)$
• $K^\pm$	$1/2(0^-)$
• K^0	$1/2(0^-)$
• K^0_S	$1/2(0^-)$
• K^0_L	$1/2(0^-)$
• $K^*(892)$	$1/2(1^-)$
• $K_1(1270)$	$1/2(1^+)$
• $K_1(1400)$	$1/2(1^+)$
• $K^*(1410)$	$1/2(1^-)$
• $K^*_0(1430)$	$1/2(0^+)$
• $K^*_2(1430)$	$1/2(2^+)$
$K(1460)$	$1/2(0^-)$
$K_2(1580)$	$1/2(2^-)$
$K_1(1650)$	$1/2(1^+)$
• $K^*(1680)$	$1/2(1^-)$
• $K_2(1770)$	$1/2(2^-)$
• $K^*_3(1780)$	$1/2(3^-)$
• $K_2(1820)$	$1/2(2^-)$
$K(1830)$	$1/2(0^-)$
$K^*_0(1950)$	$1/2(0^+)$
$K^*_2(1980)$	$1/2(2^+)$
• $K^*_4(2045)$	$1/2(4^+)$
$K_2(2250)$	$1/2(2^-)$
$K_3(2320)$	$1/2(3^+)$
$K^*_5(2380)$	$1/2(5^-)$
$K_4(2500)$	$1/2(4^-)$
$K(3100)$	$?^?(?^{??})$

CHARMED ($C = \pm1$)

• $D^\pm$	$1/2(0^-)$
• D^0	$1/2(0^-)$
• $D^*(2007)^0$	$1/2(1^-)$
• $D^*(2010)^\pm$	$1/2(1^-)$
• $D_1(2420)^0$	$1/2(1^+)$
$D_J(2440)^\pm$	$1/2(?^?)$
• $D^*_2(2460)$	$1/2(2^+)$

CHARMED, STRANGE ($C = S = \pm1$)

• $D^\pm_s$	$0(0^-)$
• $D^{*\pm}_s$	$?(?^?)$
• $D_{s1}(2536)^\pm$	$0(1^+)$
$D_{sJ}(2573)^\pm$	$?(?^?)$

BOTTOM ($B = \pm1$)

• $B^\pm$	$1/2(0^-)$
• B^0	$1/2(0^-)$
• B^*	$1/2(1^-)$

BOTTOM, STRANGE ($B = \pm1$, $S = \mp1$, $C = 0$)

	$I^G(J^{PC})$
• B^0_s	$1/2(0^-)$
B^*_s	$?(?^?)$

$c\overline{c}$

• $\eta_c(1S) = \eta_c(2980)$	$0^+(0^{-+})$
• $J/\psi(1S) = J/\psi(3097)$	$0^-(1^{--})$
• $\chi_{c0}(1P) = \chi_{c0}(3415)$	$0^+(0^{++})$
• $\chi_{c1}(1P) = \chi_{c1}(3510)$	$0^+(1^{++})$
$h_c(1P)$	$?^?(?^{??})$
• $\chi_{c2}(1P) = \chi_{c2}(3555)$	$0^+(2^{++})$
$\eta_c(2S) = \eta_c(3590)$	$?^?(?^{?+})$
• $\psi(2S) = \psi(3685)$	$0^-(1^{--})$
• $\psi(3770)$	$?^?(1^{--})$
• $\psi(4040)$	$?^?(1^{--})$
• $\psi(4160)$	$?^?(1^{--})$
• $\psi(4415)$	$?^?(1^{--})$

$b\overline{b}$

• $\Upsilon(1S) = \Upsilon(9460)$	$?^?(1^{--})$
• $\chi_{b0}(1P) = \chi_{b0}(9860)$	$?^?(0^{++})^\dagger$
• $\chi_{b1}(1P) = \chi_{b1}(9890)$	$?^?(1^{++})$
• $\chi_{b2}(1P) = \chi_{b2}(9915)$	$?^?(2^{++})$
• $\Upsilon(2S) = \Upsilon(10023)$	$?^?(1^{--})$
• $\chi_{b0}(2P) = \chi_{b0}(10235)$	$?^?(0^{++})^\dagger$
• $\chi_{b1}(2P) = \chi_{b1}(10255)$	$?^?(1^{++})^\dagger$
• $\chi_{b2}(2P) = \chi_{b2}(10270)$	$?^?(2^{++})^\dagger$
• $\Upsilon(3S) = \Upsilon(10355)$	$?^?(1^{--})$
• $\Upsilon(4S) = \Upsilon(10580)$	$?^?(1^{--})$
• $\Upsilon(10860)$	$?^?(1^{--})$
• $\Upsilon(11020)$	$?^?(1^{--})$

NON-$q\overline{q}$ CANDIDATES

Non-$q\overline{q}$ Candidates

Baryon Summary Table

This short table gives the name, the quantum numbers (where known), and the status of baryons in the Review. Only the baryons with 3- or 4-star status are included in the main Baryon Summary Table. Due to insufficient data or uncertain interpretation, the other entries in the short table are not established as baryons. The names with masses are of baryons that decay strongly. See our 1986 edition (Physics Letters **170B**) for listings of evidence for Z baryons (KN resonances).

p	P_{11}	****	$\Delta(1232)$	P_{33}	****	Λ	P_{01}	****	Σ^+	P_{11}	****	Ξ^0	P_{11}	****
n	P_{11}	****	$\Delta(1600)$	P_{33}	***	$\Lambda(1405)$	S_{01}	****	Σ^0	P_{11}	****	Ξ^-	P_{11}	****
$N(1440)$	P_{11}	****	$\Delta(1620)$	S_{31}	****	$\Lambda(1520)$	D_{03}	****	Σ^-	P_{11}	****	$\Xi(1530)$	P_{13}	****
$N(1520)$	D_{13}	****	$\Delta(1700)$	D_{33}	****	$\Lambda(1600)$	P_{01}	***	$\Sigma(1385)$	P_{13}	****	$\Xi(1620)$		*
$N(1535)$	S_{11}	****	$\Delta(1750)$	P_{31}	*	$\Lambda(1670)$	S_{01}	****	$\Sigma(1480)$		*	$\Xi(1690)$		***
$N(1650)$	S_{11}	****	$\Delta(1900)$	S_{31}	***	$\Lambda(1690)$	D_{03}	****	$\Sigma(1560)$		**	$\Xi(1820)$	D_{13}	***
$N(1675)$	D_{15}	****	$\Delta(1905)$	F_{35}	****	$\Lambda(1800)$	S_{01}	***	$\Sigma(1580)$	D_{13}	**	$\Xi(1950)$		***
$N(1680)$	F_{15}	****	$\Delta(1910)$	P_{31}	****	$\Lambda(1810)$	P_{01}	***	$\Sigma(1620)$	S_{11}	**	$\Xi(2030)$		***
$N(1700)$	D_{13}	***	$\Delta(1920)$	P_{33}	***	$\Lambda(1820)$	F_{05}	****	$\Sigma(1660)$	P_{11}	***	$\Xi(2120)$		*
$N(1710)$	P_{11}	***	$\Delta(1930)$	D_{35}	***	$\Lambda(1830)$	D_{05}	****	$\Sigma(1670)$	D_{13}	****	$\Xi(2250)$		**
$N(1720)$	P_{13}	****	$\Delta(1940)$	D_{33}	*	$\Lambda(1890)$	P_{03}	****	$\Sigma(1690)$		**	$\Xi(2370)$		**
$N(1900)$	P_{13}	*	$\Delta(1950)$	F_{37}	****	$\Lambda(2000)$		*	$\Sigma(1750)$	S_{11}	***	$\Xi(2500)$		*
$N(1990)$	F_{17}	**	$\Delta(2000)$	F_{35}	*	$\Lambda(2020)$	F_{07}	*	$\Sigma(1770)$	P_{11}	*			
$N(2000)$	F_{15}	**	$\Delta(2150)$	S_{31}	*	$\Lambda(2100)$	G_{07}	****	$\Sigma(1775)$	D_{15}	****	Ω^-		****
$N(2080)$	D_{13}	**	$\Delta(2200)$	G_{37}	*	$\Lambda(2110)$	F_{05}	***	$\Sigma(1840)$	P_{13}	*	$\Omega(2250)^-$		***
$N(2090)$	S_{11}	*	$\Delta(2300)$	H_{39}	**	$\Lambda(2325)$	D_{03}	*	$\Sigma(1880)$	P_{11}	**	$\Omega(2380)^-$		**
$N(2100)$	P_{11}	*	$\Delta(2350)$	D_{35}	*	$\Lambda(2350)$	H_{09}	***	$\Sigma(1915)$	F_{15}	****	$\Omega(2470)^-$		**
$N(2190)$	G_{17}	****	$\Delta(2390)$	F_{37}	*	$\Lambda(2585)$		**	$\Sigma(1940)$	D_{13}	***	Λ_c^+		****
$N(2200)$	D_{15}	**	$\Delta(2400)$	G_{39}	**				$\Sigma(2000)$	S_{11}	*	$\Lambda_c(2625)^+$		***
$N(2220)$	H_{19}	****	$\Delta(2420)$	$H_{3,11}$	****				$\Sigma(2030)$	F_{17}	****	$\Sigma_c(2455)$		****
$N(2250)$	G_{19}	****	$\Delta(2750)$	$I_{3,13}$	**				$\Sigma(2070)$	F_{15}	*	$\Sigma_c(2530)$		*
$N(2600)$	$I_{1,11}$	***	$\Delta(2950)$	$K_{3,15}$	**				$\Sigma(2080)$	P_{13}	**	Ξ_c^+		***
$N(2700)$	$K_{1,13}$	**							$\Sigma(2100)$	G_{17}	*	Ξ_c^0		***
									$\Sigma(2250)$		***	Ω_c^0		**
									$\Sigma(2455)$		**			
									$\Sigma(2620)$		**	Λ_b^0		***
									$\Sigma(3000)$		*			
									$\Sigma(3170)$		*			

**** Existence is certain, and properties are at least fairly well explored.

*** Existence ranges from very likely to certain, but further confirmation is desirable and/or quantum numbers, branching fractions, etc. are not well determined.

** Evidence of existence is only fair.

* Evidence of existence is poor.

Baryon Summary Table

N BARYONS
$(S = 0,\ I = 1/2)$

$p,\ N^+ = uud;\quad n,\ N^0 = udd$

p

$I(J^P) = \frac{1}{2}(\frac{1}{2}^+)$

Mass $m = 938.27231 \pm 0.00028$ MeV [a]
$= 1.007276470 \pm 0.000000012$ u
$m_{\overline{p}}/m_p = 0.99999998 \pm 0.00000004$
$|q_p + q_{\overline{p}}|/e < 2 \times 10^{-5}$
$|q_p + q_e|/e < 1.0 \times 10^{-21}$ [b]
Magnetic moment $\mu = 2.79284739 \pm 0.00000006\ \mu_N$
Electric dipole moment $d = (-4 \pm 6) \times 10^{-23}$ $e$cm
Electric polarizability $\overline{\alpha} = (10.2 \pm 0.9) \times 10^{-4}$ fm^3
Magnetic polarizability $\overline{\beta} = (4.0 \pm 0.9) \times 10^{-4}$ fm^3
Mean life $\tau > 1.6 \times 10^{25}$ years (independent of mode)
$> 10^{31} - 5 \times 10^{32}$ years [c] (mode dependent)

For N decays, p and n distinguish proton and neutron partial lifetimes. See also the "Note on Proton Mean Life Limits" in the Full Listings.

The "partial mean life" limits tabulated here are the limits on τ/B_i, where τ is the total mean life and B_i is the branching fraction for the mode in question.

p DECAY MODES	Partial mean life (10^{30} years)	Confidence level	p (MeV/c)
Antilepton + meson			
$N \to e^+\pi$	> 130 (n), > 550 (p)	90%	459
$N \to \mu^+\pi$	> 100 (n), > 270 (p)	90%	453
$N \to \nu\pi$	> 100 (n), > 25 (p)	90%	459
$p \to e^+\eta$	> 140	90%	309
$p \to \mu^+\eta$	> 69	90%	296
$n \to \nu\eta$	> 54	90%	310
$N \to e^+\rho$	> 58 (n), > 75 (p)	90%	153
$N \to \mu^+\rho$	> 23 (n), > 110 (p)	90%	119
$N \to \nu\rho$	> 19 (n), > 27 (p)	90%	153
$p \to e^+\omega$	> 45	90%	142
$p \to \mu^+\omega$	> 57	90%	104
$n \to \nu\omega$	> 43	90%	144
$N \to e^+K$	> 1.3 (n), > 150 (p)	90%	337
$p \to e^+K^0_S$	> 76	90%	337
$p \to e^+K^0_L$	> 44	90%	337
$N \to \mu^+K$	> 1.1 (n), > 120 (p)	90%	326
$p \to \mu^+K^0_S$	> 64	90%	326
$p \to \mu^+K^0_L$	> 44	90%	326
$N \to \nu K$	> 86 (n), > 100 (p)	90%	339
$p \to e^+K^*(892)^0$	> 52	90%	45
$N \to \nu K^*(892)$	> 22 (n), > 20 (p)	90%	45
Antilepton + mesons			
$p \to e^+\pi^+\pi^-$	> 21	90%	448
$p \to e^+\pi^0\pi^0$	> 38	90%	449
$n \to e^+\pi^-\pi^0$	> 32	90%	449
$p \to \mu^+\pi^+\pi^-$	> 17	90%	425
$p \to \mu^+\pi^0\pi^0$	> 33	90%	427
$n \to \mu^+\pi^-\pi^0$	> 33	90%	427
$n \to e^+K^0\pi^-$	> 18	90%	319
Lepton + meson			
$n \to e^-\pi^+$	> 65	90%	459
$n \to \mu^-\pi^+$	> 49	90%	453
$n \to e^-\rho^+$	> 62	90%	154
$n \to \mu^-\rho^+$	> 7	90%	120
$n \to e^-K^+$	> 32	90%	340
$n \to \mu^-K^+$	> 57	90%	330
Lepton + mesons			
$p \to e^-\pi^+\pi^+$	> 30	90%	448
$n \to e^-\pi^+\pi^0$	> 29	90%	449
$p \to \mu^-\pi^+\pi^+$	> 17	90%	425
$n \to \mu^-\pi^+\pi^0$	> 34	90%	427
$p \to e^-\pi^+K^+$	> 20	90%	320
$p \to \mu^-\pi^+K^+$	> 5	90%	279
Antilepton + photon(s)			
$p \to e^+\gamma$	> 460	90%	469
$p \to \mu^+\gamma$	> 380	90%	463
$n \to \nu\gamma$	> 24	90%	470
$p \to e^+\gamma\gamma$	> 100	90%	469
Three leptons			
$p \to e^+e^+e^-$	> 510	90%	469
$p \to e^+\mu^+\mu^-$	> 81	90%	457
$p \to e^+\nu\nu$	> 11	90%	469
$n \to e^+e^-\nu$	> 74	90%	470
$n \to \mu^+e^-\nu$	> 47	90%	464
$n \to \mu^+\mu^-\nu$	> 42	90%	458
$p \to \mu^+e^+e^-$	> 91	90%	464
$p \to \mu^+\mu^+\mu^-$	> 190	90%	439
$p \to \mu^+\nu\nu$	> 21	90%	463
$p \to e^-\mu^+\mu^+$	> 6	90%	457
$n \to 3\nu$	> 0.0005	90%	470
Inclusive modes			
$N \to e^+$ anything	> 0.6 (n, p)	90%	–
$N \to \mu^+$ anything	> 12 (n, p)	90%	–
$N \to e^+\pi^0$ anything	> 0.6 (n, p)	90%	–
$\Delta B = 2$ dinucleon modes			
The following are lifetime limits per iron nucleus.			
$pp \to \pi^+\pi^+$	> 0.7	90%	–
$pn \to \pi^+\pi^0$	> 2	90%	–
$nn \to \pi^+\pi^-$	> 0.7	90%	–
$nn \to \pi^0\pi^0$	> 3.4	90%	–
$pp \to e^+e^+$	> 5.8	90%	–
$pp \to e^+\mu^+$	> 3.6	90%	–
$pp \to \mu^+\mu^+$	> 1.7	90%	–
$pn \to e^+\overline{\nu}$	> 2.8	90%	–
$pn \to \mu^+\overline{\nu}$	> 1.6	90%	–
$nn \to \nu_e\overline{\nu}_e$	> 0.000012	90%	–
$nn \to \nu_\mu\overline{\nu}_\mu$	> 0.000006	90%	–

$\overline{p}$ DECAY MODES

$\overline{p}$ DECAY MODES	Partial mean life (years)	Confidence level	p (MeV/c)
$\overline{p} \to e^-\gamma$	> 1848	95%	469
$\overline{p} \to e^-\pi^0$	> 554	95%	459
$\overline{p} \to e^-\eta$	> 171	95%	309
$\overline{p} \to e^-K^0_S$	> 29	95%	337
$\overline{p} \to e^-K^0_L$	> 9	95%	337

n

$I(J^P) = \frac{1}{2}(\frac{1}{2}^+)$

Mass $m = 939.56563 \pm 0.00028$ MeV [a]
$= 1.008664904 \pm 0.000000014$ u
$m_n - m_p = 1.293318 \pm 0.000009$ MeV
$= 0.001388434 \pm 0.000000009$ u
Mean life $\tau = 887.0 \pm 2.0$ s (S = 1.3)
$c\tau = 2.659 \times 10^8$ km
Magnetic moment $\mu = -1.9130428 \pm 0.0000005\ \mu_N$
Electric dipole moment $d < 11 \times 10^{-26}$ $e$cm, CL = 95%
Electric polarizability $\alpha = (1.16^{+0.19}_{-0.23}) \times 10^{-3}$ fm^3
Charge $q = (-0.4 \pm 1.1) \times 10^{-21}$ e
Mean time for $n\overline{n}$ oscillations $> 1.2 \times 10^8$ s, CL = 90% [d]

Decay parameters [e]

$pe^-\overline{\nu}_e$	$g_A/g_V = -1.2573 \pm 0.0028$
"	$A = -0.1127 \pm 0.0011$
"	$B = 0.997 \pm 0.028$
"	$a = -0.102 \pm 0.005$
"	$\phi_{AV} = (180.07 \pm 0.18)^\circ$ [f]
"	$D = (-0.5 \pm 1.4) \times 10^{-3}$

n DECAY MODES	Fraction (Γ_i/Γ)	Confidence level	p (MeV/c)
$pe^-\overline{\nu}_e$	100 %		1.19
Charge conservation (Q) violating mode			
$p\nu_e\overline{\nu}_e$	Q < 9×10^{-24}	90%	1.29

$N(1440)$ P_{11} $I(J^P) = \frac{1}{2}(\frac{1}{2}^+)$

Mass m = 1430 to 1470 ($\approx$ 1440) MeV
Full width Γ = 250 to 450 ($\approx$ 350) MeV
p_{beam} = 0.61 GeV/c $4\pi\bar{\lambda}^2$ = 31.0 mb

N(1440) DECAY MODES	Fraction (Γ_i/Γ)	p (MeV/c)
$N\pi$	60–70 %	397
$N\pi\pi$	30–40 %	342
$\Delta\pi$	20–30 %	143
$N\rho$	<8 %	†
$N(\pi\pi)^{I=0}_{S\text{-wave}}$	5–10 %	–
$p\gamma$	0.04–0.07 %	414
$n\gamma$	0.001–0.05 %	413

$N(1520)$ D_{13} $I(J^P) = \frac{1}{2}(\frac{3}{2}^-)$

Mass m = 1515 to 1530 ($\approx$ 1520) MeV
Full width Γ = 110 to 135 ($\approx$ 120) MeV
p_{beam} = 0.74 GeV/c $4\pi\bar{\lambda}^2$ = 23.5 mb

N(1520) DECAY MODES	Fraction (Γ_i/Γ)	p (MeV/c)
$N\pi$	50–60 %	456
$N\pi\pi$	40–50 %	410
$\Delta\pi$	15–25 %	228
$N\rho$	15–25 %	†
$N(\pi\pi)^{I=0}_{S\text{-wave}}$	<8 %	–
$p\gamma$	0.45–0.53 %	470
$n\gamma$	0.34–0.48 %	470

$N(1535)$ S_{11} $I(J^P) = \frac{1}{2}(\frac{1}{2}^-)$

Mass m = 1520 to 1555 ($\approx$ 1535) MeV
Full width Γ = 100 to 250 ($\approx$ 150) MeV
p_{beam} = 0.76 GeV/c $4\pi\bar{\lambda}^2$ = 22.5 mb

N(1535) DECAY MODES	Fraction (Γ_i/Γ)	p (MeV/c)
$N\pi$	35–55 %	467
$N\eta$	30–55 %	182
$N\pi\pi$	1–10 %	422
$\Delta\pi$	<1 %	242
$N\rho$	<4 %	†
$N(\pi\pi)^{I=0}_{S\text{-wave}}$	<3 %	–
$N(1440)\pi$	<7 %	†
$p\gamma$	0.45–0.53 %	481
$n\gamma$	0.34–0.48 %	480

$N(1650)$ S_{11} $I(J^P) = \frac{1}{2}(\frac{1}{2}^-)$

Mass m = 1640 to 1680 ($\approx$ 1650) MeV
Full width Γ = 145 to 190 ($\approx$ 150) MeV
p_{beam} = 0.96 GeV/c $4\pi\bar{\lambda}^2$ = 16.4 mb

N(1650) DECAY MODES	Fraction (Γ_i/Γ)	p (MeV/c)
$N\pi$	60–80 %	547
ΛK	3–11 %	161
$N\pi\pi$	5–20 %	511
$\Delta\pi$	3–7 %	344
$N\rho$	4–14 %	†
$N(\pi\pi)^{I=0}_{S\text{-wave}}$	<4 %	–
$N(1440)\pi$	<5 %	147
$p\gamma$	0.10–0.18 %	558
$n\gamma$	0.03–0.18 %	557

$N(1675)$ D_{15} $I(J^P) = \frac{1}{2}(\frac{5}{2}^-)$

Mass m = 1670 to 1685 ($\approx$ 1675) MeV
Full width Γ = 140 to 180 ($\approx$ 150) MeV
p_{beam} = 1.01 GeV/c $4\pi\bar{\lambda}^2$ = 15.4 mb

N(1675) DECAY MODES	Fraction (Γ_i/Γ)	p (MeV/c)
$N\pi$	40–50 %	563
ΛK	<1 %	209
$N\pi\pi$	50–60 %	529
$\Delta\pi$	50–60 %	364
$N\rho$	< 1–3 %	†
$p\gamma$	0.005–0.014 %	575
$n\gamma$	0.07–0.11 %	574

$N(1680)$ F_{15} $I(J^P) = \frac{1}{2}(\frac{5}{2}^+)$

Mass m = 1675 to 1690 ($\approx$ 1680) MeV
Full width Γ = 120 to 140 ($\approx$ 130) MeV
p_{beam} = 1.01 GeV/c $4\pi\bar{\lambda}^2$ = 15.2 mb

N(1680) DECAY MODES	Fraction (Γ_i/Γ)	p (MeV/c)
$N\pi$	60–70 %	567
$N\pi\pi$	30–40 %	532
$\Delta\pi$	5–15 %	369
$N\rho$	3–15 %	†
$N(\pi\pi)^{I=0}_{S\text{-wave}}$	5–20 %	–
$p\gamma$	0.21–0.35 %	578
$n\gamma$	0.02–0.04 %	577

$N(1700)$ D_{13} $I(J^P) = \frac{1}{2}(\frac{3}{2}^-)$

Mass m = 1650 to 1750 ($\approx$ 1700) MeV
Full width Γ = 50 to 150 ($\approx$ 100) MeV
p_{beam} = 1.05 GeV/c $4\pi\bar{\lambda}^2$ = 14.5 mb

N(1700) DECAY MODES	Fraction (Γ_i/Γ)	p (MeV/c)
$N\pi$	5–15 %	580
ΛK	<3 %	250
$N\pi\pi$	85–95 %	547
$N\rho$	<35 %	†
$p\gamma$	~ 0.01 %	591

$N(1710)$ P_{11} $I(J^P) = \frac{1}{2}(\frac{1}{2}^+)$

Mass m = 1680 to 1740 ($\approx$ 1710) MeV
Full width Γ = 50 to 250 ($\approx$ 100) MeV
p_{beam} = 1.07 GeV/c $4\pi\bar{\lambda}^2$ = 14.2 mb

N(1710) DECAY MODES	Fraction (Γ_i/Γ)	p (MeV/c)
$N\pi$	10–20 %	587
ΛK	5–25 %	264
$N\pi\pi$	40–90 %	554
$\Delta\pi$	15–40 %	393
$N\rho$	5–25 %	48
$N(\pi\pi)^{I=0}_{S\text{-wave}}$	10–40 %	–

$N(1720)$ P_{13} $I(J^P) = \frac{1}{2}(\frac{3}{2}^+)$

Mass m = 1650 to 1750 ($\approx$ 1720) MeV
Full width Γ = 100 to 200 ($\approx$ 150) MeV
p_{beam} = 1.09 GeV/c $4\pi\bar{\lambda}^2$ = 13.9 mb

N(1720) DECAY MODES	Fraction (Γ_i/Γ)	p (MeV/c)
$N\pi$	10–20 %	594
ΛK	1–15 %	278
$N\pi\pi$	>70 %	561
$N\rho$	70–85 %	104
$p\gamma$	0.01–0.06 %	–

Baryon Summary Table

$N(2190)\ G_{17}$

$I(J^P) = \frac{1}{2}(\frac{7}{2}^-)$

Mass m = 2100 to 2200 (≈ 2190) MeV
Full width Γ = 350 to 550 (≈ 450) MeV
p_{beam} = 2.07 GeV/c $4\pi\lambda\!\!\!^{-2}$ = 6.21 mb

N(2190) DECAY MODES	Fraction (Γ_i/Γ)	p (MeV/c)
$N\pi$	10–20 %	888

$N(2220)\ H_{19}$

$I(J^P) = \frac{1}{2}(\frac{9}{2}^+)$

Mass m = 2180 to 2310 (≈ 2220) MeV
Full width Γ = 320 to 550 (≈ 400) MeV
p_{beam} = 2.14 GeV/c $4\pi\lambda\!\!\!^{-2}$ = 5.97 mb

N(2220) DECAY MODES	Fraction (Γ_i/Γ)	p (MeV/c)
$N\pi$	10–20 %	905

$N(2250)\ G_{19}$

$I(J^P) = \frac{1}{2}(\frac{9}{2}^-)$

Mass m = 2170 to 2310 (≈ 2250) MeV
Full width Γ = 290 to 470 (≈ 400) MeV
p_{beam} = 2.21 GeV/c $4\pi\lambda\!\!\!^{-2}$ = 5.74 mb

N(2250) DECAY MODES	Fraction (Γ_i/Γ)	p (MeV/c)
$N\pi$	5–15 %	923

$N(2600)\ I_{1,11}$

$I(J^P) = \frac{1}{2}(\frac{11}{2}^-)$

Mass m = 2550 to 2750 (≈ 2600) MeV
Full width Γ = 500 to 800 (≈ 650) MeV
p_{beam} = 3.12 GeV/c $4\pi\lambda\!\!\!^{-2}$ = 3.86 mb

N(2600) DECAY MODES	Fraction (Γ_i/Γ)	p (MeV/c)
$N\pi$	5–10 %	1126

Δ BARYONS ($S = 0$, $I = 3/2$)

$\Delta^{++} = uuu$, $\Delta^{+} = uud$, $\Delta^{0} = udd$, $\Delta^{-} = ddd$

$\Delta(1232)\ P_{33}$

$I(J^P) = \frac{3}{2}(\frac{3}{2}^+)$

Mass m = 1230 to 1234 (≈ 1232) MeV
Full width Γ = 115 to 125 (≈ 120) MeV
p_{beam} = 0.30 GeV/c $4\pi\lambda\!\!\!^{-2}$ = 94.8 mb

Δ(1232) DECAY MODES	Fraction (Γ_i/Γ)	p (MeV/c)
$N\pi$	>99 %	227
$N\gamma$	0.55–0.61 %	259

$\Delta(1600)\ P_{33}$

$I(J^P) = \frac{3}{2}(\frac{3}{2}^+)$

Mass m = 1550 to 1700 (≈ 1600) MeV
Full width Γ = 250 to 450 (≈ 350) MeV
p_{beam} = 0.87 GeV/c $4\pi\lambda\!\!\!^{-2}$ = 18.6 mb

Δ(1600) DECAY MODES	Fraction (Γ_i/Γ)	p (MeV/c)
$N\pi$	10–25 %	512
$N\pi\pi$	75–90 %	473
$\Delta\pi$	40–70 %	301
$N\rho$	<25 %	†
$N(1440)\pi$	10–35 %	74
$p\gamma$	~0 %	–

$\Delta(1620)\ S_{31}$

$I(J^P) = \frac{3}{2}(\frac{1}{2}^-)$

Mass m = 1615 to 1675 (≈ 1620) MeV
Full width Γ = 120 to 180 (≈ 150) MeV
p_{beam} = 0.91 GeV/c $4\pi\lambda\!\!\!^{-2}$ = 17.7 mb

Δ(1620) DECAY MODES	Fraction (Γ_i/Γ)	p (MeV/c)
$N\pi$	20–30 %	526
$N\pi\pi$	70–80 %	488
$\Delta\pi$	30–60 %	318
$N\rho$	7–25 %	†
$N\gamma$	0.02–0.06 %	538

$\Delta(1700)\ D_{33}$

$I(J^P) = \frac{3}{2}(\frac{3}{2}^-)$

Mass m = 1670 to 1770 (≈ 1700) MeV
Full width Γ = 200 to 400 (≈ 300) MeV
p_{beam} = 1.05 GeV/c $4\pi\lambda\!\!\!^{-2}$ = 14.5 mb

Δ(1700) DECAY MODES	Fraction (Γ_i/Γ)	p (MeV/c)
$N\pi$	10–20 %	580
$N\pi\pi$	80–90 %	547
$\Delta\pi$	30–60 %	385
$N\rho$	30–55 %	†
$N\gamma$	0.16–0.28 %	591

$\Delta(1900)\ S_{31}$

$I(J^P) = \frac{3}{2}(\frac{1}{2}^-)$

Mass m = 1850 to 1950 (≈ 1900) MeV
Full width Γ = 140 to 240 (≈ 200) MeV
p_{beam} = 1.44 GeV/c $4\pi\lambda\!\!\!^{-2}$ = 9.71 mb

Δ(1900) DECAY MODES	Fraction (Γ_i/Γ)	p (MeV/c)
$N\pi$	10–30 %	710

$\Delta(1905)\ F_{35}$

$I(J^P) = \frac{3}{2}(\frac{5}{2}^+)$

Mass m = 1870 to 1920 (≈ 1905) MeV
Full width Γ = 280 to 440 (≈ 350) MeV
p_{beam} = 1.45 GeV/c $4\pi\lambda\!\!\!^{-2}$ = 9.62 mb

Δ(1905) DECAY MODES	Fraction (Γ_i/Γ)	p (MeV/c)
$N\pi$	5–15 %	713
$N\pi\pi$	85–95 %	687
$\Delta\pi$	<25 %	542
$N\rho$	>60 %	421
$N\gamma$	0.01–0.04 %	721

$\Delta(1910)\ P_{31}$

$I(J^P) = \frac{3}{2}(\frac{1}{2}^+)$

Mass m = 1870 to 1920 (≈ 1910) MeV
Full width Γ = 190 to 270 (≈ 250) MeV
p_{beam} = 1.46 GeV/c $4\pi\lambda\!\!\!^{-2}$ = 9.54 mb

Δ(1910) DECAY MODES	Fraction (Γ_i/Γ)	p (MeV/c)
$N\pi$	15–30 %	716

$\Delta(1920)\ P_{33}$

$I(J^P) = \frac{3}{2}(\frac{3}{2}^+)$

Mass m = 1900 to 1970 (≈ 1920) MeV
Full width Γ = 150 to 300 (≈ 200) MeV
p_{beam} = 1.48 GeV/c $4\pi\lambda\!\!\!^{-2}$ = 9.37 mb

Δ(1920) DECAY MODES	Fraction (Γ_i/Γ)	p (MeV/c)
$N\pi$	5–20 %	722

$\Delta(1930)\ D_{35}$ $\quad I(J^P) = \frac{3}{2}(\frac{5}{2}^-)$

Mass m = 1920 to 1970 ($\approx$ 1930) MeV
Full width Γ = 250 to 450 ($\approx$ 350) MeV
p_{beam} = 1.50 GeV/c $\quad 4\pi\lambda\!\!\!^{-2}$ = 9.21 mb

Δ(1930) DECAY MODES	Fraction (Γ_i/Γ)	p (MeV/c)
$N\pi$	10–20 %	729

$\Delta(1950)\ F_{37}$ $\quad I(J^P) = \frac{3}{2}(\frac{7}{2}^+)$

Mass m = 1940 to 1960 ($\approx$ 1950) MeV
Full width Γ = 290 to 350 ($\approx$ 300) MeV
p_{beam} = 1.54 GeV/c $\quad 4\pi\lambda\!\!\!^{-2}$ = 8.91 mb

Δ(1950) DECAY MODES	Fraction (Γ_i/Γ)	p (MeV/c)
$N\pi$	35–40 %	741
$N\pi\pi$		716
$\quad\Delta\pi$	20–30 %	574
$\quad N\rho$	<10 %	469
$N\gamma$	0.10–0.15 %	749

$\Delta(2420)\ H_{3,11}$ $\quad I(J^P) = \frac{3}{2}(\frac{11}{2}^+)$

Mass m = 2300 to 2500 ($\approx$ 2420) MeV
Full width Γ = 300 to 500 ($\approx$ 400) MeV
p_{beam} = 2.64 GeV/c $\quad 4\pi\lambda\!\!\!^{-2}$ = 4.68 mb

Δ(2420) DECAY MODES	Fraction (Γ_i/Γ)	p (MeV/c)
$N\pi$	5–15 %	1023

Λ BARYONS
($S = -1, I = 0$)

$\Lambda^0 = uds$

Λ $\quad I(J^P) = 0(\frac{1}{2}^+)$

Mass m = 1115.684 $\pm$ 0.006 MeV
Mean life $\tau = (2.632 \pm 0.020) \times 10^{-10}$ s $\quad$ (S = 1.6)
$c\tau$ = 7.89 cm
Magnetic moment $\mu = -0.613 \pm 0.004\ \mu_N$
Electric dipole moment $d < 1.5 \times 10^{-16}$ $e$cm, CL = 95%

Decay parameters

$p\pi^-$	$\alpha_- = 0.642 \pm 0.013$	
"	$\phi_- = (-6.5 \pm 3.5)^\circ$	
"	$\gamma_- = 0.76$ [g]	
"	$\Delta_- = (8 \pm 4)^\circ$ [g]	
$n\pi^0$	$\alpha_0 = +0.65 \pm 0.05$	
$pe^-\overline{\nu}_e$	$g_A/g_V = -0.718 \pm 0.015$ [e]	

Λ DECAY MODES	Fraction (Γ_i/Γ)	p (MeV/c)
$p\pi^-$	(63.9 $\pm$0.5) %	101
$n\pi^0$	(35.8 $\pm$0.5) %	104
$n\gamma$	(1.75$\pm$0.15) $\times 10^{-3}$	162
$p\pi^-\gamma$	[h] (8.4 $\pm$1.4) $\times 10^{-4}$	101
$pe^-\overline{\nu}_e$	(8.32$\pm$0.14) $\times 10^{-4}$	163
$p\mu^-\overline{\nu}_\mu$	(1.57$\pm$0.35) $\times 10^{-4}$	131

$\Lambda(1405)\ S_{01}$ $\quad I(J^P) = 0(\frac{1}{2}^-)$

Mass m = 1407 $\pm$ 4 MeV
Full width Γ = 50.0 $\pm$ 2.0 MeV
Below $\overline{K}N$ threshold

Λ(1405) DECAY MODES	Fraction (Γ_i/Γ)	p (MeV/c)
$\Sigma\pi$	100 %	152

$\Lambda(1520)\ D_{03}$ $\quad I(J^P) = 0(\frac{3}{2}^-)$

Mass m = 1519.5 $\pm$ 1.0 MeV [i]
Full width Γ = 15.6 $\pm$ 1.0 MeV [i]
p_{beam} = 0.39 GeV/c $\quad 4\pi\lambda\!\!\!^{-2}$ = 82.8 mb

Λ(1520) DECAY MODES	Fraction (Γ_i/Γ)	p (MeV/c)
$N\overline{K}$	45 $\pm$ 1%	244
$\Sigma\pi$	42 $\pm$ 1%	267
$\Lambda\pi\pi$	10 $\pm$ 1%	252
$\Sigma\pi\pi$	0.9 $\pm$ 0.1%	152
$\Lambda\gamma$	0.8 $\pm$ 0.2%	351

$\Lambda(1600)\ P_{01}$ $\quad I(J^P) = 0(\frac{1}{2}^+)$

Mass m = 1560 to 1700 ($\approx$ 1600) MeV
Full width Γ = 50 to 250 ($\approx$ 150) MeV
p_{beam} = 0.58 GeV/c $\quad 4\pi\lambda\!\!\!^{-2}$ = 41.6 mb

Λ(1600) DECAY MODES	Fraction (Γ_i/Γ)	p (MeV/c)
$N\overline{K}$	15–30 %	343
$\Sigma\pi$	10–60 %	336

$\Lambda(1670)\ S_{01}$ $\quad I(J^P) = 0(\frac{1}{2}^-)$

Mass m = 1660 to 1680 ($\approx$ 1670) MeV
Full width Γ = 25 to 50 ($\approx$ 35) MeV
p_{beam} = 0.74 GeV/c $\quad 4\pi\lambda\!\!\!^{-2}$ = 28.5 mb

Λ(1670) DECAY MODES	Fraction (Γ_i/Γ)	p (MeV/c)
$N\overline{K}$	15–25 %	414
$\Sigma\pi$	20–60 %	393
$\Lambda\eta$	15–35 %	64

$\Lambda(1690)\ D_{03}$ $\quad I(J^P) = 0(\frac{3}{2}^-)$

Mass m = 1685 to 1695 ($\approx$ 1690) MeV
Full width Γ = 50 to 70 ($\approx$ 60) MeV
p_{beam} = 0.78 GeV/c $\quad 4\pi\lambda\!\!\!^{-2}$ = 26.1 mb

Λ(1690) DECAY MODES	Fraction (Γ_i/Γ)	p (MeV/c)
$N\overline{K}$	20–30 %	433
$\Sigma\pi$	20–40 %	409
$\Lambda\pi\pi$	~ 25 %	415
$\Sigma\pi\pi$	~ 20 %	350

$\Lambda(1800)\ S_{01}$ $\quad I(J^P) = 0(\frac{1}{2}^-)$

Mass m = 1720 to 1850 ($\approx$ 1800) MeV
Full width Γ = 200 to 400 ($\approx$ 300) MeV
p_{beam} = 1.01 GeV/c $\quad 4\pi\lambda\!\!\!^{-2}$ = 17.5 mb

Λ(1800) DECAY MODES	Fraction (Γ_i/Γ)	p (MeV/c)
$N\overline{K}$	25–40 %	528
$\Sigma\pi$	seen	493
$\Sigma(1385)\pi$	seen	345
$N\overline{K}^*(892)$	seen	†

$\Lambda(1810)\ P_{01}$ $\quad I(J^P) = 0(\frac{1}{2}^+)$

Mass m = 1750 to 1850 ($\approx$ 1810) MeV
Full width Γ = 50 to 250 ($\approx$ 150) MeV
p_{beam} = 1.04 GeV/c $\quad 4\pi\lambda\!\!\!^{-2}$ = 17.0 mb

Λ(1810) DECAY MODES	Fraction (Γ_i/Γ)	p (MeV/c)
$N\overline{K}$	20–50 %	537
$\Sigma\pi$	10–40 %	501
$\Sigma(1385)\pi$	seen	356
$N\overline{K}^*(892)$	30–60 %	†

Baryon Summary Table

$\Lambda(1820)\ F_{05}$

$I(J^P) = 0(\frac{5}{2}^+)$

Mass m = 1815 to 1825 (≈ 1820) MeV
Full width Γ = 70 to 90 (≈ 80) MeV
p_{beam} = 1.06 GeV/c $4\pi\lambda\!\!\!^{-2}$ = 16.5 mb

Λ(1820) DECAY MODES	Fraction (Γ_i/Γ)	p (MeV/c)
$N\overline{K}$	55–65 %	545
$\Sigma\pi$	8–14 %	508
$\Sigma(1385)\pi$	5–10 %	362

$\Lambda(1830)\ D_{05}$

$I(J^P) = 0(\frac{5}{2}^-)$

Mass m = 1810 to 1830 (≈ 1830) MeV
Full width Γ = 60 to 110 (≈ 95) MeV
p_{beam} = 1.08 GeV/c $4\pi\lambda\!\!\!^{-2}$ = 16.0 mb

Λ(1830) DECAY MODES	Fraction (Γ_i/Γ)	p (MeV/c)
$N\overline{K}$	3–10 %	553
$\Sigma\pi$	35–75 %	515
$\Sigma(1385)\pi$	>15 %	371

$\Lambda(1890)\ P_{03}$

$I(J^P) = 0(\frac{3}{2}^+)$

Mass m = 1850 to 1910 (≈ 1890) MeV
Full width Γ = 60 to 200 (≈ 100) MeV
p_{beam} = 1.21 GeV/c $4\pi\lambda\!\!\!^{-2}$ = 13.6 mb

Λ(1890) DECAY MODES	Fraction (Γ_i/Γ)	p (MeV/c)
$N\overline{K}$	20–35 %	599
$\Sigma\pi$	3–10 %	559
$\Sigma(1385)\pi$	seen	420
$N\overline{K}^*(892)$	seen	233

$\Lambda(2100)\ G_{07}$

$I(J^P) = 0(\frac{7}{2}^-)$

Mass m = 2090 to 2110 (≈ 2100) MeV
Full width Γ = 100 to 250 (≈ 200) MeV
p_{beam} = 1.68 GeV/c $4\pi\lambda\!\!\!^{-2}$ = 8.68 mb

Λ(2100) DECAY MODES	Fraction (Γ_i/Γ)	p (MeV/c)
$N\overline{K}$	25–35 %	751
$\Sigma\pi$	~ 5 %	704
$\Lambda\eta$	<3 %	617
ΞK	<3 %	483
$\Lambda\omega$	<8 %	443
$N\overline{K}^*(892)$	10–20 %	514

$\Lambda(2110)\ F_{05}$

$I(J^P) = 0(\frac{5}{2}^+)$

Mass m = 2090 to 2140 (≈ 2110) MeV
Full width Γ = 150 to 250 (≈ 200) MeV
p_{beam} = 1.70 GeV/c $4\pi\lambda\!\!\!^{-2}$ = 8.53 mb

Λ(2110) DECAY MODES	Fraction (Γ_i/Γ)	p (MeV/c)
$N\overline{K}$	5–25 %	757
$\Sigma\pi$	10–40 %	711
$\Lambda\omega$	seen	455
$\Sigma(1385)\pi$	seen	589
$N\overline{K}^*(892)$	10–60 %	524

$\Lambda(2350)\ H_{09}$

$I(J^P) = 0(\frac{9}{2}^+)$

Mass m = 2340 to 2370 (≈ 2350) MeV
Full width Γ = 100 to 250 (≈ 150) MeV
p_{beam} = 2.29 GeV/c $4\pi\lambda\!\!\!^{-2}$ = 5.85 mb

Λ(2350) DECAY MODES	Fraction (Γ_i/Γ)	p (MeV/c)
$N\overline{K}$	~ 12 %	915
$\Sigma\pi$	~ 10 %	867

Σ BARYONS
($S = -1, I = 1$)

$\Sigma^+ = uus$, $\Sigma^0 = uds$, $\Sigma^- = dds$

Σ^+

$I(J^P) = 1(\frac{1}{2}^+)$

Mass m = 1189.37 ± 0.07 MeV (S = 2.2)
Mean life τ = (0.799 ± 0.004) × 10^{-10} s
$c\tau$ = 2.396 cm
Magnetic moment μ = 2.458 ± 0.010 μ_N (S = 2.1)
$\Gamma(\Sigma^+ \to n\ell^+\nu)/\Gamma(\Sigma^- \to n\ell^-\overline{\nu})$ < 0.043

Decay parameters

$p\pi^0$	$\alpha_0 = -0.980^{+0.017}_{-0.015}$
"	$\phi_0 = (36 \pm 34)^\circ$
"	$\gamma_0 = 0.16$ [g]
"	$\Delta_0 = (187 \pm 6)^\circ$ [g]
$n\pi^+$	$\alpha_+ = 0.068 \pm 0.013$
"	$\phi_+ = (167 \pm 20)^\circ$ (S = 1.1)
"	$\gamma_+ = -0.97$ [g]
"	$\Delta_+ = (-73^{+133}_{-10})^\circ$ [g]
$p\gamma$	$\alpha_\gamma = -0.76 \pm 0.08$

Σ+ DECAY MODES		Fraction (Γ_i/Γ)	Confidence level	p (MeV/c)
$p\pi^0$		(51.57±0.30) %		189
$n\pi^+$		(48.30±0.30) %		185
$p\gamma$		(1.25±0.07) × 10^{-3}		225
$n\pi^+\gamma$	[h]	(4.5 ±0.5) × 10^{-4}		185
$\Lambda e^+\nu_e$		(2.0 ±0.5) × 10^{-5}		71
$\Delta S = \Delta Q$ (SQ) violating modes or $\Delta S = 1$ weak neutral current (S1) modes				
$ne^+\nu_e$	SQ	< 5 × 10^{-6}	90%	224
$n\mu^+\nu_\mu$	SQ	< 3.0 × 10^{-5}	90%	202
pe^+e^-	S1	< 7 × 10^{-6}		225

Σ^0

$I(J^P) = 1(\frac{1}{2}^+)$

J^P not measured; assumed to be the same as for the Σ^+ and Σ^-.

Mass m = 1192.55 ± 0.08 MeV (S = 1.2)
$m_{\Sigma^-} - m_{\Sigma^0}$ = 4.88 ± 0.08 MeV (S = 1.2)
$m_{\Sigma^0} - m_\Lambda$ = 76.87 ± 0.08 MeV (S = 1.2)
Mean life τ = (7.4 ± 0.7) × 10^{-20} s
$c\tau$ = 2.22 × 10^{-11} m
Transition magnetic moment $|\mu_{\Sigma\Lambda}|$ = 1.61 ± 0.08 μ_N

Σ0 DECAY MODES		Fraction (Γ_i/Γ)	Confidence level	p (MeV/c)
$\Lambda\gamma$		100 %		74
$\Lambda\gamma\gamma$		< 3 %	90%	74
Λe^+e^-	[i]	5 × 10^{-3}		74

Σ^-

$I(J^P) = 1(\frac{1}{2}^+)$

Mass m = 1197.436 ± 0.033 MeV (S = 1.2)
$m_{\Sigma^-} - m_{\Sigma^+}$ = 8.07 ± 0.08 MeV (S = 1.9)
$m_{\Sigma^-} - m_\Lambda$ = 81.752 ± 0.034 MeV (S = 1.2)
Mean life τ = (1.479 ± 0.011) × 10^{-10} s (S = 1.3)
$c\tau$ = 4.434 cm
Magnetic moment μ = −1.160 ± 0.025 μ_N (S = 1.7)

Decay parameters

$n\pi^-$	$\alpha_- = -0.068 \pm 0.008$
"	$\phi_- = (10 \pm 15)^\circ$
"	$\gamma_- = 0.98$ [g]
"	$\Delta_- = (249^{+12}_{-120})^\circ$ [g]
$ne^-\overline{\nu}_e$	$g_A/g_V = 0.340 \pm 0.017$ [e]
"	$f_2(0)/f_1(0) = 0.97 \pm 0.14$
"	$D = 0.11 \pm 0.10$
$\Lambda e^-\overline{\nu}_e$	$g_V/g_A = 0.01 \pm 0.10$ [e] (S = 1.5)
"	$g_{WM}/g_A = 2.4 \pm 1.7$ [e]

Σ^- DECAY MODES	Fraction (Γ_i/Γ)	p (MeV/c)
$n\pi^-$	(99.848 ± 0.005) %	193
$n\pi^-\gamma$	[h] $(4.6 \pm0.6) \times 10^{-4}$	193
$ne^-\overline{\nu}_e$	$(1.017\pm0.034) \times 10^{-3}$	230
$n\mu^-\overline{\nu}_\mu$	$(4.5 \pm0.4) \times 10^{-4}$	210
$\Lambda e^-\overline{\nu}_e$	$(5.73 \pm0.27) \times 10^{-5}$	79

$\Sigma(1385)\ P_{13}$

$I(J^P) = 1(\frac{3}{2}^+)$

$\Sigma(1385)^+$ mass $m = 1382.8 \pm 0.4$ MeV (S = 2.0)
$\Sigma(1385)^0$ mass $m = 1383.7 \pm 1.0$ MeV (S = 1.4)
$\Sigma(1385)^-$ mass $m = 1387.2 \pm 0.5$ MeV (S = 2.2)
$\Sigma(1385)^+$ full width $\Gamma = 35.8 \pm 0.8$ MeV
$\Sigma(1385)^0$ full width $\Gamma = 36 \pm 5$ MeV
$\Sigma(1385)^-$ full width $\Gamma = 39.4 \pm 2.1$ MeV (S = 1.7)
Below $\overline{K}N$ threshold

Σ(1385) DECAY MODES	Fraction (Γ_i/Γ)	p (MeV/c)
$\Lambda\pi$	88±2 %	208
$\Sigma\pi$	12±2 %	127

$\Sigma(1660)\ P_{11}$

$I(J^P) = 1(\frac{1}{2}^+)$

Mass $m = 1630$ to 1690 (≈ 1660) MeV
Full width $\Gamma = 40$ to 200 (≈ 100) MeV
$p_{beam} = 0.72$ GeV/c $\quad 4\pi\lambda\!\!\!^-{}^2 = 29.9$ mb

Σ(1660) DECAY MODES	Fraction (Γ_i/Γ)	p (MeV/c)
$N\overline{K}$	10–30 %	405
$\Lambda\pi$	seen	439
$\Sigma\pi$	seen	385

$\Sigma(1670)\ D_{13}$

$I(J^P) = 1(\frac{3}{2}^-)$

Mass $m = 1665$ to 1685 (≈ 1670) MeV
Full width $\Gamma = 40$ to 80 (≈ 60) MeV
$p_{beam} = 0.74$ GeV/c $\quad 4\pi\lambda\!\!\!^-{}^2 = 28.5$ mb

Σ(1670) DECAY MODES	Fraction (Γ_i/Γ)	p (MeV/c)
$N\overline{K}$	7–13 %	414
$\Lambda\pi$	5–15 %	447
$\Sigma\pi$	30–60 %	393

$\Sigma(1750)\ S_{11}$

$I(J^P) = 1(\frac{1}{2}^-)$

Mass $m = 1730$ to 1800 (≈ 1750) MeV
Full width $\Gamma = 60$ to 160 (≈ 90) MeV
$p_{beam} = 0.91$ GeV/c $\quad 4\pi\lambda\!\!\!^-{}^2 = 20.7$ mb

Σ(1750) DECAY MODES	Fraction (Γ_i/Γ)	p (MeV/c)
$N\overline{K}$	10–40 %	486
$\Lambda\pi$	seen	507
$\Sigma\pi$	<8 %	455
$\Sigma\eta$	15–55 %	81

$\Sigma(1775)\ D_{15}$

$I(J^P) = 1(\frac{5}{2}^-)$

Mass $m = 1770$ to 1780 (≈ 1775) MeV
Full width $\Gamma = 105$ to 135 (≈ 120) MeV
$p_{beam} = 0.96$ GeV/c $\quad 4\pi\lambda\!\!\!^-{}^2 = 19.0$ mb

Σ(1775) DECAY MODES	Fraction (Γ_i/Γ)	p (MeV/c)
$N\overline{K}$	37–43%	508
$\Lambda\pi$	14–20%	525
$\Sigma\pi$	2–5%	474
$\Sigma(1385)\pi$	8–12%	324
$\Lambda(1520)\pi$	17–23%	198

$\Sigma(1915)\ F_{15}$

$I(J^P) = 1(\frac{5}{2}^+)$

Mass $m = 1900$ to 1935 (≈ 1915) MeV
Full width $\Gamma = 80$ to 160 (≈ 120) MeV
$p_{beam} = 1.26$ GeV/c $\quad 4\pi\lambda\!\!\!^-{}^2 = 12.8$ mb

Σ(1915) DECAY MODES	Fraction (Γ_i/Γ)	p (MeV/c)
$N\overline{K}$	5–15 %	618
$\Lambda\pi$	seen	622
$\Sigma\pi$	seen	577
$\Sigma(1385)\pi$	<5 %	440

$\Sigma(1940)\ D_{13}$

$I(J^P) = 1(\frac{3}{2}^-)$

Mass $m = 1900$ to 1950 (≈ 1940) MeV
Full width $\Gamma = 150$ to 300 (≈ 220) MeV
$p_{beam} = 1.32$ GeV/c $\quad 4\pi\lambda\!\!\!^-{}^2 = 12.1$ mb

Σ(1940) DECAY MODES	Fraction (Γ_i/Γ)	p (MeV/c)
$N\overline{K}$	<20 %	637
$\Lambda\pi$	seen	639
$\Sigma\pi$	seen	594
$\Sigma(1385)\pi$	seen	460
$\Lambda(1520)\pi$	seen	354
$\Delta(1232)\overline{K}$	seen	410
$N\overline{K}^*(892)$	seen	320

$\Sigma(2030)\ F_{17}$

$I(J^P) = 1(\frac{7}{2}^+)$

Mass $m = 2025$ to 2040 (≈ 2030) MeV
Full width $\Gamma = 150$ to 200 (≈ 180) MeV
$p_{beam} = 1.52$ GeV/c $\quad 4\pi\lambda\!\!\!^-{}^2 = 9.93$ mb

Σ(2030) DECAY MODES	Fraction (Γ_i/Γ)	p (MeV/c)
$N\overline{K}$	17–23 %	702
$\Lambda\pi$	17–23 %	700
$\Sigma\pi$	5–10 %	657
ΞK	<2 %	412
$\Sigma(1385)\pi$	5–15 %	529
$\Lambda(1520)\pi$	10–20 %	430
$\Delta(1232)\overline{K}$	10–20 %	498
$N\overline{K}^*(892)$	<5 %	438

$\Sigma(2250)$

$I(J^P) = 1(?^?)$

Mass $m = 2210$ to 2280 (≈ 2250) MeV
Full width $\Gamma = 60$ to 150 (≈ 100) MeV
$p_{beam} = 2.04$ GeV/c $\quad 4\pi\lambda\!\!\!^-{}^2 = 6.76$ mb

Σ(2250) DECAY MODES	Fraction (Γ_i/Γ)	p (MeV/c)
$N\overline{K}$	<10 %	851
$\Lambda\pi$	seen	842
$\Sigma\pi$	seen	803

Baryon Summary Table

Ξ BARYONS
(S = −2, I = 1/2)

$\Xi^0 = uss$, $\Xi^- = dss$

Ξ^0

$I(J^P) = \frac{1}{2}(\frac{1}{2}^+)$

P is not yet measured; + is the quark model prediction.

Mass $m = 1314.9 \pm 0.6$ MeV

$m_{\Xi^-} - m_{\Xi^0} = 6.4 \pm 0.6$ MeV

Mean life $\tau = (2.90 \pm 0.09) \times 10^{-10}$ s

$c\tau = 8.71$ cm

Magnetic moment $\mu = -1.250 \pm 0.014\ \mu_N$

Decay parameters

$\Lambda\pi^0$	$\alpha = -0.411 \pm 0.022$ (S = 2.1)
"	$\phi = (21 \pm 12)^\circ$
"	$\gamma = 0.85$ [g]
"	$\Delta = (218^{+12}_{-19})^\circ$ [g]
$\Lambda\gamma$	$\alpha = 0.4 \pm 0.4$
$\Sigma^0\gamma$	$\alpha = 0.20 \pm 0.32$

Ξ^0 DECAY MODES		Fraction (Γ_i/Γ)	Confidence level	p (MeV/c)
$\Lambda\pi^0$		(99.54 ± 0.05) %		135
$\Lambda\gamma$		$(1.06 \pm 0.16) \times 10^{-3}$		184
$\Sigma^0\gamma$		$(3.5 \pm 0.4) \times 10^{-3}$		117
$\Sigma^+ e^- \overline{\nu}_e$		$< 1.1 \times 10^{-3}$	90%	120
$\Sigma^+ \mu^- \overline{\nu}_\mu$		$< 1.1 \times 10^{-3}$	90%	64
$\Delta S = \Delta Q$ (SQ) violating modes or $\Delta S = 2$ forbidden (S2) modes				
$\Sigma^- e^+ \nu_e$	SQ	$< 9 \times 10^{-4}$	90%	112
$\Sigma^- \mu^+ \nu_\mu$	SQ	$< 9 \times 10^{-4}$	90%	49
$p\pi^-$	S2	$< 4 \times 10^{-5}$	90%	299
$p e^- \overline{\nu}_e$	S2	$< 1.3 \times 10^{-3}$		323
$p \mu^- \overline{\nu}_\mu$	S2	$< 1.3 \times 10^{-3}$		309

Ξ^-

$I(J^P) = \frac{1}{2}(\frac{1}{2}^+)$

P is not yet measured; + is the quark model prediction.

Mass $m = 1321.32 \pm 0.13$ MeV

Mean life $\tau = (1.639 \pm 0.015) \times 10^{-10}$ s

$c\tau = 4.91$ cm

Magnetic moment $\mu = -0.6507 \pm 0.0025\ \mu_N$

Decay parameters

$\Lambda\pi^-$	$\alpha = -0.456 \pm 0.014$ (S = 1.8)
"	$\phi = (4 \pm 4)^\circ$
"	$\gamma = 0.89$ [g]
"	$\Delta = (188 \pm 8)^\circ$ [g]
$\Lambda e^- \overline{\nu}_e$	$g_A/g_V = -0.25 \pm 0.05$ [e]

Ξ^- DECAY MODES		Fraction (Γ_i/Γ)	Confidence level	p (MeV/c)
$\Lambda\pi^-$		(99.887 ± 0.035) %		139
$\Sigma^-\gamma$		$(1.27 \pm 0.23) \times 10^{-4}$		118
$\Lambda e^- \overline{\nu}_e$		$(5.63 \pm 0.31) \times 10^{-4}$		190
$\Lambda \mu^- \overline{\nu}_\mu$		$(3.5 {}^{+3.5}_{-2.2}) \times 10^{-4}$		163
$\Sigma^0 e^- \overline{\nu}_e$		$(8.7 \pm 1.7) \times 10^{-5}$		122
$\Sigma^0 \mu^- \overline{\nu}_\mu$		$< 8 \times 10^{-4}$	90%	70
$\Xi^0 e^- \overline{\nu}_e$		$< 2.3 \times 10^{-3}$	90%	6
$\Delta S = 2$ forbidden (S2) modes				
$n\pi^-$	S2	$< 1.9 \times 10^{-5}$	90%	303
$n e^- \overline{\nu}_e$	S2	$< 3.2 \times 10^{-3}$	90%	327
$n \mu^- \overline{\nu}_\mu$	S2	< 1.5 %	90%	314
$p\pi^-\pi^-$	S2	$< 4 \times 10^{-4}$	90%	223
$p\pi^- e^- \overline{\nu}_e$	S2	$< 4 \times 10^{-4}$	90%	304
$p\pi^- \mu^- \overline{\nu}_\mu$	S2	$< 4 \times 10^{-4}$	90%	250
$p\mu^-\mu^-$	L	$< 4 \times 10^{-4}$	90%	–

Ξ(1530) P_{13}

$I(J^P) = \frac{1}{2}(\frac{3}{2}^+)$

$\Xi(1530)^0$ mass $m = 1531.80 \pm 0.32$ MeV (S = 1.3)

$\Xi(1530)^-$ mass $m = 1535.0 \pm 0.6$ MeV

$\Xi(1530)^0$ full width $\Gamma = 9.1 \pm 0.5$ MeV

$\Xi(1530)^-$ full width $\Gamma = 9.9^{+1.7}_{-1.9}$ MeV

Ξ(1530) DECAY MODES	Fraction (Γ_i/Γ)	Confidence level	p (MeV/c)
$\Xi\pi$	100 %		152
$\Xi\gamma$	<4 %	90%	200

Ξ(1690)

$I(J^P) = \frac{1}{2}(?^?)$

Mass $m = 1690 \pm 10$ MeV [i]

Full width $\Gamma < 50$ MeV

Ξ(1690) DECAY MODES	Fraction (Γ_i/Γ)	p (MeV/c)
$\Lambda\overline{K}$	seen	240
$\Sigma\overline{K}$	seen	51
$\Xi^-\pi^+\pi^-$	possibly seen	214

Ξ(1820) D_{13}

$I(J^P) = \frac{1}{2}(\frac{3}{2}^-)$

Mass $m = 1823 \pm 5$ MeV [i]

Full width $\Gamma = 24^{+15}_{-10}$ MeV [i]

Ξ(1820) DECAY MODES	Fraction (Γ_i/Γ)	p (MeV/c)
$\Lambda\overline{K}$	large	400
$\Sigma\overline{K}$	small	320
$\Xi\pi$	small	413
$\Xi(1530)\pi$	small	234

Ξ(1950)

$I(J^P) = \frac{1}{2}(?^?)$

Mass $m = 1950 \pm 15$ MeV [i]

Full width $\Gamma = 60 \pm 20$ MeV [i]

Ξ(1950) DECAY MODES	Fraction (Γ_i/Γ)	p (MeV/c)
$\Lambda\overline{K}$	seen	522
$\Sigma\overline{K}$	possibly seen	460
$\Xi\pi$	seen	518

Ξ(2030)

$I(J^P) = \frac{1}{2}(\geq \frac{5}{2}^?)$

Mass $m = 2025 \pm 5$ MeV [i]

Full width $\Gamma = 20^{+15}_{-5}$ MeV [i]

Ξ(2030) DECAY MODES	Fraction (Γ_i/Γ)	p (MeV/c)
$\Lambda\overline{K}$	~ 20 %	589
$\Sigma\overline{K}$	~ 80 %	533
$\Xi\pi$	small	573
$\Xi(1530)\pi$	small	421
$\Lambda\overline{K}\pi$	small	501
$\Sigma\overline{K}\pi$	small	430

Ω BARYONS $(S=-3, I=0)$

$\Omega^- = sss$

Ω^-

$I(J^P) = 0(\frac{3}{2}^+)$

J^P is not yet measured; $\frac{3}{2}^+$ is the quark model prediction.

Mass $m = 1672.45 \pm 0.29$ MeV

Mean life $\tau = (0.822 \pm 0.012) \times 10^{-10}$ s

$c\tau = 2.46$ cm

Magnetic moment $\mu = -1.94 \pm 0.22\ \mu_N$

Decay parameters

ΛK^- $\alpha = -0.026 \pm 0.026$

$\Xi^0 \pi^-$ $\alpha = 0.09 \pm 0.14$

$\Xi^- \pi^0$ $\alpha = 0.05 \pm 0.21$

Ω^- DECAY MODES		Fraction (Γ_i/Γ)	Confidence level	p (MeV/c)
ΛK^-		(67.8 ± 0.7) %		211
$\Xi^0 \pi^-$		(23.6 ± 0.7) %		294
$\Xi^- \pi^0$		(8.6 ± 0.4) %		290
$\Xi^- \pi^+ \pi^-$		$(4.3^{+3.4}_{-1.3}) \times 10^{-4}$		190
$\Xi(1530)^0 \pi^-$		$(6.4^{+5.1}_{-2.0}) \times 10^{-4}$		17
$\Xi^0 e^- \overline{\nu}_e$		$(5.6 \pm 2.8) \times 10^{-3}$		319
$\Xi^- \gamma$		$< 2.2 \times 10^{-3}$	90%	314
$\Delta S = 2$ forbidden ($S2$) modes				
$\Lambda \pi^-$	$S2$	$< 1.9 \times 10^{-4}$	90%	449

$\Omega(2250)^-$

$I(J^P) = 0(?^?)$

Mass $m = 2252 \pm 9$ MeV

Full width $\Gamma = 55 \pm 18$ MeV

$\Omega(2250)^-$ DECAY MODES	Fraction (Γ_i/Γ)	p (MeV/c)
$\Xi^- \pi^+ K^-$	seen	531
$\Xi(1530)^0 K^-$	seen	437

CHARMED BARYONS $(C=+1)$

$\Lambda_c^+ = udc,\ \Sigma_c^{++} = uuc,\ \Sigma_c^+ = udc,\ \Sigma_c^0 = ddc,$

$\Xi_c^+ = usc,\ \Xi_c^0 = dsc,\ \Omega_c^0 = ssc$

Λ_c^+

$I(J^P) = 0(\frac{1}{2}^+)$

J not confirmed; $\frac{1}{2}$ is the quark model prediction.

Mass $m = 2285.1 \pm 0.6$ MeV

Mean life $\tau = (0.200^{+0.011}_{-0.010}) \times 10^{-12}$ s

$c\tau = 60.0\ \mu$m

Decay asymmetry parameters

$\Lambda \pi^+$ $\alpha = -1.03 \pm 0.29$

$\Lambda e^+ \nu_e$ $\alpha = -0.89^{+0.19}_{-0.12}$

Λ_c^+ DECAY MODES		Fraction (Γ_i/Γ)	Scale factor/ Confidence level	p (MeV/c)
Hadronic modes with a p and one K				
$p\overline{K}^0$		(2.1 ± 0.4) %		872
$pK^- \pi^+$		(4.4 ± 0.6) %		822
$p\overline{K}^*(892)^0$	[k]	(1.6 ± 0.4) %		681
$\Delta(1232)^{++} K^-$		$(7 \pm 4) \times 10^{-3}$		709
$\Lambda(1520) \pi^+$	[k]	$(3.9^{+2.0}_{-1.7}) \times 10^{-3}$		626
$pK^- \pi^+$ nonresonant		$(2.4^{+0.5}_{-0.6})$ %		822
$p\overline{K}^0 \pi^+ \pi^-$		(2.4 ± 0.8) %	S=1.3	753
$pK^- \pi^+ \pi^0$		seen		758
$pK^*(892)^- \pi^+$		seen		579
$p(K^- \pi^+)_{\text{nonresonant}} \pi^0$		(3.2 ± 0.7) %		758
$\Delta(1232) \overline{K}^*(892)$		seen		417
$pK^- \pi^+ \pi^+ \pi^-$		$(10 \pm 7) \times 10^{-4}$		670
$pK^- \pi^+ \pi^0 \pi^0$		$(7.0 \pm 3.5) \times 10^{-3}$		676
$pK^- \pi^+ \pi^0 \pi^0 \pi^0$		$(4.4 \pm 2.8) \times 10^{-3}$		573
Hadronic modes with a p and zero or two K's				
$p\pi^+ \pi^-$		$(3.0 \pm 1.6) \times 10^{-3}$		926
$pf_0(980)$	[k]	$(2.4 \pm 1.6) \times 10^{-3}$		621
$p\pi^+ \pi^+ \pi^- \pi^-$		$(1.6 \pm 1.0) \times 10^{-3}$		851
$pK^+ K^-$		$(3.0 \pm 1.1) \times 10^{-3}$		615
$p\phi$	[k]	$< 1.7 \times 10^{-3}$	CL=90%	589
Hadronic modes with a hyperon				
$\Lambda \pi^+$		$(7.9 \pm 1.8) \times 10^{-3}$		863
$\Lambda \pi^+ \pi^0$		(3.2 ± 0.9) %		843
$\Lambda \rho^0$		< 4 %	CL=95%	639
$\Lambda \pi^+ \pi^+ \pi^-$		(2.7 ± 0.6) %		806
$\Sigma^0 \pi^+$		$(8.7 \pm 2.0) \times 10^{-3}$		824
$\Sigma^0 \pi^+ \pi^0$		(1.6 ± 0.6) %		802
$\Sigma^0 \pi^+ \pi^+ \pi^-$		$(9.2 \pm 3.3) \times 10^{-3}$		762
$\Sigma^+ \pi^0$		$(8.7 \pm 2.2) \times 10^{-3}$		826
$\Sigma^+ \pi^+ \pi^-$		(3.0 ± 0.6) %		803
$\Sigma^+ \rho^0$		< 1.2 %	CL=95%	579
$\Sigma^- \pi^+ \pi^+$		(1.6 ± 0.6) %		798
$\Sigma^+ \pi^+ \pi^- \pi^0$				569
$\Sigma^+ \omega$	[k]	(2.4 ± 0.7) %		569
$\Sigma^+ \pi^+ \pi^+ \pi^- \pi^-$		$(2.6^{+3.5}_{-1.8}) \times 10^{-3}$		707
$\Sigma^+ K^+ K^-$		$(3.1 \pm 0.8) \times 10^{-3}$		346
$\Sigma^+ \phi$	[k]	$(3.0 \pm 1.3) \times 10^{-3}$		292
$\Sigma^+ K^+ \pi^-$		$(5.7^{+5.3}_{-3.1}) \times 10^{-3}$		668
$\Xi^0 K^+$		$(3.4 \pm 0.9) \times 10^{-3}$		652
$\Xi^- K^+ \pi^+$		$(3.8 \pm 1.2) \times 10^{-3}$		564
$\Xi(1530)^0 K^+$	[k]	$(2.3 \pm 0.9) \times 10^{-3}$		471
Inclusive modes				
p anything		(50 ± 16) %		–
p anything (no Λ)		(12 ± 19) %		–
n anything		(50 ± 16) %		–
n anything (no Λ)		(29 ± 17) %		–
Λ anything		(35 ± 11) %	S=1.4	–
$\Sigma^\pm$ anything	[l]	(10 ± 5) %		–
e^+ anything		(4.5 ± 1.7) %		–
pe^+ anything		(1.8 ± 0.9) %		–
Λe^+ anything		(1.4 ± 0.5) %		–
$\Lambda \mu^+$ anything		(1.5 ± 0.9) %		–

$\Lambda_c(2625)^+$

$I = 0$

Mass $m = 2625.6 \pm 0.8$ MeV

$m - m_{\Lambda_c^+} = 340.6 \pm 0.6$ MeV

Full width $\Gamma < 3.2$ MeV, CL = 90%

$\Lambda_c(2625)^+$ DECAY MODES	Fraction (Γ_i/Γ)	p (MeV/c)
$\Lambda_c^+ \pi^+ \pi^-$	seen	182
$\Sigma_c(2455)^{++} \pi^-$ + $\Sigma_c(2455)^0 \pi^+$	seen	99
$\Lambda_c^+ \pi^+ \pi^-$ nonresonant	seen	182

$\Sigma_c(2455)$

$I(J^P) = 1(\frac{1}{2}^+)$

J^P not confirmed; $\frac{1}{2}^+$ is the quark model prediction.

$\Sigma_c(2455)^{++}$ mass $m = 2453.1 \pm 0.6$ MeV

$\Sigma_c(2455)^+$ mass $m = 2453.8 \pm 0.9$ MeV

$\Sigma_c(2455)^0$ mass $m = 2452.4 \pm 0.7$ MeV (S = 1.1)

$\Sigma_c(2455)$ DECAY MODES	Fraction (Γ_i/Γ)	p (MeV/c)
$\Lambda_c^+ \pi$	100 %	91

Baryon Summary Table

Ξ_c^+

$I(J^P) = \frac{1}{2}(\frac{1}{2}^+)$

$I(J^P)$ not confirmed; $\frac{1}{2}(\frac{1}{2}^+)$ is the quark model prediction.

Mass $m = 2465.1 \pm 1.6$ MeV
Mean life $\tau = (0.35^{+0.07}_{-0.04}) \times 10^{-12}$ s
$c\tau = 106$ μm

Ξ_c^+ DECAY MODES	Fraction (Γ_i/Γ)	p (MeV/c)
$\Lambda K^- \pi^+ \pi^+$	seen	785
$\Sigma^+ K^- \pi^+$	seen	808
$\Sigma^0 K^- \pi^+ \pi^+$	seen	733
$\Xi^- \pi^+ \pi^+$	seen	850

Ξ_c^0

$I(J^P) = \frac{1}{2}(\frac{1}{2}^+)$

$I(J^P)$ not confirmed; $\frac{1}{2}(\frac{1}{2}^+)$ is the quark model prediction.

Mass $m = 2470.3 \pm 1.8$ MeV (S = 1.3)
$m_{\Xi_c^0} - m_{\Xi_c^+} = 5.2 \pm 2.2$ MeV (S = 1.1)
Mean life $\tau = (0.098^{+0.023}_{-0.015}) \times 10^{-12}$ s
$c\tau = 29$ μm

A few branching *ratios* but no absolute branching *fractions* have been measured.

Ξ_c^0 DECAY MODES	Fraction (Γ_i/Γ)	p (MeV/c)
$\Xi^- \ell^+$ anything	[m] seen	–
$\Xi^- \pi^+$	seen	875
$\Xi^- \pi^+ \pi^+ \pi^-$	seen	816
$p K^- \overline{K}^*(892)^0$	seen	406
$\Omega^- K^+$	seen	522

BOTTOM (BEAUTY) BARYON ($B = -1$)

$\Lambda_b^0 = udb$

Λ_b^0

$I(J^P) = 0(\frac{1}{2}^+)$

$I(J^P)$ not yet measured; $0(\frac{1}{2}^+)$ is the quark model prediction.

Mass $m = 5641 \pm 50$ MeV
Mean life $\tau = (1.07^{+0.19}_{-0.16}) \times 10^{-12}$ s

Λ_b^0 DECAY MODES	Fraction (Γ_i/Γ)	p (MeV/c)
$J/\psi(1S)\Lambda$	seen	1756
$p D^0 \pi^-$	seen	2383
$\Lambda_c^+ \pi^+ \pi^- \pi^-$	seen	2336
$\Lambda \ell^- X$	seen	–
$\Lambda_c^+ \ell^- X$	seen	–

NOTES

This Summary Table only includes established baryons. The Full Listings include evidence for other baryons. The masses, widths, and branching fractions for the resonances in this Table are Breit-Wigner parameters. The Full Listings also give, where available, pole parameters. See, in particular, the *Note on N and Δ Resonances.*

For most of the resonances, the parameters come from various partial-wave analyses of more or less the same sets of data, and it is not appropriate to treat the results of the analyses as independent or to average them together. Furthermore, the systematic errors on the results are not well understood. Thus, we usually only give ranges for the parameters. We then also give a best guess for the mass (as part of the name of the resonance) and for the width. The *Note on N and Δ Resonances* and the *Note on Λ and Σ Resonances* in the Full Listings review the partial-wave analyses.

When a quantity has "(S = ...)" to its right, the error on the quantity has been enlarged by the "scale factor" S, defined as $S = \sqrt{\chi^2/(N-1)}$, where N is the number of measurements used in calculating the quantity. We do this when $S > 1$, which often indicates that the measurements are inconsistent. When $S > 1.25$, we also show in the Full Listings an ideogram of the measurements. For more about S, see the Introduction.

A decay momentum p is given for each decay mode. For a 2-body decay, p is the momentum of each decay product in the rest frame of the decaying particle. For a 3-or-more-body decay, p is the largest momentum any of the products can have in this frame. For any resonance, the *nominal* mass is used in calculating p. A dagger ("†") in this column indicates that the mode is forbidden when the nominal masses of resonances are used, but is in fact allowed due to the nonzero widths of the resonances.

[a] The masses of the p and n are most precisely known in u (unified atomic mass units). The conversion factor to MeV, 1 u = 931.49432 ± 0.00028 MeV, is less well known than are the masses in u.

[b] The limit is from neutrality-of-matter experiments; it assumes $q_n = q_p + q_e$. See also the charge of the neutron.

[c] The first limit is geochemical and independent of decay mode. The second limit assumes the dominant decay modes are among those investigated. For antiprotons the best limit, inferred from the observation of cosmic ray $\overline{p}$'s is $\tau_{\overline{p}} > 10^7$ yr, the cosmic-ray storage time, but this limit depends on a number of assumptions. The best direct observation of stored antiprotons gives $\tau_{\overline{p}}/B(\overline{p} \to e^- \gamma) > 1848$ yr.

[d] There is some controversy about whether nuclear physics and model dependence complicate the analysis for bound neutrons (from which the best limit comes). For reactor experiments with free neutrons, the best limit is $> 10^7$ s.

[e] The parameters g_A, g_V, and g_{WM} for semileptonic modes are defined by $\overline{B}_f[\gamma_\lambda(g_V + g_A\gamma_5) + i(g_{WM}/m_{B_i})\,\sigma_{\lambda\nu}\,q^\nu]B_i$, and ϕ_{AV} is defined by $g_A/g_V = |g_A/g_V|e^{i\phi_{AV}}$. See the "Note on Baryon Decay Parameters" in the neutron Full Listings.

[f] Time-reversal invariance requires this to be 0° or 180°.

[g] The decay parameters γ and Δ are calculated from α and ϕ using

$$\gamma = \sqrt{1-\alpha^2}\cos\phi\,, \qquad \tan\Delta = -\frac{1}{\alpha}\sqrt{1-\alpha^2}\sin\phi\,.$$

See the "Note on Baryon Decay Parameters" in the neutron Full Listings.

[h] See the Full Listings for the pion momentum range used in this measurement.

[i] The error given here is only an educated guess. It is larger than the error on the weighted average of the published values.

[j] A theoretical value using QED.

[k] The branching fraction includes all the decay modes of the final-state resonance.

[l] The value is for the sum of the charge states indicated.

[m] ℓ indicates e or μ mode, not sum over modes.

SEARCHES FOR FREE QUARKS, MONOPOLES, SUPERSYMMETRY, COMPOSITENESS, etc.

Free Quark Searches

All searches since 1977 have had negative results.

Magnetic Monopole Searches

Isolated candidate events have not been confirmed. Most experiments obtain negative results.

Supersymmetric Particle Searches

Limits are based on the Minimal Supersymmetric Standard Model.

Assumptions include: 1) $\widetilde{\chi}_1^0$ (or $\widetilde{\gamma}$) is lightest supersymmetric particle; 2) R-parity is conserved; 3) $m_{\widetilde{f}_L} = m_{\widetilde{f}_R}$, and all scalar quarks (except $\widetilde{t}_L$ and $\widetilde{t}_R$) are degenerate in mass.

See the Full Listings for a Note giving details of supersymmetry.

$\widetilde{\chi}_i^0$ — neutralinos (mixtures of $\widetilde{\gamma}$, $\widetilde{Z}^0$, and $\widetilde{H}_i^0$)

Mass $m_{\widetilde{\gamma}}$ > 15 GeV, CL = 90% [if $m_{\widetilde{f}}$ = 100 GeV (from cosmology)]
Mass $m_{\widetilde{\chi}_1^0}$ > 18 GeV, CL = 90% [GUT relations assumed]
Mass $m_{\widetilde{\chi}_2^0}$ > 45 GeV, CL = 95% [GUT relations assumed]
Mass $m_{\widetilde{\chi}_3^0}$ > 70 GeV, CL = 95% [GUT relations assumed]
Mass $m_{\widetilde{\chi}_4^0}$ > 108 GeV, CL = 95% [GUT relations assumed]

$\widetilde{\chi}_i^{\pm}$ — charginos (mixtures of $\widetilde{W}^{\pm}$ and $\widetilde{H}_i^{\pm}$)

Mass $m_{\widetilde{\chi}_1^{\pm}}$ > 45 GeV, CL = 95% [all $m_{\widetilde{\chi}_1^0}$]
Mass $m_{\widetilde{\chi}_2^{\pm}}$ > 99 GeV, CL = 95% [GUT relations assumed]

$\widetilde{\nu}$ — scalar neutrino (sneutrino)

Mass m > 37.1 GeV, CL = 95% [one flavor]
Mass m > 41.8 GeV, CL = 95% [three degenerate flavors]

$\widetilde{e}$ — scalar electron (selectron)

Mass m > 65 GeV, CL = 95% [if $m_{\widetilde{\gamma}} = 0$]
Mass m > 50 GeV, CL = 95% [if $m_{\widetilde{\gamma}}$ < 5 GeV]
Mass m > 45 GeV, CL = 95% [if $m_{\widetilde{\chi}_1^0}$ < 41 GeV]

$\widetilde{\mu}$ — scalar muon (smuon)

Mass m > 45 GeV, CL = 95% [if $m_{\widetilde{\chi}_1^0}$ < 41 GeV]

$\widetilde{\tau}$ — scalar tau (stau)

Mass m > 45 GeV, CL = 95% [if $m_{\widetilde{\chi}_1^0}$ < 38 GeV]

$\widetilde{q}$ — scalar quark (squark)

These limits include the effects of cascade decay, for a particular choice of parameters, μ=−250 GeV, tanβ=2. The limits are weakly sensitive to these parameters over much of parameter space. Limits assume GUT relations between gaugino masses and the gauge coupling; in particular that for $|\mu|$ not small, $m_{\widetilde{\chi}_1^0} \approx m_{\widetilde{g}}/6$.

Mass m > 90 GeV, CL = 90% [any $m_{\widetilde{g}}$ <410 GeV]
Mass m > 218 GeV, CL = 90% [if $m_{\widetilde{g}} = m_{\widetilde{q}}$]

$\widetilde{g}$ — gluino

There is some controversy about a low-mass window ($1 \lesssim m_{\widetilde{g}} \lesssim 4$ GeV). Several experiments cast doubt on the existence of this window.

These limits include the effects of cascade decay, for a particular choice of parameters, μ=−250 GeV, tanβ=2. The limits are weakly sensitive to these parameters over much of parameter space. Limits assume GUT relations between gaugino masses and the gauge coupling; in particular that for $|\mu|$ not small, $m_{\widetilde{\chi}_1^0} \approx m_{\widetilde{g}}/7$.

Mass m > 100 GeV, CL = 90% [any $m_{\widetilde{q}}$]
Mass m > 218 GeV, CL = 90% [if $m_{\widetilde{q}} \leq m_{\widetilde{g}}$]

Searches for Quark and Lepton Compositeness

Scale Limits Λ for Contact Interactions (the lowest dimensional interactions with four fermions)

If the Lagrangian has the form

$$\pm \frac{g^2}{2\Lambda^2} \overline{\psi}_L \gamma_\mu \psi_L \overline{\psi}_L \gamma^\mu \psi_L$$

(with $g^2/4\pi$ set equal to 1), then we define $\Lambda \equiv \Lambda_{LL}^{\pm}$. For the full definitions and for other forms, see the Note in the Listings on Searches for Quark and Lepton Compositeness in the full *Review* and the original literature.

$\Lambda_{LL}^+(eeee)$ > 1.6 TeV, CL = 95%
$\Lambda_{LL}^-(eeee)$ > 3.6 TeV, CL = 95%
$\Lambda_{LL}^+(ee\mu\mu)$ > 2.6 TeV, CL = 95%
$\Lambda_{LL}^-(ee\mu\mu)$ > 1.9 TeV, CL = 95%
$\Lambda_{LL}^+(ee\tau\tau)$ > 1.9 TeV, CL = 95%
$\Lambda_{LL}^-(ee\tau\tau)$ > 2.9 TeV, CL = 95%
$\Lambda_{LL}^+(\ell\ell\ell\ell)$ > 3.5 TeV, CL = 95%
$\Lambda_{LL}^-(\ell\ell\ell\ell)$ > 2.8 TeV, CL = 95%
$\Lambda_{LL}^+(eeqq)$ > 1.7 TeV, CL = 95%
$\Lambda_{LL}^-(eeqq)$ > 2.2 TeV, CL = 95%
$\Lambda_{LL}^+(\mu\mu qq)$ > 1.4 TeV, CL = 95%
$\Lambda_{LL}^-(\mu\mu qq)$ > 1.6 TeV, CL = 95%
$\Lambda_{LR}^{\pm}(\nu_\mu \nu_e \mu e)$ > 3.1 TeV, CL = 90%
$\Lambda_{LL}^{\pm}(qqqq)$ > 1.4 TeV, CL = 95%

Excited Leptons

The limits from $\ell^{*+}\ell^{*-}$ do not depend on λ (where λ is the $\ell\ell^*$ transition coupling). The λ-dependent limits assume chiral coupling, except for the third limit for e^* which is for nonchiral coupling. For chiral coupling, this limit corresponds to $\lambda_\gamma = \sqrt{2}$.

$e^{*\pm}$ — excited electron

Mass m > 46.1 GeV, CL = 95% (from $e^{*+}e^{*-}$)
Mass m > 91 GeV, CL = 95% (if λ_Z > 1)
Mass m > 127 GeV, CL = 95% (if λ_γ = 1)

$\mu^{*\pm}$ — excited muon

Mass m > 46.1 GeV, CL = 95% (from $\mu^{*+}\mu^{*-}$)
Mass m > 91 GeV, CL = 95% (if λ_Z > 1)

$\tau^{*\pm}$ — excited tau

Mass m > 46.0 GeV, CL = 95% (from $\tau^{*+}\tau^{*-}$)
Mass m > 90 GeV, CL = 95% (if λ_Z > 0.18)

ν^* — excited neutrino

Mass m > 47 GeV, CL = 95% (from $\nu^*\overline{\nu}^*$)
Mass m > 91 GeV, CL = 95% (if λ_Z > 1)

q^* — excited quark

Mass m > 45.6 GeV, CL = 95% (from $q^*\overline{q}^*$)
Mass m > 88 GeV, CL = 95% (if λ_Z > 1)
Mass m > 540 GeV, CL = 95% ($p\overline{p} \to q^*X$)

Color Sextet and Octet Particles

Color Sextet Quarks (q_6)

Mass m > 84 GeV, CL = 95% (Stable q_6)

Color Octet Charged Leptons (ℓ_8)

Mass m > 86 GeV, CL = 95% (Stable ℓ_8)

Color Octet Neutrinos (ν_8)

Mass m > 110 GeV, CL = 90% ($\nu_8 \to \nu g$)

Tests of Conservation Laws

TESTS OF CONSERVATION LAWS

Revised by L. Wolfenstein and T.G. Trippe, June 1994.

In keeping with the current interest in tests of conservation laws, we collect together a Table of experimental limits on all weak and electromagnetic decays, mass differences, and moments, and on a few reactions, whose observation would violate conservation laws. The Table is given only in the full *Review of Particle Properties*, not in the Particle Physics Booklet. For the benefit of Booklet readers, we include the best limits from the Table in the following text. The Table is in two parts: "Discrete Space-Time Symmetries," *i.e.*, C, P, T, CP, and CPT; and "Number Conservation Laws," *i.e.*, lepton, baryon, hadronic flavor, and charge conservation. The references for these data can be found in the the Full Listings in the *Review*. A discussion of these tests follows.

CPT INVARIANCE

General principles of relativistic field theory require invariance under the combined transformation CPT. The simplest tests of CPT invariance are the equality of the masses and lifetimes of a particle and its antiparticle. The best test comes from the limit on the mass difference between K^0 and $\overline{K}^0$. Any such difference contributes to the CP-violating parameter ϵ. Assuming CPT invariance, ϕ_ϵ, the phase of ϵ should be very close to 44°. (See the "Note on CP Violation in K_L^0 Decay" in the Full Listings.) In contrast, if the entire source of CP violation in K^0 decays were a $K^0 - \overline{K}^0$ mass difference, ϕ_ϵ would be 44° + 90°. It is possible to deduce that [1]

$$m_{\overline{K}^0} - m_{K^0} \approx \frac{2(m_{K_L^0} - m_{K_S^0})\,|\eta|\,(\frac{2}{3}\phi_{+-} + \frac{1}{3}\phi_{00} - \phi_\epsilon)}{\sin\phi_\epsilon} .$$

Using our best values of the CP-violation parameters, we get $|(m_{\overline{K}^0} - m_{K^0})/m_{K^0}| \leq 10^{-18}$ (CL = 90%). Limits can also be placed on specific CPT-violating decay amplitudes. Given the small value of $(1 - |\eta_{00}/\eta_{+-}|)$, the value of $\phi_{00} - \phi_{+-}$ provides a measure of CPT violation in $K_L^0 \to 2\pi$ decay. Results from CERN [1] and Fermilab [2] indicate no CPT-violating effect.

CP AND T INVARIANCE

Given CPT invariance, CP violation and T violation are equivalent. So far the only evidence for CP or T violation comes from the measurements of η_{+-}, η_{00}, and the semileptonic decay charge asymmetry for K_L, *e.g.*, $|\eta_{+-}| = |A(K_L^0 \to \pi^+\pi^-)/A(K_S^0 \to \pi^+\pi^-)| = (2.269 \pm 0.023) \times 10^{-3}$ and $[\Gamma(K_L^0 \to \pi^- e^+\nu) - \Gamma(K_L^0 \to \pi^+ e^- \overline{\nu})]/[\mathrm{sum}] = (0.333 \pm 0.014)\%$. Other searches for CP or T violation divide into (a) those that involve weak interactions or parity violation, and (b) those that involve processes allowed by the strong or electromagnetic interactions. In class (a) the most sensitive is probably the search for an electric dipole moment of the neutron, measured to be $< 1.1 \times 10^{-25}$ e cm (95% CL). A nonzero value requires both P and T violation. Class (b) includes the search for C violation in η decay, believed to be an electromagnetic process, *e.g.*, as measured by $\Gamma(\eta \to \mu^+\mu^-\pi^0)/\Gamma(\eta \to \mathrm{all}) < 5 \times 10^{-6}$, and searches for T violation in a number of nuclear and electromagnetic reactions.

CONSERVATION OF LEPTON NUMBERS

Present experimental evidence and the standard electroweak theory are consistent with the absolute conservation of three separate lepton numbers: electron number L_e, muon number L_μ, and tau number L_τ. Searches for violations are of the following types:

a) $\Delta L = 2$ for one type of lepton. The best limit comes from the search for neutrinoless double beta decay $(Z, A) \to (Z+2, A) + e^- + e^-$. The best laboratory limit is $t_{1/2} > 1.4 \times 10^{24}$ yr (CL=90%) for ^{76}Ge.

b) Conversion of one lepton type to another. For purely leptonic processes, the best limits are on $\mu \to e\gamma$ and $\mu \to 3e$, measured as $\Gamma(\mu \to e\gamma)/\Gamma(\mu \to \mathrm{all}) < 5 \times 10^{-11}$ and $\Gamma(\mu \to 3e)/\Gamma(\mu \to \mathrm{all}) < 1.0 \times 10^{-12}$. For semileptonic processes, the best limit comes from the coherent conversion process in a muonic atom, $\mu^- + (Z, A) \to e^- + (Z, A)$, measured as $\Gamma(\mu^-\mathrm{Ti} \to e^-\mathrm{Ti})/\Gamma(\mu^-\mathrm{Ti} \to \mathrm{all}) < 4 \times 10^{-12}$. Of special interest is the case in which the hadronic flavor also changes, as in $K_L \to e\mu$ and $K^+ \to \pi^+ e^- \mu^+$, measured as $\Gamma(K_L \to e\mu)/\Gamma(K_L \to \mathrm{all}) < 3.3 \times 10^{-11}$ and $\Gamma(K^+ \to \pi^+ e^- \mu^+)/\Gamma(K^+ \to \mathrm{all}) < 2.1 \times 10^{-10}$. Limits on the conversion of τ into e or μ are found in τ decay and are much less stringent than those for $\mu \to e$ conversion, *e.g.*, $\Gamma(\tau \to \mu\gamma)/\Gamma(\tau \to \mathrm{all}) < 4.2 \times 10^{-6}$ and $\Gamma(\tau \to e\gamma)/\Gamma(\tau \to \mathrm{all}) < 1.2 \times 10^{-4}$.

c) Conversion of one type of lepton into another type of antilepton. The case most studied is $\mu^- + (Z, A) \to e^+ + (Z-2, A)$, the strongest limit being $\Gamma(\mu^-\mathrm{Ti} \to e^+\mathrm{Ca})/\Gamma(\mu^-\mathrm{Ti} \to \mathrm{all}) < 9 \times 10^{-11}$.

d) Relation to neutrino mass. If neutrinos have mass, then it is expected even in the standard electroweak theory that the lepton numbers are not separately conserved, as a consequence of lepton mixing analogous to Cabibbo quark mixing. However, in this case lepton-number-violating processes such as $\mu \to e\gamma$ are expected to have extremely small probability. For small neutrino masses, the lepton-number violation would be observed first in neutrino oscillations, which have been the subject of extensive experimental searches. For example, searches for $\overline{\nu}_e$ disappearance, which we label as $\overline{\nu}_e \not\to \overline{\nu}_e$, give measured limits $\Delta(m^2) < 0.0083$ eV^2 for $\sin^2(2\theta) = 1$, and $\sin^2(2\theta) < 0.14$ for large $\Delta(m^2)$, where θ is the neutrino mixing angle. Searches for $\nu_\mu \to \nu_e$ set limits $\Delta(m^2) < 0.09$ eV^2 for $\sin^2(2\theta) = 1$, and $\sin^2(2\theta) < 0.0025$ for large $\Delta(m^2)$. For larger neutrino masses ($\gg 1$ keV), lepton-number violation is searched for by looking for anomalous decays such as $\pi \to e\nu_x$, where ν_x is a massive neutrino. If the $\Delta L = 2$ type of violation occurs, it is expected that neutrinos will have a nonzero mass of the Majorana type.

CONSERVATION OF HADRONIC FLAVORS

In strong and electromagnetic interactions, hadronic flavor is conserved, *i.e.* the conversion of a quark of one flavor (d, u, s, c, b, t) into a quark of another flavor is forbidden. In the Standard Model, the weak interactions violate these conservation laws in a manner described by the Cabibbo-Kobayashi-Maskawa mixing (see the section "Cabibbo-Kobayashi-Maskawa Mixing Matrix"). The way in which these conservation laws are violated is tested as follows:

a) $\Delta S = \Delta Q$ rule. In the semileptonic decay of strange particles, the strangeness change equals the change in charge of the hadrons. Tests come from limits on decay rates such as $\Gamma(\Sigma^+ \to n e^+ \nu)/\Gamma(\Sigma^+ \to \mathrm{all}) < 5 \times 10^{-6}$, and from a detailed analysis of $K_L \to \pi e \nu$, which yields the parameter x, measured to be $(\mathrm{Re}\,x, \mathrm{Im}\,x) = (0.006 \pm 0.018, -0.003 \pm 0.026)$. Corresponding rules are $\Delta C = \Delta Q$ and $\Delta B = \Delta Q$.

b) Change of flavor by two units. In the Standard Model this occurs only in second-order weak interactions. The classic example is $\Delta S = 2$ via $K^0 - \overline{K}^0$ mixing, which is directly measured by $m(K_S) - m(K_L) = (3.510 \pm 0.018) \times 10^{-12}$ MeV. There is now evidence for $B^0 - \overline{B}^0$ mixing ($\Delta B = 2$), with the corresponding mass difference between the eigenstates ($m_{B^0_H} - m_{B^0_L}$) $= (0.71 \pm 0.06)\ \Gamma_B = (3.1 \pm 0.4) \times 10^{-10}$ MeV. No evidence exists for $D^0 - \overline{D}^0$ mixing, which is expected to be much smaller in the Standard Model.

c) Flavor-changing neutral currents. In the Standard Model the neutral-current interactions do not change flavor. The low rate $\Gamma(K_L \to \mu^+\mu^-)/\Gamma(K_L \to \text{all}) = (7.4 \pm 0.4) \times 10^{-9}$ puts limits on such interactions; the nonzero value for this rate is attributed to a combination of the weak and electromagnetic interactions. The best test should come from a limit on $K^+ \to \pi^+\nu\overline{\nu}$, which occurs in the Standard Model only as a second-order weak process with a branching fraction of $(1 \text{ to } 8) \times 10^{-10}$. The current limit is $\Gamma(K^+ \to \pi^+\nu\overline{\nu})/\Gamma(K^+ \to \text{all}) < 5.2 \times 10^{-9}$. Limits for charm-changing or bottom-changing neutral currents are much less stringent: $\Gamma(D^0 \to \mu^+\mu^-)/\Gamma(D^0 \to \text{all}) < 1.1 \times 10^{-5}$ and $\Gamma(B^0 \to \mu^+\mu^-)/\Gamma(B^0 \to \text{all}) < 5.9 \times 10^{-6}$. One cannot isolate flavor-changing neutral current (FCNC) effects in non leptonic decays. For example, the FCNC transition $s \to d + (\overline{u} + u)$ is equivalent to the charged-current transition $s \to u + (\overline{u} + d)$. Tests for FCNC are therefore limited to hadron decays into lepton pairs. Such decays are expected only in second-order in the electroweak coupling in the Standard Model.

References

1. R. Carosi *et al.*, Phys. Lett. **B237**, 303 (1990).
2. M. Karlsson *et al.*, Phys. Rev. Lett. **64**, 2976 (1990); L.K. Gibbons *et al.*, Phys. Rev. Lett. **70**, 1199 (1993).

TESTS OF DISCRETE SPACE-TIME SYMMETRIES

CHARGE CONJUGATION (C)

$\Gamma(\pi^0 \to 3\gamma)/\Gamma_{total}$		$<3.1 \times 10^{-8}$, CL = 90%
$\Gamma((e^+e^-)_{J=0} \to 3\gamma)/\Gamma((e^+e^-)_{J=0} \to 2\gamma)$	[a]	$<1 \times 10^{-5}$, CL = 90%
$\Gamma((e^+e^-)_{J=1} \to 4\gamma)/\Gamma((e^+e^-)_{J=1} \to 3\gamma)$	[a]	$<1 \times 10^{-5}$, CL = 90%
η C-nonconserving decay parameters		
$\pi^+\pi^-\pi^0$ left-right asymmetry parameter		$(0.09 \pm 0.17) \times 10^{-2}$
$\pi^+\pi^-\pi^0$ sextant asymmetry parameter		$(0.18 \pm 0.16) \times 10^{-2}$
$\pi^+\pi^-\pi^0$ quadrant asymmetry parameter		$(-0.17 \pm 0.17) \times 10^{-2}$
$\pi^+\pi^-\gamma$ left-right asymmetry parameter		$(0.9 \pm 0.4) \times 10^{-2}$
$\pi^+\pi^-\gamma$ parameter β (D-wave)		0.05 ± 0.06 (S = 1.5)
$\Gamma(\eta \to 3\gamma)/\Gamma_{total}$		$<5 \times 10^{-4}$, CL = 95%
$\Gamma(\eta \to \pi^0 e^+ e^-)/\Gamma_{total}$	[b]	$<4 \times 10^{-5}$, CL = 90%
$\Gamma(\eta \to \pi^0 \mu^+ \mu^-)/\Gamma_{total}$	[b]	$<5 \times 10^{-6}$, CL = 90%

PARITY (P)

e electric dipole moment	$(-0.3 \pm 0.8) \times 10^{-26}$ $e$cm
μ electric dipole moment	$(3.7 \pm 3.4) \times 10^{-19}$ $e$cm
τ electric dipole moment	$<5 \times 10^{-17}$ $e$cm, CL = 95%
$\Gamma(\eta \to \pi^+\pi^-)/\Gamma_{total}$	$<1.5 \times 10^{-3}$
p electric dipole moment	$(-4 \pm 6) \times 10^{-23}$ $e$cm
n electric dipole moment	$<11 \times 10^{-26}$ $e$cm, CL = 95%
Λ electric dipole moment	$<1.5 \times 10^{-16}$ $e$cm, CL = 95%

Limits are given at the 90% confidence level, while errors are given as ± 1 standard deviation.

TIME REVERSAL (T)

e electric dipole moment		$(-0.3 \pm 0.8) \times 10^{-26}$ $e$cm
μ electric dipole moment		$(3.7 \pm 3.4) \times 10^{-19}$ $e$cm
μ decay parameters		
transverse e^+ polarization normal to plane of μ spin, e^+ momentum		0.007 ± 0.023
α'/A		$(0 \pm 4) \times 10^{-3}$
β'/A		$(2 \pm 6) \times 10^{-3}$
τ electric dipole moment		$<5 \times 10^{-17}$ $e$cm, CL = 95%
Im(ξ) in $K^{\pm}_{\mu 3}$ decay (from transverse μ pol.)		-0.017 ± 0.025
Im(ξ) in $K^0_{\mu 3}$ decay (from transverse μ pol.)		-0.007 ± 0.026
p electric dipole moment		$(-4 \pm 6) \times 10^{-23}$ $e$cm
n electric dipole moment		$<11 \times 10^{-26}$ $e$cm, CL = 95%
$n \to pe^-\nu$ decay parameters		
ϕ_{AV}, phase of g_A relative to g_V	[c]	$(180.07 \pm 0.18)^\circ$
triple correlation coefficient D		$(-0.5 \pm 1.4) \times 10^{-3}$
Λ electric dipole moment		$<1.5 \times 10^{-16}$ $e$cm, CL = 95%
triple correlation coefficient D for $\Sigma^- \to ne^-\overline{\nu}_e$		0.11 ± 0.10

CHARGE CONJUGATION TIMES PARITY (CP)

τ weak dipole moment		$<3.7 \times 10^{-17}$ $e$cm, CL = 95%
$\Gamma(\eta \to \pi^+\pi^-)/\Gamma_{total}$		$<1.5 \times 10^{-3}$
$K^{\pm} \to \pi^{\pm}\pi^+\pi^-$ rate difference/average		$(0.07 \pm 0.12)\%$
$K^{\pm} \to \pi^{\pm}\pi^0\pi^0$ rate difference/average		$(0.0 \pm 0.6)\%$
$K^{\pm} \to \pi^{\pm}\pi^0\gamma$ rate difference/average		$(0.9 \pm 3.3)\%$
$(g_{\tau^+} - g_{\tau^-}) / (g_{\tau^+} + g_{\tau^-})$ for $K^{\pm} \to \pi^{\pm}\pi^+\pi^-$		$(-0.7 \pm 0.5)\%$
CP-violation parameters in K^0_S decay		
$\text{Im}(\eta_{+-0})^2 = \Gamma(K^0_S \to \pi^+\pi^-\pi^0, CP\text{-violating}) / \Gamma(K^0_L \to \pi^+\pi^-\pi^0)$		<0.12, CL = 90%
$\text{Im}(\eta_{000})^2 = \Gamma(K^0_S \to 3\pi^0) / \Gamma(K^0_L \to 3\pi^0)$		<0.1, CL = 90%
charge asymmetry j for $K^0_L \to \pi^+\pi^-\pi^0$		0.0011 ± 0.0008
$\lvert\eta_{+-\gamma}\rvert = \lvert A(K^0_L \to \pi^+\pi^-\gamma)/A(K^0_S \to \pi^+\pi^-\gamma)\rvert$		$(2.15 \pm 0.26 \pm 0.20) \times 10^{-3}$
$\phi_{+-\gamma}$ = phase of $\eta_{+-\gamma}$		$(72 \pm 23 \pm 17)^\circ$
$\lvert\epsilon'_{+-\gamma}\rvert/\epsilon$		<0.3, CL = 90%
$\Gamma(K^0_L \to \pi^0\mu^+\mu^-)/\Gamma_{total}$	[d]	$<5.1 \times 10^{-9}$, CL = 90%
$\Gamma(K^0_L \to \pi^0 e^+ e^-)/\Gamma_{total}$	[d]	$<4.3 \times 10^{-9}$, CL = 90%
$\Gamma(K^0_L \to \pi^0\nu\overline{\nu})/\Gamma_{total}$	[e]	$<2.2 \times 10^{-4}$, CL = 90%
$[\Gamma(D^0 \to K^+K^-) - \Gamma(\overline{D}^0 \to K^+K^-)]/\text{sum}$		<0.45, CL = 90%
$\lvert\text{Re}(\epsilon_{B^0})\rvert$		<0.045
$[\alpha_-(\Lambda) + \alpha_+(\overline{\Lambda})] / [\alpha_-(\Lambda) - \alpha_+(\overline{\Lambda})]$		-0.03 ± 0.06

CHARGE CONJUGATION TIMES PARITY (CP) VIOLATION OBSERVED

charge asymmetry in $K^0_{\ell 3}$ decays		
$\delta(\mu) = [\Gamma(\pi^-\mu^+\nu_\mu) - \Gamma(\pi^+\mu^-\overline{\nu}_\mu)]/\text{sum}$		$(0.304 \pm 0.025)\%$
$\delta(e) = [\Gamma(\pi^- e^+\nu_e) - \Gamma(\pi^+ e^-\overline{\nu}_e)]/\text{sum}$		$(0.333 \pm 0.014)\%$
parameters for $K^0_L \to 2\pi$ decay		
$\lvert\eta_{00}\rvert = \lvert A(K^0_L \to 2\pi^0) / A(K^0_S \to 2\pi^0)\rvert$		$(2.259 \pm 0.023) \times 10^{-3}$ (S = 1.1)
$\lvert\eta_{+-}\rvert = \lvert A(K^0_L \to \pi^+\pi^-) / A(K^0_S \to \pi^+\pi^-)\rvert$		$(2.269 \pm 0.023) \times 10^{-3}$ (S = 1.1)
$\epsilon'/\epsilon \approx \text{Re}(\epsilon'/\epsilon) = (1 - \lvert\eta_{00}/\eta_{+-}\rvert)/3$	[f]	$(1.5 \pm 0.8) \times 10^{-3}$ (S = 1.8)
ϕ_{+-}, phase of η_{+-}		$(44.3 \pm 0.8)^\circ$
ϕ_{00}, phase of η_{00}		$(43.3 \pm 1.3)^\circ$
$\Gamma(K^0_L \to \pi^+\pi^-)/\Gamma_{total}$		$(2.03 \pm 0.04) \times 10^{-3}$ (S = 1.2)
$\Gamma(K^0_L \to \pi^0\pi^0)/\Gamma_{total}$		$(9.14 \pm 0.34) \times 10^{-4}$ (S = 1.8)

Tests of Conservation Laws

CPT INVARIANCE

$(m_{W^+} - m_{W^-}) / m_{\text{average}}$		-0.002 ± 0.007
$(m_{e^+} - m_{e^-}) / m_{\text{average}}$		$<4 \times 10^{-8}$, CL = 90%
$\lvert q_{e^+} + q_{e^-}\rvert/e$		$<4 \times 10^{-8}$
$(g_{e^+} - g_{e^-}) / g_{\text{average}}$		$(-0.5 \pm 2.1) \times 10^{-12}$
$\tau_{\mu^+}/\tau_{\mu^-}$ mean life ratio		1.00002 ± 0.00008
$(\tau_{\mu^+} - \tau_{\mu^-}) / \tau_{\text{average}}$		$(2 \pm 8) \times 10^{-5}$
$(g_{\mu^+} - g_{\mu^-}) / g_{\text{average}}$		$(-2.6 \pm 1.6) \times 10^{-8}$
$(m_{\pi^+} - m_{\pi^-}) / m_{\text{average}}$		$(2 \pm 5) \times 10^{-4}$
$(\tau_{\pi^+} - \tau_{\pi^-}) / \tau_{\text{average}}$		$(6 \pm 7) \times 10^{-4}$
$(m_{K^+} - m_{K^-}) / m_{\text{average}}$		$(-0.6 \pm 1.8) \times 10^{-4}$
$(\tau_{K^+} - \tau_{K^-}) / \tau_{\text{average}}$		$(0.11 \pm 0.09)\%$ (S = 1.2)
$K^\pm \to \mu^\pm \nu_\mu$ rate difference/average		$(-0.5 \pm 0.4)\%$
$K^\pm \to \pi^\pm \pi^0$ rate difference/average	[g]	$(0.8 \pm 1.2)\%$
$\lvert m_{K^0} - m_{\overline{K}^0}\rvert / m_{\text{average}}$	[h]	$<9 \times 10^{-19}$
phase difference $\phi_{00} - \phi_{+-}$		$(-1.0 \pm 1.0)^\circ$
$(m_p - m_{\overline{p}}) / m_{\text{average}}$		$(2 \pm 4) \times 10^{-8}$
$\lvert q_p + q_{\overline{p}}\rvert/e$		$<2 \times 10^{-5}$
$(\mu_p - \lvert\mu_{\overline{p}}\rvert) / \lvert\mu_{\text{average}}\rvert$		$(-2.6 \pm 2.9) \times 10^{-3}$
$(m_n - m_{\overline{n}}) / m_{\text{average}}$		$(9 \pm 5) \times 10^{-5}$
$(m_\Lambda - m_{\overline{\Lambda}}) / m_\Lambda$		$(-1.0 \pm 0.9) \times 10^{-5}$
$(\tau_\Lambda - \tau_{\overline{\Lambda}}) / \tau_{\text{average}}$		0.04 ± 0.09
$(\mu_{\Sigma^+} - \lvert\mu_{\overline{\Sigma}^-}\rvert) / \lvert\mu\rvert_{\text{average}}$		0.014 ± 0.015
$(m_{\Xi^-} - m_{\overline{\Xi}^+}) / m_{\text{average}}$		$(1.1 \pm 2.7) \times 10^{-4}$
$(\tau_{\Xi^-} - \tau_{\overline{\Xi}^+}) / \tau_{\text{average}}$		0.02 ± 0.18
$(m_{\Omega^-} - m_{\overline{\Omega}^+}) / m_{\text{average}}$		$(0 \pm 5) \times 10^{-4}$

TESTS OF NUMBER CONSERVATION LAWS

LEPTON FAMILY NUMBER

Lepton family number conservation means separate conservation of each of L_e, L_μ, L_τ.

$\Gamma(Z \to e^\pm \mu^\mp)/\Gamma_{\text{total}}$	[i]	$<6 \times 10^{-6}$, CL = 95%
$\Gamma(Z \to e^\pm \tau^\mp)/\Gamma_{\text{total}}$	[i]	$<1.3 \times 10^{-5}$, CL = 95%
$\Gamma(Z \to \mu^\pm \tau^\mp)/\Gamma_{\text{total}}$	[i]	$<1.9 \times 10^{-5}$, CL = 95%
limit on $\mu^- \to e^-$ conversion		
$\sigma(\mu^- {}^{32}\text{S} \to e^- {}^{32}\text{S})$ / $\sigma(\mu^- {}^{32}\text{S} \to \nu_\mu {}^{32}\text{P}^*)$		$<7 \times 10^{-11}$, CL = 90%
$\sigma(\mu^- \text{Ti} \to e^- \text{Ti})$ / $\sigma(\mu^- \text{Ti} \to \text{capture})$		$<4.3 \times 10^{-12}$, CL = 90%
limit on muonium $\to$ antimuonium conversion $R_g = G_C / G_F$		<0.13, CL = 90%
$\Gamma(\mu^- \to e^- \nu_e \overline{\nu}_\mu)/\Gamma_{\text{total}}$	[j]	$<1.2 \times 10^{-2}$, CL = 90%
$\Gamma(\mu^- \to e^- \gamma)/\Gamma_{\text{total}}$		$<4.9 \times 10^{-11}$, CL = 90%
$\Gamma(\mu^- \to e^- e^+ e^-)/\Gamma_{\text{total}}$		$<1.0 \times 10^{-12}$, CL = 90%
$\Gamma(\mu^- \to e^- 2\gamma)/\Gamma_{\text{total}}$		$<7.2 \times 10^{-11}$, CL = 90%
$\Gamma(\tau^- \to e^- \gamma)/\Gamma_{\text{total}}$		$<1.2 \times 10^{-4}$, CL = 90%
$\Gamma(\tau^- \to \mu^- \gamma)/\Gamma_{\text{total}}$		$<4.2 \times 10^{-6}$, CL = 90%
$\Gamma(\tau^- \to e^- \pi^0)/\Gamma_{\text{total}}$		$<1.4 \times 10^{-4}$, CL = 90%
$\Gamma(\tau^- \to \mu^- \pi^0)/\Gamma_{\text{total}}$		$<4.4 \times 10^{-5}$, CL = 90%
$\Gamma(\tau^- \to e^- K^0)/\Gamma_{\text{total}}$		$<1.3 \times 10^{-3}$, CL = 90%
$\Gamma(\tau^- \to \mu^- K^0)/\Gamma_{\text{total}}$		$<1.0 \times 10^{-3}$, CL = 90%
$\Gamma(\tau^- \to e^- \eta)/\Gamma_{\text{total}}$		$<6.3 \times 10^{-5}$, CL = 90%
$\Gamma(\tau^- \to \mu^- \eta)/\Gamma_{\text{total}}$		$<7.3 \times 10^{-5}$, CL = 90%
$\Gamma(\tau^- \to e^- \rho^0)/\Gamma_{\text{total}}$		$<1.9 \times 10^{-5}$, CL = 90%
$\Gamma(\tau^- \to \mu^- \rho^0)/\Gamma_{\text{total}}$		$<2.9 \times 10^{-5}$, CL = 90%
$\Gamma(\tau^- \to e^- K^*(892)^0)/\Gamma_{\text{total}}$		$<3.8 \times 10^{-5}$, CL = 90%
$\Gamma(\tau^- \to \mu^- K^*(892)^0)/\Gamma_{\text{total}}$		$<4.5 \times 10^{-5}$, CL = 90%
$\Gamma(\tau^- \to \ell^- \ell^- \ell^+)/\Gamma_{\text{total}}$	[k]	$<3.4 \times 10^{-5}$, CL = 90%
$\Gamma(\tau^- \to e^- e^+ e^-)/\Gamma_{\text{total}}$		$<1.3 \times 10^{-5}$, CL = 90%
$\Gamma(\tau^- \to (e\mu\mu)^-)/\Gamma_{\text{total}}$		$<2.7 \times 10^{-5}$, CL = 90%
$\Gamma(\tau^- \to e^- \mu^+ \mu^-)/\Gamma_{\text{total}}$		$<1.9 \times 10^{-5}$, CL = 90%
$\Gamma(\tau^- \to e^+ \mu^- \mu^-)/\Gamma_{\text{total}}$		$<1.6 \times 10^{-5}$, CL = 90%
$\Gamma(\tau^- \to (\mu ee)^-)/\Gamma_{\text{total}}$		$<2.7 \times 10^{-5}$, CL = 90%
$\Gamma(\tau^- \to \mu^- e^+ e^-)/\Gamma_{\text{total}}$		$<1.4 \times 10^{-5}$, CL = 90%
$\Gamma(\tau^- \to \mu^- \mu^+ \mu^-)/\Gamma_{\text{total}}$		$<1.7 \times 10^{-5}$, CL = 90%
$\Gamma(\tau^- \to \ell^\pm \pi^\mp \pi^-)/\Gamma_{\text{total}}$	[i,k]	$<6.3 \times 10^{-5}$, CL = 90%
$\Gamma(\tau^- \to e^\mp \pi^\pm \pi^-)/\Gamma_{\text{total}}$	[i]	$<6.0 \times 10^{-5}$, CL = 90%
$\Gamma(\tau^- \to e^- \pi^+ \pi^-)/\Gamma_{\text{total}}$		$<2.7 \times 10^{-5}$, CL = 90%
$\Gamma(\tau^- \to \mu^\mp \pi^\pm \pi^-)/\Gamma_{\text{total}}$	[i]	$<3.9 \times 10^{-5}$, CL = 90%
$\Gamma(\tau^- \to \mu^- \pi^+ \pi^-)/\Gamma_{\text{total}}$		$<3.6 \times 10^{-5}$, CL = 90%
$\Gamma(\tau^- \to \ell^\pm \pi^\mp K^-)/\Gamma_{\text{total}}$	[i,k]	$<1.2 \times 10^{-4}$, CL = 90%
$\Gamma(\tau^- \to (e\pi K)^-$, all charged$)/\Gamma_{\text{total}}$		$<7.7 \times 10^{-5}$, CL = 90%
$\Gamma(\tau^- \to e^- \pi^\pm K^\mp)/\Gamma_{\text{total}}$	[i]	$<5.8 \times 10^{-5}$, CL = 90%
$\Gamma(\tau^- \to e^- \pi^+ K^-)/\Gamma_{\text{total}}$		$<2.9 \times 10^{-5}$, CL = 90%
$\Gamma(\tau^- \to e^- \pi^- K^+)/\Gamma_{\text{total}}$		$<5.8 \times 10^{-5}$, CL = 90%
$\Gamma(\tau^- \to (\mu\pi K)^-$, all charged$)/\Gamma_{\text{total}}$		$<7.7 \times 10^{-5}$, CL = 90%
$\Gamma(\tau^- \to \mu^- \pi^\pm K^\mp)/\Gamma_{\text{total}}$	[i]	$<7.7 \times 10^{-5}$, CL = 90%
$\Gamma(\tau^- \to \mu^- \pi^+ K^-)/\Gamma_{\text{total}}$		$<7.7 \times 10^{-5}$, CL = 90%
$\Gamma(\tau^- \to \mu^- \pi^- K^+)/\Gamma_{\text{total}}$		$<7.7 \times 10^{-5}$, CL = 90%
$\Gamma(\tau^- \to e^-$ light spinless boson$)/\Gamma_{\text{total}}$		$<3.2 \times 10^{-3}$, CL = 95%
$\Gamma(\tau^- \to \mu^-$ light spinless boson$)/\Gamma_{\text{total}}$		$<6 \times 10^{-3}$, CL = 95%

ν oscillations. (For other lepton mixing effects in particle decays, see the Full Listings.)

$\overline{\nu}_e \not\to \overline{\nu}_e$		
$\Delta(m^2)$ for $\sin^2(2\theta) = 1$		<0.0083 eV^2, CL = 90%
$\sin^2(2\theta)$ for "Large" $\Delta(m^2)$		<0.14, CL = 68%
$\nu_e \to \nu_\tau$		
$\Delta(m^2)$ for $\sin^2(2\theta) = 1$		<9 eV^2, CL = 90%
$\sin^2(2\theta)$ for "Large" $\Delta(m^2)$		<0.12, CL = 90%
$\overline{\nu}_e \to \overline{\nu}_\tau$		
$\sin^2(2\theta)$ for "Large" $\Delta(m^2)$		<0.7, CL = 90%
$\nu_\mu \to \nu_e$		
$\Delta(m^2)$ for $\sin^2(2\theta) = 1$		<0.09 eV^2, CL = 90%
$\sin^2(2\theta)$ for "Large" $\Delta(m^2)$		$<2.5 \times 10^{-3}$, CL = 90%
$\overline{\nu}_\mu \to \overline{\nu}_e$		
$\Delta(m^2)$ for $\sin^2(2\theta) = 1$		<0.14 eV^2, CL = 90%
$\sin^2(2\theta)$ for "Large" $\Delta(m^2)$		<0.004, CL = 95%
$\nu_\mu(\overline{\nu}_\mu) \to \nu_e(\overline{\nu}_e)$		
$\Delta(m^2)$ for $\sin^2(2\theta) = 1$		<0.075 eV^2, CL = 90%
$\sin^2(2\theta)$ for "Large" $\Delta(m^2)$		$< 3 \times 10^{-3}$, CL = 90%
$\nu_\mu \to \nu_\tau$		
$\Delta(m^2)$ for $\sin^2(2\theta) = 1$		<0.9 eV^2, CL = 90%
$\sin^2(2\theta)$ for "Large" $\Delta(m^2)$		<0.004, CL = 90%
$\overline{\nu}_\mu \to \overline{\nu}_\tau$		
$\Delta(m^2)$ for $\sin^2(2\theta) = 1$		<2.2 eV^2, CL = 90%
$\sin^2(2\theta)$ for "Large" $\Delta(m^2)$		$<4.4 \times 10^{-2}$, CL = 90%
$\nu_\mu(\overline{\nu}_\mu) \to \nu_\tau(\overline{\nu}_\tau)$		
$\Delta(m^2)$ for $\sin^2(2\theta) = 1$		<1.5 eV^2, CL = 90%
$\sin^2(2\theta)$ for "Large" $\Delta(m^2)$		$<8 \times 10^{-3}$, CL = 90%
$\nu_e \not\to \nu_e$		
$\Delta(m^2)$ for $\sin^2(2\theta) = 1$		<2.3 eV^2
$\sin^2(2\theta)$ for "Large" $\Delta(m^2)$		$<7 \times 10^{-2}$, CL = 90%
$\nu_\mu \not\to \nu_\mu$		
$\Delta(m^2)$ for $\sin^2(2\theta) = 1$		<0.23 or >1500 eV^2
$\sin^2(2\theta)$ for $\Delta(m^2) = 100$eV^2	[l]	<0.02, CL = 90%
$\overline{\nu}_\mu \not\to \overline{\nu}_\mu$		
$\Delta(m^2)$ for $\sin^2(2\theta) = 1$		<7 or >1200 eV^2
$\sin^2(2\theta)$ for 190 eV^2 $< \Delta(m^2) <$ 320 eV^2	[m]	<0.02, CL = 90%
$\Gamma(\pi^+ \to \mu^+ \nu_e)/\Gamma_{\text{total}}$	[n]	$<8.0 \times 10^{-3}$, CL = 90%
$\Gamma(\pi^+ \to \mu^- e^+ e^+ \nu)/\Gamma_{\text{total}}$		$<1.6 \times 10^{-6}$, CL = 90%
$\Gamma(\pi^0 \to \mu^+ e^- + e^- \mu^+)/\Gamma_{\text{total}}$		$<1.72 \times 10^{-8}$, CL = 90%
$\Gamma(K^+ \to \mu^- \nu e^+ e^+)/\Gamma_{\text{total}}$		$<2.0 \times 10^{-8}$, CL = 90%
$\Gamma(K^+ \to \mu^+ \nu_e)/\Gamma_{\text{total}}$	[n]	$<4 \times 10^{-3}$, CL = 90%
$\Gamma(K^+ \to \pi^+ \mu^+ e^-)/\Gamma_{\text{total}}$		$<2.1 \times 10^{-10}$, CL = 90%
$\Gamma(K^+ \to \pi^+ \mu^- e^+)/\Gamma_{\text{total}}$		$<7 \times 10^{-9}$, CL = 90%
$\Gamma(K^0_L \to e^\pm \mu^\mp)/\Gamma_{\text{total}}$	[i]	$<3.3 \times 10^{-11}$, CL = 90%
$\Gamma(D^+ \to \pi^+ e^\pm \mu^\mp)/\Gamma_{\text{total}}$	[i]	$<3.8 \times 10^{-3}$, CL = 90%
$\Gamma(D^+ \to \pi^+ e^+ \mu^-)/\Gamma_{\text{total}}$		$<3.3 \times 10^{-3}$, CL = 90%
$\Gamma(D^+ \to \pi^+ e^- \mu^+)/\Gamma_{\text{total}}$		$<3.3 \times 10^{-3}$, CL = 90%
$\Gamma(D^+ \to K^+ e^+ \mu^-)/\Gamma_{\text{total}}$		$<3.4 \times 10^{-3}$, CL = 90%
$\Gamma(D^+ \to K^+ e^- \mu^+)/\Gamma_{\text{total}}$		$<3.4 \times 10^{-3}$, CL = 90%
$\Gamma(D^0 \to \mu^\pm e^\mp)/\Gamma_{\text{total}}$	[i]	$<1.0 \times 10^{-4}$, CL = 90%
$\Gamma(B^+ \to \pi^+ e^+ \mu^-)/\Gamma_{\text{total}}$		$<6.4 \times 10^{-3}$, CL = 90%
$\Gamma(B^+ \to \pi^+ e^- \mu^+)/\Gamma_{\text{total}}$		$<6.4 \times 10^{-3}$, CL = 90%
$\Gamma(B^+ \to K^+ e^+ \mu^-)/\Gamma_{\text{total}}$		$<6.4 \times 10^{-3}$, CL = 90%

Limits are given at the 90% confidence level, while errors are given as ± 1 standard deviation.

Tests of Conservation Laws

$\Gamma(B^+ \to K^+ e^- \mu^+)/\Gamma_{\text{total}}$		$<6.4 \times 10^{-3}$, CL = 90%
$\Gamma(B^0 \to e^\pm \mu^\mp)/\Gamma_{\text{total}}$	[i]	$<5.9 \times 10^{-6}$, CL = 90%
$\Gamma(B^0 \to e^\pm \tau^\mp)/\Gamma_{\text{total}}$	[i]	$<5.3 \times 10^{-4}$, CL = 90%
$\Gamma(B^0 \to \mu^\pm \tau^\mp)/\Gamma_{\text{total}}$	[i]	$<8.3 \times 10^{-4}$, CL = 90%

TOTAL LEPTON NUMBER

Violation of total lepton number conservation also implies violation of lepton family number conservation.

limit on $\mu^- \to e^+$ conversion		
$\sigma(\mu^- {}^{32}\text{S} \to e^+ {}^{32}\text{Si}^*)$ / $\sigma(\mu^- {}^{32}\text{S} \to \nu_\mu {}^{32}\text{P}^*)$		$<9 \times 10^{-10}$, CL = 90%
$\sigma(\mu^- {}^{127}\text{I} \to e^+ {}^{127}\text{Sb}^*)$ / $\sigma(\mu^- {}^{127}\text{I} \to \text{anything})$		$<3 \times 10^{-10}$, CL = 90%
$\sigma(\mu^- \text{Ti} \to e^+ \text{Ca})$ / $\sigma(\mu^- \text{Ti} \to \text{capture})$		$<8.9 \times 10^{-11}$, CL = 90%
$\Gamma(\tau^- \to \pi^- \gamma)/\Gamma_{\text{total}}$		$<2.8 \times 10^{-4}$, CL = 90%
$\Gamma(\tau^- \to \pi^- \pi^0)/\Gamma_{\text{total}}$		$<3.7 \times 10^{-4}$, CL = 90%
$\Gamma(\tau^- \to \mu^+ e^- e^-)/\Gamma_{\text{total}}$		$<1.4 \times 10^{-5}$, CL = 90%
$\Gamma(\tau^- \to \ell^\pm \pi^\mp \pi^-)/\Gamma_{\text{total}}$	[i,k]	$<6.3 \times 10^{-5}$, CL = 90%
$\Gamma(\tau^- \to e^\mp \pi^\pm \pi^-)/\Gamma_{\text{total}}$	[i]	$<6.0 \times 10^{-5}$, CL = 90%
$\Gamma(\tau^- \to e^+ \pi^- \pi^-)/\Gamma_{\text{total}}$		$<1.7 \times 10^{-5}$, CL = 90%
$\Gamma(\tau^- \to \mu^\mp \pi^\pm \pi^-)/\Gamma_{\text{total}}$	[i]	$<3.9 \times 10^{-5}$, CL = 90%
$\Gamma(\tau^- \to \mu^+ \pi^- \pi^-)/\Gamma_{\text{total}}$		$<3.9 \times 10^{-5}$, CL = 90%
$\Gamma(\tau^- \to \ell^\pm \pi^\mp K^-)/\Gamma_{\text{total}}$	[i,k]	$<1.2 \times 10^{-4}$, CL = 90%
$\Gamma(\tau^- \to (e \pi K)^-$, all charged$)/\Gamma_{\text{total}}$		$<7.7 \times 10^{-5}$, CL = 90%
$\Gamma(\tau^- \to e^+ \pi^- K^-)/\Gamma_{\text{total}}$		$<2.0 \times 10^{-5}$, CL = 90%
$\Gamma(\tau^- \to (\mu \pi K)^-$, all charged$)/\Gamma_{\text{total}}$		$<7.7 \times 10^{-5}$, CL = 90%
$\Gamma(\tau^- \to \mu^+ \pi^- K^-)/\Gamma_{\text{total}}$		$<4.0 \times 10^{-5}$, CL = 90%
$\Gamma(\tau^- \to \overline{p} \gamma)/\Gamma_{\text{total}}$		$<2.9 \times 10^{-4}$, CL = 90%
$\Gamma(\tau^- \to \overline{p} \pi^0)/\Gamma_{\text{total}}$		$<6.6 \times 10^{-4}$, CL = 90%
$\Gamma(\tau^- \to \overline{p} \eta)/\Gamma_{\text{total}}$		$<1.30 \times 10^{-3}$, CL = 90%
$\nu_e \to (\overline{\nu}_e)_L$		
$\alpha\Delta(m^2)$ for $\sin^2(2\theta) = 1$		<0.14 eV^2, CL = 90%
$\alpha^2\sin^2(2\theta)$ for "Large" $\Delta(m^2)$		<0.032, CL = 90%
$\nu_\mu \to (\overline{\nu}_e)_L$		
$\alpha\Delta(m^2)$ for $\sin^2(2\theta) = 1$		<0.16 eV^2, CL = 90%
$\alpha^2\sin^2(2\theta)$ for "Large" $\Delta(m^2)$		<0.001, CL = 90%
$\Gamma(\pi^+ \to \mu^+ \overline{\nu}_e)/\Gamma_{\text{total}}$	[n]	$<1.5 \times 10^{-3}$, CL = 90%
$\Gamma(K^+ \to \pi^- \mu^+ e^+)/\Gamma_{\text{total}}$		$<7 \times 10^{-9}$, CL = 90%
$\Gamma(K^+ \to \pi^- e^+ e^+)/\Gamma_{\text{total}}$		$<1.0 \times 10^{-8}$, CL = 90%
$\Gamma(K^+ \to \pi^- \mu^+ \mu^+)/\Gamma_{\text{total}}$		$<1.5 \times 10^{-4}$, CL = 90%
$\Gamma(K^+ \to \mu^+ \overline{\nu}_e)/\Gamma_{\text{total}}$	[n]	$<3.3 \times 10^{-3}$, CL = 90%
$\Gamma(K^+ \to \pi^0 e^+ \overline{\nu}_e)/\Gamma_{\text{total}}$	[n]	$<3 \times 10^{-3}$, CL = 90%
$\Gamma(D^+ \to \pi^- e^+ e^+)/\Gamma_{\text{total}}$		$<4.8 \times 10^{-3}$, CL = 90%
$\Gamma(D^+ \to \pi^- \mu^+ \mu^+)/\Gamma_{\text{total}}$		$<6.8 \times 10^{-3}$, CL = 90%
$\Gamma(D^+ \to \pi^- e^+ \mu^+)/\Gamma_{\text{total}}$		$<3.7 \times 10^{-3}$, CL = 90%
$\Gamma(D^+ \to K^- e^+ e^+)/\Gamma_{\text{total}}$		$<9.1 \times 10^{-3}$, CL = 90%
$\Gamma(D^+ \to K^- \mu^+ \mu^+)/\Gamma_{\text{total}}$		$<4.3 \times 10^{-3}$, CL = 90%
$\Gamma(D^+ \to K^- e^+ \mu^+)/\Gamma_{\text{total}}$		$<4.0 \times 10^{-3}$, CL = 90%
$\Gamma(B^+ \to \pi^- e^+ e^+)/\Gamma_{\text{total}}$		$<3.9 \times 10^{-3}$, CL = 90%
$\Gamma(B^+ \to \pi^- \mu^+ \mu^+)/\Gamma_{\text{total}}$		$<9.1 \times 10^{-3}$, CL = 90%
$\Gamma(B^+ \to \pi^- e^+ \mu^+)/\Gamma_{\text{total}}$		$<6.4 \times 10^{-3}$, CL = 90%
$\Gamma(B^+ \to K^- e^+ e^+)/\Gamma_{\text{total}}$		$<3.9 \times 10^{-3}$, CL = 90%
$\Gamma(B^+ \to K^- \mu^+ \mu^+)/\Gamma_{\text{total}}$		$<9.1 \times 10^{-3}$, CL = 90%
$\Gamma(B^+ \to K^- e^+ \mu^+)/\Gamma_{\text{total}}$		$<6.4 \times 10^{-3}$, CL = 90%
$\Gamma(\Xi^- \to p \mu^- \mu^-)/\Gamma_{\text{total}}$		$<4 \times 10^{-4}$, CL = 90%

BARYON NUMBER

$\Gamma(\tau^- \to \overline{p} \gamma)/\Gamma_{\text{total}}$		$<2.9 \times 10^{-4}$, CL = 90%
$\Gamma(\tau^- \to \overline{p} \pi^0)/\Gamma_{\text{total}}$		$<6.6 \times 10^{-4}$, CL = 90%
$\Gamma(\tau^- \to \overline{p} \eta)/\Gamma_{\text{total}}$		$<1.30 \times 10^{-3}$, CL = 90%
p mean life		$>1.6 \times 10^{25}$ years

A few examples of proton or bound neutron decay follow. For limits on many other nucleon decay channels, see the Baryon Summary Table.

$\tau(N \to e^+ \pi)$		> 130 (n), > 550 (p) $\times 10^{30}$ years, CL = 90%
$\tau(N \to \mu^+ \pi)$		> 100 (n), > 270 (p) $\times 10^{30}$ years, CL = 90%
$\tau(N \to e^+ K)$		> 1.3 (n), > 150 (p) $\times 10^{30}$ years, CL = 90%
$\tau(N \to \mu^+ K)$		> 1.1 (n), > 120 (p) $\times 10^{30}$ years, CL = 90%
mean time for $n\overline{n}$ transition in vacuum	[o]	$>1.2 \times 10^{8}$ s, CL = 90%

ELECTRIC CHARGE (Q)

e mean life / branching fraction	[p]	$>2.7 \times 10^{23}$ yr, CL = 68%
$\Gamma(n \to p \nu_e \overline{\nu}_e)/\Gamma_{\text{total}}$		$<9 \times 10^{-24}$, CL = 90%

$\Delta S = \Delta Q$ RULE

Allowed in second-order weak interactions.

$\Gamma(K^+ \to \pi^+ \pi^+ e^- \overline{\nu}_e)/\Gamma_{\text{total}}$		$<1.2 \times 10^{-8}$, CL = 90%
$\Gamma(K^+ \to \pi^+ \pi^+ \mu^- \overline{\nu}_\mu)/\Gamma_{\text{total}}$		$<3.0 \times 10^{-6}$, CL = 95%
$x = \text{A}(\overline{K}^0 \to \pi^- \ell^+ \nu)/\text{A}(K^0 \to \pi^- \ell^+ \nu) = \text{A}(\Delta S{=}-\Delta Q)/\text{A}(\Delta S{=}\Delta Q)$		
real part of x		0.006 ± 0.018 (S = 1.3)
imaginary part of x		-0.003 ± 0.026 (S = 1.2)
$\Gamma(\Sigma^+ \to n \ell^+ \nu)/\Gamma(\Sigma^- \to n \ell^- \overline{\nu})$		<0.043
$\Gamma(\Sigma^+ \to n e^+ \nu_e)/\Gamma_{\text{total}}$		$<5 \times 10^{-6}$, CL = 90%
$\Gamma(\Sigma^+ \to n \mu^+ \nu_\mu)/\Gamma_{\text{total}}$		$<3.0 \times 10^{-5}$, CL = 90%
$\Gamma(\Xi^0 \to \Sigma^- e^+ \nu_e)/\Gamma_{\text{total}}$		$<9 \times 10^{-4}$, CL = 90%
$\Gamma(\Xi^0 \to \Sigma^- \mu^+ \nu_\mu)/\Gamma_{\text{total}}$		$<9 \times 10^{-4}$, CL = 90%

$\Delta S = 2$ FORBIDDEN

Allowed in second-order weak interactions.

$\Gamma(\Xi^0 \to p \pi^-)/\Gamma_{\text{total}}$		$<4 \times 10^{-5}$, CL = 90%
$\Gamma(\Xi^0 \to p e^- \overline{\nu}_e)/\Gamma_{\text{total}}$		$<1.3 \times 10^{-3}$
$\Gamma(\Xi^0 \to p \mu^- \overline{\nu}_\mu)/\Gamma_{\text{total}}$		$<1.3 \times 10^{-3}$
$\Gamma(\Xi^- \to n \pi^-)/\Gamma_{\text{total}}$		$<1.9 \times 10^{-5}$, CL = 90%
$\Gamma(\Xi^- \to n e^- \overline{\nu}_e)/\Gamma_{\text{total}}$		$<3.2 \times 10^{-3}$, CL = 90%
$\Gamma(\Xi^- \to n \mu^- \overline{\nu}_\mu)/\Gamma_{\text{total}}$		$<1.5 \times 10^{-2}$, CL = 90%
$\Gamma(\Xi^- \to p \pi^- \pi^-)/\Gamma_{\text{total}}$		$<4 \times 10^{-4}$, CL = 90%
$\Gamma(\Xi^- \to p \pi^- e^- \overline{\nu}_e)/\Gamma_{\text{total}}$		$<4 \times 10^{-4}$, CL = 90%
$\Gamma(\Xi^- \to p \pi^- \mu^- \overline{\nu}_\mu)/\Gamma_{\text{total}}$		$<4 \times 10^{-4}$, CL = 90%
$\Gamma(\Omega^- \to \Lambda \pi^-)/\Gamma_{\text{total}}$		$<1.9 \times 10^{-4}$, CL = 90%

$\Delta S = 2$ VIA MIXING

Allowed in second-order weak interactions, e.g. mixing.

$m_{K^0_L} - m_{K^0_S}$		$(0.5333 \pm 0.0027) \times 10^{10}\ \hbar\ \text{s}^{-1}$ (S = 1.2)
$m_{K^0_L} - m_{K^0_S}$		$(3.510 \pm 0.018) \times 10^{-12}$ MeV (S = 1.3)

$\Delta C = 2$ VIA MIXING

Allowed in second-order weak interactions, e.g. mixing.

$\lvert m_{D^0_1} - m_{D^0_2} \rvert$	[q]	$<20 \times 10^{10}\ \hbar\ \text{s}^{-1}$, CL = 90%
$\Gamma(K^+ \pi^-$ (via $\overline{D}^0$)$)/\Gamma(K^- \pi^+)$		<0.0037, CL = 90%
$\Gamma(\mu^-$ anything (via $\overline{D}^0$)$)/\Gamma(\mu^+$ anything$)$		<0.0056, CL = 90%

Limits are given at the 90% confidence level, while errors are given as ± 1 standard deviation.

Tests of Conservation Laws

$\Delta B = 2$ VIA MIXING

Allowed in second-order weak interactions, e.g. mixing.

χ_d		0.156 ± 0.024
$\Delta m_{B^0} = m_{B^0_H} - m_{B^0_L}$		$(0.51 \pm 0.06) \times 10^{12}\ \hbar\ s^{-1}$
$x_d = \Delta m_{B^0}/\Gamma_{B^0}$		0.71 ± 0.06
χ_s		0.62 ± 0.13
$\Delta m_{B^0_s} = m_{B^0_{sH}} - m_{B^0_{sL}}$		$>1.8 \times 10^{12}\ \hbar\ s^{-1}$, CL = 95%
$x_s = \Delta m_{B^0_s}/\Gamma_{B^0_s}$		>2.0, CL = 95%

$\Delta S = 1$ WEAK NEUTRAL CURRENT FORBIDDEN

Allowed by higher-order electroweak interactions.

$\Gamma(K^+ \to \pi^+ e^+ e^-)/\Gamma_{total}$		$(2.74 \pm 0.23) \times 10^{-7}$
$\Gamma(K^+ \to \pi^+ \mu^+ \mu^-)/\Gamma_{total}$		$<2.3 \times 10^{-7}$, CL = 90%
$\Gamma(K^+ \to \pi^+ \nu\overline{\nu})/\Gamma_{total}$		$<5.2 \times 10^{-9}$, CL = 90%
$\Gamma(K^0_S \to \mu^+ \mu^-)/\Gamma_{total}$		$<3.2 \times 10^{-7}$, CL = 90%
$\Gamma(K^0_S \to e^+ e^-)/\Gamma_{total}$		$<1.0 \times 10^{-5}$, CL = 90%
$\Gamma(K^0_S \to \pi^0 e^+ e^-)/\Gamma_{total}$		$<1.1 \times 10^{-6}$, CL = 90%
$\Gamma(K^0_L \to \mu^+ \mu^-)/\Gamma_{total}$		$(7.4 \pm 0.4) \times 10^{-9}$
$\Gamma(K^0_L \to \mu^+ \mu^- \gamma)/\Gamma_{total}$		$(2.8 \pm 2.8) \times 10^{-7}$
$\Gamma(K^0_L \to e^+ e^-)/\Gamma_{total}$		$<4.1 \times 10^{-11}$, CL = 90%
$\Gamma(K^0_L \to e^+ e^- \gamma)/\Gamma_{total}$		$(9.1 \pm 0.5) \times 10^{-6}$
$\Gamma(K^0_L \to e^+ e^- \gamma\gamma)/\Gamma_{total}$	[r]	$(6.6 \pm 3.2) \times 10^{-7}$
$\Gamma(K^0_L \to \pi^+ \pi^- e^+ e^-)/\Gamma_{total}$		$<2.5 \times 10^{-6}$, CL = 90%
$\Gamma(K^0_L \to \mu^+ \mu^- e^+ e^-)/\Gamma_{total}$		$<4.9 \times 10^{-6}$, CL = 90%
$\Gamma(K^0_L \to e^+ e^- e^+ e^-)/\Gamma_{total}$	[s]	$(3.9 \pm 0.7) \times 10^{-8}$
$\Gamma(K^0_L \to \pi^0 \mu^+ \mu^-)/\Gamma_{total}$		$<5.1 \times 10^{-9}$, CL = 90%
$\Gamma(K^0_L \to \pi^0 e^+ e^-)/\Gamma_{total}$		$<4.3 \times 10^{-9}$, CL = 90%
$\Gamma(K^0_L \to \pi^0 \nu\overline{\nu})/\Gamma_{total}$		$<2.2 \times 10^{-4}$, CL = 90%
$\Gamma(\Sigma^+ \to p e^+ e^-)/\Gamma_{total}$		$<7 \times 10^{-6}$

$\Delta C = 1$ WEAK NEUTRAL CURRENT FORBIDDEN

Allowed by higher-order electroweak interactions.

$\Gamma(D^+ \to \pi^+ e^+ e^-)/\Gamma_{total}$		$<2.5 \times 10^{-3}$, CL = 90%
$\Gamma(D^+ \to \pi^+ \mu^+ \mu^-)/\Gamma_{total}$		$<2.9 \times 10^{-3}$, CL = 90%
$\Gamma(D^0 \to e^+ e^-)/\Gamma_{total}$		$<1.3 \times 10^{-4}$, CL = 90%
$\Gamma(D^0 \to \mu^+ \mu^-)/\Gamma_{total}$		$<1.1 \times 10^{-5}$, CL = 90%
$\Gamma(D^0 \to \rho^0 e^+ e^-)/\Gamma_{total}$		$<4.5 \times 10^{-4}$, CL = 90%
$\Gamma(D^0 \to \rho^0 \mu^+ \mu^-)/\Gamma_{total}$		$<8.1 \times 10^{-4}$, CL = 90%

$\Delta B = 1$ WEAK NEUTRAL CURRENT FORBIDDEN

Allowed by higher-order electroweak interactions.

$\Gamma(B^+ \to \pi^+ e^+ e^-)/\Gamma_{total}$		$<3.9 \times 10^{-3}$, CL = 90%
$\Gamma(B^+ \to \pi^+ \mu^+ \mu^-)/\Gamma_{total}$		$<9.1 \times 10^{-3}$, CL = 90%
$\Gamma(B^+ \to K^+ e^+ e^-)/\Gamma_{total}$		$<6 \times 10^{-5}$, CL = 90%
$\Gamma(B^+ \to K^+ \mu^+ \mu^-)/\Gamma_{total}$		$<1.7 \times 10^{-4}$, CL = 90%
$\Gamma(B^+ \to K^*(892)^+ e^+ e^-)/\Gamma_{total}$		$<6.9 \times 10^{-4}$, CL = 90%
$\Gamma(B^+ \to K^*(892)^+ \mu^+ \mu^-)/\Gamma_{total}$		$<1.2 \times 10^{-3}$, CL = 90%
$\Gamma(\overline{b} \to e^+ e^-\ \text{anything})/\Gamma_{total}$	[t]	$<2.4 \times 10^{-3}$, CL = 90%
$\Gamma(\overline{b} \to \mu^+ \mu^-\ \text{anything})/\Gamma_{total}$	[t]	$<5.0 \times 10^{-5}$, CL = 90%
$\Gamma(B^0 \to e^+ e^-)/\Gamma_{total}$		$<5.9 \times 10^{-6}$, CL = 90%
$\Gamma(B^0 \to \mu^+ \mu^-)/\Gamma_{total}$		$<5.9 \times 10^{-6}$, CL = 90%
$\Gamma(B^0 \to K^0 e^+ e^-)/\Gamma_{total}$		$<3.0 \times 10^{-4}$, CL = 90%
$\Gamma(B^0 \to K^0 \mu^+ \mu^-)/\Gamma_{total}$		$<3.6 \times 10^{-4}$, CL = 90%
$\Gamma(B^0 \to K^*(892)^0 e^+ e^-)/\Gamma_{total}$		$<2.9 \times 10^{-4}$, CL = 90%
$\Gamma(B^0 \to K^*(892)^0 \mu^+ \mu^-)/\Gamma_{total}$		$<2.3 \times 10^{-5}$, CL = 90%

Limits are given at the 90% confidence level, while errors are given as ± 1 standard deviation.

NOTES

In this Summary Table:

When a quantity has "(S = ...)" to its right, the error on the quantity has been enlarged by the "scale factor" S, defined as $S = \sqrt{\chi^2/(N-1)}$, where N is the number of measurements used in calculating the quantity. We do this when $S > 1$, which often indicates that the measurements are inconsistent. When $S > 1.25$, we also show in the Full Listings an ideogram of the measurements. For more about S, see the Introduction.

[a] Positronium data are from A.P. Mills and S. Berko, Physical Review Letters **18** 420 (1967); and K. Marko and A. Rich, Physical Review Letters **33** 980 (1974). Values for 90% confidence level, are from A.P. Mills, private communication.

[b] C parity forbids this to occur as a single-photon process.

[c] Time-reversal invariance requires this to be 0° or 180°.

[d] Allowed by higher-order electroweak interactions.

[e] Violates CP in leading order. Test of direct CP violation since the indirect CP-violating and CP-conserving contributions are expected to be suppressed.

[f] ϵ'/ϵ is derived from $|\eta_{00}/\eta_{+-}|$ measurements using theoretical input on phases.

[g] Neglecting photon channels. See, *e.g.*, A. Pais and S.B. Treiman, Phys. Rev. **D12**, 2744 (1975).

[h] Derived from measured values of ϕ_{+-}, ϕ_{00}, $|\eta|$, $\tau_{K^0_S}$, and $|m_{K^0_L} - m_{K^0_S}|$, as described in the introduction to this Table.

[i] The value is for the sum of the charge states indicated.

[j] A test of additive vs. multiplicative lepton family number conservation.

[k] ℓ means a sum over e and μ modes.

[l] $\Delta(m^2) = 100\ \text{eV}^2$.

[m] $190\ \text{eV}^2 < \Delta(m^2) < 320\ \text{eV}^2$.

[n] Derived from an analysis of neutrino-oscillation experiments.

[o] There is some controversy about whether nuclear physics and model dependence complicate the analysis for bound neutrons (from which the best limit comes). For reactor experiments with free neutrons, the best limit is $> 10^7$ s.

[p] This is the best "electron disappearance" limit. The best limit for the mode $e^- \to \nu\gamma$ is $> 2.35 \times 10^{25}$ yr (CL=68%).

[q] The D^0_1-D^0_2 limits are inferred from the limit on $D^0 \to \overline{D}{}^0 \to K^+\pi^-$.

[r] See the K^0_L Full Listings for the energy limits used in this measurement.

[s] $m_{e^+e^-} > 470$ MeV.

[t] B^0, B^+, and B^0_s not separated.

8

TABLES OF ALPHA-PARTICLE EMITTERS: ENERGIES AND EMISSION PROBABILITIES

A.L. Nichols

Table of contents

The recommended half-lives, branching ratios, alpha-particle energies and emission probabilities have been compiled from the most up-to-date data to be found within:

Nuclear Data Sheets, Academic Press, Inc., New York,
Decay Data of the Transactinium Nuclides, Technical Reports Series No.261, IAEA, Vienna, 1986,
Recommended Energy and Intensity Values of Alpha Particles from Radioactive Decay, *At. Data Nucl. Data Tables*, 47, 205, 1991,
UK Heavy Element and Actinide Decay Data Files (UKHEDD-2), 1991.

A large majority of the alpha-particle-emitting radionuclides are identified with the heavier elements, which are generally more unstable with respect to ^{4}He emissions than the other decay modes. It should be noted that the resulting tabulations are not comprehensive; they do not include all known alpha-particle-emitting radionuclides nor all of the alpha-particle emissions identified with each individual radionuclide. Only the major alpha-particle emissions are listed (i.e. generally emission probabilities greater than 0.01), along with indications of the commonly used energy and emission probability standards.

I. INTRODUCTION

Alpha-particle spectroscopy can be used to determine the composition of a mixture of heavy elements that are predominantly unstable to alpha decay.

Semiconductor detectors are used extensively to measure alpha-particle spectra,[1,2] although it is stressed that the best energy resolution and quantification can only be achieved if the samples are prepared correctly with minimal thickness to avoid spectral attenuation. Such studies require a sound knowledge of the half-lives of these radionuclides, and the energies and emission probabilities per decay of their alpha-particle emissions.

A number of evaluations and compilations have been produced to assist in defining the radioactive characteristics of radionuclides that undergo alpha decay.[3–10] The relevant data from the most recent and comprehensive of these data files have been brought together within Tables 1 and 2 to give suitable listings to assist in the calibration of alpha spectrometers and determine the composition and quantities of alpha-emitting radionuclides. Inevitably these data can be expected to undergo further modifications as more precise measurements and evaluations are undertaken (e.g. various experimental programmes are under way to resolve discrepancies between previous studies of the alpha, conversion-electron and gamma-ray decay of ^{237}Np).

II. HALF-LIVES AND BRANCHING RATIOS

Recommended half-lives and branching ratios were obtained by a judicious combination of the data evaluated in references 5, 8 and 9 (Table 1). Decay modes include

α	alpha decay
β^-	beta decay
EC	electron capture decay
SF	spontaneous fission decay

with the following half-life units:

s	second
m	minute
h	hour
d	day
y	year (= 365.2422 days).

The branching ratios are expressed as a fraction of the decay. All data have been listed with 1 sigma uncertainty. Although some of the radionuclides have extremely short half-lives (e.g. ^{211}Po, ^{212}Po, ^{213}Po, ^{214}Po and ^{220}Rn), they are included because they form part of the natural decay chains and undergo significant alpha-particle emission.

III. ALPHA-PARTICLE ENERGIES AND EMISSION PROBABILITIES

Both the alpha-particle energies and emission probabilities are given in Table 2, as compiled from references 5, 7, 8 and 9. Energies and their uncertainties have been defined in units of keV, while the emission probabilities are listed as fraction per decay:

5340.54(15) = 5340.54 ± 0.15 keV
0.267(2) = 0.267 ± 0.002 (or (26.7 ± 0.2)% when expressed as percentage).

All data have been listed with 1 sigma uncertainty. Some of the emission probabilities are known to a high degree of accuracy because of the simplicity of the alpha-particle decay to only one or two nuclear levels of the daughter nuclide (e.g. ^{210}Po, ^{213}Po and ^{222}Rn). Energy and emission probability standards are denoted by an asterisk in Table 2, and constitute the recommended data for detector calibration.

REFERENCES

1. **Siegbahn, K.**, *Alpha-, Beta- and Gamma-ray Spectroscopy*, Vol. 1, North-Holland Pub. Co., Amsterdam, 1968.
2. **Glover, K.M.**, Alpha-particle Spectrometry and Its Applications, *Int. J. Appl. Radiat. Isot.* **35**, 239, 1984.
3. **Rogers, F.J.G.**, A Listing of Alpha Particle Energies and Intensities Arranged (a) by Nuclides in Ascending Order of Z and A, and (b) in order of Energy, UKAEA Harwell Report AERE-R 8005, 1976.
4. **Westmeier, W., Merklin, A.**, Catalog of Alpha Particles from Radioactive Decay, Physik Daten/Physics Data, 29-1, Fachinformationszentrum Energie, Physik-Mathematik GmbH, Karlsruhe, 1985.
5. International Atomic Energy Agency, Decay Data of the Transactinium Nuclides, Technical Reports Series No.261, International Atomic Energy Agency, Vienna, 1986.
6. **Tepel, J.W., Müller, H.W.**, GAMCAT — A Personal Computer Database on Alpha Particles and Gamma Rays from Radioactive Decay, *Nucl. Instrum. Methods Phys. Res.* **A286**, 443, 1990.
7. **Rytz, A.**, Recommended Energy and Intensity Values of Apha Particles from Radioactive Decay, *At. Data Nucl. Data Tables* **47**, 205, 1991.
8. **Tobias, A., Davies, B.S.J., Nichols, A.L., James, M.F.**, The UKCNDC Radioactive Decay Data Libraries, *Nucl. Energy* **22**, 445, 1983; **Nichols, A.L.**, Heavy Element and Actinide Decay Data: UKHEDD-2 Data Files, AEA Technology Report AEA-RS-5219, 1991.
9. Nuclear Data Sheets, Produced by the National Nuclear Data Center for the International Network for Nuclear Structure Data Evaluation, Academic Press, Inc., New York.
10. **Nordborg, C., Gruppelaar, H., Salvatores, M.**, Status of the JEF and EFF Projects, Int. Conf. Nuclear Data for Science and Technology, 13–17 May 1991, Jülich, Germany.

Table 1 Selected alpha-particle emitters — half-lives and branching ratios

Nuclide	Decay mode	Half-life uncertainty			Branching ratio
		Value	Exponent	Units	
62-Sm-146	α	1.03 ± 0.05	E+08	y	1.0
62-Sm-147	α	1.06 ± 0.02	E+11	y	1.0
64-Gd-148	α	75 ± 3		y	1.0
66-Dy-154	α	3.0 ± 1.5	E+06	y	1.0
83-Bi-211	α	2.17 ± 0.04		m	0.99727 ± 0.00004
	β^-	13.25 ± 0.28		h	0.00273 ± 0.00004
83-Bi-212	Total	60.55 ± 0.06		m	
	α	168.5 ± 0.3		m	0.3594 ± 0.0006
	β^-	94.52 ± 0.13		m	0.6406 ± 0.0006
84-Po-208	α	2.898 ± 0.002		y	1.0
84-Po-209	α	102 ± 5		y	0.9974 ± 0.0001
	EC	3.92 ± 0.24	E+04	y	0.0026 ± 0.0001
84-Po-210	α	138.4 ± 0.2		d	1.0
84-Po-211	α	0.516 ± 0.003		s	1.0
84-Po-212	α	0.298 ± 0.003	E−06	s	1.0
84-Po-213	α	4.2 ± 0.8	E−03	s	1.0
84-Po-214	α	0.165 ± 0.003	E−03	s	1.0
84-Po-215	α	1.780 ± 0.004	E−03	s	1.0
84-Po-216	α	0.145 ± 0.002		s	1.0
84-Po-218	α	3.05 ± 0.09		m	0.9998 ± 0.0001
	β^-	11 ± 6		d	0.0002 ± 0.0001
85-At-217	α	3.23 ± 0.04		s	0.99988 ± 0.00004
	β^-	7 ± 3		h	0.00012 ± 0.00004
86-Rn-219	α	3.96 ± 0.01		s	1.0
86-Rn-220	α	55.6 ± 0.1		s	1.0

continued over leaf

Table 1 Continued

Nuclide	Decay mode	Half-life uncertainty			Branching ratio
		Value	Exponent	Units	
86-Rn-222	α	3.8235 ± 0.0003		d	1.0
87-Fr-221	α	4.9 ± 0.2		m	1.0
88-Ra-223	α	11.43 ± 0.02		d	1.0
88-Ra-224	α	3.66 ± 0.04		d	1.0
88-Ra-226	α	1600 ± 7		y	1.0
89-Ac-225	α	10.0 ± 0.1		d	1.0
90-Th-227	α	18.72 ± 0.02		d	1.0
90-Th-228	α	1.913 ± 0.002		y	1.0
90-Th-229	α	7.34 ± 0.16	E+03	y	1.0
90-Th-230	α	7.54 ± 0.03	E+04	y	1.0
90-Th-232	α	1.405 ± 0.006	E+10	y	1.0
91-Pa-231	α	3.276 ± 0.011	E+04	y	1.0
92-U-232	α	69.8 ± 0.5		y	1.0
92-U-233	α	1.592 ± 0.002	E+05	y	1.0
92-U-234	α	2.457 ± 0.003	E+05	y	1.0
92-U-235	α	7.037 ± 0.007	E+08	y	1.0
92-U-236	α	2.342 ± 0.005	E+07	y	1.0
92-U-238	α	4.468 ± 0.005	E+09	y	1.0
93-Np-237	α	2.14 ± 0.01	E+06	y	1.0
94-Pu-236	α	2.9 ± 0.1		y	1.0
94-Pu-238	α	87.7 ± 0.3		y	1.0
94-Pu-239	α	2.411 ± 0.003	E+04	y	1.0
94-Pu-240	α	6.563 ± 0.007	E+03	y	0.9999943 ± 0.0000001
	SF	1.16 ± 0.04	E+11	y	0.0000057 ± 0.0000001
94-Pu-242	α	3.735 ± 0.011	E+05	y	0.999451 ± 0.000008
	SF	6.8 ± 0.1	E+10	y	0.000549 ± 0.000008

Table 1 Continued

Nuclide	Decay mode	Half-life uncertainty			Branching ratio
		Value	Exponent	Units	
94-Pu-244	α	8.00 ± 0.09	E+07	y	0.99881 ± 0.00006
	SF	6.7 ± 0.3	E+10	y	0.00119 ± 0.00006
95-Am-241	α	432.7 ± 0.5		y	1.0
95-Am-243	α	7.370 ± 0.015	E+03	y	1.0
96-Cm-242	α	162.94 ± 0.06		d	0.99999367 ± 0.00000013
	SF	7.05 ± 0.14	E+06	y	0.00000633 ± 0.00000013
96-Cm-243	Total	28.5 ± 0.2		y	
	α	28.6 ± 0.2		y	0.9976 ± 0.0003
	EC	1.19 ± 0.15	E+04	y	0.0024 ± 0.0003
96-Cm-244	α	18.10 ± 0.02		y	0.9998653 ± 0.0000007
	SF	1.344 ± 0.007	E+07	y	0.0001347 ± 0.0000007
96-Cm-246	α	4.73 ± 0.15	E+03	y	0.9997386 ± 0.0000005
	SF	1.81 ± 0.06	E+07	y	0.0002614 ± 0.0000005
96-Cm-247	α	1.60 ± 0.05	E+07	y	1.0
96-Cm-248	Total	3.40 ± 0.04	E+05	y	
	α	3.71 ± 0.04	E+05	y	0.9174 ± 0.0003
	SF	4.12 ± 0.05	E+06	y	0.0826 ± 0.0003
98-Cf-249	α	351 ± 2		y	1.0
98-Cf-250	α	13.08 ± 0.09		y	0.99923 ± 0.00003
	SF	1.70 ± 0.07	E+04	y	0.00077 ± 0.00003
98-Cf-251	α	898 ± 44		y	1.0
98-Cf-252	Total	2.645 ± 0.008		y	
	α	2.73 ± 0.01		y	0.96908 ± 0.00008
	SF	85.5 ± 0.3		y	0.03092 ± 0.00008

continued overleaf

Table 1 Continued

Nuclide	Decay mode	Half-life uncertainty			Branching ratio
		Value	Exponent	Units	
99-Es-253	α	20.47 ± 0.03		d	1.0
99-Es-254	α	275.7 ± 0.5		d	1.0

Table 2 Alpha-particle energies and emission probabilities (recommended standards are denoted by an asterisk (*))

Nuclide	Energy (keV)	Emission probability
62-Sm-146	2455(4)	1.000
62-Sm-147	2235(3)	1.000
64-Gd-148	3182.68(2)*	1.000*
66-Dy-154	2870(5)	1.000
83-Bi-211	6278.8(7)	0.164(4)
	6623.1(6)	0.834(4)
83-Bi-212	6050.776(26)*	0.252(1)*
	6089.879(34)*	0.0967(5)*
84-Po-208	5114.9(14)	1.000
84-Po-209	4880(3)	0.794(10)
	4883(3)	0.199(10)
84-Po-210	5304.33(7)	1.000
84-Po-211	7450(3)	0.98919(19)
84-Po-212	8784.86(12)*	1.000(1)*
84-Po-213	8376(3)	0.99997(1)
84-Po-214	7686.82(7)	0.999895(6)
84-Po-215	7386.1(8)	1.000
84-Po-216	6778.3(5)	0.999981(3)
84-Po-218	6002.55(10)	0.999789(1)
85-At-217	7066.9(16)	0.999(1)
86-Rn-219	6425.0(10)	0.075(6)
	6552.6(10)	0.129(6)
	6819.1(3)	0.794(10)
86-Rn-220	6288.27(10)	0.99874(12)
86-Rn-222	5489.66(30)*	0.99928(21)*
87-Fr-221	6127(2)	0.151(2)
	6341.1(5)	0.834(8)
88-Ra-223	5540.0(2)	0.091(2)
	5606.9(2)	0.242(4)
	5716.4(2)	0.525(8)
	5747.2(2)	0.095(2)
88-Ra-224	5448.84(16)	0.0506(4)
	5685.53(15)	0.9492(4)
88-Ra-226	4601.9(5)	0.0555(5)
	4784.50(25)	0.9445(5)
89-Ac-225	5581(3)	0.012(1)
	5608(3)	0.011(1)
	5637(3)	0.045(3)
	5682(3)	0.014(2)
	5723(3)	0.029(5)
	5731(3)	0.100(1)
	5791(2)	0.086(9)
	5793(2)	0.181(20)
	5829(2)	0.507(15)

Table 2 Continued

Nuclide	Energy (keV)	Emission probability
90-Th-227	5667.9(15)	0.0206(12)
	5701.4(16)	0.0333(50)
	5709(1)	0.0820(3)
	5713(1)	0.0489(80)
	5757.06(15)	0.203(20)
	5807.5(15)	0.0127(2)
	5959.6(15)	0.030(10)
	5977.92(10)	0.234(20)
	6008.7(15)	0.029(4)
	6038.21(15)	0.245(30)
90-Th-228	5340.54(15)*	0.267(2)*
	5423.33(22)*	0.727(10)*
90-Th-229	4814.6(12)	0.0930(8)
	4837(2)	0.048(5)
	4845.3(12)	0.562(2)
	4901.0(12)	0.1020(8)
	4967.5(12)	0.0597(6)
	4978.5(12)	0.0317(4)
	5050(2)	0.052(5)
	5052(2)	0.016(2)
90-Th-230	4620.5(15)*	0.234(1)*
	4687.0(15)*	0.763(3)*
90-Th-232	3954(8)	0.23(2)
	4013(3)	0.77(2)
91-Pa-231	4678.6(9)	0.0161(7)
	4710.7(9)	0.0100(7)
	4734.7(8)	0.0847(40)
	4851.8(9)	0.014(1)
	4934.4(9)	0.030(4)
	4951.0(9)	0.229(5)
	4986.2(8)	0.014(2)
	5013.5(8)	0.254(5)
	5029.6(8)	0.20(2)
	5032.2(8)	0.032(3)
	5059.1(8)	0.110(5)
92-U-232	5263.41(9)*	0.317(4)*
	5320.17(14)*	0.680(4)*
92-U-233	4729(2)	0.0185(5)
	4783.5(12)*	0.149(2)*
	4824.2(12)*	0.827(3)*
92-U-234	4722.6(9)*	0.2842(2)*
	4774.9(8)*	0.7137(2)*
92-U-235	4218(2)	0.057(6)
	4325.0(9)	0.044(5)
	4365.4(9)	0.17(2)

Table 2 Continued

Nuclide	Energy (keV)	Emission probability
	4400(2)	0.55(3)
	4414(4)	0.021(2)
	4502(2)	0.017(2)
	4556(2)	0.042(3)
	4599(2)	0.050(5)
92-U-236	4445(5)	0.259(40)
	4494(3)	0.738(40)
92-U-238	4147(5)*	0.23(4)*
	4196(5)*	0.77(4)*
93-Np-237	4639.5(20)	0.0618(12)
	4664.1(20)	0.0332(10)
	4766.1(15)	0.08(3)
	4771.1(15)	0.25(6)
	4788.1(15)	0.47(9)
	4817.4(20)	0.025(4)
	4873.1(20)	0.026(2)
94-Pu-236	5720.87(10)	0.3076(33)
	5767.53(8)	0.6914(33)
94-Pu-238	5456.5(4)*	0.2884(6)*
	5499.21(20)*	0.7104(6)*
94-Pu-239	5008.76(20)	0.00019(2)
	5055.42(20)	0.00030(4)
	5076.35(20)	0.00036(4)
	5105.5(8)	0.118(2)
	5143.8(8)	0.150(2)
	5156.70(14)	0.731(7)
94-Pu-240	5123.45(23)*	0.2639(21)*
	5168.13(15)*	0.7351(36)*
94-Pu-242	4856.2(12)	0.2348(17)
	4900.5(12)	0.7649(17)
94-Pu-244	4542(2)	0.194(8)
	4589(1)	0.805(8)
95-Am-241	5442.90(13)*	0.128(2)*
	5485.60(12)*	0.852(8)*
95-Am-243	5181(1)	0.011(1)
	5233.4(10)*	0.106(11)*
	5275.3(10)*	0.88(8)*
96-Cm-242	6069.42(12)*	0.250(5)*
	6112.72(8)*	0.740(5)*
96-Cm-243	5686(3)	0.016(1)
	5742.2(10)	0.106(2)
	5785.1(10)	0.735(10)
	5992.2(10)	0.065(2)
	6010(3)	0.010(1)
	6058(3)	0.047(5)

Table 2 Continued

Nuclide	Energy (keV)	Emission probability
	6067(3)	0.015(2)
96-Cm-244	5762.16(3)*	0.2300(5)*
	5804.77(5)*	0.7698(5)*
96-Cm-246	5343.5(10)	0.178(12)
	5386.5(10)	0.822(12)
96-Cm-247	4818(4)	0.047(3)
	4868(4)	0.710(10)
	4941(4)	0.016(2)
	4983(4)	0.020(2)
	5145(4)	0.012(2)
	5210(4)	0.057(5)
	5265(4)	0.138(7)
96-Cm-248	5034.89(25)	0.165(2)
	5078.38(25)	0.751(4)
98-Cf-249	5759.7(10)	0.0366(21)
	5813.5(10)	0.844(30)
	5849.5(10)	0.0104(11)
	5903.4(10)	0.0279(44)
	5946.2(10)	0.040(5)
	6139.5(7)	0.0111(5)
	6194.0(7)	0.0217(5)
98-Cf-250	5988.9(6)	0.151(12)
	6030.22(20)	0.845(12)
98-Cf-251	5566(2)	0.015(2)
	5632(1)	0.046(11)
	5648(1)	0.036(14)
	5677(1)	0.36(2)
	5738(7)	0.010(3)
	5762(3)	0.039(5)
	5793(1)	0.020(3)
	5814(4)	0.043(5)
	5852(1)	0.28(2)
	6014(3)	0.118(7)
	6074(3)	0.028(4)
98-Cf-252	6075.64(11)*	0.152(3)*
	6118.10(4)*	0.816(3)*
99-Es-253	6590.5(14)	0.0661(1)
	6632.51(5)	0.898(2)
99-Es-254	6358.6(20)	0.026(3)
	6429.3(23)	0.931(1)

9

X-RAY AND GAMMA-RAY STANDARDS FOR DETECTOR CALIBRATION

A. Lorenz and A.L. Nichols

Table of contents

The recommended half-life and X- and gamma-ray emission probability data presented here were prepared by members of a Coordinated Research Programme (CRP) on X-ray and Gamma-ray Standards for Detector Calibration under the auspices of the International Atomic Energy Agency (IAEA). This concerted measurement and evaluation effort has resulted in a significant improvement in the accuracy and consistency of the decay parameters required for the efficiency calibration of X- and gamma-ray detectors.

The following representatives from major national and standards laboratories participated in this IAEA Coordinated Research Programme:

W. Bambynek, CEC-JRC Central Bureau for Nuclear Measurements (CBNM), Geel, Belgium

T. Barta and R. Jedlovszky, National Office of Measures (OMH), Budapest, Hungary

P. Christmas, National Physical Laboratory (NPL), Teddington, Middlesex, United Kingdom

N. Coursol, Laboratoire de Métrologie des Rayonnements Ionisants (LMRI)*, Gif-sur-Yvette, France

K. Debertin, Physikalisch-Technische Bundesanstalt (PTB), Braunschweig, Germany

R. G. Helmer, Idaho National Engineering Laboratory (INEL), Idaho Falls, Idaho, USA

---

*now Laboratoire Primaire des Rayonnements Ionisants (LPRI).

A. L. Nichols, AEA Technology, Winfrith Technology Centre, Dorchester, Dorset, United Kingdom
F. J. Schima, National Institute of Standards and Technology (NIST), Gaithersburg, Maryland, USA
Y. Yoshizawa, Faculty of Science, Hiroshima University, Hiroshima-Shi, Japan.

At the same time, other laboratories in Japan, Europe and the USA were engaged in measurements relevant to the objectives of the CRP and contributed to this effort. Valuable work was also undertaken within multinational decay data and measurement techniques intercomparison projects organized by the International Committee for Radionuclide Metrology (ICRM) and the Bureau International des Poids et Mesures (BIPM), respectively.

The programme was coordinated by A. Lorenz and H. D. Lemmel of the IAEA Nuclear Data Section. A detailed account of this project has been published[1] including listings of the recommended decay data for the selected radionuclides (i.e. X-rays from 5 to 90 keV and gamma-rays from 30 to about 3000 keV).

I. INTRODUCTION

A major objective of the IAEA nuclear data programme is to promote improvements in the quality of nuclear data used in science and technology. For many years various groups around the world have engaged in the compilation and evaluation of decay data for radionuclides. Generally, these evaluators have operated independently and arrived at different values for the same quantity. Such disagreements in the recommended data are particularly critical when attempting to define the decay characteristics of radionuclides used as standards to determine the efficiencies of X- and gamma-ray detectors to a high degree of accuracy. Not only do such discrepancies impede comparisons of measurements made by different experimentalists, but the differences can also be propagated into the subsequent measurements of decay data for other radionuclides.

Although various factors, such as source preparation and source–detector geometry, may affect the quality of measurements made with gamma-ray spectrometers, these measurements depend invariably upon the accuracy of the efficiency versus energy calibration curve derived from radionuclidic decay data.

The IAEA CRP on X-ray and Gamma-ray Standards for Detector Calibration was given the task to establish a data file that would be internationally accepted so as to improve the uniformity and validity of gamma-ray measurements. It is hoped that the evaluated data presented in these tables will serve as internationally recognized reference standards for gamma-ray spectroscopists worldwide.

II. HALF-LIVES AND GAMMA-RAY EMISSION PROBABILITIES

Based on the initial assessment of the existing half-life and gamma-ray emission probability data, members of the CRP concluded that certain half-lives and emission probabilities needed to be remeasured, and that a far greater consistency and uniformity needed to be observed in the evaluation of these data. As a consequence of the latter conclusion, the CRP adopted an evaluation procedure for the half-life data, which could also be applied to gamma-ray emission probabilities. This methodology, as outlined below, has also been described elsewhere.[2]

> *The recommended value consisted of the weighted average of the published values in which the weights were taken to be the inverse of the squares of the overall uncertainties. A set of data was considered self-consistent if the reduced Chi-squared value was approximately 1.0 or less. When the data in a set were inconsistent, and there were three or more such values, the method of limitation of the relative weight proposed by Zijp[3] was recommended. The sum of the individual weights was then computed; if any one weight contributed over 50% of the total, the corresponding uncertainty was increased so that the contribution to the value of the sum of the weights would be less than 50%. The weighted average was then recalculated and used if the reduced Chi-squared value for this average was less than 2. If the reduced Chi-squared value was larger than 2, the weighted or unweighted mean was chosen according to whether or not the 1 sigma uncertainty on each mean value included the other term.*

Photon energies of these decay data need only be known to the nearest 1 to 0.1 keV; it was therefore deemed unnecessary to carry out additional evaluations of the gamma-ray energies. Most energies quoted in this report were taken from Helmer *et al.*[4] Original references were cited when the data were not available from this source.

Internal conversion coefficients are often used in the evaluation of gamma-ray emission probabilities, either directly in the determination of a particular emission probability or in testing the consistency of the decay scheme. Theoretical values of this quantity were taken from the compilation published by Rösel *et al*;[5] when necessary, the required data were obtained by computer interpolation.[6]

All recommended half-lives and gamma-ray emission probabilities presented in this report are based on re-evaluations of these quantities, incorporating both old and newly measured values and implementing the methodology outlined above. The recommended data are listed in Tables 1 and 2, respectively.

The calibration radionuclides included in this report facilitate the precise determination of the efficiency of a gamma-ray detector from approximately 14 keV to 2.7 MeV with either a ^{24}Na or ^{228}Th source, or to 3.6 MeV with a

Table 1 Half-lives of radionuclides used for detector calibration (all values given in days)

Nuclide	Decay mode	Half-life uncertainty	
		Value	Exponent
11-Na-22	EC	950.8 ± 0.9	
11-Na-24	β^-	0.62356 ± 0.00017	
21-Sc-46	β^-	83.79 ± 0.04	
24-Cr-51	EC	27.706 ± 0.007	
25-Mn-54	EC	312.3 ± 0.4	
26-Fe-55	EC	999 ± 8	
27-Co-56	EC	77.31 ± 0.19	
27-Co-57	EC	271.79 ± 0.09	
27-Co-58	EC	70.86 ± 0.07	
27-Co-60	β^-	1925.5 ± 0.5	
30-Zn-65	EC	244.26 ± 0.26	
34-Se-75	EC	119.64 ± 0.24	
38-Sr-85	EC	64.849 ± 0.004	
39-Y-88	EC	106.630 ± 0.025	
41-Nb-93m	IT	5890 ± 50	
41-Nb-94	β^-	7.3 ± 0.9	E+06
41-Nb-95	β^-	34.975 ± 0.007	
48-Cd-109	EC	462.6 ± 0.7	
49-In-111	EC	2.8047 ± 0.0005	
50-Sn-113	EC	115.09 ± 0.04	
51-Sb-125	β^-	1007.7 ± 0.6	
53-I-125	EC	59.43 ± 0.06	
55-Cs-134	β^-	754.28 ± 0.22	
55-Cs-137	β^-	1.102 ± 0.006	E+04
56-Ba-133	EC	3862 ± 15	
58-Ce-139	EC	137.640 ± 0.023	
63-Eu-152	EC	4933 ± 11	
63-Eu-154	β^-	3136.8 ± 2.9	
63-Eu-155	β^-	1770 ± 50	
79-Au-198	β^-	2.6943 ± 0.0008	
80-Hg-203	β^-	46.595 ± 0.013	
83-Bi-207	EC	1.16 ± 0.07	E+04
90-Th-228	α	698.2 ± 0.6	
93-Np-239	β^-	2.350 ± 0.004	
95-Am-241	α	1.5785 ± 0.0024	E+05
95-Am-243	α	2.690 ± 0.008	E+06

Table 2 Gamma-ray energies and emission probabilities of radionuclides used for detector calibration

Nuclide	Energy (keV)	Emission probability
11-Na-22	1274.542(7)	0.99935(15)
11-Na-24	1368.633(6)	0.999936(15)
11-Na-24	2754.030(14)	0.99855(5)
21-Sc-46	889.277(3)	0.999844(16)
21-Sc-46	1120.545(4)	0.999874(11)
24-Cr-51	320.0842(9)	0.0986(5)
25-Mn-54	834.843(6)	0.999758(24)
27-Co-56	846.764(6)	0.99933(7)
27-Co-56	1037.844(4)	0.1413(5)
27-Co-56	1175.099(8)	0.02239(11)
27-Co-56	1238.287(6)	0.6607(19)
27-Co-56	1360.206(6)	0.04256(15)
27-Co-56	1771.350(15)	0.1549(5)
27-Co-56	2015.179(11)	0.03029(13)
27-Co-56	2034.759(11)	0.07771(27)
27-Co-56	2598.460(10)	0.1696(6)
27-Co-56	3201.954(14)	0.0313(9)
27-Co-56	3253.417(14)	0.0762(24)
27-Co-56	3272.998(14)	0.0178(6)
27-Co-56	3451.154(13)	0.0093(4)
27-Co-56	3548.27(10)	0.00178(9)
27-Co-57	14.4127(4)	0.0916(15)
27-Co-57	122.0614(3)	0.8560(17)
27-Co-57	136.4743(5)	0.1068(8)
27-Co-58	810.775(9)	0.9945(1)
27-Co-60	1173.238(4)	0.99857(22)
27-Co-60	1332.502(5)	0.99983(6)
30-Zn-65	1115.546(4)	0.5060(24)
34-Se-75	96.7344(10)	0.0341(4)
34-Se-75	121.1171(14)	0.171(1)
34-Se-75	136.0008(6)	0.588(3)
34-Se-75	264.6580(17)	0.590(2)
34-Se-75	279.5431(22)	0.250(1)
34-Se-75	400.6593(13)	0.115(1)
38-Sr-85	514.0076(22)	0.984(4)
39-Y-88	898.042(4)	0.940(3)
39-Y-88	1836.063(13)	0.9936(3)
41-Nb-94	702.645(6)	0.9979(5)
41-Nb-94	871.119(4)	0.9986(5)
41-Nb-95	765.807(6)	0.9981(3)
48-Cd-109	88.0341(11)	0.0363(2)
49-In-111	171.28(3)	0.9078(10)
49-In-111	245.35(4)	0.9416(6)
50-Sn-113	391.702(4)	0.6489(13)

Table 2 Continued

Nuclide	Energy (keV)	Emission probability
51-Sb-125	176.313(1)	0.0685(7)
51-Sb-125	380.452(8)	0.01518(16)
51-Sb-125	427.875(6)	0.297(3)
51-Sb-125	463.365(5)	0.1048(11)
51-Sb-125	600.600(4)	0.1773(18)
51-Sb-125	606.718(3)	0.0500(5)
51-Sb-125	635.954(5)	0.1121(12)
53-I-125	35.4919(5)	0.0658(8)
55-Cs-134	475.364(3)	0.0149(2)
55-Cs-134	563.240(4)	0.0836(3)
55-Cs-134	569.328(3)	0.1539(6)
55-Cs-134	604.720(3)	0.9763(6)
55-Cs-134	795.859(5)	0.854(3)
55-Cs-134	801.948(5)	0.0869(3)
55-Cs-134	1038.610(7)	0.00990(5)
55-Cs-134	1167.968(5)	0.01792(7)
55-Cs-134	1365.185(7)	0.03016(11)
55-Cs-137 (a)	661.660(3)	0.851(2)
56-Ba-133	80.998(5)	0.3411(28)
56-Ba-133	276.398(1)	0.07147(30)
56-Ba-133	302.853(1)	0.1830(6)
56-Ba-133	356.017(2)	0.6194(14)
56-Ba-133	383.851(3)	0.08905(29)
58-Ce-139	165.857(6)	0.7987(6)
63-Eu-152	121.7824(4)	0.2837(13)
63-Eu-152	244.6989(10)	0.0753(4)
63-Eu-152	344.2811(19)	0.2657(11)
63-Eu-152	411.126(3)	0.02238(10)
63-Eu-152 (b)	443.965(4)	0.03125(14)
63-Eu-152	778.903(6)	0.1297(6)
63-Eu-152	867.390(6)	0.04214(25)
63-Eu-152	964.055(4)	0.1463(6)
63-Eu-152	1085.842(4)	0.1013(5)
63-Eu-152	1089.767(14)	0.01731(9)
63-Eu-152	1112.087(6)	0.1354(6)
63-Eu-152	1212.970(13)	0.01412(8)
63-Eu-152	1299.152(9)	0.01626(11)
63-Eu-152	1408.022(4)	0.2085(9)
63-Eu-154	123.071(1)	0.412(5)
63-Eu-154	247.930(1)	0.0695(9)
63-Eu-154	591.762(5)	0.0499(6)
63-Eu-154	692.425(4)	0.0180(3)
63-Eu-154	723.305(5)	0.202(2)
63-Eu-154	756.804(5)	0.0458(6)
63-Eu-154	873.190(5)	0.1224(15)

Table 2 Continued

Nuclide	Energy (keV)	Emission probability
63-Eu-154	996.262(6)	0.1048(13)
63-Eu-154	1004.725(7)	0.182(2)
63-Eu-154	1274.436(6)	0.350(4)
63-Eu-154	1494.048(9)	0.0071(2)
63-Eu-154	1596.495(18)	0.0181(2)
79-Ag-198	411.8044(11)	0.9557(47)
80-Hg-203	279.1967(12)	0.8148(8)
83-Bi-207	569.702(2)	0.9774(3)
83-Bi-207	1063.662(4)	0.745(2)
83-Bi-207	1770.237(9)	0.0687(4)
90-Th-228	84.373(3)	0.0122(2)
90-Th-228 (c)	238.632(2)	0.435(4)
90-Th-228 (c)	240.987(6)	0.0410(5)
90-Th-228 (c)	277.358(10)	0.0230(3)
90-Th-228 (c)	300.094(10)	0.0325(3)
90-Th-228 (c)	510.77(10)	0.0818(10)
90-Th-228 (c)	583.191(2)	0.306(2)
90-Th-228 (c)	727.330(9)	0.0669(9)
90-Th-228 (c)	860.564(5)	0.0450(4)
90-Th-228 (c)	1620.735(10)	0.0149(5)
90-Th-228 (c)	2614.533(13)	0.3586(6)
93-Np-239	106.123(2)	0.267(4)
93-Np-239	228.183(1)	0.1112(15)
93-Np-239	277.599(2)	0.1431(20)
95-Am-241	26.345(1)	0.024(1)
95-Am-241	59.537(1)	0.360(4)
95-Am-243	43.53(1)	0.0594(11)
95-Am-243	74.66(1)	0.674(10)

(a) Gamma-ray belongs to daughter radionuclide 56-Ba-137 m (2.55 min half-life) in secular equilibrium with the 55-Cs-137 parent radionuclide, which has a half-life of 11020 days.
(b) Gamma-ray has been recognized to be an unresolved doublet.
(c) 90-Th-228 (698.2 d half-life) gamma-rays are in secular equilibrium with parent

Table 3 X-ray energies and emission probabilities of radionuclides used for detector calibration

Nuclide	Trans	E_{min} (keV)	E_{max} (keV)	Emission probability
24-Cr-51	VKα	4.95		0.201(3)
24-Cr-51	VKβ	5.43		0.027(1)
24-Cr-51	VKX	4.95	5.43	0.228(3)
25-Mn-54	CrKα	5.41		0.226(7)
25-Mn-54	CrKβ	5.95		0.030(1)
25-Mn-54	CrKX	5.41	5.95	0.256(8)
26-Fe-55	MnKα	5.89		0.249(9)
26-Fe-55	MnKβ	6.49		0.034(1)
26-Fe-55	MnKX	5.89	6.49	0.283(10)
27-Co-57	FeKα	6.40		0.510(7)
27-Co-57	FeKβ	7.06		0.069(1)
27-Co-57	FeKX	6.40	7.06	0.579(8)
27-Co-58	FeKα	6.40		0.235(3)
27-Co-58	FeKβ	7.06		0.032(1)
27-Co-58	FeKX	6.40	7.06	0.267(3)
30-Zn-65	CuKα	8.03	8.05	0.341(6)
30-Zn-65	CuKβ	8.91		0.046(1)
30-Zn-65	CuKX	8.03	8.91	0.387(6)
34-Se-75	AsKα	10.51	10.54	0.493(11)
34-Se-75	AsKβ	11.72	11.95	0.075(2)
34-Se-75	AsKX	10.51	11.95	0.568(13)
38-Sr-85	RbKα	13.34	13.40	0.500(3)
38-Sr-85	RbKβ	14.96	15.29	0.087(2)
38-Sr-85	RbKX	13.34	15.29	0.587(4)
39-Y-88	SrKα	14.10	14.17	0.522(6)
39-Y-88	SrKβ	15.83	16.19	0.094(2)
39-Y-88	SrKX	14.10	16.19	0.616(7)
41-Nb-93m	NbKα	16.52	16.62	0.0925(30)
41-Nb-93m	NbKβ	18.62	19.07	0.0179(7)
41-Nb-93m	NbKX	16.52	19.07	0.1104(35)
48-Cd-109	AgKα	21.99	22.16	0.821(9)
48-Cd-109	AgKβ	24.93	25.60	0.173(3)
48-Cd-109	AgKX	21.99	25.60	0.994(10)
49-In-111	CdKα	22.98	23.17	0.684(5)
49-In-111	CdKβ	26.09	26.80	0.146(3)
49-In-111	CdKX	22.98	26.80	0.830(5)
50-Sn-113	InKα	24.00	24.21	0.796(6)
50-Sn-113	InKβ	27.27	28.02	0.172(3)
50-Sn-113	InKX	24.00	28.02	0.968(6)
53-I-125	TeKα	27.20	27.47	1.135(21)
53-I-125	TeKβ	30.98	31.88	0.255(6)
53-I-125	TeKX	27.20	31.88	1.390(25)

Table 3 Continued

Nuclide	Trans	E_{min} (keV)	E_{max} (keV)	Emission probability
55-Cs-137	BaKα	31.82	32.19	0.0566(16)
55-Cs-137	BaKβ	36.36	37.45	0.0134(4)
55-Cs-137	BaKX	31.82	37.45	0.0700(20)
56-Ba-133	CsKα	30.63	30.97	0.980(14)
56-Ba-133	CsKβ	34.97	36.01	0.230(5)
56-Ba-133	CsKX	30.63	36.01	1.210(16)
58-Ce-139	LaKα	33.03	33.44	0.643(18)
58-Ce-139	LaKβ	37.78	38.93	0.154(5)
58-Ce-139	LaKX	33.03	38.93	0.792(22)
63-Eu-152	SmKα	39.52	40.12	0.591(12)
63-Eu-152	GdKα	42.31	43.00	0.00648(22)
63-Eu-152	SmKβ	45.38	46.82	0.149(3)
63-Eu-152	GdKβ	48.65	50.21	0.00176(18)
63-Eu-152	SmKX	39.52	46.82	0.740(12)
63-Eu-152	GdKX	42.31	50.21	0.00824(28)
63-Eu-152	(Sm+Gd)KX	39.52	50.21	0.748(12)
63-Eu-154	GdKα	42.31	43.00	0.205(6)
63-Eu-154	GdKβ	48.65	50.21	0.051(2)
63-Eu-154	GdKX	42.31	50.21	0.256(6)
79-Au-198	HgKα	68.89	70.82	0.0219(8)
79-Au-198	HgKβ	80.12	82.78	0.0061(3)
79-Au-198	HgKX	68.89	82.78	0.0280(10)
80-Hg-203	T1LX	8.95	14.40	0.060(12)
80-Hg-203	T1Kα_2	70.83		0.038(2)
80-Hg-203	T1Kα_1	72.87		0.064(2)
80-Hg-203	T1Kβ'_1	82.43		0.022(1)
80-Hg-203	T1Kβ'_2	85.19		0.0063(3)
80-Hg-203	T1KX	70.83	85.19	0.130(4)
83-Bi-207	PbLX	9.19	14.91	0.325(13)
83-Bi-207	PbKα_2	72.80		0.226(12)
83-Bi-207	PbKα_1	74.97		0.382(20)
83-Bi-207	PbKβ'_1	84.79		0.130(10)
83-Bi-207	PbKβ'_2	87.63		0.039(3)
83-Bi-207	PbKX	72.80	87.63	0.777(26)
95-Am-241	NpL$_1$	11.871		0.0085(3)
95-Am-241	NpLα	13.927		0.132(4)
95-Am-241	NpLβ_η	17.611		0.194(6)
95-Am-241	NpL$_\gamma$	20.997		0.049(2)
95-Am-241	NpLX	11.871	20.997	0.3835(75)

^{56}Co source. Even though data above 3.6 MeV were not evaluated by this CRP, certain methods and sources can be used to extend the efficiency calibration to energies above 10 MeV. These range from ^{66}Ga, which has a relatively short half-life of 9.5 hours with calibration energies ranging from 3.8 to 4.8 MeV, to gamma-ray generation methods involving nuclear reactions induced by thermal neutrons, protons and alpha particles. A more extensive discussion of detector efficiency calibration at high energies is given in the original IAEA publication.[1]

III. X-RAY EMISSION PROBABILITIES

The relevant data required for the evaluation of the X-ray emission probabilities were derived as follows:

- Electron-capture probabilities were calculated from the electron wave functions of Mann and Waber;[7] exchange and overlap corrections were made according to Bahcall[8] and Vatai[9] as recalculated by Bambynek *et al.*[10] for $Z < 54$; corrections by Suslov[11] and Martin and Blichert-Toft[12] were made for $Z > 54$. The method of calculation and the input data have been described by Bambynek *et al.*[10]
- Fluorescence yields were deduced from recent evaluations by Bambynek.[13]
- Internal conversion data were obtained from compilations of experimental values,[14,15] and from evaluations of specific transitions.[16,17]
- Relative X-ray emission rates were taken from Salem *et al.*[18] allowing for the contribution of radiative Auger satellites to these ratios.[19]

All resulting X-ray data were critically assessed against measured X-ray or gamma-ray emission probabilities and adjusted when necessary. X-ray energies were not evaluated, but taken from Browne and Firestone.[20] Uncertainties were estimated according to the recommendations of BIPM, the Bureau International des Poids et Mesures.[21] The recommended X-ray data covering the energy range from 4.95 to 87 keV are listed in Table 3.

REFERENCES

1. International Atomic Energy Agency, X-Ray and Gamma-Ray Standards for Detector Efficiency Calibration, Report IAEA-TECDOC-619, International Atomic Energy Agency, Vienna, 1991.
2. **Woods, M.J., Munster, A.S.**, Evaluation of Half-Life Data, Report RS(EXT)95, National Physical Laboratory, Teddington, 1988.
3. **Zijp, W.L.**, On the Statistical Evaluation of Inconsistent Measurement Results Illustrated by the Example of the ^{90}Sr Half-life, Report ECN-179, Netherlands Energy Research Foundation ECN, Petten, 1985.

4. **Helmer, R.G., Van Assche, P.H.M., van der Leun, C.**, *At. Data Nucl. Data Tables* 24, 39, 1979.
5. **Rösel, F., Fries, H.M., Alder, K., Pauli, H.C.**, *At. Data Nucl. Data Tables* 21, 91, 1978.
6. **Coursol, N.**, Report RI-LPRI-102, Département des Applications et de la Métrologie des Rayonnements Ionisants, Gif-sur-Yvette, 1990.
7. **Mann, J.B., Waber, J.T.**, *At. Data* 5, 201, 1973.
8. **Bahcall, J.N.**, *Phys. Rev.* 129, 2963, 1963;
   ibid, 131, 1756, 1963;
   ibid, 132, 362, 1963.
9. **Vatai, E.**, *Nucl. Phys.* A156, 541, 1970.
10. **Bambynek, W., Behrens, H., Chen, M.H., Crasemann, B., Fitzpatrick, M.L., Ledingham, K.W.D., Genz, H., Mutterer, M., Intemann, R.L.**, *Rev. Mod. Phys.* 49, 77, 1977.
11. **Suslov, Yu.P.**, *Izv. Akad. Nauk, SSSR Ser. Fiz.* 34, 79, 1970. English translation: *Bull. Acad. Sci. USSR, Phys. Ser.* 34, 91, 1983.
12. **Martin, M.J., Blichert-Toft, P.H.**, *Nucl. Data* A8, 1, 1970.
13. **Bambynek, W.**, in *Proc. Conf. on X-ray and Inner-shell Processes in Atoms, Molecules and Solids*, Leipzig, Meisel, A., Ed., VEB Druckerei Thomas Münzer, Langensalza, Germany, paper P1, 1984.
14. **Hansen, H.H.**, Report Physics Data 17-1, Fachinformationszentrum Karlsruhe, 1981.
15. **Hansen, H.H.**, Report Physics Data 17-2, Fachinformationszentrum Karlsruhe, 1985.
16. **Hansen, H.H.**, *Eur. Appl. Res. — Nucl. Sci. Technol.* 6, 777, 1985.
17. **Lagoutine, F., Coursol, N., Legrand, J.**, *Table de Radionucléides*, Volume 4, Département des Applications et de la Métrologie des Rayonnements Ionisants, Gif-sur-Yvette, 1987.
18. **Salem, S.I., Panossian, S.L., Krause, R.A.**, *At. Data Nucl. Data Tables* 14, 91, 1974.
19. **Campbell, J.L., Perujo, A., Teesdale, W.J., Millman, B.M.**, *Phys. Rev.* A33, 240, 1986.
20. **Browne, E., Firestone, R.B.**, *Table of Radioactive Isotopes*, Shirley, V.S., Ed., John Wiley & Sons, New York, 1986.
21. **Giacomo, P.**, in *Quantum Metrology and Fundamental Physical Constants*, Cuttler, P.H., Lukas, A.A., Eds., Plenum Publ. Co., New York, 623, 1983; **Giacomo, P.**, *Metrologia* 17, 69, 1981.

10
Table of nuclides*

Jagdish K. Tuli†

Table of contents

I. Introduction

This table presents half–life, abundance, decay modes, spin, parity, and mass excess for all known nuclides and some of their isomeric states. The data given here are from the adopted properties of the various nuclides as given in the *Evaluated Nuclear Structure Data File* (ENSDF)[1]. The data in ENSDF are based on experimental results as reported in *Nuclear Data Sheets*[2] for A=45 to 266 and in *Nuclear Physics*[3,4] for A<45. For those nuclides for which either there are no data in ENSDF or more recent data are available, the half–life and decay modes are taken from the *Chart of the Nuclides*, 14th Edition[5]. The ground–state mass excesses are from the mass adjustments by G. Audi and A. H. Wapstra[6]. The isotopic abundances are those of N. E. Holden[7].

For other references, experimental data, and information on the data measurements please refer to the original evaluations[1–4].

---

*The Nuclear Properties given here are based upon the author's pocket–size handbook *Nuclear Wallet Cards* (1990) published and distributed free by the National Nuclear Data Center, Brookhaven National Laboratory, Upton, NY 11973, USA. *Nuclear Wallet Cards* also contains a number of useful appendices.

†This research was supported by the Office of Basic Energy Sciences, US Department of Energy at Brookhaven National Laboratory, which is operated by the Associated Universities, Inc., under contract with the US Department of Energy.

II. Explanation of table

Column 1, Isotope (*Z*, *El*, *A*)

Nuclides are listed in order of increasing atomic number (Z), and are subordered by increasing mass number (A). Included are all isotopic species, all isomers with half-life ≥1 s, and other selected well-known isomers. A nuclide is included even if the only datum is its mass estimate.

Isomeric states are denoted by the symbol "m" after the mass number and are given in the order of increasing excitation energy. More than one entry for a nuclide, without the symbol "m" for any, indicates that their relative excitation energies are not known.

The ^{235}U fission products, with fractional yields $>10^{-6}$, are *italicized* in the table. The fission product information is taken from reference[8].

The symbols Rf (Rutherfordium) and Ha (Hahnium) have been used for elements with Z=104 and 105, respectively. These, however, have not been accepted internationally owing to conflicting claims of their discovery.

Column 2, $J\pi$

Spin and parity assignments without and with parentheses are based upon strong and weak arguments, respectively. See the introductory pages of any issue of *Nuclear Data Sheets*[2] for a description of strong and weak arguments for $J\pi$ assignments. A colon (:) in the field refers to range of values.

Column 3, Mass Excess, Δ

Mass excesses, $M-A$, are given in MeV with $\Delta(^{12}C)=0$, by definition. For isomers the values are obtained by adding the excitation energy to the Δ(g.s.) values. Wherever the excitation energy is not known, the mass excess for the next lower isomer (or g.s.) is given. An appended "s" denotes that the value is from systematics.

Column 4, $T_{1/2}$ or Abundance

The half–life and the abundance are given followed by units ("%" symbol in case of abundance) which are followed by the uncertainty in *italics*. The uncertainty given is in the last significant figures. For example, 8.1 s *10* means $T_{1/2}$=8.1±1.0 s. For some very short–lived nuclei, level widths rather than half–lives are given. There also, the width is followed by the units (e.g. eV, keV, or MeV) which are followed by the uncertainty in *italics*, if known.

Column 5, Decay Mode:

Decay modes are given followed by the percentage branching, if known ("w" indicates a weak branch). The decay modes are given in decreasing strength from left to right. The percentage branching is omitted where there is no competing mode of decay.

The various modes of decay are given below:

$\beta-$	β^- decay
ε	ε (electron capture), or $\varepsilon+\beta^+$, or β^+ decay
IT	isomeric transition (through γ or conversion–electron decay)
n, p, α, ...	neutron, proton, alpha, ... decay
SF	spontaneous fission
$2\beta-$, 3α, ...	double β^- decay ($\beta^-\beta^-$), decay through emission of 3 α's, ...
$\beta-$n, $\beta-$p, $\beta-\alpha$, ...	delayed n, p, α, ... emission following β^- decay
εp, $\varepsilon\alpha$, εSF, ...	delayed p, α, SF, ... decay following ε or β^+ decay

Table of Nuclides

Isotope Z	El	A	Jπ	Δ (MeV)	T1/2 or Abundance	Decay Mode
0	n	1	1/2+	8.071	10.4 m *2*	β−
1	H	1	1/2+	7.289	99.985% *1*	
		2	1+	13.136	0.015% *1*	
		3	1/2+	14.950	12.33 y *6*	β−
		4?	2−	26.002		
2	He	3	1/2+	14.931	0.000137% *3*	
		4	0+	2.425	99.999863% *3*	
		5	3/2−	11.386	0.60 MeV *2*	α, n
		6	0+	17.594	806.7 ms *15*	β−
		7	(3/2)−	26.110	160 keV *30*	n
		8	0+	31.598	119.0 ms *15*	β−, β−n 16%
		9		40.819	very short	n
3	Li	4	2−	25.320		
		5	3/2−	11.679	≈1.5 MeV	α, p
		6	1+	14.086	7.5% *2*	
		7	3/2−	14.908	92.5% *2*	
		8	2+	20.945	838 ms *6*	β−, β−2α
		9	3/2−	24.954	178.3 ms *4*	β−, β−n 49.5%, β−n2α
		10		33.445	1.2 MeV *3*	n
		11	3/2−	40.788	8.5 ms *2*	β−, β−nα 0.027%, β−n
4	Be	6	0+	18.375	92 keV *6*	2p, α
		7	3/2−	15.769	53.29 d *7*	ε
		8	0+	4.942	6.8 eV *17*	2α
		9	3/2−	11.348	100%	
		10	0+	12.607	1.51×10^6 y *6*	β−
		11	1/2+	20.174	13.81 s *8*	β−, β−α 3.1%
		12	0+	25.076	23.6 ms *9*	β−, β−n<1%
		13	(1/2,5/2)+	35.158		n
		14	0+	39.882	4.35 ms *17*	β−, β−n 81%, β−2n 5%
5	B	7	(3/2−)	27.868	1.4 MeV *2*	α, p
		8	2+	22.921	770 ms *3*	εα, ε, ε2α
		9	3/2−	12.416	0.54 keV *21*	2α, p
		10	3+	12.051	19.9% *2*	
		11	3/2−	8.668	80.1% *2*	
		12	1+	13.369	20.20 ms *2*	β−, β−3α 1.58%
		13	3/2−	16.562	17.36 ms *16*	β−
		14	2−	23.664	13.8 ms *10*	β−
		15		28.967	10.5 ms *3*	β−
		16	(0−)	37.139s		n
		17	(3/2−)	43.716	5.08 ms *5*	β−, β−
		18		52.322s		
6	C	8	0+	35.094	230 keV *50*	α, p
		9	(3/2−)	28.914	126.5 ms *9*	ε, εp, ε2α
		10	0+	15.699	19.255 s *53*	ε
		11	3/2−	10.650	20.39 m *2*	ε
		12	0+		98.90% *3*	
		13	1/2−	3.125	1.10% *3*	
		14	0+	3.020	5730 y *40*	β−
		15	1/2+	9.873	2.449 s *5*	β−
		16	0+	13.694	0.747 s *8*	β−
		17		21.037	193 ms *13*	β−, β−n 32%
		18	(0+)	24.924	66 ms *+25−15*	β−
		19		32.833		β−
		20	0+	37.560		β−

Isotope Z El	A	Jπ	Δ (MeV)	T1/2 or Abundance	Decay Mode
7 N	10		39.699s		
	11	(1/2+)	24.961		p
	11m	1/2-	24.961	0.74 MeV *10*	p
	12	1+	17.338	11.000 ms *16*	ε, ε3α 3.44%
	13	1/2-	5.345	9.965 m *4*	ε
	14	1+	2.863	99.63% *2*	
	15	1/2-	0.101	0.37% *2*	
	16	2-	5.682	7.13 s *2*	β-
	17	1/2-	7.871	4.173 s *4*	β-
	18	1-	13.117	624 ms *12*	β-
	19		15.860	0.27 s *6*	β-, β-n≈33%
	20		21.767	100 ms *+30-20*	β-, β-n≈61%
	21		25.232	95 ms *13*	β-n 84%, β-
	22		32.081	24 ms *7*	β-n 35%, β-
8 O	12	0+	32.063	0.40 MeV *25*	p
	13	(3/2-)	23.111	8.58 ms *5*	ε
	14	0+	8.007	70.606 s *18*	ε
	15	1/2-	2.855	122.24 s *16*	ε
	16	0+	-4.737	99.76% *1*	
	17	5/2+	-0.809	0.038% *3*	
	18	0+	-0.782	0.20% *1*	
	19	5/2+	3.332	26.91 s *8*	β-
	20	0+	3.797	13.51 s *5*	β-
	21	(1/2:5/2)+	8.062	3.42 s *10*	β-
	22	0+	9.284	2.25 s *15*	β-
	23		14.616	82 ms *37*	β-, β-n 31%
	24	0+	18.975	61 ms *26*	β-, β-n 58%
9 F	14	(2-)	33.608s		p
	15	(1/2+)	16.777	1.0 MeV *2*	p
	16	0-	10.680	40 keV *20*	p
	17	5/2+	1.952	64.49 s *16*	ε
	18	1+	0.873	109.77 m *5*	ε
	19	1/2+	-1.487	100.%	
	20	2+	-0.017	11.00 s *2*	β-
	21	5/2+	-0.048	4.158 s *20*	β-
	22	4+,(3+)	2.794	4.23 s *4*	β-
	23	(3/2,5/2)+	3.329	2.23 s *14*	β-
	24	(1,2,3)+	7.544	0.34 s *8*	β-
	25		11.266		
	26		18.288		
	27		25.050		
10 Ne	16	0+	23.989	122 keV *37*	p
	17	1/2-	16.485	109.2 ms *6*	ε
	18	0+	5.319	1672 ms *8*	ε
	19	1/2+	1.751	17.34 s *9*	ε
	20	0+	-7.042	90.48% *3*	
	21	3/2+	-5.732	0.27% *1*	
	22	0+	-8.024	9.25% *3*	
	23	5/2+	-5.154	37.24 s *12*	β-
	24	0+	-5.948	3.38 m *2*	β-
	25	(1/2,3/2)+	-2.059	602 ms *8*	β-
	26	0+	0.430	0.23 s *6*	β-
	27		7.094		
	28	0+	11.279		β-
11 Na	18		25.318s		p
	19		12.929		ε, εp
	20	2+	6.845	447.9 ms *23*	ε
	21	3/2+	-2.184	22.49 s *4*	ε
	22	3+	-5.182	2.6019 y *4*	ε
	23	3/2+	-9.530	100%	

Isotope Z El	A	Jπ	Δ (MeV)	T1/2 or Abundance	Decay Mode
11 Na	24	4+	-8.418	14.9590 h *12*	β-
	24m	1+	-7.946	20.20 ms *7*	IT, β-≈0.05%
	25	5/2+	-9.358	59.1 s *6*	β-
	26	3+	-6.903	1.072 s *9*	β-
	27	5/2+	-5.581	301 ms *6*	β-, β-n 0.08%
	28	1+	-1.034	30.5 ms *4*	β-, β-n 0.58%
	29		2.619	44.9 ms *12*	β-, β-n 21.5%
	30	2+	8.594	48 ms *2*	β-, β-n 30%, β-2n 1.17%, β-α 5.5×10^{-5}%
	31		12.664	17.0 ms *4*	β-, β-n 37%, β-2n 0.9%
	32		18.304	13.2 ms *4*	β-, β-n 24%, β-2n 8%
	33		25.510	8.2 ms *4*	β-, β-n 52%, β- 12%
	34		32.509s	5.5 ms *10*	β-n 115%, β-
	35		41.153s	1.5 ms *5*	β-, β-n
12 Mg	20	0+	17.570	95 ms *+80-50*	ε, εp≥3%
	21	(3/2,5/2)+	10.912	122 ms *3*	ε, εp 29.3%
	22	0+	-0.397	3.857 s *9*	ε
	23	3/2+	-5.473	11.317 s *11*	ε
	24	0+	-13.933	78.99% *3*	
	25	5/2+	-13.193	10.00% *1*	
	26	0+	-16.215	11.01% *2*	
	27	1/2+	-14.587	9.458 m *12*	β-
	28	0+	-15.019	20.91 h *3*	β-
	29	3/2+	-10.661	1.30 s *12*	β-
	30	0+	-8.882	335 ms *17*	β-
	31		-3.215	230 ms *20*	β-, β-n 1.7%
	32	0+	-0.796	120 ms *20*	β-, β-n 2.4%
	33		5.204	90 ms *20*	β-, β-n 17%
	34	0+	8.451	20 ms *10*	β-, β-n
13 Al	22		18.183s	70 ms *+50-35*	ε, εp>0%, ε2p>0%
	23		6.767	0.47 s *3*	ε, εp
	24	4+	-0.055	2.053 s *4*	ε, εα 0.04%
	24m	1+	0.371	131.3 ms *25*	IT 82%, ε 18%, εα 0.03%
	25	5/2+	-8.916	7.183 s *12*	ε
	26	5+	-12.210	7.4×10^{5} y *3*	ε
	26m	0+	-11.982	6.3452 s *19*	ε
	27	5/2+	-17.197	100%	
	28	3+	-16.851	2.2414 m *12*	β-
13 Al	29	5/2+	-18.215	6.56 m *6*	β-
	30	3+	-15.872	3.60 s *6*	β-
	31	(3/2,5/2)+	-14.954	644 ms *25*	β-
	32	1+	-11.062	33 ms *4*	β-
	33		-8.505		
	34		-2.862	60 ms *18*	β-, β-n 27%
	35		-0.058	150 ms *50*	β-, β-n 65%
	36	0+	5.916		
	37		9.604		
14 Si	22	0+	32.164s	6 ms *3*	ε, εp
	24	0+	10.755	102 ms *35*	ε, εp≈7%
	25	5/2+	3.825	220 ms *3*	ε, εp
	26	0+	-7.145	2.234 s *13*	ε
	27	5/2+	-12.385	4.16 s *2*	ε
	28	0+	-21.493	92.23% *1*	
	29	1/2+	-21.895	4.67% *1*	
	30	0+	-24.433	3.10% *1*	
	31	3/2+	-22.949	157.3 m *3*	β-

Isotope Z El	A	Jπ	Δ (MeV)	T1/2 or Abundance	Decay Mode
14 Si	32	0+	-24.081	172 y *4*	β-
	33		-20.492	6.18 s *18*	β-
	34	0+	-19.957	2.77 s *20*	β-
	35		-14.360	0.78 s *12*	β-
	36	0+	-12.401	0.45 s *6*	β-, β-n < 10%
	37		-6.524		
	38	0+	-3.745		
15 P	25		18.872s		
	26	(3+)	10.973s	20 ms *+35-15*	ε, εp 2%
	27		-0.753	260 ms *80*	ε, εp 6%
	28	3+	-7.161	270.3 ms *5*	ε
	29	1/2+	-16.952	4.142 s *15*	ε
	30	1+	-20.201	2.498 m *4*	ε
	31	1/2+	-24.441	100%	
	32	1+	-24.305	14.262 d *14*	β-
	33	1/2+	-26.338	25.34 d *12*	β-
	34	1+	-24.558	12.43 s *8*	β-
	35	1/2+	-24.858	47.3 s *7*	β-
	36		-20.251	5.6 s *3*	β-
	37		-18.995	2.31 s *13*	β-
	38		-14.466	0.64 s *14*	β-, β-n < 10%
	39		-12.650	0.16 s *+30-10*	β-, β-n 41%
	40		-8.337	260 ms *80*	β-, β-n 30%
	41		-4.844	120 ms *20*	β-, β-n 30%
	42		0.084s	110 ms *30*	β-, β-n 50%
	45				
16 S	27		17.507s		ε
	28	0+	4.073	125 ms *10*	ε, εp > 0%
	29		-3.159	187 ms *4*	ε
	30	0+	-14.063	1.178 s *5*	ε
	31	1/2+	-19.045	2.572 s *13*	ε
	32	0+	-26.016	95.02% *9*	
	33	3/2+	-26.586	0.75% *1*	
	34	0+	-29.932	4.21% *8*	
	35	3/2+	-28.846	87.51 d *12*	β-
	36	0+	-30.664	0.02% *1*	
	37	7/2-	-26.896	5.05 m *2*	β-
	38	0+	-26.861	170.3 m *7*	β-
	39	(3/2:7/2)-	-23.161	11.5 s *5*	β-
	40	0+	-22.850	8.8 s *22*	β-
	41		-18.602		
	42	0+	-17.242		
	43		-12.482	220 ms *65*	β-, β-n 40%
	44	0+	-10.880s	200 ms *40*	β-, β-n 30%
17 Cl	29		13.143s		
	30		4.443s		
	31		-7.064	150 ms *25*	ε, εp 0.44%
	32	1+	-13.331	298 ms *1*	ε, εα 0.01%, εp 7.0×10^{-3}%
	33	3/2+	-21.003	2.511 s *3*	ε
	34	0+	-24.440	1.5264 s *14*	ε
	34m	3+	-24.294	32.00 m *4*	ε 55.4%, IT 44.6%
	35	3/2+	-29.014	75.77% *5*	
	36	2+	-29.522	3.01×10^{5} y *2*	β- 98.1%, ε 1.9%
	37	3/2+	-31.761	24.23% *5*	
	38	2-	-29.798	37.24 m *5*	β-
	38m	5-	-29.127	715 ms *3*	IT
	39	3/2+	-29.800	55.6 m *2*	β-
	40	2-	-27.558	1.35 m *2*	β-
	41	(1/2,3/2)+	-27.339	38.4 s *8*	β-
	42		-24.987	6.8 s *3*	β-

Isotope Z El	A	Jπ	Δ (MeV)	T1/2 or Abundance	Decay Mode
17 Cl	43		-24.029	3.3 s *2*	β-
	44		-19.991		
	45		-18.909	400 ms *43*	β-, β-n 24%
18 Ar	32	0+	-2.179	98 ms *2*	ε, εp
	33	1/2+	-9.381	173.0 ms *20*	ε, εp 38.7%
	34	0+	-18.378	844.5 ms *34*	ε
	35	3/2+	-23.048	1.775 s *4*	ε
	36	0+	-30.230	0.337% *3*	
	37	3/2+	-30.948	35.04 d *4*	ε
	38	0+	-34.715	0.063% *1*	
	39	7/2-	-33.242	269 y *3*	β-
	40	0+	-35.040	99.600% *3*	
	41	7/2-	-33.067	109.34 m *12*	β-
	42	0+	-34.422	32.9 y *11*	β-
	43	(3/2,5/2)	-31.978	5.37 m *6*	β-
	44	0+	-32.262	11.87 m *5*	β-
	45		-29.719	21.48 s *15*	β-
	46	0+	-29.721	8.4 s *6*	β-
	47		-25.908		
19 K	33		6.763s		
	34		-1.481s		
	35	3/2+	-11.167	190 ms *30*	ε, εp 0.37%
	36	2+	-17.425	342 ms *2*	ε, εp 0.05%, εα 3.4×10^{-3}%
	37	3/2+	-24.799	1.226 s *7*	ε
	38	3+	-28.802	7.636 m *18*	ε
	38	0+	-28.672	923.9 ms *6*	ε
	39	3/2+	-33.807	93.2581% *30*	
	40	4-	-33.535	1.277×10^{9} y *8* 0.0117% *1*	β- 89.28%, ε 10.72%
	41	3/2+	-35.559	6.7302% *30*	
	42	2-	-35.021	12.360 h *3*	β-
	43	3/2+	-36.593	22.3 h *1*	β-
	44	2-	-35.810	22.13 m *19*	β-
	45	3/2+	-36.608	17.3 m *6*	β-
	46	(2-)	-35.419	105 s *10*	β-
	47	1/2+	-35.697	17.5 s *3*	β-
	48	(2-)	-32.125	6.8 s *2*	β-, β-n 1.14%
	49	(1/2+,3/2+)	-30.320	1.26 s *5*	β-, β-n 86%
	50	(0-,1,2-)	-25.353	472 ms *4*	β-, β-n 29%
	51	(1/2+,3/2+)		365 ms *5*	β-, β-n 47%
	52			105 ms *5*	β-, β-n>88%
	53	(3/2+)		30 ms *5*	β-, β-n 85%
	54			10 ms *5*	β-, β-n
20 Ca	35		4.439s	50 ms *30*	ε, ε2p
	36	0+	-6.439	100 ms *+90-40*	ε, εp≈20%
	37	3/2+	-13.161	175 ms *3*	ε, εp 76%
	38	0+	-22.059	440 ms *8*	ε
	39	3/2+	-27.276	859.6 ms *14*	ε
	40	0+	-34.846	96.941% *13*	
	41	7/2-	-35.138	1.03×10^{5} y *4*	ε
	42	0+	-38.547	0.647% *3*	
	43	7/2-	-38.408	0.135% *3*	
	44	0+	-41.469	2.086% *5*	
	45	7/2-	-40.813	163.8 d *18*	β-
	46	0+	-43.135	0.004% *3*	
	47	7/2-	-42.340	4.536 d *2*	β-
	48	0+	-44.215	$>6\times10^{18}$ y 0.187% *3*	β-, 2β-
	49	3/2-	-41.290	8.715 m *23*	β-
	50	0+	-39.571	13.9 s *6*	β-

Isotope Z El	A	Jπ	Δ (MeV)	T1/2 or Abundance	Decay Mode
20 Ca	51	(3/2−)	−35.905	10.0 s 8	β−, β-n
	52	0+	−32.509	4.6 s 3	β−
	53	(3/2−,5/2−)	−27.898s	90 ms 15	β−, β-n>30%
21 Sc	38		−4.937s		
	39		−14.168		
	40	4−	−20.526	182.3 ms 7	ε, εp 0.44%, εα 0.02%
	41	7/2−	−28.642	596.3 ms 17	ε
	42	0+	−32.121	681.3 ms 7	ε
	42m	7+(5+,6+)	−31.505	61.7 s 4	ε
	43	7/2−	−36.188	3.891 h 12	ε
	44	2+	−37.816	3.927 h 8	ε
	44m	6+	−37.545	58.6 h 1	IT 98.8%, ε 1.2%
	45	7/2−	−41.069	100%	
	45m	3/2+	−41.057	318 ms 7	IT
	46	4+	−41.759	83.79 d 4	β−
	46m	1−	−41.616	18.75 s 4	IT
	47	7/2−	−44.332	3.345 d 3	β−
	48	6+	−44.493	43.67 h 9	β−
	49	7/2−	−46.552	57.2 m 2	β−
	50	5+	−44.538	102.5 s 5	β−
	50m	2+,3+	−44.281	0.35 s 4	IT>97.5%, β−<2.5%
	51	(7/2)−	−43.219	12.4 s 1	β−
	52	3+	−40.456	8.2 s 2	β−
	53		−37.968s		
22 Ti	40	0+	−8.850	50 ms 15	ε, εp
	41	3/2+	−15.713s	80 ms 2	ε, εp≈100%
	42	0+	−25.121	199 ms 6	ε
	43	7/2−	−29.320	509 ms 5	ε
	44	0+	−37.548	49 y 3	ε
	45	7/2−	−39.007	184.8 m 5	ε
	46	0+	−44.125	8.0% 1	
	47	5/2−	−44.932	7.3% 1	
	48	0+	−48.487	73.8% 1	
	49	7/2−	−48.558	5.5% 1	
	50	0+	−51.426	5.4% 1	
	51	3/2−	−49.727	5.76 m 1	β−
	52	0+	−49.464	1.7 m 1	β−
	53	(3/2)−	−46.825	32.7 s 9	β−
	54	0+	−45.608		
23 V	42		−8.169s		
	43		−18.024s		
	44		−23.846s	90 ms 25	ε, εα
	45	7/2−	−31.874	547 ms 6	ε
	46	0+	−37.074	422.37 ms 20	ε
	46m	3+	−36.273	1.02 ms 7	IT
	47	3/2−	−42.004	32.6 m 3	ε
	48	4+	−44.475	15.9735 d 25	ε
	49	7/2−	−47.956	338 d 5	ε
	50	6+	−49.218	1.4×10^{17} y +4−3, 0.250% 2	ε 83%, β− 17%
	51	7/2−	−52.198	99.750% 2	
	52	3+	−51.438	3.75 m 1	β−
	53	7/2−	−51.845	1.61 m 4	β−
	54	3+	−49.887	49.8 s 5	β−
	55	(7/2−)	−49.148	6.54 s 15	β−
	56		−46.157		
	59		−37.912		β−
24 Cr	44	0+	−13.535s		
	45		−19.412s	50 ms 6	ε, εp>27%

Isotope Z El	A	Jπ	Δ (MeV)	T1/2 or Abundance	Decay Mode
24 Cr	46	0+	-29.471	0.26 s *6*	ε
	47	3/2-	-34.553	508 ms *10*	ε
	48	0+	-42.815	21.56 h *3*	ε
	49	5/2-	-45.326	42.3 m *1*	ε
	50	0+	-50.255	>1.8×10^17 y 4.345% *9*	
	51	7/2-	-51.445	27.702 d *4*	ε
	52	0+	-55.413	83.79% *1*	
	53	3/2-	-55.281	9.50% *1*	
	54	0+	-56.929	2.365% *5*	
	55	3/2-	-55.104	3.497 m *3*	β-
	56	0+	-55.289	5.94 m *10*	β-
	57	3/2-:7/2-	-52.393	21.1 s *10*	β-
	58	0+	-51.894	7.0 s *3*	β-
	59		-47.773	0.74 s *24*	β-
	60	0+	-46.826	0.57 s *6*	β-
	61		-42.765		β-
25 Mn	46		-12.370s		
	47		-22.263s		ε, εp
	48	4+	-29.286s	158.1 ms *22*	ε, εp 0.28%, εα<6.0×10^-4%
	49	5/2-	-37.611	384 ms *17*	ε
	50	0+	-42.622	283.07 ms *36*	ε
	50m	5+	-42.393	1.75 m *3*	ε
	51	5/2-	-48.237	46.2 m *1*	ε
	52	6+	-50.701	5.591 d *3*	ε
	52m	2+	-50.323	21.1 m *2*	ε 98.25%, IT 1.75%
	53	7/2-	-54.684	3.74×10^6 y *4*	ε
	54	3+	-55.552	312.12 d *10*	ε, β-<0.001%
	55	5/2-	-57.707	100%	
	56	3+	-56.906	2.5785 h *2*	β-
	57	5/2-	-57.485	85.4 s *18*	β-
	58	3+	-55.902	65.3 s *7*	β-
	58m	+	-55.902	3.0 s *1*	β-
	59	3/2-,5/2-	-55.473	4.6 s *1*	β-
	60	0+	-52.775	51 s *6*	β-
	60m	3+	-52.503	1.77 s *2*	β- 88.5%, IT 11.5%
	61	(5/2)-	-51.570	0.71 s *1*	β-
	62	(3+)	-48.466	0.88 s *15*	β-
	63		-46.752	0.25 s *4*	β-
	64		-43.100		
	65		-40.893		β-
26 Fe	48	0+	-18.108s	≥200 ns	
	49	(7/2-)	-24.582s	75 ms *10*	ε, εp≤60%
	50	0+	-34.472	150 ms *20*	ε, εp≈0%
	51	(5/2-)	-40.217	305 ms *5*	ε
	52	0+	-48.329	8.275 h *8*	ε
	52m	(12+)	-41.509	45.9 s *6*	ε
	53	7/2-	-50.941	8.51 m *2*	ε
	53m	19/2-	-47.901	2.58 m *4*	IT
	54	0+	-56.249	5.9% *2*	
	55	3/2-	-57.475	2.73 y *3*	ε
	56	0+	-60.601	91.72% *15*	
	57	1/2-	-60.176	2.1% *1*	
	58	0+	-62.149	0.28% *2*	
	59	3/2-	-60.659	44.503 d *6*	β-
	60	0+	-61.407	1.5×10^6 y *3*	β-
	61	3/2-,5/2-	-58.918	5.98 m *6*	β-
	62	0+	-58.898	68 s *2*	β-

Isotope Z El	A	Jπ	Δ (MeV)	T1/2 or Abundance	Decay Mode
26 Fe	63	(5/2)-	-55.513	6.1 s *6*	β-
	64	0+	-54.902	2.0 s *2*	β-
	65		-51.288	0.4 s *2*	β-
	66	0+	-50.319		
	67		-46.575		β-
27 Co	50		-17.503s		
	51		-27.470s		
	52		-34.316s		ε, εp
	53	(7/2-)	-42.639	240 ms *20*	ε
	53m	(19/2-)	-39.449	247 ms *12*	ε≈98.5%, p≈1.5%
	54	0+	-48.006	193.24 ms *14*	ε
	54m	(7)+	-47.806	1.48 m *2*	ε
	55	7/2-	-54.024	17.53 h *3*	ε
	56	4+	-56.035	77.27 d *3*	ε
	57	7/2-	-59.340	271.79 d *9*	ε
	58	2+	-59.842	70.82 d *3*	ε
	58m	5+	-59.817	9.15 h *10*	IT
	59	7/2-	-62.224	100%	
	60	5+	-61.645	5.2714 y *5*	β-
	60m	2+	-61.585	10.467 m *6*	IT 99.76%, β- 0.24%
	61	7/2-	-62.895	1.650 h *5*	β-
	62	2+	-61.428	1.50 m *4*	β-
	62m	5+	-61.406	13.91 m *5*	β- >99%, IT<1%
	63	(7/2)-	-61.837	27.4 s *5*	β-
	64	1+	-59.790	0.30 s *3*	β-
	65	(7/2)-	-59.165	1.20 s *6*	β-
	66	(3+)	-56.052	0.23 s *2*	β-
	67	(7/2-)	-55.321	0.42 s *7*	β-
	70		-46.752s		β-
	71		-44.963s		β-
28 Ni	52		-22.654s		ε, εp
	53	(7/2-)	-29.379s	45 ms *15*	ε
	54	0+	-39.206		ε
	55	7/2-	-45.330	212.1 ms *38*	ε
	56	0+	-53.900	5.9 d *1*	ε
	57	3/2-	-56.076	35.60 h *6*	ε
	58	0+	-60.223	68.077% *5*	
	59	3/2-	-61.151	7.6×10^4 y *5*	ε
	60	0+	-64.468	26.223% *5*	
	61	3/2-	-64.217	1.140% *1*	
	62	0+	-66.743	3.634% *1*	
	63	1/2-	-65.509	100.1 y *20*	β-
	64	0+	-67.096	0.926% *1*	
	65	5/2-	-65.123	2.5172 h *3*	β-
	66	0+	-66.029	54.6 h *4*	β-
	67	(1/2-)	-63.743	21 s *1*	β-
	68	0+	-63.486	19 s *5*	β-
	69		-60.378	11.4 s *3*	β-
	70	0+	-59.485		β-
	71		-55.890	1.86 s *35*	β-
	73		-50.329s	0.90 s *15*	β-
29 Cu	55		-32.118s	>200 ns	ε, εp
	56		-38.601s		ε, εp
	57	3/2-	-47.306	199.4 ms *32*	ε
	58	1+	-51.660	3.204 s *7*	ε
	59	3/2-	-56.352	81.5 s *5*	ε
	60	2+	-58.341	23.7 m *4*	ε
	61	3/2-	-61.980	3.333 h *5*	ε
	62	1+	-62.795	9.74 m *2*	ε
	63	3/2-	-65.576	69.17% *2*	

Isotope Z El	A	Jπ	Δ (MeV)	T1/2 or Abundance	Decay Mode
29 Cu	64	1+	-65.421	12.700 h *2*	ε 61%, β- 39%
	65	3/2-	-67.260	30.83% *2*	
	66	1+	-66.254	5.088 m *11*	β-
	67	3/2-	-67.300	61.83 h *12*	β-
	68	1+	-65.542	31.1 s *15*	β-
	68m	(6-)	-64.820	3.75 m *5*	IT 84%, β- 16%
	69	3/2-	-65.740	2.85 m *15*	β-
	70	1+	-62.961	4.5 s *10*	β-
	70m	3-,4-,5-	-62.821	47 s *5*	β-
	71	(3/2-)	-62.764	19.5 s *16*	β-
	72		-59.904s	6.6 s *1*	β-
	73		-59.159s	3.9 s *3*	β-
	75		-54.576s	1.3 s *1*	β-, β-n 3.5%
	76		-50.738s	0.61 s *10*	β-, β-n
	79		-42.709s	188 ms *25*	β-, β-n 55%
30 Zn	57	(7/2-)	-32.686s	40 ms *10*	ε, εp≥65%
	58	0+	-42.293		ε
	59	3/2-	-47.258	182.0 ms *18*	ε, εp 0.1%
	60	0+	-54.183	2.38 m *5*	ε
	61	3/2-	-56.343	89.1 s *2*	ε
	62	0+	-61.168	9.186 h *13*	ε
	63	3/2-	-62.210	38.47 m *5*	ε
	64	0+	-66.000	48.6% *3*	
	65	5/2-	-65.908	244.26 d *26*	ε
	66	0+	-68.897	27.9% *2*	
	67	5/2-	-67.877	4.1% *1*	
	68	0+	-70.004	18.8% *4*	
	69	1/2-	-68.415	56.4 m *9*	β-
	69m	9/2+	-67.976	13.76 h *2*	IT 99.97%, β- 0.03%
	70	0+	-69.560	>5×10^{14} y 0.6% *1*	
	71	1/2-	-67.322	2.45 m *10*	β-
	71m	9/2+	-67.164	3.96 h *5*	β-, IT≤0.05%
	72	0+	-68.126	46.5 h *1*	β-
	73	(1/2)-	-65.410	23.5 s *10*	β-
	*73*m	(7/2+)	-65.214	5.8 s *8*	β-, IT
	74	0+	-65.709	96 s *1*	β-
	75	(7/2+)	-62.468	10.2 s *2*	β-
	76	0+	-62.043	5.7 s *3*	β-
	77	(7/2+)	-58.604	2.08 s *5*	β-
	*77*m	(1/2-)	-57.832	1.05 s *10*	β-<50%, IT>50%
	78	0+	-57.222	1.47 s *15*	β-
	79	(9/2+)	-53.398s	0.995 s *19*	β-, β-n 1.3%
	80	0+	-51.777	0.545 s *16*	β-, β-n 1%
	81		-46.128s	0.29 s *5*	β-, β-n 7.5%
31 Ga	61	(3/2-)	-47.348s		ε
	62	0+	-51.997	116.12 ms *23*	ε
	63	3/2-,5/2-	-56.690	32.4 s *5*	ε
	64	0+	-58.835	2.630 m *11*	ε
	65	3/2-	-62.653	15.2 m *2*	ε
	66	0+	-63.722	9.49 h *7*	ε
	67	3/2-	-66.877	3.2612 d *6*	ε
	68	1+	-67.083	67.629 m *24*	ε
	69	3/2-	-69.321	60.108% *6*	
	70	1+	-68.905	21.14 m *3*	β- 99.59%, ε 0.41%
	71	3/2-	-70.135	39.892% *6*	
	72	3-	-68.584	14.10 h *2*	β-
	73	3/2-	-69.704	4.86 h *3*	β-
	74	(3-)	-68.054	8.12 m *12*	β-

Isotope Z El	A	Jπ	Δ (MeV)	T1/2 or Abundance	Decay Mode
31 Ga	74m	(0)	-67.994	9.5 s *10*	IT 75%, β-<50%
	75	3/2-	-68.464	126 s *2*	β-
	76	(3-)	-66.203	29.1 s *7*	β-
	77	(3/2-)	-65.874	13.2 s *2*	β-
	78	(3+)	-63.662	5.09 s *5*	β-
	79	(3/2-)	-62.488	2.847 s *3*	β-, β-n 0.09%
	80	(3)	-59.068	1.697 s *11*	β-, β-n 0.79%
	81	(5/2-)	-57.982	1.221 s *5*	β-, β-n 11.4%
	82	(1,2,3)	-52.946s	0.602 s *6*	β-, β-n 19.8%
	83		-49.490s	0.31 s *1*	β-, β-n 54%
32 Ge	61	(3/2-)	-33.729s	40 ms *15*	ε, εp≈80%
	63		-46.910s		
	64	0+	-54.425	63.7 s *25*	ε
	65	(3/2)-	-56.411	30.9 s *5*	ε
	66	0+	-61.622	2.26 h *5*	ε
	67	1/2-	-62.654	18.9 m *3*	ε
	68	0+	-66.977	270.82 d *27*	ε
	69	5/2-	-67.094	39.05 h *10*	ε
	70	0+	-70.561	21.23% *4*	
	71	1/2-	-69.905	11.43 d *3*	ε
	71m	9/2+	-69.707	20.40 ms *17*	IT
	72	0+	-72.585	27.66% *3*	
	73	9/2+	-71.297	7.73% *1*	
	73m	1/2-	-71.230	0.499 s *11*	IT
	74	0+	-73.422	35.94% *2*	
	75	1/2-	-71.856	82.78 m *4*	β-
	75m	7/2+	-71.716	47.7 s *5*	IT 99.97%, β- 0.03%
	76	0+	-73.213	7.44% *2*	
	77	7/2+	-71.214	11.30 h *1*	β-
	77m	1/2-	-71.054	52.9 s *6*	β- 79%, IT 21%
	78	0+	-71.862	88.0 m *10*	β-
	79	(1/2)-	-69.488	18.98 s *3*	β-
	79m	(7/2+)	-69.302	39.0 s *10*	β- 96%, IT 4%
	80	0+	-69.448	29.5 s *4*	β-
	81	(9/2+)	-66.302	7.6 s *6*	β-
	81m	(1/2+)	-65.623	7.6 s *6*	β-
	82	0+	-65.539	4.60 s *35*	β-
	83	(5/2+)	-61.004s	1.85 s *6*	β-
	84	0+	-58.395s	1.2 s *3*	β-
33 As	65		-47.056s	0.19 s +*11*-*7*	ε
	66		-51.822s	95.77 ms *23*	ε
	67	(5/2-)	-56.644	42.5 s *12*	ε
	68	3	-58.877	151.6 s *8*	ε
	69	5/2-	-63.081	15.2 m *2*	ε
	70	4(+)	-64.341	52.6 m *3*	ε
	71	5/2-	-67.893	65.28 h *15*	ε
	72	2-	-68.229	26.0 h *1*	ε
	73	3/2-	-70.956	80.30 d *6*	ε
	74	2-	-70.859	17.77 d *2*	ε 66%, β- 34%
	75	3/2-	-73.032	100%	
	76	2-	-72.289	26.32 h *7*	β-, ε<0.02%
	77	3/2-	-73.916	38.83 h *5*	β-
	78	2-	-72.816	90.7 m *2*	β-
	79	3/2-	-73.636	9.01 m *15*	β-
	80	1+	-72.118	15.2 s *2*	β-
	81	3/2-	-72.533	33.3 s *8*	β-
	82	(1+)	-70.239	19.1 s *5*	β-
	82m	(5-)	-70.239	13.6 s *4*	β-
	83	(5/2-,3/2-)	-69.880	13.4 s *3*	β-
	84	0(-),1(-),2-	-66.080s	5.5 s *3*	β-, β-n 0.08%

Isotope Z El	A	Jπ	Δ (MeV)	T1/2 or Abundance	Decay Mode
33 As	84m		-66.080s	0.65 s	β-
	85	(3/2-)	-63.519s	2.028 s 12	β-, β-n 23%
	86		-59.401s	0.9 s 2	β-, β-n 12%
	87	(3/2-)	-56.281s	0.73 s 6	β-, β-n 44%
34 Se	65		-32.919s		ε
	67		-46.491s		
	68	0+	-54.148s	1.6 m 4	ε
	69	(3/2-)	-56.298	27.4 s 2	ε, εp 0.05%
	70	0+	-61.941s	41.1 m 3	ε
	71	5/2-	-63.093s	4.74 m 5	ε
	72	0+	-67.894	8.40 d 8	ε
	73	9/2+	-68.216	7.15 h 8	ε
	73m	3/2-	-68.190	39.8 m 13	IT 72.6%, ε 27.4%
	74	0+	-72.213	0.89% 2	
	75	5/2+	-72.169	119.779 d 4	ε
	76	0+	-75.251	9.36% 12	
	77	1/2-	-74.599	7.63% 5	
	77m	7/2+	-74.437	17.36 s 5	IT
	78	0+	-77.025	23.78% 15	
	79	7/2+	-75.917	≤6.5×10^5 y	β-
	79m	1/2-	-75.821	3.92 m 1	IT 99.94%, β- 0.06%
	80	0+	-77.759	49.61% 31	
	81	1/2-	-76.389	18.45 m 12	β-
	81m	7/2+	-76.286	57.28 m 2	IT 99.95%, β- 0.05%
	82	0+	-77.593	1.4×10^{20} y 4 8.73% 6	2β-
	83	9/2+	-75.340	22.3 m 3	β-
	83m	1/2-	-75.112	70.1 s 4	β-
	84	0+	-75.950	3.1 m 1	β-
	85	(5/2+)	-72.426	31.7 s 9	β-
	86	0+	-70.536	15.3 s 9	β-
	87	(5/2+)	-66.578	5.85 s 15	β-, β-n 0.18%
	88	0+	-63.874	1.52 s 3	β-, β-n 0.94%
	89	(5/2+)	-59.597s	0.41 s 4	β-, β-n 5%
	91		-50.888s	0.27 s 5	β-, β-n 21%
35 Br	69		-46.679s		
	70		-51.591s	79.1 ms 8	ε
	70m		-51.591s	2.2 s 2	ε
	71	(5/2)-	-56.593s	21.4 s 6	ε
	72	(3)+	-59.180	78.6 s 24	ε
	72m		-59.180	7.2 s 5	εp
	72m	(1)-	-59.079	10.6 s 3	ε?, IT
	73	1/2-	-63.560	3.4 m 2	ε
	74	(0-,1)	-65.306	25.4 m 3	ε
	74m	4(-)	-65.306	46 m 2	ε
	75	3/2-	-69.139	96.7 m 13	ε
	76	1-	-70.288	16.2 h 2	ε
	76m	(4)+	-70.186	1.31 s 2	IT>99.4%, ε<0.6%
	77	3/2-	-73.234	57.036 h 6	ε
	77m	9/2+	-73.128	4.28 m 10	IT
	78	1+	-73.452	6.46 m 4	ε≥99.99%, β-≤0.01%
	79	3/2-	-76.068	50.69% 5	
	79m	9/2+	-75.860	4.86 s 4	IT
	80	1+	-75.889	17.68 m 2	β- 91.7%, ε 8.3%
	80m	5-	-75.803	4.4205 h 8	IT
	81	3/2-	-77.974	49.31% 5	
	82	5-	-77.496	35.30 h 2	β-

Isotope Z El	A	Jπ	Δ (MeV)	T1/2 or Abundance	Decay Mode
35 Br	82m	2-	-77.450	6.13 m *5*	IT 97.6%, β- 2.4%
	83	3/2-	-79.008	2.40 h *2*	β-
	84	2-	-77.775	31.80 m *8*	β-
	*84*m	(5-,6-)	-77.455	6.0 m *2*	β-
	85	3/2-	-78.608	2.90 m *6*	β-
	86	(2-)	-75.635	55.1 s *4*	β-
	87	3/2-	-73.853	55.60 s *15*	β-, β-n 2.57%
	88	(1,2-)	-70.728	16.5 s *1*	β-, β-n 6.4%
	89	(3/2-,5/2-)	-68.563	4.40 s *3*	β-, β-n 13%
	90		-64.609	1.92 s *2*	β-, β-n 24.6%
	91		-61.551	0.541 s *5*	β-, β-n 18.3%
	92	(2-)	-56.623	0.343 s *15*	β-, β-n 33%
	93	(5/2-)	-53.002s	102 ms	β-, β-n 77%
	94			70 ms *20*	β-, β-n 30%
36 Kr	71		-46.100s	97 ms *9*	ε
	72	0+	-54.140	17.2 s *3*	ε
	73	5/2-	-56.885	27.0 s *12*	ε, εp 0.68%
	74	0+	-62.168	11.50 m *11*	ε
	75	(5/2)+	-64.240	4.3 m *2*	ε
	76	0+	-68.977	14.8 h *1*	ε
	77	5/2+	-70.170	74.4 m *6*	ε
	78	0+	-74.158	0.35% *2*	
	79	1/2-	-74.442	35.04 h *10*	ε
	79m	7/2+	-74.312	50 s *3*	IT
	80	0+	-77.893	2.25% *2*	
	81	7/2+	-77.693	2.29×10^{5} y *11*	ε
	81m	1/2-	-77.502	13.10 s *3*	IT, ε 2.5×10^{-3}%
	82	0+	-80.588	11.6% *1*	
	83	9/2+	-79.981	11.5% *1*	
	*83*m	1/2-	-79.939	1.83 h *2*	IT
	84	0+	-82.430	57.0% *3*	
	85	9/2+	-81.478	10.756 y *18*	β-
	*85*m	1/2-	-81.173	4.480 h *8*	β- 78.6%, IT 21.4%
	86	0+	-83.261	17.3% *2*	
	87	5/2+	-80.706	76.3 m *6*	β-
	88	0+	-79.688	2.84 h *3*	β-
	89	(3/2+,5/2+)	-76.717	3.15 m *4*	β-
	90	0+	-74.960	32.32 s *9*	β-
	91	(5/2+)	-71.353	8.57 s *4*	β-
	92	0+	-68.827	1.840 s *8*	β-, β-n 0.03%
	93	(1/2+)	-64.102	1.286 s *10*	β-, β-n 2.01%
	94	0+	-61.220s	0.20 s *1*	β-, β-n 5.7%
	95		-56.141s	0.78 s *3*	β-
	97?			<0.1 s	β-
37 Rb	73		-46.261s		
	74	(0+)	-51.724	64.9 ms *5*	ε
	75	(3/2-,5/2-)	-57.220	19.0 s *12*	ε
	76	1	-60.477	39.1 s *6*	ε
	77	3/2-	-64.826	3.75 m *8*	ε
	78	0(+)	-66.934	17.66 m *8*	ε
	78m	4(-)	-66.831	5.74 m *5*	ε 90%, IT 10%
	79	5/2+	-70.793	22.9 m *5*	ε
	80	1+	-72.170	34 s *4*	ε
	81	3/2-	-75.455	4.576 h *5*	ε
	81m	9/2+	-75.369	30.5 m *3*	IT 97.6%, ε 2.4%
	82	1+	-76.187	1.273 m *2*	ε
	82m	5-	-76.107	6.472 h *6*	ε, IT<0.33%
	83	5/2-	-79.071	86.2 d *1*	ε
	84	2-	-79.748	32.77 d *14*	ε 96.2%, β- 3.8%
	84m	6-	-79.284	20.26 m *4*	IT

Isotope Z	El	A	Jπ	Δ (MeV)	T1/2 or Abundance	Decay Mode
37	Rb	85	5/2-	-82.165	72.17% 1	
		86	2-	-82.745	18.631 d 18	β- 99.99%, ε 0.0052%
		86m	6-	-82.189	1.017 m 3	IT
		87	3/2-	-84.593	4.75×10^{10} y 4 27.83% 1	β-
		88	2-	-82.602	17.78 m 11	β-
		89	3/2-	-81.703	15.15 m 12	β-
		90	0-	-79.351	158 s 5	β-
		90m	3-	-79.244	258 s 4	β- 97.4%, IT 2.6%
		91	3/2(-)	-77.788	58.4 s 4	β-
		92	0-	-74.814	4.51 s 2	β-, β-n 9.9×10^{-3}%
		93	5/2-	-72.702	5.84 s 2	β-, β-n 1.35%
		94	3(-)	-68.530	2.702 s 5	β-, β-n 10.4%
		95	5/2-	-65.863	0.377 s 9	β-, β-n 8.62%
		96	2+	-61.227	0.199 s 3	β-
		96	2+	-61.227	202.8 ms 33	β-, β-n 14%
		97	3/2+	-58.375	169.9 ms 7	β-, β-n 25.1%
		98	(1,0)	-54.266	114 ms 5	β-, β-n 13.6%, β- 0.05%
		98m	(4,5)	-53.996	96 ms 3	β-, β-n?
		99	(5/2+)	-50.926	59 ms 1	β-, β-n 15%
		100		-46.696s	51 ms 8	β-, β-n 6%
		101		-43.598	32 ms 4	β-, β-n 31%
38	Sr	77	(5/2+,7/2+)	-57.973	9.0 s 2	ε, εp<0.25%
		78	0+	-63.172	2.5 m 3	ε
		79	3/2(-)	-65.475	2.25 m 10	ε
		80	0+	-70.302	106.3 m 15	ε
		81	1/2-	-71.524	22.3 m 4	ε
		82	0+	-76.007	25.55 d 15	ε
		83	7/2+	-76.795	32.41 h 3	ε
		83m	1/2-	-76.536	4.95 s 12	IT
		84	0+	-80.643	0.56% 1	
		85	9/2+	-81.100	64.84 d 2	ε
		85m	1/2-	-80.861	67.63 m 4	IT 86.6%, ε 13.4%
		86	0+	-84.519	9.86% 1	
		87	9/2+	-84.876	7.00% 1	
		87m	1/2-	-84.487	2.803 h 3	IT 99.7%, ε 0.3%
		88	0+	-87.918	82.58% 1	
		89	5/2+	-86.205	50.53 d 7	β-
		90	0+	-85.941	28.78 y 4	β-
		91	5/2+	-83.649	9.63 h 5	β-
		92	0+	-82.920	2.71 h 1	β-
		93	5/2+	-80.162	7.423 m 24	β-
		94	0+	-78.837	75.3 s 2	β-
		95	1/2+	-75.159	23.90 s 14	β-
		96	0+	-72.983	1.07 s 1	β-
		97	1/2+	-68.795	429 ms 5	β-, β-n≤0.05%
		98	0+	-66.610	0.653 s 2	β-, β-n 0.23%
		99	(3/2+)	-62.173	0.271 s 4	β-, β-n 0.32%
		100	0+	-60.220	202 ms 3	β-, β-n 0.73%
		101	(5/2)	-55.408	118 ms 3	β-, β-n 2.37%
		102	0+	-53.078	69 ms 6	β-, β-n 4.8%
39	Y	79	(5/2+)	-58.354	14.8 s 6	ε, εp
		80	(3,4,5)	-61.162s	35 s 2	ε
		81	(5/2+)	-66.013	72.4 s 13	ε
		82	1+	-68.190	9.5 s 3	ε
		83	(9/2+)	-72.329	7.08 m 6	ε
		83m	(3/2-)	-72.267	2.85 m 2	ε 60%, IT 40%
		84	1+	-74.233	4.6 s 2	ε
		84m	(5-)	-73.733	40 m 1	ε

Isotope Z El	A	Jπ	Δ (MeV)	T1/2 or Abundance	Decay Mode
39 Y	85	(1/2)-	-77.845	2.68 h 5	ε
	85m	9/2+	-77.825	4.86 h 13	ε, IT<2.0×10^{-3}%
	86	4-	-79.279	14.74 h 2	ε
	86m	(8+)	-79.061	48 m 1	IT 99.31%, ε 0.69%
	87	1/2-	-83.015	79.8 h 3	ε
	87m	9/2+	-82.634	13.37 h 3	IT 98.43%, ε 1.57%
	88	4-	-84.295	106.65 d 4	ε
	89	1/2-	-87.701	100%	
	*89*m	9/2+	-86.793	16.06 s 4	IT
	90	2-	-86.487	64.10 h 8	β-
	*90*m	7+	-85.805	3.19 h 1	IT, β- 1.8×10^{-3}%
	91	1/2-	-86.349	58.51 d 6	β-
	*91*m	9/2+	-85.793	49.71 m 4	IT, β-<1.5%
	92	2-	-84.831	3.54 h 1	β-
	93	1/2-	-84.245	10.18 h 8	β-
	*93*m	7/2+	-83.486	0.82 s 4	IT
	94	2-	-82.348	18.7 m 1	β-
	95	1/2-	-81.239	10.3 m 1	β-
	96	0-	-78.355	5.34 s 5	β-
	*96*m	(8)+	-78.355	9.6 s 2	β-
	97	(1/2-)	-76.262	3.75 s 3	β-, β-n 0.06%
	*97*m	(9/2)+	-75.594	1.17 s 3	β->99.3%, IT<0.7%, β-n<0.08%
	*97*m	(27/2-)	-72.739	142 ms 8	IT>80%, β-<20%
	98	(0)-	-72.436	0.548 s 2	β-, β-n 0.24%
	*98*m	(4,5)	-72.436	2.0 s 2	β- 90%, IT<20%, β-n 3.4%
	99	(5/2+)	-70.204	1.47 s 2	β-, β-n 0.96%
	100	1-,2-	-67.295	735 ms 7	β-, β-n 0.81%
	*100*m	(3,4,5)	-67.295	0.94 s 3	β-
	101	(5/2+)	-64.913	448 ms 19	β-, β-n 1.94%
40 Zr	80	0+	-55.340s		
	81		-58.853	15 s 5	ε, εp
	82	0+	-64.190	32 s 5	ε
	83	(1/2-)	-66.461	44 s 1	ε, εp
	84	0+	-71.492s	25.9 m 8	ε
	85	7/2+	-73.152	7.86 m 4	ε
	85m	(1/2-)	-72.860	10.9 s 3	IT≤92%, ε>8%
	86	0+	-77.806	16.5 h 1	ε
	87	(9/2)+	-79.349	1.68 h 1	ε
	87m	(1/2)-	-79.013	14.0 s 2	IT
	88	0+	-83.625	83.4 d 3	ε
	89	9/2+	-84.869	78.41 h 12	ε
	89m	1/2-	-84.281	4.18 m 1	IT 93.77%, ε 6.23%
	90	0+	-88.769	51.45% 2	
	*90*m	5-	-86.450	809.2 ms 20	IT
	91	5/2+	-87.893	11.22% 3	
	92	0+	-88.456	17.15% 2	
	93	5/2+	-87.119	1.53×10^{6} y 10	β-
	94	0+	-87.268	17.38% 3	
	95	5/2+	-85.659	64.02 d 5	β-
	96	0+	-85.441	2.80% 1	
	97	1/2+	-82.950	16.90 h 5	β-
	98	0+	-81.266	30.7 s 4	β-
	99	(1/2)+	-77.771	2.1 s 1	β-
	100	0+	-76.605	7.1 s 4	β-
	101	(3/2+)	-73.458	2.1 s 3	β-

Isotope Z El A	Jπ	Δ (MeV)	T1/2 or Abundance	Decay Mode
40 Zr *102*	0+	-71.743	2.9 s *2*	β-
103	(5/2)	-68.375	1.3 s *1*	β-
104	0+	-66.341s	1.2 s *3*	β-
41 Nb 83	(5/2+)	-58.961	4.1 s *3*	ε
84	(3+)	-61.879s	12 s *3*	ε, εp
85	(9/2+)	-67.152	20.9 s *7*	ε
86	(5+)	-69.828	88 s *1*	ε
87	(9/2+)	-74.180	2.6 m *1*	ε
87m	(1/2-)	-74.180	3.7 m *1*	ε
88	(8+)	-76.425s	14.5 m *1*	ε
88m	(4-)	-76.425s	7.8 m *1*	ε
89	(1/2)-	-80.579	1.18 h *10*	ε
89m	(9/2+)	-80.579	1.9 h *2*	ε
90	8+	-82.658	14.60 h *5*	ε
90m	4-	-82.533	18.81 s *6*	IT
91	9/2+	-86.639	6.8×10^2 y *13*	ε
91m	1/2-	-86.535	60.86 d *22*	IT 93%, ε 7%
92	(7)+	-86.450	3.47×10^7 y *24*	ε, β-<0.05%
92m	(2)+	-86.315	10.15 d *2*	ε
93	9/2+	-87.210	100%	
*93*m	1/2-	-87.179	16.13 y *14*	IT
94	(6)+	-86.366	2.03×10^4 y *16*	β-
94m	3+	-86.325	6.263 m *4*	IT 99.5%, β- 0.5%
95	9/2+	-86.783	34.975 d *7*	β-
*95*m	1/2-	-86.547	86.6 h *8*	IT 94.4%, β- 5.6%
96	6+	-85.605	23.35 h *5*	β-
97	9/2+	-85.608	72.1 m *7*	β-
*97*m	1/2-	-84.865	52.7 s *18*	IT
98	1+	-83.527	2.86 s *6*	β-
*98*m	(5+)	-83.443	51.3 m *4*	β- 99.9%, IT<0.2%
99	9/2+	-82.328	15.0 s *2*	β-
*99*m	1/2-	-81.963	2.6 m *2*	β-, IT<2.5%
100	1+	-79.940	1.5 s *2*	β-
*100*m	(4+,5+)	-79.460	2.99 s *11*	β-
101	+	-78.943	7.1 s *3*	β-
*102*m	1+	-76.348	1.3 s *2*	β-
*102*m		-76.348	4.3 s *4*	β-
103	(5/2+)	-75.320	1.5 s *2*	β-
104	(1+)	-72.229	4.8 s *4*	β-, β-n 0.71%
*104*m		-72.014	0.92 s *4*	β-
105		-70.856	2.95 s *6*	β-
106		-66.984s	1.02 s *5*	β-
107		-65.037s	330 ms *50*	β-
42 Mo 86		-65.018s		
87	(7/2+)	-67.694	13.4 s *4*	ε, εp>0%
88	0+	-72.701	8.0 m *2*	ε
89	(9/2+)	-75.004	2.04 m *11*	ε
89m	(1/2-)	-74.617	190 ms *15*	IT
90	0+	-80.169	5.67 h *5*	ε
91	9/2+	-82.205	15.49 m *1*	ε
91m	1/2-	-81.552	65.0 s *7*	IT 50.1%, ε 49.9%
92	0+	-86.806	14.84% *4*	
93	5/2+	-86.805	4.0×10^3 y *8*	ε
93m	21/2+	-84.380	6.85 h *7*	IT 99.88%, ε 0.12%
94	0+	-88.411	9.25% *2*	
95	5/2+	-87.709	15.92% *4*	
96	0+	-88.792	16.68% *4*	
97	5/2+	-87.542	9.55% *2*	
98	0+	-88.113	24.13% *6*	

Isotope Z El	A	Jπ	Δ (MeV)	T1/2 or Abundance	Decay Mode
42 Mo	*99*	1/2+	-85.967	65.94 h *1*	β-
	100	0+	-86.185	9.63% *2*	
	101	1/2+	-83.512	14.61 m *3*	β-
	102	0+	-83.558	11.3 m *2*	β-
	103	(3/2+)	-80.850	67.5 s *15*	β-
	104	0+	-80.334	60 s *2*	β-
	105	(3/2+)	-77.341	35.6 s *16*	β-
	106	0+	-76.257	8.4 s *5*	β-
	107		-72.941	3.5 s *5*	β-
	108	0+	-71.300s	1.5 s *4*	β-
	109		-67.356s	0.53 s *6*	β-
43 Tc	88		-62.568s		
	89		-67.494		
	90	1+	-71.029s	8.7 s *2*	ε
	90m	4,5,6	-70.529s	49.2 s *4*	ε
	91	(9/2)+	-75.985	3.14 m *2*	ε
	91m	(1/2)-	-75.635	3.3 m *1*	ε, IT<1%
	92	(8)+	-78.936	4.23 m *15*	ε
	93	9/2+	-83.604	2.75 h *5*	ε
	93m	1/2-	-83.212	43.5 m *10*	IT 76.7%, ε 23.3%
	94	7+	-84.155	293 m *1*	ε
	94m	(2)+	-84.080	52.0 m *10*	ε, IT<0.1%
	95	9/2+	-86.018	20.0 h *1*	ε
	95m	1/2-	-85.979	61 d *2*	ε 96.12%, IT 3.88%
	96	7+	-85.819	4.28 d *7*	ε
	96m	4+	-85.785	51.5 m *10*	IT 98%, ε 2%
	97	9/2+	-87.221	2.6×10^6 y *4*	ε
	97m	1/2-	-87.124	90.1 d *10*	IT, ε<0.34%
	98	(6)+	-86.429	4.2×10^6 y *3*	β-
	99	9/2+	-87.324	2.111×10^5 y *12*	β-
	99m	1/2-	-87.181	6.01 h *1*	IT, β- 0.004%
	100	1+	-86.017	15.8 s *1*	β-
	101	(9/2)+	-86.337	14.22 m *1*	β-
	102	1+	-84.568	5.28 s *15*	β-
	102m	(4,5)	-84.568	4.35 m *7*	β- 98%, IT 2%
	103	5/2+	-84.600	54.2 s *8*	β-
	104	(3+)	-82.489	18.3 m *3*	β-
	105	(5/2+)	-82.291	7.6 m *1*	β-
	106	(1,2)	-79.777	36 s *1*	β-
	107		-79.101	21.2 s *2*	β-
	108	(2-,3)	-75.935	5.17 s *7*	β-
	109		-74.867s	0.87 s *4*	β-
	110		-71.362s	0.92 s *3*	β-
	111		-69.815s	0.30 s *3*	β-
	113		-63.966s	130 ms *50*	β-
44 Ru	90	0+	-65.409s	13 s *5*	ε
	91	(9/2+)	-68.580	9 s *1*	ε
	91m	(1/2-)	-68.580	7.6 s *8*	ε>0%, εp>0%, IT
	92	0+	-74.408s	3.65 m *5*	ε
	93	(9/2)+	-77.267	59.7 s *6*	ε
	93m	(1/2)-	-76.533	10.8 s *3*	ε 78%, IT 22%, εp 0.01%
	94	0+	-82.563	51.8 m *6*	ε
	95	5/2+	-83.445	1.643 h *14*	ε
	96	0+	-86.067	5.54% *2*	
	97	5/2+	-86.107	2.9 d *1*	ε
	98	0+	-88.225	1.86% *2*	
	99	5/2+	-87.618	12.7% *1*	
	100	0+	-89.219	12.6% *1*	
	101	5/2+	-87.950	17.1% *1*	

Isotope Z El A		Jπ	Δ (MeV)	T1/2 or Abundance	Decay Mode
44 Ru	101	11/2−	−87.423	17.5 μs 4	IT
	102	0+	−89.099	31.6% 2	
	103	3/2+	−87.260	39.26 d 2	β−
	103	11/2−	−87.022	1.69 ms 7	IT
	104	0+	−88.092	18.6% 2	
	105	3/2+	−85.931	4.44 h 2	β−
	106	0+	−86.324	373.59 d 15	β−
	107	(5/2)+	−83.921	3.75 m 5	β−
	108	0+	−83.655	4.55 m 5	β−
	109	(5/2+)	−80.852	34.5 s 10	β−
	110	0+	−80.140	14.6 s 10	β−
	111		−76.792s	2.12 s 7	β−
	112	0+	−75.867s	1.75 s 7	β−
	113		−72.154s	0.80 s 5	β−
	114	0+	−70.794s	0.57 s 5	β−
	115		−66.779s	0.40 s 10	β−, β-n
45 Rh	92		−63.360s		
	93		−69.173s		
	94m	(8+)	−72.933s	25.8 s 2	ε
	94m	(3+)	−72.933s	70.6 s 6	ε
	95	(9/2)+	−78.335	5.02 m 10	ε
	95m	(1/2)−	−77.792	1.96 m 4	IT 88%, ε 12%
	96	(6+)	−79.620	9.90 m 10	ε
	96m	(3+)	−79.568	1.51 m 2	IT 60%, ε 40%
	97	9/2+	−82.584	30.7 m 6	ε
	97m	1/2−	−82.325	46.2 m 16	ε 94.4%, IT 5.6%
	98	(2)+	−83.167	8.7 m 2	ε
	98m	(5+)	−83.167	3.5 m 3	ε > 0%, IT
	99	(1/2−)	−85.515	16.1 d 2	ε
	99m	9/2+	−85.451	4.7 h 1	ε, IT < 0.16%
	100	1−	−85.590	20.8 h 1	ε
	100m	(5+)	−85.590	4.6 m 2	IT≈98.3%, ε≈1.7%
	101	1/2−	−87.409	3.3 y 3	ε
	101m	9/2+	−87.252	4.34 d 1	ε 93.6%, IT 6.4%
	102	(1−,2−)	−86.776	207 d 3	ε 80%, β− 20%
	102m	6(+)	−86.635	≈2.9 y	ε 99.73%, IT 0.23%
	103	1/2−	−88.023	100%	
	103m	7/2+	−87.983	56.114 m 9	IT
	104	1+	−86.951	42.3 s 4	β− 99.55%, ε 0.45%
	104m	5+	−86.822	4.34 m 3	IT 99.87%, β− 0.13%
	105	7/2+	−87.848	35.36 h 6	β−
	105m	1/2−	−87.718	≈40 s	IT
	106	1+	−86.363	29.80 s 8	β−
	106m	(6)+	−86.226	130 m 2	β−
	107	7/2+	−86.861	21.7 m 4	β−
	108m	1+	−85.016	16.8 s 5	β−
	108m	(5+)	−85.016	6.0 m 3	β−
	109	7/2+	−85.012	80 s 2	β−
	110m	1+	−82.950	3.2 s 2	β−
	110m	(≥4)	−82.950	28.5 s 15	β−
	111	(7/2+)	−82.288s	11 s 1	β−
	112m	1+	−79.537s	3.8 s 6	β−
	112m	≥4	−79.537s	6.8 s 2	β−
	113		−78.786s	2.72 s 22	β−
	114	(1+)	−75.594s	1.85 s 5	β−
	114m	(≥4)	−75.594s	1.85 s 5	β−
	115	(7/2+)	−74.403	0.99 s 5	β−

Isotope Z El A	Jπ	Δ (MeV)	T1/2 or Abundance	Decay Mode
45 Rh *116*	1+	-71.053s	0.68 s *6*	β-
116	5,6,7	-71.053s	0.9 s *4*	β-
117	(7/2+)	-69.536s	0.44 s *4*	β-
46 Pd 93			60 s *20*	
94	0+	-66.350s	9.0 s *5*	ε
95		-70.151s		
95m	(21/2+)	-68.151s	13.3 s *3*	ε≥91.3%, IT≤9.7%, εp 0.9%
96	0+	-76.170	122 s *2*	ε
97	(5/2+)	-77.794	3.10 m *9*	ε
98	0+	-81.295	17.7 m *3*	ε
99	(5/2)+	-82.149	21.4 m *2*	ε
100	0+	-85.227	3.63 d *9*	ε
101	(5/2+)	-85.429	8.47 h *6*	ε
102	0+	-87.926	1.02% *1*	
103	5/2+	-87.480	16.991 d *19*	ε
104	0+	-89.392	11.14% *8*	
105	5/2+	-88.414	22.33% *8*	
106	0+	-89.905	27.33% *3*	
107	5/2+	-88.372	6.5×10^{6} y *3*	β-
*107*m	11/2-	-88.157	21.3 s *5*	IT
108	0+	-89.521	26.46% *9*	
109	5/2+	-87.603	13.7012 h *24*	β-
*109*m	11/2-	-87.414	4.696 m *3*	IT
110	0+	-88.350	11.72% *9*	
111	5/2+	-86.029	23.4 m *2*	β-
*111*m	11/2-	-85.857	5.5 h *1*	IT 73%, β- 27%
112	0+	-86.337	21.03 h *5*	β-
113	(5/2)+	-83.693	93 s *5*	β-
*113*m		-83.693	≥100 s	
114	0+	-83.494	2.42 m *6*	β-
115	(5/2+)	-80.403	25 s *2*	β-
*115*m	(11/2-)	-80.314	50 s *3*	β- 92%, IT 8%
116	0+	-79.953	11.8 s *4*	β-
117	(5/2+)	-76.532s	4.3 s *3*	β-
*117*m	(11/2-)	-76.329s	19.1 ms *7*	IT
118	0+	-75.544	2.4 s *4*	β-
119		-72.023s	0.92 s *13*	β-
47 Ag 96	(8+,9+)	-64.571s	5.1 s *4*	ε, εp 8%
97	(9/2+)	-70.794s	19 s *2*	ε
98	(5+)	-72.875	46.7 s *9*	ε
99	(9/2)+	-76.719	124 s *3*	ε
99m	(1/2-)	-76.213	10.5 s *5*	IT
100	(5)+	-78.153	2.01 m *9*	ε
100m	(2)+	-78.137	2.24 m *13*	ε, IT
101	9/2+	-81.225	11.1 m *3*	ε
101m	1/2-	-80.951	3.10 s *10*	IT
102	5+	-82.003	12.9 m *3*	ε
102m	2+	-81.994	7.7 m *5*	ε 51%, IT 49%
103	7/2+	-84.792	65.7 m *7*	ε
103m	1/2-	-84.658	5.7 s *3*	IT
104	5+	-85.113	69.2 m *10*	ε
104m	2+	-85.106	33.5 m *20*	ε 99.93%, IT
105	1/2-	-87.069	41.29 d *7*	ε
105m	7/2+	-87.044	7.23 m *16*	IT 99.66%, ε 0.34%
106	1+	-86.939	23.96 m *4*	ε 99.5%, β- <1%
106m	6+	-86.849	8.46 d *10*	ε
107	1/2-	-88.405	51.839% *5*	
*107*m	7/2+	-88.312	44.3 s *2*	IT
108	1+	-87.603	2.37 m *1*	β- 97.15%,

Isotope Z El	A	Jπ	Δ (MeV)	T1/2 or Abundance	Decay Mode
47 Ag					ε 2.85%
	108m	6+	-87.494	418 y 21	ε 91.3%, IT 8.7%
	109	1/2-	-88.719	48.161% 5	
	109m	7/2+	-88.631	39.6 s 2	IT
	110	1+	-87.457	24.6 s 2	β- 99.7%, ε 0.3%
	110m	6+	-87.339	249.79 d 20	β- 98.64%, IT 1.36%
	111	1/2-	-88.217	7.45 d 1	β-
	111m	7/2+	-88.157	64.8 s 8	IT 99.3%, β- 0.7%
	112	2(-)	-86.625	3.130 h 9	β-
	113	1/2-	-87.033	5.37 h 5	β-
	113m	7/2+	-86.990	68.7 s 16	IT 64%, β- 36%
	114	1+	-84.945	4.6 s 1	β-
	115	1/2-	-84.987	20.0 m 5	β-
	115m	7/2+	-84.946	18.0 s 7	β- 79%, IT 21%
	116	(2-)	-82.560	2.68 m 10	β-
	116m	(5+)	-82.478	8.6 s 3	β- 94%, IT 6%
	117	(1/2-)	-82.243	72.8 s +20-7	β-≈100%
	117m	(7/2+)	-82.214	5.34 s 5	β- 94%, IT 6%
	118	(1)	-79.644	3.76 s 15	β-
	118m		-79.516	2.0 s 2	β- 59%, IT 41%
	119m	(7/2+)	-78.555	2.1 s 1	β-
	119m	(1/2-)	-78.555	6.0 s 5	β-
	120		-75.773	1.23 s 3	β-
	120m		-75.570	0.32 s 4	β-≈63%, IT≈37%
	121	(7/2+)	-74.547	0.78 s 1	β-, β-n 0.08%
	122	(3+)	-71.427s	0.56 s 3	β-, β-n
	122m		-71.427s	1.5 s 5	β-, β-n
	123	(7/2+)	-69.955s	0.309 s 15	β-, β-n
	124	(1,2,3)+	-66.574s	0.22 s 3	β-, β-n
48 Cd	97m			3 s +4-2	ε, E
	98	0+	-67.455s	9.2 s 3	ε
	99	(5/2+)	-69.853s	16 s 3	ε, εp 0.17%, εα<1×10^{-4}%
	100	0+	-74.263	49.1 s 5	ε
	101	(5/2+)	-75.748	1.2 m 2	ε
	102	0+	-79.416	5.5 m 5	ε
	103	(5/2)+	-80.650	7.3 m 1	ε
	104	0+	-83.976	57.7 m 10	ε
	105	5/2+	-84.330	55.5 m 4	ε
	106	0+	-87.134	1.25% 4	
	107	5/2+	-86.988	6.50 h 2	ε
	108	0+	-89.253	0.89% 2	
	109	5/2+	-88.506	462.6 d 4	ε
	109m	1/2+	-88.446	12 μs 2	IT
	109m	11/2-	-88.043	10.9 μs 5	IT
	110	0+	-90.349	12.49% 12	
	111	1/2+	-89.254	12.80% 8	
	111m	11/2-	-88.858	48.54 m 5	IT
	112	0+	-90.581	24.13% 14	
	113	1/2+	-89.049	9.3×10^{15} y 19 12.22% 8	β-
	113m	11/2-	-88.785	14.1 y 5	β- 99.86%, IT 0.14%
	114	0+	-90.021	28.73% 28	
	115	1/2+	-88.090	53.46 h 10	β-
	115m	11/2-	-87.910	44.6 d 3	β-
	116	0+	-88.719	7.49% 12	
	117	1/2+	-86.425	2.49 h 4	β-
	117m	(11/2)-	-86.289	3.36 h 5	β-
	118	0+	-86.709	50.3 m 2	β-

Isotope Z El A	Jπ	Δ (MeV)	T1/2 or Abundance	Decay Mode
48 Cd *119*	3/2+	-83.905	2.69 m *2*	β-
*119*m	(11/2-)	-83.758	2.20 m *2*	β-
120	0+	-83.973	50.80 s *21*	β-
121	(3/2+)	-80.947	13.5 s *3*	β-
*121*m	(11/2-)	-80.732	8.3 s *8*	β-
122	0+	-80.574s	5.3 s *1*	β-
123	(3/2)+	-77.313	2.10 s *2*	β-
*123*m	(11/2-)	-76.996	1.82 s *3*	β-, IT
124	0+	-76.710	1.24 s *5*	β-
125	(3/2+)	-73.321	0.65 s *2*	β-
*125*m	(11/2-)	-73.272	0.57 s *9*	β-
126	0+	-72.327	0.506 s *15*	β-
127	(3/2+)	-68.526	0.4 s *1*	β-
128	0+	-67.290	0.28 s *4*	β-
129			0.27 s *4*	β-
130	0+		0.20 s *4*	β-, β-n≈4%
49 In 100		-63.733s		ε, εp
101		-68.409s	16 s *3*	ε≈100%, εp
102	(5)	-70.516	24 s *4*	ε
103	(9/2)+	-74.600	65 s *7*	ε
104	(6+)	-76.067	1.8 m *2*	ε
104m	(3+)	-75.973	15.7 s *5*	IT 80%, ε 20%
105	(9/2+)	-79.481	5.07 m *7*	ε
105m	(1/2-)	-78.807	48 s *6*	IT
106	7+	-80.613	6.2 m *1*	ε
106m	(3)+	-80.584	5.2 m *1*	ε
107	9/2+	-83.562	32.4 m *3*	ε
107m	1/2-	-82.884	50.4 s *6*	IT
108	7+	-84.105	58.0 m *12*	ε
108m	2+	-84.075	39.6 m *7*	ε
109	9/2+	-86.485	4.2 h *1*	ε
109m	1/2-	-85.835	1.34 m *7*	IT
109m	(19/2+)	-84.383	0.21 s *1*	IT
110	7+	-86.471	4.9 h *1*	ε
110m	2+	-86.409	69.1 m *5*	ε
111	9/2+	-88.388	2.8049 d *1*	ε
111m	1/2-	-87.851	7.7 m *2*	IT
112	1+	-87.994	14.97 m *10*	ε 56%, β- 44%
112m	4+	-87.837	20.56 m *6*	IT
113	9/2+	-89.365	4.3% *2*	
*113*m	1/2-	-88.973	1.6582 h *6*	IT
114	1+	-88.568	71.9 s	β- 99.5%, ε 0.5%
114m	5+	-88.378	49.51 d *1*	IT 95.6%, ε 4.4%
114m	8-	-88.066	43.1 ms *6*	IT
115	9/2+	-89.536	4.41×10^{14} y *25* 95.7% *2*	β-
*115*m	1/2-	-89.200	4.486 h *4*	IT 95%, β- 5%
116	1+	-88.249	14.10 s *3*	β- 99.94%, ε<0.06%
116m	5+	-88.122	54.41 m *3*	β-
116m	8-	-87.959	2.18 s *4*	IT
117	9/2+	-88.941	43.2 m *3*	β-
*117*m	1/2-	-88.626	116.2 m *3*	β- 52.9%, IT 47.1%
118	1+	-87.228	5.0 s *5*	β-
*118*m	5+	-87.168	4.45 m *5*	β-
*118*m	8-	-87.028	8.5 s *3*	IT 98.6%, β- 1.4%
119	9/2+	-87.702	2.4 m *1*	β-
*119*m	1/2-	-87.391	18.0 m *3*	β- 94.4%, IT 5.6%
120	1+	-85.731	3.08 s *8*	β-
120	(3,4,5)+	-85.731	46.2 s *8*	β-

Isotope Z El A	Jπ	Δ (MeV)	T1/2 or Abundance	Decay Mode
49 In *120*	(8-)	-85.731	47.3 s *5*	β-
121	9/2+	-85.837	23.1 s *6*	β-
*121*m	1/2-	-85.523	3.88 m *10*	β- 98.8%, IT 1.2%
122	1+	-83.575	1.5 s *3*	β-
*122*m	(4,5)+	-83.575	10.3 s *6*	β-
*122*m	8-	-83.355	10.8 s *4*	β-
123	9/2+	-83.428	5.98 s *6*	β-
*123*m	1/2-	-83.101	47.8 s *5*	β-
124	3+	-80.876	3.17 s *5*	β-
*124*m	(8-)	-80.686	3.4 s *5*	β-
125	9/2(+)	-80.480	2.36 s *4*	β-
*125*m	1/2(-)	-80.120	12.2 s *2*	β-
126	3(+)	-77.813	1.60 s *10*	β-
*126*m	7,8,9	-77.711	1.64 s *5*	β-
127	(9/2+)	-76.994	1.15 s *5*	β-
*127*m	(1/2-)	-76.834	3.76 s *3*	β-, β-n
128	3+	-74.360	0.80 s *1*	β-, β-n
*128*m	8-	-74.280	0.7 s *1*	β-, β-n
129	(9/2+)	-72.975	0.63 s *4*	β-, β-n
*129*m	(1/2-)	-72.775	1.23 s *2*	β-, β-n
130	1(-)	-69.993	0.32 s *2*	β-, β-n 0.9%
*130*m	(10-)	-69.943	0.55 s *1*	β-, β-n<1.67%
*130*m	(5+)	-69.593	0.55 s *1*	β-, β-n<1.67%
131	(9/2+)	-68.199	0.27 s *2*	β-, β-n
*131*m	(21/2+)	-68.199	0.32 s *6*	β-
*131*m	1/2-	-68.199	0.35 s *5*	β-
132	(7-)	-63.020	0.201 s *13*	β-, β-n 6.2%
133			180 ms *20*	β-, β-n
50 Sn 102	0+	-64.748s		
103		-66.946s	7 s *3*	ε
104	0+	-71.552	20.8 s *5*	ε
105		-73.233	31 s *6*	ε, εp
106	0+	-77.428	2.10 m *15*	ε
107	(5/2+)	-78.563	2.90 m *5*	ε
108	0+	-82.013	10.30 m *8*	ε
109	5/2(+)	-82.635	18.0 m *2*	ε
110	0+	-85.834	4.11 h *10*	ε
111	7/2+	-85.943	35.3 m *6*	ε
112	0+	-88.658	0.97% *1*	
113	1/2+	-88.329	115.09 d *4*	ε
113m	7/2+	-88.252	21.4 m *4*	IT 91.1%, ε 8.9%
114	0+	-90.557	0.65% *1*	
115	1/2+	-90.031	0.36% *1*	
116	0+	-91.523	14.53% *11*	
117	1/2+	-90.397	7.68% *7*	
*117*m	11/2-	-90.082	13.60 d *4*	IT
118	0+	-91.652	24.22% *11*	
119	1/2+	-90.066	8.58% *4*	
*119*m	11/2-	-89.976	293.1 d 7	IT
120	0+	-91.102	32.59% *10*	
121	3/2+	-89.201	27.06 h *4*	β-
*121*m	11/2-	-89.195	55 y *5*	IT 77.6%, β- 22.4%
122	0+	-89.944	4.63% *3*	
123	11/2-	-87.819	129.2 d *4*	β-
*123*m	3/2+	-87.794	40.06 m *1*	β-
124	0+	-88.236	5.79% *5*	
125	11/2-	-85.898	9.64 d *3*	β-
*125*m	3/2+	-85.870	9.52 m *5*	β-
126	0+	-86.020	$\approx 1\times10^5$ y	β-
127	(11/2-)	-83.508	2.10 h *4*	β-

Isotope Z El A	Jπ	Δ (MeV)	T1/2 or Abundance	Decay Mode
50 Sn *127*m	(3/2+)	-83.503	4.13 m *3*	β-
128	0+	-83.336	59.1 m *5*	β-
*128*m	(7-)	-81.245	6.5 s *5*	IT
129	(3/2+)	-80.630	2.4 m *1*	β-
*129*m	(11/2-)	-80.595	6.9 m *1*	β-, IT 0.0002%
130	0+	-80.242	3.72 m *4*	β-
*130*m	(7-)	-78.295	1.7 m *1*	β-
131		-77.383	39 s *2*	β-
*131*m	(3/2+)	-77.383	61 s *2*	β-
132	0+	-76.620	39.7 s *5*	β-
133	(7/2-)	-71.126	1.44 s *4*	β-, β-n 0.08%
134	0+	-67.226s	1.04 s *2*	β-, β-n 17%
51 Sb 104		-59.029s		
105		-63.910s		
106		-66.353s		
107		-70.654s		
108	(4+)	-72.507s	7.0 s *5*	ε
109	(5/2+)	-76.255	17.0 s *7*	ε
110	3+	-77.534s	23.0 s *4*	ε
111	(5/2+)	-80.843s	75 s *1*	ε
112	3+	-81.603	51.4 s *10*	ε
113	5/2+	-84.424	6.67 m *7*	ε
114	3+	-84.676	3.49 m *3*	ε
115	5/2+	-87.001	32.1 m *3*	ε
116	3+	-86.816	15.8 m *8*	ε
116m	8-	-86.433	60.3 m *6*	ε
117	5/2+	-88.640	2.80 h *1*	ε, ε 1.7%
118	1+	-87.995	3.6 m *1*	ε
118m	8-	-87.783	5.00 h *2*	ε
119	5/2+	-89.472	38.19 h *22*	ε
120	1+	-88.421	15.89 m *4*	ε
120m	8-	-88.421	5.76 d *2*	ε
121	5/2+	-89.589	57.36% *15*	
122	2-	-88.324	2.70 d *1*	β- 97.6%, ε 2.4%
122m	8-	-88.160	4.21 m *2*	IT
123	7/2+	-89.222	42.64% *15*	
124	3-	-87.618	60.20 d *3*	β-
124m	5+	-87.607	93 s *5*	IT 75%, β- 25%
124m	8-	-87.581	20.2 m *2*	IT
125	7/2+	-88.262	2.7582 y *11*	β-
126	(8)-	-86.398	12.46 d *3*	β-
*126*m	(5)+	-86.380	19.15 m *8*	β- 86%, IT 14%
*126*m	(3)-	-86.358	≈11 s	
127	7/2+	-86.709	3.85 d *5*	β-
128	8-	-84.610	9.01 h *3*	β-
*128*m	5+	-84.590	10.4 m *2*	β- 96.4%, IT 3.6%
129		-84.626	17.7 m	β-
129	7/2+	-84.626	4.40 h *1*	β-
130	(8-)	-82.393	39.5 m *8*	β-
*130*m	(5)+	-82.393	6.3 m *2*	β-
131	(7/2+)	-82.021	23 m *2*	β-
132	(4+)	-79.923	2.79 m *5*	β-
*132*m	(8-)	-79.923	4.10 m *5*	β-
133	(7/2+)	-78.956	2.5 m *1*	β-
134	(0-)	-73.977	0.85 s *10*	β-
134	(7-)	-73.977	10.43 s *14*	β-, β-n 0.1%
135	(7/2+)	-69.705	1.71 s *2*	β-, β-n 16.4%
136		-65.083s	0.82 s *2*	β-, β-n 24%
52 Te 106	0+	-58.000s	70 μs *20*	α
107		-60.519s	3.6 ms *+6-4*	α 70%, ε 30%
108	0+	-65.686	2.1 s *1*	α 68%, ε 32%

Isotope Z	El	A	Jπ	Δ (MeV)	T1/2 or Abundance	Decay Mode
52	Te	109		-67.582	4.6 s 3	ε 96%, α 4%
		110	0+	-72.280	18.6 s 8	ε≈100%, α≈3.0×10^{-3}%
		111		-73.475	19.3 s 4	ε, εp
		112	0+	-77.259	2.0 m 2	ε
		113	(7/2+)	-78.324s	1.7 m 2	ε
		114	0+	-81.933s	15.2 m 7	ε
		115	7/2+	-82.363	5.8 m 2	ε
		115m	(1/2)+	-82.343	6.7 m 4	ε≤100%, IT
		116	0+	-85.316	2.49 h 4	ε
		117	1/2+	-85.105	62 m 2	ε, ε 25%
		117m	(11/2-)	-84.809	103 ms 3	IT
		118	0+	-87.717	6.00 d 2	ε
		119	1/2+	-87.179	16.03 h 5	ε
		119m	11/2-	-86.918	4.70 d 4	ε, IT 8.0×10^{-3}%
		120	0+	-89.399	0.095% 5	
		121	1/2+	-88.553	16.78 d 35	ε
		121m	11/2-	-88.259	154 d 7	IT 88.6%, ε 11.4%
		122	0+	-90.303	2.59% 1	
		123	1/2+	-89.171	>1×10^{13} y 0.905% 5	ε
		123m	11/2-	-88.923	119.7 d 1	IT
		124	0+	-90.524	4.79% 2	
		125	1/2+	-89.028	7.12% 2	
		125m	11/2-	-88.883	57.40 d 15	IT
		126	0+	-90.071	18.93% 3	
		127	3/2+	-88.290	9.35 h 7	β-
		127m	11/2-	-88.202	109 d 2	IT 97.6%, β- 2.4%
		128	0+	-88.993	>8.×10^{24} y 31.70% 2	2β-
		129	3/2+	-87.005	69.6 m 2	β-
		129m	11/2-	-86.900	33.6 d 1	IT 64%, β- 36%
		130	0+	-87.353	≤1.25×10^{21} y 33.87% 7	
		131	3/2+	-85.211	25.0 m 1	β-
		131m	11/2-	-85.029	30 h 2	β- 77.8%, IT 22.2%
		132	0+	-85.209	3.204 d 13	β-
		133	(3/2+)	-82.959	12.5 m 3	β-
		133m	(11/2-)	-82.625	55.4 m 4	β- 82.5%, IT 17.5%
		134	0+	-82.394	41.8 m 8	β-
		135	(7/2-)	-77.825	19.0 s 2	β-
		136	0+	-74.423	17.5 s 2	β-, β-n 1.1%
		137	(7/2-)	-69.559	2.49 s 5	β-, β-n 2.7%
		138	0+	-65.931s	1.4 s 4	β-, β-n 6.3%
53	I	108		-52.569s	≈50 ms	
		109		-57.576	100 μs 5	P
		110		-60.347s	0.65 s 2	ε 83%, α 17%, εp 11%, εα 1.1%
		111		-64.951s	2.5 s 2	ε 99.9%, α≈0.1%
		112		-67.096s	3.42 s 11	ε, α≈0.0012%, εα, εp
		113	5/2+	-71.124	6.6 s 2	ε, α 3.3×10^{-7}%
		114		-72.796s	2.1 s 2	ε, εp
		115	(5/2+)	-76.403s	1.3 m 2	ε
		116	(7-)	-77.571	3.27 μs 16	
		116	1+	-77.571	2.91 s 15	ε
		117	(5/2)+	-80.452	2.22 m 4	ε
		118	2-	-80.673	13.7 m 5	ε
		118m	(7-)	-80.673	8.5 m 5	ε

Isotope Z El A	Jπ	Δ (MeV)	T1/2 or Abundance	Decay Mode
53 I 119	5/2+	-83.665	19.1 m 4	ε
120	2-	-83.784	81.0 m 6	ε
120m	>3	-83.784	53 m 4	ε
121	5/2+	-86.282	2.12 h 1	ε
122	1+	-86.069	3.63 m 6	ε
123	5/2+	-87.929	13.27 h 8	ε
124	2-	-87.364	4.18 d 2	ε
125	5/2+	-88.842	59.408 d 8	ε
126	2-	-87.916	13.11 d 5	ε 56.3%, β- 43.7%
127	5/2+	-88.988	100%	
128	1+	-87.743	24.99 m 2	β- 93.1%, ε 6.9%
129	7/2+	-88.503	1.57×10^{7} y 4	β-
130	5+	-86.932	12.36 h 3	β-
130m	2+	-86.892	9.0 m 1	IT 84%, β- 16%
131	7/2+	-87.444	8.04 d 1	β-
132	4+	-85.702	2.295 h 13	β-
132m	(8-)	-85.582	1.387 h 15	IT 86%, β- 14%
133	7/2+	-85.877	20.8 h 1	β-
133m	(19/2-)	-84.243	9 s 2	IT
134	4+	-83.954	52.6 m 4	β-
134m	8-	-83.638	3.69 m 7	IT 97.7%, β- 2.3%
135	7/2+	-83.787	6.57 h 2	β-
136	2-	-79.498	83.4 s 10	β-
136m	(6-)	-78.858	46.9 s 10	β-
137	(7/2+)	-76.501	24.5 s 2	β-, β-n 7.1%
138	(2-)	-72.299	6.49 s 7	β-, β-n 5.5%
139	(7/2+)	-68.843	2.29 s 2	β-, β-n 9.9%
140	(3)	-64.236s	0.86 s 4	β-, β-n 9.4%
141		-60.482s	0.43 s 2	β-, β-n 21.2%
142			≈0.2 s	β-
54 Xe 110	0+	-51.689s		α
111		-54.381s	0.74 s 20	
112	0+	-59.944	2.7 s 8	ε 99.16%, α 0.84%
113		-62.061	2.74 s 8	ε 99.97%, εp 4.2%, α 0.04%
114	0+	-66.935s	10.0 s 4	ε
115	(5/2+)	-68.444s	18 s 4	ε, εp
116	0+	-72.911s	56 s 2	ε >0%
117	5/2(+)	-74.006	61 s 2	ε, εp 2.9×10^{-3}%
118	0+	-77.728	3.8 m 9	ε
119	(5/2+)	-78.664	5.8 m 3	ε
120	0+	-81.824	40 m 1	ε
121	5/2(+)	-82.550	40.1 m 20	ε
122	0+	-85.174	20.1 h 1	ε
123	(1/2)+	-85.253	2.08 h 2	ε
124	0+	-87.658	0.10% 1	
125	(1/2)+	-87.190	16.9 h 2	ε
125m	(9/2)-	-86.937	57 s 1	IT
126	0+	-89.174	0.09% 1	
127	(1/2+)	-88.325	36.4 d 1	ε
127m	(9/2-)	-88.028	69.2 s 9	IT
128	0+	-89.861	1.91% 3	
129	1/2+	-88.697	26.4% 6	
129m	11/2-	-88.461	8.89 d 2	IT
130	0+	-89.881	4.1% 1	
131	3/2+	-88.415	21.2% 4	
131m	11/2-	-88.251	11.9 d 1	IT
132	0+	-89.279	26.9% 5	
132m	(10+)	-86.527	8.39 ms 11	IT
133	3/2+	-87.648	5.243 d 1	β-
133m	11/2-	-87.415	2.19 d 1	IT

Isotope Z El A	Jπ	Δ (MeV)	T1/2 or Abundance	Decay Mode
54 Xe *134*	0+	-88.124	10.4% *2*	
*134*m	7-	-86.165	290 ms *17*	IT
135	3/2+	-86.435	9.14 h *2*	β-
*135*m	11/2-	-85.908	15.29 m *5*	IT, β- 0.004%
136	0+	-86.424	$\geq 2.36\times10^{21}$ y 8.9% *1*	2β-
137	(7/2)-	-82.378	3.818 m *13*	β-
138	0+	-80.119	14.08 m *8*	β-
139	3/2-	-75.649	39.68 s *14*	β-
140	0+	-72.999	13.60 s *10*	β-
141	5/2+	-68.321	1.73 s *1*	β-, β-n 0.04%
142	0+	-65.478	1.22 s *2*	β-
143	5/2-	-60.398s	0.30 s *3*	β-
144	0+	-57.250s	1.15 s *20*	β-
145			0.9 s *3*	β-, β-n
146	0+			β-
55 Cs 113		-51.677	33 μs *7*	p≈100%
114	(1+)	-54.565s	0.57 s *2*	ε≈100%, εp 7%, εα 0.16%, α 0.02%
115		-59.677s	1.4 s *8*	ε, εp≈0.07%
116	(1+)	-62.434	0.70 s *4*	ε, εα, εp
116m	≥5+	-62.434	3.84 s *16*	ε, εα>0%, εp>0%
117m		-66.482	6.5 s *4*	ε
117m		-66.482	8.4 s *6*	ε
118m	2	-68.428	14 s *2*	εα, ε, εp
118m	8,7,6	-68.428	17 s *3*	εα, ε, εp
119	9/2+	-72.336	43.0 s *2*	ε
119m	3/2(+)	-72.336	30.4 s *1*	ε
120	high	-73.902	57 s *6*	ε, εp≤ 1.0×10^{-5}%
120	2	-73.902	64 s *3*	ε
121	3/2(+)	-77.150	155 s *4*	ε
121m	9/2(+)	-77.082	122 s *3*	ε 83%, IT 17%
122	1+	-78.120	21.0 s *7*	ε
122m		-78.120	0.36 s *2*	IT
122m	8-	-78.120	4.5 m *2*	ε
123	1/2+	-81.052	5.94 m *4*	ε
123m	(11/2)-	-80.896	1.64 s *12*	IT
124	1+	-81.741	30.8 s *5*	ε
124m	(7)+	-81.278	6.3 s *2*	IT
125	(1/2+)	-84.098	45 m *1*	ε
126	1+	-84.348	1.64 m *2*	ε
127	1/2+	-86.245	6.25 h *10*	ε
128	1+	-85.931	3.62 m *2*	ε
129	1/2+	-87.502	32.06 h *6*	ε
130	1+	-86.898	29.21 m *4*	ε 98.4%, β- 1.6%
130m	5-	-86.735	3.46 m *6*	IT 99.84%, ε 0.16%
131	5/2+	-88.063	9.69 d *1*	ε
132	2+	-87.160	6.479 d *7*	ε 98.13%, β- 1.87%
133	7/2+	-88.075	100%	
134	4+	-86.896	2.062 y *5*	β-, ε 0.0003%
134m	8-	-86.757	2.91 h *1*	IT
135	7/2+	-87.586	2.3×10^{6} y *3*	β-
*135*m	19/2-	-85.953	53 m *2*	IT
136	5+	-86.343	13.16 d *3*	β-
*136*m	8-	-86.343	19 s *2*	β-?, IT
137	7/2+	-86.550	30.1 y *2*	β-
138	3-	-82.893	33.41 m *18*	β-
*138*m	6-	-82.813	2.91 m *8*	IT 81%, β- 19%
139	7/2+	-80.706	9.27 m *5*	β-

Isotope Z El	A	Jπ	Δ (MeV)	T1/2 or Abundance	Decay Mode
55 Cs	*140*	1-	-77.059	63.7 s *3*	β-
	141	7/2+	-74.471	24.94 s *6*	β-, β-n 0.03%
	142	0-	-70.518	1.70 s *2*	β-, β-n 0.28%
	143	3/2+	-67.705	1.78 s *1*	β-, β-n 1.62%
	144	1	-63.316	1.01 s *1*	β-, β-n 3.17%
	*144*m	(≥4)	-63.316	<1 s	β-
	145	3/2+	-60.164	0.594 s *13*	β-, β-n 13.8%
	146	1-	-55.662	0.343 s *7*	β-, β-n 13.2%
	147	(3/2+)	-52.232	0.225 s *5*	β-, β-n 43%
	148		-47.523	158 ms *7*	β-
56 Ba	117	(3/2)	-56.962s	1.75 s *7*	ε, εα>0%, εp>0%
	118	0+	-62.000s		
	119	(5/2+)	-64.239	5.4 s *3*	ε, εp>0%
	120	0+	-68.902	32 s *5*	ε
	121	5/2(+)	-70.335	29.7 s *15*	ε, εp 0.02%
	122	0+	-74.277s	1.95 m *15*	ε
	123	5/2+	-75.591s	2.7 m *4*	ε
	124	0+	-79.094	11.9 m *10*	ε
	125	1/2(+)	-79.538	3.5 m *4*	ε
	126	0+	-82.675	100 m *2*	ε
	127	(1/2+)	-82.795	12.7 m *4*	ε
	128	0+	-85.409	2.43 d *5*	ε
	129	1/2+	-85.068	2.23 h *11*	ε
	129m	(7/2)+	-85.060	2.17 h *4*	ε
	130	0+	-87.271	0.106% *2*	
	130m	8-	-84.796	11 ms *2*	IT
	131	1/2+	-86.693	11.8 d *2*	ε
	131m	9/2-	-86.505	14.6 m *2*	IT
	132	0+	-88.439	0.101% *3*	
	133	1/2+	-87.558	10.52 y *13*	ε
	133m	11/2-	-87.270	38.9 h *1*	IT 99.99%, ε 0.01%
	134	0+	-88.954	2.42% *4*	
	135	3/2+	-87.855	6.593% *24*	
	*135*m	11/2-	-87.587	28.7 h *2*	IT
	136	0+	-88.891	7.85% *5*	
	*136*m	7-	-86.860	0.3084 s *19*	IT
	137	3/2+	-87.726	11.23% *5*	
	*137*m	11/2-	-87.064	2.552 m *1*	IT
	138	0+	-88.266	71.70% *9*	
	139	7/2-	-84.918	83.06 m *28*	β-
	140	0+	-83.278	12.752 d *3*	β-
	141	3/2-	-79.726	18.27 m *7*	β-
	142	0+	-77.825	10.6 m *2*	β-
	143	5/2-	-73.948	14.33 s *8*	β-
	144	0+	-71.780	11.5 s *2*	β-, β-n 3.6%
	145	5/2-	-68.051	4.31 s *16*	β-
	146	0+	-65.039	2.22 s *7*	β-
	147	(3/2-)	-61.485	0.893 s *1*	β-, β-n 0.02%
	148	0+	-58.048	0.607 s *25*	β-, β-n≤0.02%
	149		-53.961s	0.356 s *8*	β-, β-n 0.43%
57 La	120		-57.687s	2.8 s *2*	ε, εp
	121		-62.401s	5.3 s *2*	ε
	122		-64.543s	8.7 s *7*	ε, εp
	123		-68.707s	17 s *3*	ε
	124	(7+)	-70.300s	29 s *3*	ε
	125	(11/2-)	-73.895s	76 s *6*	ε
	126		-75.106s	54 s *2*	ε>0%
	127	(3/2+)	-78.096s	3.8 m *5*	ε
	127m	(11/2-)	-78.096s	5.0 m *5*	
	128	4-,5-	-78.759	5.0 m *3*	ε

Isotope Z El A	Jπ	Δ (MeV)	T1/2 or Abundance	Decay Mode
57 La 129	(3/2+)	-81.348	11.6 m *2*	ε
129m	11/2-	-81.176	0.56 s *5*	IT
130	3(+)	-81.673s	8.7 m *1*	ε
131	3/2+	-83.733	59 m *2*	ε
132	2-	-83.731	4.8 h *2*	ε
132m	6-	-83.543	24.3 m *5*	IT 76%, ε 24%
133	5/2+	-85.328	3.912 h *8*	ε
134	1+	-85.241	6.45 m *16*	ε
135	5/2+	-86.655	19.5 h *2*	ε
136	1+	-86.021	9.87 m *3*	ε
136m		-85.791	114 ms *3*	IT
137	7/2+	-87.126	6×10^{4} y *2*	ε
138	5+	-86.529	1.05×10^{11} y *2* 0.0902% *2*	ε 66.4%, β- 33.6%
139	7/2+	-87.235	99.9098% *2*	
140	3-	-84.325	1.6781 d *3*	β-
141	(7/2+)	-82.942	3.92 h *3*	β-
142	2-	-80.037	91.1 m *5*	β-
143	(7/2)+	-78.191	14.2 m *1*	β-
144	(3-)	-74.899	40.8 s *4*	β-
145		-72.982	24.8 s *20*	β-
146	2-	-69.157	6.27 s *10*	β-
*146*m	(6-)	-69.157	10.0 s *1*	β-
147	(3/2+,5/2+)	-67.235	4.015 s *8*	β-, β-n 0.04%
148	(2-)	-63.163	1.05 s *1*	β-, β-n 0.11%
149		-61.292s	1.2 s *4*	β-, β-n
150		-57.156s		
58 Ce 123	(5/2)	-60.072s	3.8 s	ε, εp
124	0+	-64.720s	6 s *2*	ε
125	(5/2+)	-66.565s	9.0 s *6*	ε, εp
126	0+	-70.700s	50 s *3*	ε >0%
127		-71.958s	32 s *4*	ε
128	0+	-75.572s	6 m *2*	ε
129		-76.298s	3.5 m *5*	ε
130	0+	-79.466s	25 m *2*	ε
131		-79.713	10 m *1*	ε
131		-79.713	5 m *1*	ε
132	0+	-82.447s	3.51 h *11*	ε
133	9/2-	-82.391s	4.9 h *4*	ε
133m	1/2+	-82.391s	97 m *4*	ε
134	0+	-84.741	75.9 h *9*	ε
135	1/2(+)	-84.629	17.7 h *2*	ε
135m	11/2(-)	-84.183	20 s *1*	IT
136	0+	-86.494	0.19% *1*	
137	3/2+	-85.904	9.0 h *3*	ε
137m	11/2-	-85.650	34.4 h *3*	IT 99.22%, ε 0.78%
138	0+	-87.573	0.25% *1*	
138m	7-	-85.444	8.65 ms *20*	IT
139	3/2+	-86.957	137.640 d *23*	ε
139m	11/2-	-86.203	54.8 s *10*	IT
140	0+	-88.087	88.43% *10*	
141	7/2-	-85.444	32.501 d *5*	β-
142	0+	-84.542	$>5\times10^{16}$ y 11.13% *10*	
143	3/2-	-81.616	33.039 h *6*	β-
144	0+	-80.441	284.893 d *8*	β-
145	(3/2-)	-77.099	3.01 m *6*	β-
146	0+	-75.704	13.52 m *13*	β-
147	(5/2-)	-72.180	56.4 s *10*	β-
148	0+	-70.425	56 s *1*	β-

Isotope Z El A	Jπ	Δ (MeV)	T1/2 or Abundance	Decay Mode
58 Ce *149*		-66.798	5.2 s *3*	β-
150	0+	-64.993	4.0 s *6*	β-
151		-61.460s	1.02 s *6*	β-
152	0+	-59.047s	3.1 s *3*	β-
59 Pr 121			1.4 s *8*	ε
124		-53.021s	1.2 s *2*	ε, εp
126		-60.258s	3.1 s *3*	ε>0%, εp
128		-66.322s	3.2 s *5*	ε, εp
129		-69.992s	24 s *5*	ε
130		-71.371s	40.0 s *4*	ε
131		-74.463	1.7 m *4*	ε
132		-75.339s	1.6 m *3*	ε
133	5/2(+)	-78.059s	6.5 m *3*	ε
134	2-	-78.534s	17 m *2*	ε
134m	(5-)	-78.534s	11 m	ε
135	3/2(+)	-80.909	24 m *2*	ε
136	2+	-81.368	13.1 m *1*	ε
137	5/2+	-83.202	1.28 h *3*	ε
138	1+	-83.136	1.45 m *5*	ε
138m	7-	-82.772	2.12 h *4*	ε
139	5/2+	-84.828	4.41 h *4*	ε
140	1+	-84.699	3.39 m *1*	ε
141	5/2+	-86.025	100%	
142	2-	-83.797	19.12 h *4*	β- 99.98%, ε 0.02%
59 Pr 142m	5-	-83.793	14.6 m *5*	IT
143	7/2+	-83.077	13.57 d *2*	β-
144	0-	-80.759	17.28 m *5*	β-
*144*m	3-	-80.700	7.2 m *3*	IT 99.93%, β- 0.07%
145	7/2+	-79.636	5.984 h *10*	β-
146	(2)-	-76.739	24.15 m *18*	β-
147	(3/2+)	-75.470	13.4 m *4*	β-
148	1-	-72.485	2.27 m *4*	β-
*148*m	(4)	-72.395	2.0 m *1*	β-
149	(5/2+)	-70.988	2.26 m *7*	β-
150	1(-)	-68.003	6.19 s *16*	β-
151	1/2-:5/2-	-66.786	18.90 s *7*	β-
152		-63.463s	3.24 s *19*	β-
153		-61.544s	4.3 s *2*	β-
154	(3+,2+)	-57.697s	2.3 s *1*	β-
60 Nd 127	(5/2)	-55.424s	1.8 s *4*	ε, εp
128		-60.184s	4 s *2*	ε, εp
129	(5/2-)	-62.168s	4.9 s *2*	ε, εp
130	0+	-66.341s	28 s *3*	ε
131		-67.903	24 s *3*	ε, εp
131		-67.903	25 s	
132	0+	-71.613s	1.75 m *17*	ε
133		-72.461s	70 s *10*	ε
133m	(9/2-)	-72.461s	<2 m	ε
134	0+	-75.764s	8.5 m *15*	ε
135	9/2(-)	-76.159s	12.4 m *6*	ε
135m		-76.159s	5.5 m *5*	ε
136	0+	-79.157	50.65 m *33*	ε
137	1/2+	-79.512	38.5 m *15*	ε
137m	11/2-	-78.992	1.60 s *15*	IT
138	0+	-82.036s	5.04 h *9*	ε
139	3/2+	-82.042	29.7 m *5*	ε
139m	11/2-	-81.811	5.50 h *20*	ε 88.2%, IT 11.8%
140	0+	-84.477	3.37 d *2*	ε
141	3/2+	-84.202	2.49 h *3*	ε

Isotope Z El A	Jπ	Δ (MeV)	T1/2 or Abundance	Decay Mode
60 Nd 141m	11/2-	-83.445	62.0 s 8	IT, ε<0.05%
142	0+	-85.959	27.13% 10	
143	7/2-	-84.011	12.18% 5	
144	0+	-83.757	2.29×10^{15} y 16	α
			23.80% 10	
145	7/2-	-81.441	8.30% 5	
146	0+	-80.935	17.19% 8	
147	5/2-	-78.156	10.98 d 1	β-
148	0+	-77.417	5.76% 3	
149	5/2-	-74.385	1.72 h 1	β-
150	0+	-73.693	$>1\times10^{18}$ y	2β-
			5.64% 3	
151	(3/2)+	-70.956	12.44 m 7	β-
152	0+	-70.157	11.4 m 2	β-
153	(1/2:5/2)	-67.068s	28.9 s 4	β-
154	0+	-65.614s	25.9 s 2	β-
155		-62.010s	8.9 s 2	β-
156	0+	-60.109s	5.47 s 11	β-
61 Pm 130		-55.470s	2.2 s 5	ε, εp
132	(3+)	-61.711s	6.3 s 7	ε, εp≈5.0×10^{-5}%
133		-65.465s	12 s 3	ε
134	(5+)	-66.881s	24 s 2	ε
135	(11/2-)	-70.141s	49 s 7	ε
136	(3+)	-71.307	≈107 s	ε
136	5(+),6-	-71.307	107 s 6	ε
137	(11/2)-	-73.932s	2.4 m 1	ε
138	1+	-75.136s	10 s 2	ε
138m	(3+)	-75.136s	3.24 m 5	ε
138m	(5-)	-75.136s	3.24 m	ε
139	(5/2)+	-77.520	4.15 m 5	ε
139m	(11/2)-	-77.331	180 ms 20	IT, ε?
140	1+	-78.388	9.2 s 2	ε
140m	7-	-78.388	5.95 m 5	ε
141	5/2+	-80.487	20.90 m 5	ε
142	1+	-81.085	40.5 s 5	ε
143	5/2+	-82.970	265 d 7	ε
144	5-	-81.425	363 d 14	ε
145	5/2+	-81.278	17.7 y 4	ε, α 3×10^{-7}%
146	3-	-79.463	5.53 y 5	ε 66%, β- 34%
147	7/2+	-79.052	2.6234 y 2	β-
148	1-	-76.878	5.370 d 9	β-
148m	6-	-76.740	41.29 d 11	β- 95%, IT 5%
149	7/2+	-76.075	53.08 h 5	β-
150	(1-)	-73.607	2.68 h 2	β-
151	5/2+	-73.399	28.40 h 4	β-
152	1+	-71.268	4.1 m 1	β-
152m	4-	-71.098	7.52 m 8	β-
152m	(8)	-71.098	13.8 m 2	β-, IT
153	5/2-	-70.668	5.4 m 2	β-
154	(0,1)	-68.411	1.73 m 10	β-
154m	(3,4)	-68.411	2.68 m 7	β-
155	(5/2-)	-67.030s	48 s 4	β-
156	4(-)	-64.217	26.70 s 10	β-
157		-62.224s	10.90 s 20	β-
158		-58.973s	4.8 s 5	β-
62 Sm 131			1.2 s 2	ε, εp,
132	0+		4.0 s 3	ε, εp
133	(5/2+)	-57.073s	2.9 s 2	ε, εp
134	0+	-61.460s	11 s 2	ε
135	(7/2+)	-63.016s	10 s 2	ε, εp
136	0+	-66.788s	43 s 3	ε

Isotope Z El A	Jπ	Δ (MeV)	T1/2 or Abundance	Decay Mode
62 Sm 137	(9/2-)	-67.878s	45 s *1*	ε
138	0+	-71.222s	3.1 m *2*	ε
139	(1/2)+	-72.060	2.57 m *10*	ε
139m	(11/2)-	-71.602	10.7 s *6*	IT 93.7%, ε 6.3%
140	0+	-75.368s	14.82 m *10*	ε
141	1/2+	-75.944	10.2 m *2*	ε
141m	11/2-	-75.768	22.6 m *2*	ε 99.69%, IT 0.31%
142	0+	-78.987	72.49 m *5*	ε
143	3/2+	-79.527	8.83 m *1*	ε
143m	11/2-	-78.773	66 s *2*	IT 99.76%, ε 0.24%
143m	23/2-	-76.732	30 ms *3*	IT
144	0+	-81.975	3.1% *1*	
145	7/2-	-80.661	340 d *3*	ε
146	0+	-81.005	10.3×10^{7} y *5*	α
147	7/2-	-79.276	1.06×10^{11} y *2* 15.0% *2*	α
148	0+	-79.346	7×10^{15} y *3* 11.3% *1*	α
149	7/2-	-77.146	$>2\times10^{15}$ y 13.8% *1*	
150	0+	-77.061	7.4% *1*	
151	5/2-	-74.586	90 y *8*	β-
152	0+	-74.772	26.7% *2*	
153	3/2+	-72.569	46.27 h *1*	β-
154	0+	-72.465	22.7% *2*	
155	3/2-	-70.201	22.3 m *2*	β-
156	0+	-69.372	9.4 h *2*	β-
157		-66.771	8.07 m *12*	β-
158	0+	-65.270s	5.51 m *9*	β-
159		-62.224s	11.2 s *2*	β-
160	0+	-60.286s	9.6 s *3*	β-
63 Eu 134			0.5 s *2*	ε, εp
135		-54.287s	1.5 s *2*	ε
136	(1+)	-56.355s	3.9 s *5*	ε, εp
136m	(7+)	-56.355s	≈3.2 s	
137	(11/2-)	-60.351s	11 s *2*	ε
138	(6-)	-61.991s	12.1 s *6*	ε
139	(11/2)-	-65.382s	17.9 s *6*	ε
140	1(-)	-66.968s	1.54 s *13*	ε
140m		-66.968s	0.125 s *2*	ε
141	5/2+	-70.394	40.0 s *7*	ε
141m	11/2-	-70.298	2.7 s *3*	IT 87%, ε 13%
142	1+	-71.627	2.4 s *2*	ε
142m	8-	-71.627	1.22 m *2*	ε
143	5/2+	-74.358	2.63 m *5*	ε
144	1+	-75.646	10.2 s *1*	ε
145	5/2+	-78.001	5.93 d *4*	ε
146	4-	-77.127	4.59 d *3*	ε
147	5/2+	-77.554	24.1 d *6*	ε, α 2.2×10^{-3}%
148	5-	-76.239	54.5 d *5*	ε, α 9.4×10^{-7}%
149	5/2+	-76.454	93.1 d *4*	ε
150	5(-)	-74.800	35.8 y *10*	ε
150m	0(-)	-74.758	12.8 h *1*	β- 89%, ε 11%
151	5/2+	-74.663	47.8% *5*	
152	3-	-72.898	13.542 y *10*	ε 72.08%, β- 27.92%
152m	0-	-72.852	9.274 h *9*	β- 72%, ε 28%
152m	8-	-72.750	96 m *1*	IT
153	5/2+	-73.377	52.2% *5*	

Isotope Z	El	A	Jπ	Δ (MeV)	T1/2 or Abundance	Decay Mode
63	Eu	154	3−	−71.748	8.593 y 4	β− 99.98%, ε 0.02%
		154m	(8−)	−71.603	46.3 m 4	IT
		155	5/2+	−71.828	4.68 y 5	β−
		156	0+	−70.094	15.19 d 8	β−
		157	5/2+	−69.471	15.18 h 3	β−
		158	(1−)	−67.215	45.9 m 2	β−
		159	5/2+	−66.057	18.1 m 1	β−
		160	1(−)	−63.372s	38 s 4	β−
		161		−61.777s	26 s 3	β−
		162		−58.647s	10.6 s 10	β−
64	Gd	137		−51.558s	7 s 3	ε, εp
		138	0+	−55.918s		
		139		−57.678s	4.9 s 10	ε, εp
		140	0+	−61.508s	16 s 1	ε
		141	(1/2+)	−63.146s	14 s 4	ε, εp 0.03%
		141m	(11/2−)	−62.768s	24.5 s 5	ε 89%, IT 11%
		142	0+	−67.127s	70.2 s 6	ε
		143	(1/2)+	−68.351	39 s 2	ε
		143m	(11/2−)	−68.198	112 s 2	ε
		144	0+	−71.907s	4.5 m 1	ε
		145	1/2+	−72.947	23.0 m 4	ε
		145m	11/2−	−72.198	85 s 3	IT 94.3%, ε 5.7%
		146	0+	−76.097	48.27 d 10	ε
		147	7/2−	−75.367	38.06 h 12	ε
		148	0+	−76.279	74.6 y 30	α
		149	7/2(−)	−75.135	9.4 d 3	ε, α
		150	0+	−75.771	1.79×10^{6} y 8	α
		151	7/2−	−74.199	124 d 1	ε, α 1.0×10^{-6}%
		152	0+	−74.716	1.08×10^{14} y 8 0.20% 1	α
		153	3/2−	−72.892	241.6 d 2	ε
		154	0+	−73.716	2.18% 3	
		155	3/2−	−72.080	14.80% 5	
		156	0+	−72.545	20.47% 4	
		157	3/2−	−70.834	15.65% 3	
		158	0+	−70.700	24.84% 12	
		159	3/2−	−68.572	18.56 h 8	β−
		160	0+	−67.952	21.86% 4	
		161	5/2−	−65.516	3.66 m	β−
		162	0+	−64.290	8.4 m 2	β−
		163	(5/2−)	−61.488s	68 s 3	β−
		164	0+	−59.746s	45 s 3	β−
65	Tb	140		−50.708s	2.4 s 4	ε, εp
		141	(5/2−)	−54.809s	3.5 s 2	ε
		141m		−54.809s	7.9 s 6	ε
		142	1+	−57.067s	597 ms 17	ε, εp≈3.0×10^{-7}%
		142m	(5−)	−56.787s	303 ms 7	ε, εp, IT
		143	(11/2−)	−60.957s	12 s 1	ε
		143m	(5/2+)	−60.957s	<21 s	
		144	(1+)	−62.988s	≈1 s	ε
		144m	(6−)	−62.591s	4.25 s 15	IT 66%, ε 34%
		145	(1/2+)	−66.438		
		145m	(11/2−)	−66.438	29.5 s 15	ε
		146	1+	−68.022	8 s 4	ε
		146m	5−	−68.022	23 s 2	ε
		147	(1/2+)	−70.755	1.7 h 1	ε
		147m	(11/2)−	−70.704	1.83 m 6	ε
		148	2−	−70.586	60 m 1	ε
		148m	9+	−70.496	2.20 m 5	ε
		149	1/2+	−71.499	4.13 h 2	ε 84.2%, α 15.8%

Isotope Z El A	Jπ	Δ (MeV)	T1/2 or Abundance	Decay Mode
65 Tb 149m	11/2-	-71.463	4.16 m *4*	ε, α
150m	(8+,9+)	-71.115	5.8 m *2*	ε
150m	(2-)	-71.115	3.48 h *16*	ε, α<0.05%
151	1/2(+)	-71.633	17.609 h *1*	ε, α 0.0095%
151m	(11/2-)	-71.533	25 s *3*	IT 93.8%, ε 6.2%
152	2-	-70.726	17.5 h *1*	ε, α<7.0×10^{-7}%
152m	8+	-70.224	4.2 m *1*	IT 78.9%, ε 21.1%
153	5/2+	-71.322	2.34 d *1*	ε
154	0	-70.154	21.5 h *4*	ε, β-<0.1%
154m	3-	-70.154	9.4 h *4*	ε 78.2%, IT 21.8%, β-<0.1%
154m	7-	-70.154	22.7 h *5*	ε 98.2%, IT 1.8%
155	3/2+	-71.259	5.32 d *6*	ε
156	3-	-70.101	5.35 d *10*	ε, β-
156m	(7-)	-70.051	24.4 h *10*	IT
156m	(0+)	-70.013	5.3 h *2*	ε, IT
157	3/2+	-70.774	99 y *10*	ε
158	3-	-69.480	180 y *11*	ε 83.4%, β- 16.6%
158m	0-	-69.370	10.5 s *2*	IT, β-<0.6%, ε<0.01%
159	3/2+	-69.542	100%	
160	3-	-67.846	72.3 d *2*	β-
161	3/2+	-67.471	6.88 d *3*	β-
162	1-	-65.684	7.60 m *15*	β-
163	3/2+	-64.605	19.5 m *3*	β-
164	(5+)	-62.087	3.0 m *1*	β-
165	(3/2+)	-60.659s	2.11 m *10*	β-
66 Dy 141	(9/2-)	-45.466s	0.9 s *2*	ε, εp
142	0+	-50.167s	2.3 s *3*	ε, εp≈8.0×10^{-5}%
143		-52.192s	3.9 s *4*	ε, εp
144	0+	-56.756s	9.1 s *4*	ε, εp
145		-58.718s		
145m	(11/2-)	-58.718s	13.6 s *10*	ε
146	0+	-62.862	29 s *3*	ε
146m	(10+)	-62.862	150 ms *20*	IT
147	1/2+	-64.383	40 s *10*	ε, εp>0%
147m	11/2-	-63.632	55 s *1*	ε 65%, IT 35%
148	0+	-67.908	3.1 m *1*	ε
149	(7/2-)	-67.687	4.23 m *18*	ε
150	0+	-69.321	7.17 m *5*	ε 64%, α 36%
151	7/2(-)	-68.762	17.9 m *3*	ε 94.4%, α 5.6%
152	0+	-70.128	2.38 h *2*	ε 99.9%, α 0.1%
153	7/2(-)	-69.151	6.4 h *1*	ε 99.99%, α 9.4×10^{-3}%
154	0+	-70.400	3.0×10^{6} y *15*	α
155	3/2-	-69.164	10.0 h *3*	ε
156	0+	-70.534	0.06% *1*	
157	3/2-	-69.432	8.14 h *4*	ε
157m	11/2-	-69.233	20.2 ms *14*	IT
158	0+	-70.417	0.10% *1*	
159	3/2-	-69.177	144.4 d *2*	ε
160	0+	-69.682	2.34% *5*	
161	5/2+	-68.065	18.9% *1*	
162	0+	-68.190	25.5% *2*	
163	5/2-	-66.390	24.9% *2*	
164	0+	-65.977	28.2% *2*	
165	7/2+	-63.621	2.334 h *1*	β-
165m	1/2-	-63.513	1.257 m *6*	IT 97.76%, β- 2.24%
166	0+	-62.593	81.6 h *1*	β-
167	(1/2-)	-59.942	6.20 m *8*	β-

Isotope Z El A	Jπ	Δ (MeV)	T1/2 or Abundance	Decay Mode
66 Dy 168	0+	-58.470s	8.5 m *5*	β-
169	(5/2-)	-55.606	39 s *8*	β-
67 Ho 144		-45.000s	0.7 s *1*	ε, εp
145		-49.611s		
146	(10+)	-52.182s	3.6 s *3*	ε, εp
147	(11/2-)	-56.234s	5.8 s *4*	ε, εp
148	1+	-58.508s	2.2 s *11*	ε
148m	6-	-58.508s	9.59 s *15*	ε, εp 0.08%
149	(1/2+)	-61.673	>30 s	ε
149m	(11/2-)	-61.673	21.4 s *18*	ε
150	(9+)	-62.081s	26 s *2*	ε
150	(2-)	-62.081s	72 s *4*	ε
151	(11/2-)	-63.635	35.2 s *1*	ε 78%, α 22%
151m	(1/2+)	-63.594	47.2 s *10*	α>40%
152	2-	-63.654	161.8 s *3*	ε 88%, α 12%
152m	9+	-63.494	49.5 s *3*	ε 89.2%, α 10.8%
153	11/2-	-65.023	2.0 m *1*	ε 99.95%, α 0.05%
153m	1/2+	-64.955	9.3 m *5*	ε 99.82%, α 0.18%
154	(2)-	-64.648	11.76 m *19*	ε 99.98%, α 0.02%
154m	8+	-64.328	3.10 m *14*	ε, α<1.0×10^{-3}%, IT≈0%
155	5/2+	-66.062	48 m *1*	α, ε
156	(4+)	-65.474s	56 m *1*	ε
157	7/2-	-66.892	12.6 m *2*	ε
158	5+	-66.180	11.3 m *4*	ε
158m	2-	-66.113	27 m *2*	IT>81%, ε<19%
158m	(9+)	-66.000	21.3 m *23*	ε
159	7/2-	-67.339	33.05 m *1*	ε
159m	1/2+	-67.133	8.30 s	IT
160	5+	-66.390	25.6 m *3*	ε
160m	2-	-66.330	5.02 h *5*	IT 65%, ε
160m	(9+)	-66.221	3 s	IT
161	7/2-	-67.206	2.48 h *5*	ε
161m	1/2+	-66.995	6.76 s *7*	IT
162	1+	-66.050	15.0 m *10*	ε
162m	6-	-65.944	67.0 m *7*	IT 62%, ε 38%
163	7/2-	-66.387	4570 y *25*	ε
163m	1/2+	-66.089	1.09 s *3*	IT
164	1+	-64.990	29 m *1*	ε 60%, β- 40%
164m	6-	-64.850	37.5 m *+15-5*	IT
165	7/2-	-64.907	100%	
166	0-	-63.080	26.83 h *2*	β-
166m	(7)-	-63.074	1.20×10^{3} y *18*	β-
167	7/2-	-62.292	3.1 h *1*	β-
168	3+	-60.084	2.99 m *7*	β-
169	7/2-	-58.806	4.7 m *1*	β-
170	(6+)	-56.248	2.76 m *5*	β-
170m	1(+)	-56.128	43 s *2*	β-
171	(7/2-)	-54.528	53 s *2*	β-
68 Er 146		-44.758s		
147	(11/2-)	-47.134s	2.5 s *2*	ε, εp>0%
147m	(1/2+)	-47.134s	≈2.5 s	ε, εp>0%
148	0+	-51.754s	4.6 s *2*	ε
149	(1/2+)	-53.939s	10.7 s *4*	ε, εp
149m	(11/2-)	-53.197s	10.8 s *6*	ε, εp, IT
150	0+	-57.973s	18.5 s *7*	ε
151	(7/2-)	-58.414s	23.5 s *13*	ε
151m	(27/2-)	-55.829s	0.58 s *2*	IT 95.3%, ε 4.7%
152	0+	-60.549	10.3 s *1*	α 90%, ε 10%
153	(7/2-)	-60.459	37.1 s *2*	α 53%, ε 47%
154	0+	-62.617	3.73 m *9*	ε 99.53%, α 0.47%

Isotope Z El A	Jπ	Δ (MeV)	T1/2 or Abundance	Decay Mode
68 Er 155	(7/2-)	-62.219	5.3 m *3*	ε 99.98%, α 0.02%
156	0+	-64.104s	19.5 m *10*	ε
157	3/2-	-63.422	18.65 m *10*	ε, α?
158	0+	-65.280s	2.24 h *7*	ε
159	3/2-	-64.571	36 m *1*	ε
160	0+	-66.062	28.58 h *9*	ε
161	3/2-	-65.203	3.21 h *3*	ε
162	0+	-66.345	0.14% *1*	
163	5/2-	-65.177	75.0 m *4*	ε
164	0+	-65.952	1.61% *1*	
165	5/2-	-64.531	10.36 h *4*	ε
166	0+	-64.934	33.6% *2*	
167	7/2+	-63.299	22.95% *13*	
167m	1/2-	-63.091	2.269 s *6*	IT
168	0+	-62.999	26.8% *2*	
169	1/2-	-60.931	9.40 d *2*	β-
170	0+	-60.118	14.9% *1*	
171	5/2-	-57.728	7.516 h *2*	β-
172	0+	-56.493	49.3 h *3*	β-
173	(7/2-)	-53.663s	1.4 m *1*	β-
174	0+	-52.117s	3.3 m *2*	β-
69 Tm 147	(11/2-)	-36.408s	0.56 s *4*	ε≈90%, p≈10%
148m	(10+)	-39.756s	0.7 s *2*	ε
149	(11/2-)	-44.367s	0.9 s *2*	ε
150	(6-)	-47.143s	2.3 s *4*	ε
151m	(11/2-)	-50.884s	4.13 s *11*	ε
151m	(1/2+)	-50.884s	5.2 s *20*	ε
152	(9+)	-51.884s	5.2 s *6*	ε
152m	(2-)	-51.884s	8.0 s *10*	ε
153	(11/2-)	-54.000	1.48 s *1*	α 91%, ε 9%
153m	(1/2+)	-53.957	2.5 s *2*	α 95%, ε 5%
154	(2-)	-54.563s	8.1 s *3*	ε 56%, α 44%
154m	(9+)	-54.563s	3.30 s *7*	α 90%, ε 10%, IT
155		-56.640	32 s *7*	ε>94%, α<6%
156	2-	-56.885	83.8 s *18*	ε 99.94%, α 0.06%
156m		-56.885	19 s *3*	α
157	1/2+	-58.942	3.5 m *2*	ε
158	2-	-58.750s	4.02 m *10*	ε
158m	(5+)	-58.750s	≈20 s	
159	5/2(+)	-60.721	9.15 m *17*	ε
160	1-	-60.172	9.4 m *3*	ε
160m	5	-60.102	74.5 s *15*	IT 85%, ε 15%
161	7/2+	-62.039	33 m *3*	ε
162	1-	-61.536	21.70 m *19*	ε
162m	5+	-61.469	24.3 s *17*	IT 82%, ε 18%
163	1/2+	-62.738	1.810 h *5*	ε
164	1+	-61.990	2.0 m *1*	ε
164	6-	-61.990	5.1 m *1*	IT≈80%, ε≈20%
165	1/2+	-62.938	30.06 h *3*	ε
166	2+	-61.894	7.70 h *3*	ε
167	1/2+	-62.551	9.25 d *2*	ε
168	3(+)	-61.320	93.1 d *2*	ε 99.99%, β- 0.01%
169	1/2+	-61.282	100%	
170	1-	-59.804	128.6 d *3*	β- 99.85%, ε 0.15%
171	1/2+	-59.219	1.92 y *1*	β-
172	2-	-57.383	63.6 h *2*	β-
173	(1/2+)	-56.262	8.24 h *8*	β-
174	(4)-	-53.873	5.4 m *1*	β-
175	(1/2+,3/2+)	-52.319	15.2 m *5*	β-

Isotope Z El A	Jπ	Δ (MeV)	T1/2 or Abundance	Decay Mode
69 Tm 176	(4+)	-49.617s	1.9 m *1*	β-
177	(1/2+)	-47.804s	85 s *+10-15*	β-
70 Yb 151	(1/2+)	-41.684s	≈1.6 s	ε, εp
151m	(11/2-)	-41.684s	≈1.6 s	ε, εp
152	0+	-46.419s	3.1 s *2*	ε
153		-47.311s	4.2 s *1*	α 50%, ε 50%
154	0+	-50.074s	0.404 s *14*	α 92.8%, ε 7.2%
155		-50.652s	1.72 s *12*	α 84%, ε 16%
156	0+	-53.312	26.1 s *7*	ε 90%, α 10%
157	(7/2-)	-53.412	38.6 s *10*	ε 99.5%, α 0.5%
158	0+	-56.021	1.57 m *9*	ε, α≈0.003%
159	(5/2)	-55.671	1.40 m *20*	ε
160	0+	-58.162s	4.8 m *2*	ε
161	3/2-	-57.889s	4.2 m *2*	ε
162	0+	-59.848s	18.87 m *19*	ε
163	3/2-	-59.368	11.05 m *25*	ε
164	0+	-60.994s	75.8 m *17*	ε
165	5/2-	-60.176	9.9 m *3*	ε
166	0+	-61.590	56.7 h *1*	ε
167	5/2-	-60.596	17.5 m *2*	ε
168	0+	-61.577	0.13% *1*	
169	7/2+	-60.373	32.026 d *5*	ε
169m	1/2-	-60.348	46 s *2*	IT
170	0+	-60.772	3.05% *5*	
171	1/2-	-59.315	14.3% *2*	
172	0+	-59.264	21.9% *3*	
173	5/2-	-57.560	16.12% *18*	
174	0+	-56.953	31.8% *4*	
175	7/2-	-54.704	4.185 d *1*	β-
176	0+	-53.497	12.7% *1*	
176m	(8-)	-52.447	11.4 s *3*	IT≥90%, β-<10%
177	(9/2+)	-50.992	1.911 h *3*	β-
177m	(1/2-)	-50.661	6.41 s *3*	IT
178	0+	-49.701	74 m *3*	β-
179		-46.714s	8.1 m *8*	β-
180			2.4 m *5*	β-
71 Lu 150		-25.124s		
151		-30.683s	85 ms *10*	p
152	(6-,5-)	-34.074s	0.7 s *1*	ε
153		-38.480s		
154m	(7+)	-39.961s	1.12 s *8*	ε≈100%
155		-42.688s	70 ms *6*	α 79%, ε 21%
155m		-40.890s	2.60 ms *7*	α
156		-43.866s	≈0.5 s	α≈75%, ε≈25%
156m		-43.866s	0.18 s *2*	α≈95%, ε≈5%
157		-46.479	5.4 s *2*	ε 94%, α 6%
158		-47.348s	10.4 s *1*	ε>98.5%, α<1.5%
159		-49.683	12.3 s *1*	ε, α 0.04%
160		-50.282s	36.1 s *3*	ε, α≤1.0×10⁻⁴%
160m		-50.282s	40 s *1*	ε≤100%, α
161	(5/2+)	-52.589s	72 s	ε
161m	(9/2-)	-52.453s	7.3 ms *4*	IT
162	(1-)	-52.628s	1.37 m *2*	ε
162m	(4-)	-52.628s	1.5 m	ε≤100%
162m		-52.628s	1.9 m	ε≤100%
163	(1/2-)	-54.768	238 s *8*	ε
164		-54.744s	3.14 m *3*	ε
165	(7/2+)	-56.257	10.74 m *10*	ε
165	1/2+	-56.257	12 m	
166	(6-)	-56.111	2.65 m *10*	ε
166m	(3-)	-56.076	1.41 m *10*	ε 58%, IT 42%

Isotope Z El A	Jπ	Δ (MeV)	T1/2 or Abundance	Decay Mode
71 Lu 166m	(0-)	-56.068	2.12 m *10*	ε>80%, IT<20%
167	7/2+	-57.466	51.5 m *10*	ε
168	(6-)	-57.102	5.5 m *1*	ε
168m	3+	-56.882	6.7 m *4*	ε>95%, IT<5%
169	7/2+	-58.079	34.06 h *5*	ε
169m	1/2-	-58.050	160 s *10*	IT
170	0+	-57.313	2.00 d *3*	ε
170m	4-	-57.220	0.67 s *10*	IT
171	7/2+	-57.836	8.24 d *3*	ε
171m	1/2-	-57.765	79 s *2*	IT
172	4-	-56.744	6.70 d *3*	ε
172m	1-	-56.702	3.7 m *5*	IT
173	7/2+	-56.889	1.37 y *1*	ε
174	(1)-	-55.579	3.31 y *5*	ε
174m	(6)-	-55.408	142 d *2*	IT 99.38%, ε 0.62%
175	7/2+	-55.174	97.41% *2*	
176	7-	-53.391	3.78×10^{10} y *2* 2.59% *2*	β-
176m	1-	-53.268	3.635 h *3*	β- 99.91%, ε 0.1%
177	7/2+	-52.392	6.734 d *12*	β-
177m	23/2-	-51.422	160.4 d *3*	β- 78.3%, IT 21.7%
178	1(+)	-50.346	28.4 m *2*	β-
178m	(9-)	-50.126	23.1 m *3*	β-
179	(7/2+)	-49.067	4.59 h *6*	β-
180	3+,4+,5+	-46.687	5.7 m *1*	β-
181	(7/2+)	-44.926s	3.5 m *3*	β-
182	(0,1,2)		2.0 m *2*	β-
183	(7/2+)		58 s *4*	β-
184			≈20 s	β-
72 Hf 154	0+	-33.301s	2 s *1*	ε≈100%, α≈0%
155		-34.689s	0.89 s *12*	α, ε
156	0+	-37.961s	25 ms *4*	α≥81%
157		-39.004s	110 ms *6*	α 91%, ε 9%
158	0+	-42.245s	2.9 s *2*	ε 54%, α 46%
159		-43.004s	5.6 s *5*	ε 88%, α 12%
160	0+	-45.985	13.0 s *15*	ε 97.7%, α 2.3%
161		-46.265	17 s *2*	α, ε
162	0+	-49.179	37.6 s *8*	ε 99.99%, α 8.7×10^{-3}%
163		-49.318s	40.0 s *6*	ε
164	0+	-51.770s	111 s *8*	ε
165	(5/2-)	-51.661s	76 s *4*	ε
166	0+	-53.794s	6.77 m *30*	ε
167	(5/2-)	-53.468s	2.05 m *5*	ε
168	0+	-55.303s	25.95 m *20*	ε
169	(5/2)-	-54.810	3.24 m *4*	ε
170	0+	-56.216s	16.01 h *13*	ε
171	(7/2+)	-55.433s	12.1 h *4*	ε
172	0+	-56.394	1.87 y *3*	ε
173	1/2-	-55.284s	23.6 h *1*	ε
174	0+	-55.851	2.0×10^{15} y *4* 0.162% *2*	α
175	5/2-	-54.488	70 d *2*	ε
176	0+	-54.582	5.206% *4*	
177	7/2-	-52.890	18.606% *3*	
177m	23/2+	-51.575	1.08 s *6*	IT
177m	37/2-	-50.150	51.4 m *5*	IT
178	0+	-52.445	27.297% *3*	
178m	8-	-51.298	4.0 s *2*	IT

Isotope Z El A	Jπ	Δ (MeV)	T1/2 or Abundance	Decay Mode
72 Hf 178m	16+	-49.999	31 y *1*	IT
179	9/2+	-50.473	13.629% *5*	
179m	1/2-	-50.098	18.67 s *3*	IT
179m	25/2-	-49.367	25.1 d *3*	IT
180	0+	-49.790	35.100% *6*	
180m	8-	-48.647	5.5 h *1*	IT≥98.6%, β-<1.4%
181	1/2-	-47.414	42.39 d *6*	β-
182	0+	-46.060	9×10^6 y *2*	β-
182m	8-	-44.887	61.5 m *15*	β- 58%, IT 42%
183	(3/2-)	-43.286	1.067 h *17*	β-
184	0+	-41.500	4.12 h *5*	β-
73 Ta 156		-26.371s		
157		-29.673s	5.3 ms *18*	α>77%
158		-31.328s	36.8 ms *16*	α 93%, ε 7%
159		-34.517s	0.57 s *18*	α 80%, ε 20%
160		-35.896s	1.5 s *2*	ε 66%, α 34%
161		-38.774	2.7 s *2*	ε≈95%, α≈5%
162		-39.916s	3.52 s *12*	ε 99.93%, α 0.07%
163		-42.509	11.0 s *8*	ε≈99.8%, α≈0.2%
164	(3+)	-43.249s	14.2 s *3*	ε
165		-45.813s	31.0 s *15*	ε
166	(2)+	-46.137s	34.4 s *5*	ε
167		-48.464s	1.4 m *3*	ε
168	(3+)	-48.633s	2.44 m *35*	ε
169	(5/2-)	-50.375s	4.9 m *4*	ε
170	(3+)	-50.217s	6.76 m *6*	ε
171	(5/2-)	-51.735s	23.3 m *3*	ε
172	(3-)	-51.474	36.8 m *3*	ε
173	5/2(-)	-52.494s	3.14 h *13*	ε
174	3(+)	-52.006	1.05 h *3*	ε
175	7/2+	-52.490s	10.5 h *2*	ε
176	(1)-	-51.472	8.09 h *5*	ε
177	7/2+	-51.724	56.56 h *6*	ε
178	1+	-50.533	9.31 m *3*	ε
178	(7)-	-50.533	2.36 h *8*	ε
179	7/2+	-50.362	1.79 y *8*	ε
180	1+	-48.936	8.152 h *6*	ε 86%, β- 14%
180m	9-	-48.861	$>1.2\times10^{15}$ y 0.012% *2*	
181	7/2+	-48.441	99.988% *2*	
182	3-	-46.433	114.43 d *3*	β-
182m	5+	-46.417	283 ms *3*	IT
182m	10-	-45.913	15.84 m *10*	IT
183	7/2+	-45.296	5.1 d *1*	β-
184	(5-)	-42.840	8.7 h *1*	β-
185	(7/2+)	-41.397	49 m *2*	β-
186	2,3	-38.611	10.5 m *5*	β-
74 W 158	0+	-24.276s	≈1.4 ms	α
159		-25.821s	7.3 ms *27*	α
160	0+	-29.464s	81 ms *15*	α≥54%
161		-30.656s	410 ms *40*	α≈82%, ε≈18%
162	0+	-34.146s	1.39 s *4*	ε 53%, α 47%
163		-35.059s	2.75 s *25*	ε 59%, α 41%
164	0+	-38.281	6.4 s *8*	ε 97.4%, α 2.6%
165		-38.809	5.1 s *5*	ε, α<0.2%
166	0+	-41.898	18.8 s *4*	ε 99.97%, α 0.04%
167		-42.224s	19.9 s *5*	α, ε
168	0+	-44.839s	53 s *2*	ε
169	(5/2-)	-44.936s	76 s *6*	ε
170	0+	-47.238s	4 m *1*	ε

Isotope Z El A	Jπ	Δ (MeV)	T1/2 or Abundance	Decay Mode
74 W 171	(5/2-)	-47.162s	2.38 m *4*	ε
172	0+	-48.974s	6.7 m *10*	ε
173m	(5/2-)	-48.494s	7.97 m *27*	ε
174	0+	-50.152s	31 m *1*	ε
175	(1/2-)	-49.583s	35.2 m *6*	ε
176	0+	-50.683s	2.5 h *1*	ε
177	(7/2+)	-49.723s		
177	(1/2-)	-49.723s	135 m *3*	ε
178	0+	-50.442	21.6 d *3*	ε
179	(7/2-)	-49.303	37.5 m *5*	ε
179m	(1/2-)	-49.081	6.4 m *1*	IT 99.72%, ε 0.28%
180	0+	-49.644	0.12% *3*	
181	9/2+	-48.253	121.2 d *2*	ε
182	0+	-48.246	26.3% *2*	
183	1/2-	-46.366	$>1.1\times10^{17}$ y 14.28% *5*	
183m	11/2+	-46.057	5.2 s *3*	IT
184	0+	-45.706	$>3\times10^{17}$ y 30.7% *2*	
185	3/2-	-43.389	75.1 d *3*	β-
185m	11/2+	-43.192	1.67 m *3*	IT
186	0+	-42.512	28.6% *2*	
187	3/2-	-39.907	23.72 h *6*	β-
188	0+	-38.669	69.4 d *5*	β-
189	(3/2-)	-35.479	11.5 m *3*	β-
190	0+	-34.288	30.0 m *15*	β-
75 Re 160		-17.247s	0.79 ms *16*	p 91%, α 9%
161		-20.809s	10 ms *+15-5*	α
162		-22.629s	0.10 s *3*	ε<97%, α>3%
163		-26.025s	260 ms *40*	α 64%, ε 36%
164		-27.548s	0.88 s *24*	α≈58%, ε≈42%
165		-30.692	2.4 s *6*	ε 87%, α 13%
166		-31.854s	2.8 s *3*	α
167		-34.841s	6.1 s *2*	ε≈99.3%, α≈0.7%
168		-35.761s	6.9 s *8*	α, ε
168m		-35.761s	6.6 s *15*	α, ε
169		-38.350s	?	α, ε
169m		-38.350s	12.9 s *11*	α≈0.2%
170	(5)	-38.971s	8.0 s *5*	ε
171	(9/2-)	-41.492s	15.2 s *4*	ε
172	(5)	-41.648s	15 s *3*	α, ε
172m	(2)	-41.648s	55 s *5*	α, ε, IT
173		-43.722s	1.98 m *26*	ε
174		-43.675s	2.40 m *4*	ε
175	(5/2-)	-45.277s	5.89 m *5*	ε
176	3(+)	-45.112s	5.3 m *3*	ε
177	(5/2-)	-46.323s	14 m *1*	ε
178	(3)	-45.782	13.2 m *2*	ε
179	(5/2+)	-46.592	19.5 m *1*	ε
180	(1)-	-45.841	2.44 m *6*	ε
181	5/2+	-46.515	19.9 h *7*	ε
182	7+	-45.446	64.0 h *5*	ε
182m	2+	-45.446	12.7 h *2*	ε
183	5/2+	-45.810	70.0 d *14*	ε
183m	(25/2)+	-43.904	1.04 ms *5*	IT
184	3(-)	-44.223	38.0 d *5*	ε
184m	8(+)	-44.035	169 d *8*	IT 75.4%, ε 24.6%
185	5/2+	-43.822	37.40% *2*	
186	1-	-41.930	90.64 h *9*	β- 93.1%, ε 6.9%
186m	(8+)	-41.781	2.0×10^{5} y *5*	IT, β-<10%

Isotope Z El A	Jπ	Δ (MeV)	T1/2 or Abundance	Decay Mode
75 Re 187	5/2+	-41.218	4.35×10^{10} y *13* 62.60% *2*	β-
187	5/2+	-41.218	4.35×10^{10} y *13*	$\alpha<1.0\times10^{-4}$%
188	1-	-39.018	16.98 h *2*	β-
188m	(6)-	-38.846	18.6 m *1*	IT
189	5/2+	-37.979	24.3 h	β-
190	(2)-	-35.558	3.1 m *3*	β-
190m	(6-)	-35.439	3.2 h *2*	β- 54.4%, IT 45.6%
191	(3/2+,1/2+)	-34.350	9.8 m *5*	β-
192		-31.708s	16 s *1*	β-
76 Os 162	0+	-15.072s	1.9 ms *7*	α
163		-16.722s	?	α, ε
164	0+	-20.561s	41 ms *20*	α≈98%, ε≈2%
165		-21.914s	65 ms *+70-30*	α>60%, ε<40%
166	0+	-25.591s	181 ms *38*	α 72%, ε 18%
167		-26.655s	0.83 s *12*	α 67%, ε 33%
168	0+	-30.038	2.2 s *1*	ε 51%, α 49%
169		-30.666	3.4 s *2*	ε 89%, α 11%
170	0+	-33.932	7.1 s *2*	ε 88%, α 12%
171	(5/2-)	-34.429s	8.0 s *7*	ε 98.3%, α 1.7%
172	0+	-37.187s	19 s *2*	ε 99.8%, α 0.2%
173		-37.454s	16 s *5*	ε 99.98%, α 0.02%
174	0+	-39.941s	44 s *4*	ε 99.98%, α 0.02%
175	(5/2-)	-40.024s	1.4 m *1*	ε
176	0+	-41.948s	3.6 m *5*	ε
177	(1/2-)	-41.850s	2.8 m *3*	ε
178	0+	-43.447	5.0 m *4*	ε
179	(1/2-)	-42.914s	6.5 m *3*	ε
180	0+	-44.375s	21.5 m *4*	ε
181	1/2-	-43.585s	105 m *3*	ε
181m	(7/2)-	-43.536s	2.7 m *1*	ε
182	0+	-44.538	22.10 h *25*	ε
183	9/2+	-43.678s	13.0 h *5*	ε
183m	1/2-	-43.507s	9.9 h *3*	ε 85%, IT 15%
184	0+	-44.255	$>5.6\times10^{13}$ y 0.02% *1*	
185	1/2-	-42.809	93.6 d *5*	ε
186	0+	-43.000	2.0×10^{15} y *11* 1.58% *10*	α
187	1/2-	-41.221	1.6% *1*	
188	0+	-41.139	13.3% *2*	
189	3/2-	-38.988	16.1% *3*	
189m	9/2-	-38.957	5.8 h *1*	IT
190	0+	-38.708	26.4% *4*	
190m	(10)-	-37.003	9.9 m *1*	IT
191	9/2-	-36.396	15.4 d *1*	β-
191m	3/2-	-36.322	13.10 h *5*	IT
192	0+	-35.882	41.0% *3*	
192m	(10-)	-33.867	5.9 s *1*	IT>87%, β-<13%
193	3/2-	-33.396	30.5 h *4*	β-
194	0+	-32.436	6.0 y *2*	β-
195		-29.693	6.5 m	β-
196	0+	-28.297	34.9 m *2*	β-
77 Ir 166		-13.501s	>5 ms	α 99%
167		-17.057s	>5 ms	α
168		-18.712s	?	α
169		-21.991	0.4 s *1*	α≈100%, ε, p
170		-23.256s	1.05 s *15*	α 75%, ε 25%
171		-26.257s	1.5 s *1*	α≈100%, ε, p
172		-27.346s	2.1 s *1*	ε≈97%, α≈3%

Isotope Z El A	Jπ	Δ (MeV)	T1/2 or Abundance	Decay Mode
77 Ir 173		-30.080s	3.0 s *10*	ε 97.98%, α 2.02%
174		-30.922s	4 s *1*	ε 99.53%, α 0.47%
175	(5/2-)	-33.447s	9 s *2*	ε 99.15%, α 0.85%
176		-33.986s	8 s *1*	ε 97.9%, α 2.1%
177	(5/2-)	-36.170s	30 s *2*	ε 99.94%, α 0.06%
178		-36.250s	12 s *2*	ε
179		-38.052s	4 m *1*	ε
180		-37.958s	1.5 m *1*	ε
181	(5/2)-	-39.516s	4.90 m *15*	ε
182m	(5-)	-38.929	15 m *1*	ε
183	5/2-	-40.228s	57 m *4*	ε
184	5-	-39.693	3.09 h *3*	ε
185	5/2-	-40.436s	14.4 h *1*	ε
186	5+	-39.168	16.64 h *3*	ε
186m	2-	-39.168	2.0 h *1*	ε, IT
187	3/2+	-39.718	10.5 h *3*	ε
188	1-	-38.329	41.5 h *5*	ε
189	3/2+	-38.456	13.2 d *1*	ε
189m	11/2-	-38.084	13.3 ms *3*	IT
189m	(25/2+)	-36.123	3.7 ms *2*	IT
190	(4)+	-36.708	11.78 d *10*	ε
190m	(7)+	-36.682	1.2 h	IT
190m	(11)-	-36.533	3.25 h *20*	ε 94.4%, IT 5.6%
191	3/2+	-36.709	37.3% *5*	
191m	11/2-	-36.539	4.94 s *3*	IT
191m		-34.662	5.5 s *7*	IT
192	4(+)	-34.836	73.831 d *8*	β- 95.24%, ε 4.76%
192m	1(-)	-34.779	1.45 m *5*	IT 99.98%, β- 0.02%
192m	(9)	-34.681	241 y *9*	IT
193	3/2+	-34.537	62.7% *5*	
193m	11/2-	-34.457	10.53 d *4*	IT
194	1-	-32.532	19.15 h *3*	β-
194m	(10,11)	-32.342	171 d *11*	β-
195	3/2+	-31.693	2.5 h *2*	β-
195m	11/2-	-31.593	3.8 h *2*	β- 95%, IT 5%
196	(0-)	-29.454	52 s *2*	β-
196m	(10,11-)	-29.044	1.40 h *2*	β-
197	3/2+	-28.284	5.8 m *5*	β-
197m	11/2-	-28.169	8.9 m *3*	β- 99.75%, IT 0.25%
198		-25.821s	8 s *1*	β-
78 Pt 168	0+	-11.146s	?	α
169		-12.649s	2.5 ms *+25-10*	α≤100%
170	0+	-16.462s	6 ms *+5-2*	α
171		-17.623s	25 ms *9*	α≈99%, ε≈1%
172	0+	-21.148	0.10 s *1*	α 98%, ε 2%
173		-21.888	342 ms *18*	α 84%, ε 16%
174	0+	-25.324	0.90 s *1*	α 83%, ε 17%
175		-25.825s	2.52 s *8*	α 64%, ε 36%
176	0+	-28.876s	6.33 s *15*	ε 62%, α 38%
177	(5/2-)	-29.386s	11 s *1*	ε 94.4%, α 5.6%
178	0+	-31.942s	21.0 s *6*	ε 92.3%, α 7.7%
179	1/2-	-32.318s	43 s *10*	ε 99.76%, α 0.24%
180	0+	-34.266s	52 s *3*	ε, α≈0.3%
181	1/2-	-34.292s	51 s *5*	ε, α≈0.06%
182	0+	-36.079	2.2 m *1*	ε≈99.98%, α≈0.02%
183	1/2-	-35.650s	6.5 m *10*	ε, α≈1.3×10^{-3}%
183m	(7/2)-	-35.615s	43 s *5*	ε, α<4.0×10^{-4}%, IT

Isotope Z El	A	Jπ	Δ (MeV)	T1/2 or Abundance	Decay Mode
78 Pt	184	0+	-37.360s	17.3 m *2*	ε, α≈0.001%
	184m	8-	-35.521s	1.01 ms *5*	IT
	185	(9/2+)	-36.618s	70.9 m *24*	ε
	185m	(1/2-)	-36.515s	33.0 m *8*	ε 99%, IT<2%
	186	0+	-37.790	2.0 h *1*	ε, α≈1.4×10^{-4}%
	187	3/2-	-36.610s	2.35 h *3*	ε
	188	0+	-37.823	10.2 d *3*	ε, α 2.6×10^{-5}%
	189	3/2-	-36.485	10.87 h *12*	ε
	190	0+	-37.325	6.5×10^{11} y *3* 0.01% *1*	α
	191	3/2-	-35.691	2.9 d *1*	ε
	192	0+	-36.296	0.79% *5*	
	193	1/2-	-34.480	50 y *9*	ε
	193m	13/2+	-34.330	4.33 d *3*	IT
	194	0+	-34.779	32.9% *5*	
	195	1/2-	-32.813	33.8% *5*	
	195m	13/2+	-32.554	4.02 d *1*	IT
	196	0+	-32.664	25.3% *5*	
	197	1/2-	-30.439	18.3 h *3*	β-
	197m	13/2+	-30.039	95.41 m *18*	IT 96.7%, β- 3.3%
	198	0+	-29.924	7.2% *2*	
	199	5/2-	-27.409	30.8 m *4*	β-
	199m	(13/2)+	-26.985	13.6 s *4*	IT
	200	0+	-26.619	12.5 h *3*	β-
	201	(5/2-)	-23.744	2.5 m *1*	β-
79 Au	173		-12.669	59 ms *+45-18*	α
	174		-14.049s	120 ms *20*	α>0%
	175		-17.054s	200 ms *22*	α 94%, ε 6%
	176		-18.379s	1.25 s *30*	α, ε
	177		-21.229s	1.18 s *7*	α≤40%
	178		-22.379s	2.6 s *5*	ε≤60%, α≥40%
	179		-24.941s	7.5 s *4*	ε 78%, α 22%
	180		-25.713s	8.1 s *3*	ε≤98.2%, α≥1.8%
	181	5/2-	-27.993s	11.4 s *5*	ε 98.5%, α 1.5%
	182		-28.299s	21 s *1*	ε, α 0.038%
	183	(5/2)-	-30.161s	42.0 s *12*	ε 99.64%, α 0.36%
	184	3+	-30.233s	53.0 s *14*	ε, α 0.02%
	185	5/2-	-31.911s	4.3 m *1*	ε 99.9%, α 0.1%
	185m		-31.911s	6.8 m *3*	ε, IT
	186	3-	-31.749	10.7 m *5*	ε
	187	1/2+	-33.010s	8.4 m *3*	ε, α 3.0×10^{-3}%
	187m	9/2-	-32.889s	2.3 s *1*	IT
	188	1(-)	-32.523s	8.84 m *6*	ε
	189	1/2+	-33.635s	28.7 m *3*	ε, α<3.0×10^{-5}%
	189m	11/2-	-33.388s	4.59 m *11*	ε, IT>0%
	190	1-	-32.883	42.8 m *10*	ε, α<1.0×10^{-6}%
	191	3/2+	-33.861	3.18 h *8*	ε
	191m	(11/2-)	-33.595	0.92 s *11*	IT
	192	1-	-32.780	4.94 h *9*	ε
	193	3/2+	-33.412	17.65 h *15*	ε
	193m	11/2-	-33.122	3.9 s *3*	IT 99.97%, ε≈0.03%
	194	1-	-32.287	38.02 h *10*	ε
	195	3/2+	-32.586	186.09 d *4*	ε
	195m	11/2-	-32.267	30.5 s *2*	IT
	196	2-	-31.158	6.183 d *10*	ε 92.5%, β- 7.5%
	196m	5+	-31.073	8.1 s *2*	IT
	196m	12-	-30.562	9.7 h *1*	IT
	197	3/2+	-31.157	100%	
	197m	11/2-	-30.749	7.73 s *6*	IT
	198	2-	-29.598	2.6935 d *4*	β-
	198m	(12-)	-28.786	2.30 d *4*	IT

Isotope Z El	A	Jπ	Δ (MeV)	T1/2 or Abundance	Decay Mode
79 Au	199	3/2+	-29.111	3.139 d *7*	β-
	200	1(-)	-27.276	48.4 m *3*	β-
	200m	12-	-26.286	18.7 h *5*	β- 82%, IT 18%
	201	3/2+	-26.404	26 m *1*	β-
	202	(1-)	-24.416	28.8 s *19*	β-
	203	3/2+	-23.145	53 s *2*	β-
	204	(2-)	-20.907s	39.8 s *9*	β-
80 Hg	175		-8.159s	20 ms *+40-13*	α
	176	0+	-11.799	34 ms *+18-9*	α≈100%
	177		-12.725	0.130 s *5*	α 85%, ε 15%
	178	0+	-16.321	0.26 s *3*	α≈50%, ε≈50%
	179		-16.969s	1.09 s *4*	α≈53%, ε≈47%, εp≈0.15%
	180	0+	-20.193s	3.0 s *3*	ε 51%, α 49%
	181	1/2(-)	-20.674s	3.6 s *3*	ε 64%, α 36%
	182	0+	-23.519s	11.3 s *5*	ε 84.8%, α 15.2%
	183	1/2-	-23.854s	8.8 s *5*	ε 74.5%, α 25.5%, εp 0.06%
	184	0+	-26.179s	30.6 s *3*	ε 98.89%, α 1.11%
	185	1/2-	-26.089s	49 s *1*	ε 94%, α 6%
	185m	13/2+	-25.990s	21 s *1*	IT 54%, ε 46%, α≈0.03%
	186	0+	-28.449	1.38 m *7*	ε 99.98%, α 0.02%
	187	13/2+	-28.145s	2.4 m *3*	ε, α>1.2×10^{-4}%
	187m	3/2-	-28.145s	1.9 m *3*	ε, α>2.5×10^{-4}%
	188	0+	-30.225s	3.25 m *15*	ε, α 3.7×10^{-5}%
	189	3/2-	-29.685s	7.6 m *1*	ε, α<3.0×10^{-5}%
	189m	13/2+	-29.685s	8.6 m *1*	ε, α<3.0×10^{-5}%
	190	0+	-31.409s	20.0 m *5*	ε, α<5.0×10^{-5}%
	191	(3/2-)	-30.681	49 m *10*	ε
	191m	13/2+	-30.681	50.8 m *15*	ε
	192	0+	-32.068s	4.85 h *20*	ε
	193	3/2-	-31.071	3.80 h *15*	ε
	193m	13/2+	-30.930	11.8 h *2*	ε 92.9%, IT 7.1%
	194	0+	-32.247	520 y *32*	ε
	195	1/2-	-31.076	9.9 h *5*	ε
	195m	13/2+	-30.900	41.6 h *8*	IT 54.2%, ε 45.8%
	196	0+	-31.844	0.15% *1*	
	197	1/2-	-30.558	64.14 h *5*	ε
	197m	13/2+	-30.259	23.8 h *1*	IT 93%, ε 7%
	198	0+	-30.971	9.97% *8*	
	199	1/2-	-29.563	16.87% *10*	
	199m	13/2+	-29.031	42.6 m *2*	IT
	200	0+	-29.520	23.10% *16*	
	201	3/2-	-27.679	13.10% *8*	
	202	0+	-27.362	29.86% *20*	
	203	5/2-	-25.284	46.612 d *18*	β-
	204	0+	-24.708	6.87% *4*	
	205	1/2-	-22.304	5.2 m *1*	β-
	206	0+	-20.959	8.15 m *10*	β-
	207	(9/2+)	-16.270	2.9 m *2*	β-
81 Tl	179		-7.769s	0.16 s *+9-4*	α
	179m		-7.769s	1.4 ms *5*	α
	180		-9.135s		
	181		-12.204s		
	182		-13.404s		α
	183	(1/2+)	-16.208s	6.9 s *7*	ε>0%
	183m	(9/2-)	-15.658s	60 ms *15*	α
	184		-16.990s	11 s *1*	ε 97.9%, α 2.1%
	185	(1/2+)	-19.468s		
	185m	(9/2-)	-19.014s	1.8 s *1*	α, IT

Isotope Z El	A	Jπ	Δ (MeV)	T1/2 or Abundance	Decay Mode
81 Tl	186m	(7+)	-19.983s	27.5 s *10*	ε, α 6.0×10^{-4}%
	186m	(10-)	-19.609s	2.9 s *2*	IT
	187	(1/2+)	-22.197s	≈51 s	ε<100%, α>0%
	187m	(9/2-)	-21.862s	15.60 s *12*	ε<100%, IT<100%, α>0%
	188m	(2-)	-22.430s	71 s *2*	ε
	188m	(7+)	-22.430s	71 s *1*	ε
	189	(1/2+)	-24.509s	2.3 m *2*	ε
	189m	(9/2-)	-24.228s	1.4 m *1*	ε, IT<4%
	190m	(2)-	-24.409s	2.6 m *3*	ε
	190m	(7+)	-24.409s	3.7 m *3*	ε
	191	(1/2+)	-26.188s		
	191m	9/2(-)	-25.889s	5.22 m *16*	ε
	192m	(2-)	-25.948s	9.6 m *4*	ε
	192m	(7+)	-25.948s	10.8 m *2*	ε
	193	1/2(+)	-27.432s	21.6 m *8*	ε
	193m	(9/2-)	-27.067s	2.11 m *15*	IT 75%, ε 25%
	194	2-	-26.965s	33.0 m *5*	ε, α<1.0×10^{-7}%
	194m	(7+)	-26.965s	32.8 m *2*	ε
	195	1/2+	-28.274s	1.16 h *5*	ε
	195m	9/2-	-27.791s	3.6 s *4*	IT
	196	2-	-27.467s	1.84 h *3*	ε
	196m	(7+)	-27.072s	1.41 h *2*	ε 95.5%, IT 4.5%
	197	1/2+	-28.375	2.84 h *4*	ε
	197m	9/2-	-27.767	0.54 s *1*	IT
	198	2-	-27.511	5.3 h *5*	ε
	198m	7+	-26.967	1.87 h *3*	ε 56%, IT 44%
	198m	(10-)	-26.769	32.1 ms *10*	IT
	199	1/2+	-28.119	7.42 h *8*	ε
	199m	9/2-	-27.369	28.4 ms *2*	IT
	200	2-	-27.064	26.1 h *1*	ε
	200m	7+	-26.310	34.3 ms *10*	IT
	201	1/2+	-27.197	72.912 h *17*	ε
	201m	(9/2-)	-26.277	2.035 ms *7*	IT
	202	2-	-25.998	12.23 d *2*	ε
	203	1/2+	-25.776	29.524% *9*	
	204	2-	-24.360	3.78 y *2*	β- 97.43%, ε 2.57%
	205	1/2+	-23.835	70.476% *9*	
	206	0-	-22.268	4.199 m *15*	β-
	206m	(12-)	-19.625	3.74 m *3*	IT
	207	1/2+	-21.045	4.77 m *2*	β-
	207m	11/2-	-19.697	1.33 s *11*	IT
	208	5(+)	-16.763	3.053 m *4*	β-
	209	(1/2+)	-13.648	2.20 m *7*	β-
	210	(5+)	-9.258	1.30 m *3*	β-, β-n 7.0×10^{-3}%
82 Pb	182	0+	-6.820	55 ms *+40-35*	α
	183	(1/2-)	-7.517s	300 ms *80*	α≈94%, ε≈6%
	184	0+	-10.993s	0.55 s *6*	α
	185		-11.569s	4.1 s *3*	α
	186	0+	-14.624s	4.79 s *5*	α
	187m		-15.034s	15.2 s *3*	α, ε
	187m	(13/2+)	-15.034s	18.3 s *3*	ε 98%, α 2%
	188	0+	-17.642s	24.2 s *10*	ε 78%, α 22%
	189m		-17.810s	51 s *3*	ε>99%, α≈0.4%
	190	0+	-20.326	1.2 m *1*	ε 99.1%, α 0.9%
	191		-20.307s	1.33 m *8*	ε 99.99%, α 0.01%
	191m	(13/2+)	-20.307s	2.18 m *8*	ε
	192	0+	-22.579s	3.5 m *1*	ε 99.99%, α 5.7×10^{-3}%
	193	(3/2-)	-22.281s	?	ε

Isotope Z El A	Jπ	Δ (MeV)	T1/2 or Abundance	Decay Mode
82 Pb 193m	(13/2+)	-22.181s	5.8 m *2*	ε
194	0+	-24.247s	12.0 m *5*	ε, α 7.3×10^{-6}%
195	3/2-	-23.780s	≈15 m	ε
195m	13/2+	-23.579s	15.0 m *12*	ε
196	0+	-25.420s	37 m *3*	ε, α<0.0001%
197	3/2-	-24.796s	8 m *2*	ε
197m	13/2+	-24.477s	43 m *1*	ε 81%, IT 19%, α<3.0×10^{-4}%
198	0+	-26.100s	2.40 h *10*	ε
199	3/2-	-25.236	90 m *10*	ε
199m	13/2+	-24.806	12.2 m *3*	IT 93%, ε 7%
200	0+	-26.254	21.5 h *4*	ε
201	5/2-	-25.294	9.33 h *3*	ε
201m	13/2+	-24.665	61 s *2*	IT>99%, ε<1%
202	0+	-25.948	52.5×10^{3} y *28*	ε, α<1%
202m	9-	-23.778	3.53 h *1*	IT 90.5%, ε 9.5%
203	5/2-	-24.801	51.873 h *9*	ε
203m	13/2+	-23.976	6.3 s *2*	IT
203m	29/2-	-21.852	0.48 s *2*	IT
204	0+	-25.124	≥1.4×10^{17} y 1.4% *1*	
204m	9-	-22.938	67.2 m *3*	IT
205	5/2-	-23.784	1.53×10^{7} y *7*	ε
205m	13/2+	-22.770	5.54 ms *10*	IT
206	0+	-23.801	24.1% *1*	
207	1/2-	-22.467	22.1% *1*	
207m	13/2+	-20.834	0.806 s *6*	IT
208	0+	-21.764	52.4% *1*	
209	9/2+	-17.628	3.253 h *14*	β-
210	0+	-14.742	22.3 y *2*	β-, α 1.9×10^{-6}%
211	9/2+	-10.496	36.1 m *2*	β-
212	0+	-7.557	10.64 h *1*	β-
213	(9/2+)	-3.171s	10.2 m *3*	β-
214	0+	-0.189	26.8 m *9*	β-
83 Bi 186		-3.279s		
187	(9/2-)	-6.094s	35 ms *4*	α>50%
187m	(1/2+)	-6.034s	8 ms *6*	α
188m		-7.291s	44 ms *3*	α, ε
188m		-7.291s	0.21 s *9*	α, ε
189	(9/2-)	-9.774s	680 ms *30*	α>50%, ε<50%
189m	(1/2+)	-9.678s	≈5 ms	α>50%, ε<50%
190m		-10.695s	6.2 s *1*	α 68%, ε 32%
190m		-10.695s	6.3 s *1*	α 82%, ε 18%
191	(9/2-)	-12.991s	12 s *1*	α 60%, ε 40%
191m	(1/2+)	-12.749s	150 ms *15*	α>50%, ε<50%
192	(2+,3+)	-13.629s	37 s *3*	ε 82%, α 18%
192m	(10-)	-13.524s	39.6 s *4*	ε 90.8%, α 9.2%
193	(9/2-)	-15.779s	67 s *3*	ε 95%, α 5%
193m	(1/2+)	-15.472s	3.2 s *7*	α 90%, ε≈10%
194	(2+,3+)	-16.066s	106 s *3*	ε
194m	(6+,7+)	-16.066s	92 s *5*	ε 99.93%, α 0.07%
194m	(10-)	-16.066s	125 s *2*	ε 99.79%, α 0.21%
195	(9/2-)	-17.930s	183 s *4*	ε 99.97%, α 0.03%
195m	(1/2+)	-17.529s	87 s *1*	ε 67%, α 33%
196m	(10-)	-18.063	4.6 m *5*	ε, IT
196m		-18.063	5 m	ε
197	(9/2-)	-19.620		ε, α 1.0×10^{-4}%
197m	(1/2+)	-19.120	5.2 m *6*	α 55%, ε 45%, IT<0.3%
198	(7+)	-19.540	11.85 m *18*	ε
198m	(10-)	-19.291	7.7 s *5*	IT

Isotope Z El A	Jπ	Δ (MeV)	T1/2 or Abundance	Decay Mode
83 Bi 199	9/2-	-20.891	27 m *1*	ε
199m	(1/2+)	-20.211	24.70 m *15*	ε 99%, IT≤2%, α≈0.01%
200	7+	-20.362	36.4 m *5*	ε
200m	(2+)	-20.162	31 m *2*	ε>90%, IT<10%
200m	(10-)	-19.934	0.40 s *5*	IT
201	9/2-	-21.450	108 m *3*	ε, α<1×10⁻⁴%
201m	1/2+	-20.604	59.1 m *6*	ε>93%, IT≤6.8%, α≈0.3%
202	5+	-20.793	1.72 h *5*	ε, α<1×10⁻⁵%
203	9/2-	-21.548	11.76 h *5*	ε, α≈1.0×10⁻⁵%
203m	1/2+	-20.450	303 ms *5*	IT
204	6+	-20.686	11.22 h *10*	ε
204m	(17+)	-17.853	1.07 ms *3*	IT
205	9/2-	-21.076	15.31 d *4*	ε
206	6(+)	-20.043	6.243 d *3*	ε
207	9/2-	-20.069	31.55 y *5*	ε
208	(5)+	-18.884	3.68×10^5 y *4*	ε
208m	(10)-	-17.313	2.58 ms *4*	IT
209	9/2-	-18.272	100%	
210	1-	-14.806	5.013 d *5*	β-, α 1.3×10⁻⁴%
210m	9-	-14.535	3.04×10^6 y *6*	α
211	9/2-	-11.869	2.14 m *2*	α 99.72%, β- 0.28%
212	1(-)	-8.131	60.55 m *6*	β- 64.06%, α 35.94%, β-α 0.014%
212m	(9-)	-7.881	25.0 m *2*	α 67%, β- 33%
212m		-6.221	7.0 m *3*	β-≈100%
213	9/2-	-5.241	45.59 m *6*	β- 97.91%, α 2.09%
214	1-	-1.212	19.9 m *4*	β- 99.98%, α 0.02%
215		1.707	7.6 m *2*	β-
216		5.774s		
84 Po 192	0+	-7.898s	0.034 s *3*	α≈99%, ε≈1%
193m		-8.289s	260 ms *20*	α
193m		-8.289s	360 ms *50*	α
194	0+	-10.914	0.44 s *6*	α
195	(3/2-)	-11.136s	4.5 s *5*	α 75%, ε 25%
195m	(13/2+)	-10.906s	2.0 s *2*	α≈90%, ε≈10%, IT<0.01%
196	0+	-13.498s	5.5 s *5*	α, ε
197	(3/2-)	-13.445s	56 s *3*	ε 56%, α 44%
197m	(13/2+)	-13.241s	26 s *2*	α 84%, ε 16%, IT 0.01%
198	0+	-15.513s	1.76 m *3*	α 70%, ε 30%
199	3/2-	-15.281s	5.2 m *1*	ε 88%, α 12%
199m	13/2+	-14.971s	4.2 m *1*	ε 59%, α 39%, IT 2.1%
200	0+	-17.014s	11.5 m *1*	ε 85%, α 15%
201	3/2-	-16.572s	15.3 m *2*	ε 98.4%, α 1.6%
201m	13/2+	-16.148s	8.9 m *2*	ε 57%, IT 40%, α≈2.9%
202	0+	-17.975s	44.7 m *5*	ε 98%, α 2%
203	5/2-	-17.315	36.7 m *5*	ε 99.89%, α 0.11%
203m	13/2+	-16.674	45 s *2*	IT≈100%, α≈0.04%
204	0+	-18.344	3.53 h *2*	ε 99.34%, α 0.66%
205	5/2-	-17.545	1.66 h *2*	ε 99.96%, α 0.04%
206	0+	-18.197	8.8 d *1*	ε 94.55%, α 5.45%

Isotope Z El A	Jπ	Δ (MeV)	T1/2 or Abundance	Decay Mode
84 Po 207	5/2-	-17.160	5.80 h *2*	ε 99.98%, α 0.02%
207m	19/2-	-15.777	2.79 s *8*	IT
208	0+	-17.483	2.898 y *2*	α, ε
209	1/2-	-16.380	102 y *5*	α 99.52%, ε 0.48%
210	0+	-15.969	138.376 d *2*	α
211	9/2+	-12.448	0.516 s *3*	α
211m	(25/2+)	-10.986	25.2 s *6*	α 99.98%, IT 0.02%
212	0+	-10.385	0.299 μs *2*	α
212m	(18+)	-7.463	45.1 s *6*	α 99.93%, IT 0.07%
213	9/2+	-6.667	4.2 μs *8*	α
214	0+	-4.484	164.3 μs *20*	α
215	9/2+	-0.545	1.781 ms *4*	α, β- 2.3×10^{-4}%
216	0+	1.774	0.145 s *2*	α
217		5.914s	<10 s	α>95%, β-<5%
218	0+	8.351	3.10 m *1*	α 99.98%, β- 0.02%
85 At 194		-0.770s	0.18 s *8*	α
195		-3.166s	?	α>75%, ε<25%
196		-4.002s	0.3 s *1*	α
197	(9/2-)	-6.251s	0.35 s *4*	α 96%, ε 4%
197m	(1/2+)	-6.199s	3.7 s *25*	α≤100%, ε
198		-6.748s	4.9 s *5*	α, ε
198m		-6.648s	1.5 s *3*	α, ε, IT
199	(9/2-)	-8.726s	7.2 s *5*	α 90%, ε 10%
200	(5+)	-9.041	43 s *2*	ε 65%, α 35%
200m	(10-)	-8.751	4.3 s *3*	IT≈80%, α≈10%, ε≈10%
201	(9/2-)	-10.722	89 s *3*	α 71%, ε 29%
202	(5+)	-10.762	181 s *3*	ε 88%, α 12%
202m	(10-)	-10.371	≤1.5 s	IT
203	9/2-	-12.256	7.4 m *2*	ε 69%, α 31%
204	(7)+	-11.867	9.2 m *2*	ε 95.7%, α 4.3%
204m	(10)-	-11.867	108 ms *10*	IT
205	9/2-	-13.005	26.2 m *5*	ε 90%, α 10%
206	(5)+	-12.479	30.0 m *6*	ε 99.11%, α 0.89%
207	9/2-	-13.250	1.80 h *4*	ε 91.4%, α 8.6%
208	6+	-12.510	1.63 h *3*	ε 99.45%, α 0.55%
209	9/2-	-12.894	5.41 h *5*	ε 95.9%, α 4.1%
210	(5)+	-11.987	8.1 h *4*	ε 99.82%, α 0.18%
211	9/2-	-11.661	7.214 h *7*	ε 58.2%, α 41.8%
212	(1-)	-8.630	0.314 s *2*	α, ε<0.03%, β-<2.0×10^{-6}%
212m	(9-)	-8.408	0.119 s *3*	α>99%, IT<1%
213	9/2-	-6.594	125 ns *6*	α
214	1-	-3.394	558 ns *10*	α
214m		-3.335	265 ns *30*	α
214m	9-	-3.162	760 ns *15*	α
215	9/2-	-1.266	0.10 ms *2*	α
216	1(-)	2.243	0.30 ms *3*	α, ε<0.006%, β-<3×10^{-7}%
217	9/2-	4.386	32.3 ms *4*	α 99.99%, β- 0.01%
218	(2-)	8.087	1.6 s *4*	α 99.9%, β- 0.1%
219		10.523	56 s *3*	α≈97%, β-≈3%
220		14.252s		
221			2.3 m *2*	β-
223			50 s *7*	β-
86 Rn 198	0+	-1.145		α, ε
198m	0+	-1.145	50 ms *9*	α, ε, IT

Isotope Z El A	Jπ	Δ (MeV)	T1/2 or Abundance	Decay Mode
86 Rn 199	(3/2-)	-1.575s	0.62 s *3*	α 95%, ε 5%
199m	(13/2+)	-1.575s	0.3 s *1*	α, ε
200	0+	-4.029s	1.06 s *2*	α≈98%, ε≈2%
201	(3/2-)	-4.159s	7.0 s *4*	α≈80%, ε≈20%
201m	(13/2+)	-3.879s	3.8 s *4*	α≈90%, ε≈10%, IT
202	0+	-6.314s	9.85 s *20*	ε<30%, α
203	(3/2,5/2)-	-6.227s	45 s *3*	α 66%, ε 34%
203m	(13/2+)	-5.866s	28 s *2*	α≈80%, ε≈20%, IT<0.1%
204	0+	-8.043s	1.24 m *3*	α 68%, ε 32%
205	5/2-	-7.761s	2.8 m *1*	ε 77%, α 23%
206	0+	-9.166s	5.67 m *17*	α 62%, ε 38%
207	5/2-	-8.639	9.25 m *17*	ε 79%, α 21%
208	0+	-9.659	24.35 m *14*	α 62%, ε 38%
209	5/2-	-8.965	28.5 m *10*	ε 83%, α 17%
210	0+	-9.613	2.4 h *1*	α 96%, ε 4%
211	1/2-	-8.770	14.6 h *2*	ε 72.7%, α 27.4%
212	0+	-8.674	23.9 m *12*	α
213	(9/2+)	-5.712	25.0 ms *2*	α
214	0+	-4.335	0.27 μs *2*	α
214m	6+	-2.892	0.7 ns *3*	α
214m	8+	-2.710	6.5 ns *30*	α
215	9/2+	-1.184	2.30 μs *10*	α
216	0+	0.240	45 μs *5*	α
217	9/2+	3.647	0.54 ms *5*	α
218	0+	5.204	35 ms *5*	α
219	5/2+	8.826	3.96 s *1*	α
220	0+	10.604	55.6 s *1*	α
221	7/2(+)	14.486s	25 m *2*	β- 78%, α 22%
222	0+	16.366	3.8235 d *3*	α
223	7/2		23.2 m *4*	β-
224	0+		107 m *3*	β-
225	7/2-		4.5 m *3*	β-
226	0+		6.0 m *5*	β-
227			22.5 s *7*	β-
228	0+		65 s *2*	β-
87 Fr 201	(9/2-)	3.712s	48 ms *15*	α, ε<1%
202		3.065s	0.34 s *4*	α≈97%, ε≈3%
203	(9/2-)	0.976s	0.55 s *2*	α≈95%, ε≈5%
204	(5+,6+)	0.554	2.1 s *2*	α≈80%, ε≈20%
205	(9/2-)	-1.242	3.85 s *10*	α, ε<1%
206	(5+)	-1.412	15.9 s *2*	α 88%, ε 12%
206m		-0.881	0.7 s *1*	α
207	9/2-	-2.930	14.8 s *1*	α 95%, ε 5%
208	7+	-2.671	59.1 s *3*	α 90%, ε 10%
209	9/2-	-3.803	50.0 s *3*	α 89%, ε 11%
210	6+	-3.351	3.18 m *6*	α 60%, ε 40%
211	9/2-	-4.165	3.10 m *2*	α>80%, ε<20%
212	5+	-3.556	20.0 m *6*	ε 57%, α 43%
213	9/2-	-3.563	34.6 s *3*	α 99.45%, ε 0.55%
214	(1-)	-0.975	5.0 ms *2*	α
214m	(9-)	-0.853	3.35 ms *5*	α
215	9/2-	0.304	86 ns *5*	α
216	(1-)	2.970	0.70 μs *2*	α, ε<2×10^{-7}%
217	9/2-	4.301	22 μs *5*	α
218	(1-)	7.046	1.0 ms *6*	α
218m		7.132	22.0 ms *5*	α≤100%
219	9/2-	8.608	20 ms *2*	α
220	1+	11.469	27.4 s *3*	α 99.65%, β- 0.35%
221	5/2-	13.269	4.9 m *2*	α, β-<0.1%

Isotope Z El A	Jπ	Δ (MeV)	T1/2 or Abundance	Decay Mode
87 Fr 222	2-	16.341	14.2 m *3*	β-
223	3/2(-)	18.379	21.8 m *4*	β- 99.99%, α 6.0×10^{-3}%
224	1(-)	21.637	3.30 m *10*	β-
225	3/2-	23.851	4.0 m *2*	β-
226	1	27.295	48 s *1*	β-
227	1/2+	29.658	2.47 m *3*	β-
228	2-	33.273	39 s *1*	β-
229			50 s *20*	β-
230			19.1 s *5*	β-
231			17.5 s *8*	β-
88 Ra 204	0+	6.046s		
205		5.772s	0.22 s *6*	α, ε
206	0+	3.526s	0.24 s *2*	α≈100%
207	(5/2-,3/2-)	3.471s	1.3 s *2*	α≈90%, ε≈10%
207m	(13/2+)	3.941s	55 ms *10*	IT 85%, α 15%, ε≈0.35%
208	0+	1.655s	1.3 s *2*	α 95%, ε 5%
209	5/2-	1.811s	4.6 s *2*	α≈90%, ε≈10%
210	0+	0.416s	3.7 s *2*	α≈96%, ε≈4%
211	5/2(-)	0.832	13 s *2*	α>93%, ε<7%
212	0+	-0.202	13.0 s *2*	α≈90%, ε≈15%
213	1/2-	0.320	2.74 m *6*	α 80%, ε 20%
213m		2.090	2.1 ms *1*	IT≈99%, α≈1%
214	0+	0.084	2.46 s *3*	α 99.94%, ε 0.06%
215	(9/2+)	2.519	1.59 ms *9*	α
216	0+	3.277	182 ns *10*	α, ε
217	(9/2+)	5.874	1.6 μs *2*	α
218	0+	6.636	25.6 μs *11*	α
219	(7/2)+	9.371	10 ms *3*	α
220	0+	10.260	25 ms *5*	α
221	5/2+	12.957	28 s *2*	α
222	0+	14.309	38.0 s *5*	α, ^{14}C 3×10^{-6}%
223	3/2+	17.230	11.435 d *4*	α, ^{14}C 6.4×10^{-8}%
224	0+	18.818	3.66 d *4*	α, ^{12}C 4.3×10^{-9}%
225	1/2+	21.986	14.9 d *2*	β-
226	0+	23.661	1600 y *7*	α, ^{14}C 3×10^{-9}%
227	3/2+	27.171	42.2 m *5*	β-
228	0+	28.935	5.75 y *3*	β-
229	5/2(+)	32.434	4.0 m *2*	β-
230	0+	34.543	93 m *2*	β-
231	(7/2-,1/2+)		103 s *3*	β-
232	0+		250 s *50*	β-
89 Ac 209	(9/2-)	8.917	0.10 s *5*	α≈99%, ε≈1%
210		8.621	0.35 s *5*	α≈96%, ε≈4%
211		7.120	0.25 s *5*	α≈100%
212		7.275	0.93 s *5*	α≈97%, ε≈3%
213		6.125	0.80 s *5*	α≤100%
214		6.424	8.2 s *2*	α≥89%, ε≤11%
215	9/2-	6.008	0.17 s *1*	α 99.91%, ε 0.09%
216	(1-)	8.112	≈0.33 ms	α
216m	(9-)	8.112	0.33 ms *2*	α
217	9/2-	8.693	69 ns *4*	α, ε≤2%
218		10.828	1.12 μs *11*	α
219	9/2-	11.556	11.8 μs *15*	α
220		13.742	26.1 ms *5*	α, ε 5×10^{-4}%
221		14.508	52 ms *2*	α
222	(1-)	16.599	5.0 s *5*	α 99%, ε≤2%
222m		16.599	63 s *4*	α≥88%, IT≤10%, ε≤2%
223	(5/2-)	17.816	2.10 m *5*	α 99%, ε 1%

Isotope Z El A	Jπ	Δ (MeV)	T1/2 or Abundance	Decay Mode
89 Ac 224	0−	20.221	2.9 h *2*	ε 90.9%, α 9.1%, β−<1.6%
225	(3/2−)	21.629	10.0 d *1*	α
226	(1)	24.302	29.4 h *1*	β− 83%, ε 17%, α 0.006%
227	3/2−	25.846	21.773 y *3*	β− 98.62%, α 1.38%
228	3(+)	28.889	6.15 h *2*	β−, α 5.5×10^{-6}%
229	(3/2+)	30.674	62.7 m *5*	β−
230	(1+)	33.556	122 s *3*	β−
231	(1/2+)	35.910	7.5 m *1*	β−
232	(1+)	39.143	119 s *5*	β−
233	(1/2+)		145 s *10*	β−
234	(1+)		44 s *7*	β−
90 Th 212	0+	12.032s	30 ms *+20−10*	α, ε≈0.3%
213		12.074s	140 ms *25*	α≤100%
214	0+	10.666s	100 ms *25*	α
215	(1/2−)	10.922	1.2 s *2*	α
216	0+	10.293	0.028 s *2*	α, ε≈0.006%
216	(8+,11−)	12.321	0.18 ms *4*	IT≈97%, α≈3%
217	(9/2+)	12.169	0.252 ms *7*	α
218	0+	12.358	109 ns *13*	α
219		14.458	1.05 μs *3*	α
220	0+	14.655	9.7 μs *6*	α, ε 2×10^{-7}%
221	(7/2+)	16.926	1.68 ms *6*	α
222	0+	17.190	2.8 ms *3*	α
223	(5/2)+	19.363	0.60 s *2*	α
224	0+	19.989	1.05 s *2*	α
225	(3/2)+	22.304	8.72 m *4*	α≈90%, ε≈10%
226	0+	23.185	30.6 m *1*	α
227	(1/2+)	25.802	18.72 d *2*	α
228	0+	26.763	1.9131 y *9*	α
229	5/2+	29.579	7340 y *160*	α
230	0+	30.856	7.538×10^{4} y *30*	α, SF≤$5.\times10^{-11}$%
231	5/2+	33.810	25.52 h *1*	β−, α≈1.0×10^{-8}%
232	0+	35.443	1.405×10^{10} y *6* 100%	α
232	0+	35.443	1.405×10^{10} y *6*	SF<1.0×10^{-9}%
233	1/2+	38.728	22.3 m *1*	β−
234	0+	40.610	24.10 d *3*	β−
235	(1/2+)	44.251	7.1 m *2*	β−
236	0+		37.5 m *2*	β−
91 Pa 215		17.712	14 ms *+20−3*	α
216		17.713	0.20 s *4*	α≈80%, ε≈20%
217		17.038	4.9 ms *6*	α
217m		18.892	1.6 ms *10*	α≤100%
218		18.643	0.12 ms *+4−2*	α
219	9/2−	18.518	53 ns *10*	α
220		20.366		
221	9/2−	20.366	5.9 μs *17*	α
222		22.050s	≈4.3 ms	α
223		22.323	6.5 ms *10*	α
224		23.862	0.95 s *15*	α 99.9%, ε 0.1%
225		24.326	1.7 s *2*	α
226		26.011	1.8 m *2*	α 74%, ε 26%
227	(5/2−)	26.821	38.3 m *3*	α 85%, ε 15%
228	(3+)	28.874	22 h *1*	ε 98.15%, α 1.85%
229	(5/2+)	29.895	1.50 d *5*	ε 99.52%, α 0.48%
230	(2−)	32.166	17.4 d *5*	ε 91.6%, β− 8.4%, α 3.2×10^{-3}%
231	3/2−	33.420	32760 y *110*	α, SF≤$3.0\times10^{-}$%

Isotope Z El A	Jπ	Δ (MeV)	T1/2 or Abundance	Decay Mode
91 Pa 232	(2-)	35.938	1.31 d *2*	β-, ε 3.0×10^{-3}%
233	3/2-	37.483	26.967 d *2*	β-
234	4(+)	40.337	6.70 h *5*	β-
234m	(0-)	40.411	1.17 m *3*	β- 99.87%, IT 0.13%
235	(3/2-)	42.323	24.5 m *2*	β-
236	1(-)	45.340	9.1 m *1*	β-
237	(1/2+)	47.635	8.7 m *2*	β-
238	(3-)	50.764	2.3 m *1*	β-
92 U 222	0+	24.283s	1.0 μs *+10-4*	α
225		27.371	50 ms *30*	α
226	0+	27.321	0.5 s *2*	α
227	(3/2+)	28.999	1.1 m *1*	α
228	0+	29.217	9.1 m *2*	α>95%, ε<5%
229	(3/2+)	31.204	58 m *3*	ε≈80%, α≈20%
230	0+	31.603	20.8 d	α
231	(5/2-)	33.778	4.2 d *1*	ε, α≈5.5×10^{-3}%
232	0+	34.601	68.9 y *4*	α
233	5/2+	36.912	1.592×10^{5} y *2*	α, SF<6.0×10^{-9}%
234	0+	38.140	2.45×10^{5} y *2* 0.0055% *5*	α, SFw
235	7/2-	40.913	703.8×10^{6} y *5* 0.720% *1*	α, SF 7.0×10^{-9}%
235m	1/2+	40.913	≈25 m	IT
236	0+	42.440	2.342×10^{7} y *3*	α, SF 9.6×10^{-8}%
237	1/2+	45.385	6.75 d *1*	β-
238	0+	47.305	4.468×10^{9} y *3* 99.2745% *15*	α, SF 0.0001%
239	5/2+	50.570	23.45 m *2*	β-
240	0+	52.708	14.1 h *1*	β-, α
242	0+		16.8 m *5*	β-
93 Np 227		32.564	0.51 s *6*	α
228		33.701s	1.00 m *8*	SF?
229		33.763	4.0 m *2*	α>50%, ε<50%
230		35.214	4.6 m *3*	ε≤97%, α≥3%
231	(5/2)	35.614	48.8 m *2*	ε 98%, α 2%
232	(4+)	37.299s	14.7 m *3*	ε
233	(5/2+)	38.146s	36.2 m *1*	ε, α≤1.0×10^{-3}%
234	(0+)	39.950	4.4 d *1*	ε
235	5/2+	41.037	396.1 d *12*	ε, α 2.6×10^{-3}%
236	(6-)	43.380	154×10^{3} y *6*	ε 87.3%, β- 12.5%, α 0.16%
236m	1	43.440	22.5 h *4*	ε 52%, β- 48%
237	5/2+	44.867	2.14×10^{6} y *1*	α, SF≤2×10^{-1}%
238	2+	47.450	2.117 d *2*	β-
239	5/2+	49.304	2.3565 d *4*	β-
240	(5+)	52.320	61.9 m *2*	β-
240m	1(+)	52.320	7.22 m *2*	β- 99.89%, IT 0.11%
241	(5/2+)	54.255	13.9 m *2*	β-
242	(1+)	57.412	2.2 m *2*	β-
242	(6)	57.412	5.5 m *1*	β-
243	(5/2-)	59.919	1.85 m *15*	β-
94 Pu 230	0+	36.921		α≤100%
231		38.424s		
232	0+	38.358	34.1 m *7*	ε 80%, α 20%
233		40.045	20.9 m *4*	ε 99.88%, α 0.12%
234	0+	40.338	8.8 h *1*	ε 94%, α 6%
235	(5/2+)	42.204s	25.3 m *5*	ε, α 2.7×10^{-3}%
236	0+	42.893	2.858 y *8*	α, SF 1.4×10^{-7}%
237	7/2-	45.087	45.2 d *1*	ε, α 0.004%

Isotope Z El A	Jπ	Δ (MeV)	T1/2 or Abundance	Decay Mode
94 Pu 237m	1/2+	45.233	0.18 s *2*	IT
238	0+	46.158	87.74 y *4*	α, SF w
239	1/2+	48.583	24110 y *30*	α, SF 3×10^{-1}%
240	0+	50.120	6564 y *11*	α, SF 5.7×10^{-6}%
241	5/2+	52.950	14.35 y *10*	β−, α
242	0+	54.712	3.733×10^{5} y *12*	α, SF 5.5×10^{-4}%
243	7/2+	57.749	4.956 h *3*	β−
244	0+	59.799	8.08×10^{7} y *10*	α 99.88%, SF 0.12%
245	(9/2−)	63.097	10.5 h *1*	β−
246	0+	65.388	10.84 d *2*	β−
247			2.27 d *23*	β−
95 Am 232			79 s *2*	ε≈98%, α≈2%
233		43.288s		
234		44.511s	2.6 m *2*	α?, ε
235		44.739s	?	
236		46.173s		α, ε
237	5/2(−)	46.817s	73.0 m *10*	ε 99.98%, α 0.02%
238	1+	48.416	98 m *2*	ε>99.99%, α 0.0001%
239	(5/2)−	49.386	11.9 h *1*	ε 99.99%, α 0.01%
240	(3−)	51.499	50.8 h *3*	ε, α 1.9×10^{-4}%
241	5/2−	52.929	432.7 y *6*	α, SF
242	1−	55.463	16.02 h *2*	β− 82.7%, ε 17.3%
242m	5−	55.512	141 y *2*	IT 99.54%, α 0.46%, SF
243	5/2−	57.167	7370 y *40*	α, SF 3.7×10^{-9}%
244	(6−)	59.875	10.1 h *1*	β−
244m	1+	59.963	≈26 m	β− 99.96%, ε 0.04%
245	(5/2)+	61.893	2.05 h *1*	β−
246	(7−)	64.988	39 m *3*	β−
246m	2(−)	64.988	25.0 m *2*	β−, IT<0.01%
247	(5/2)	67.227s	23.0 m *13*	β−
248		70.485s	?	β−
96 Cm 235		48.049s	?	
236	0+	47.883s		α, ε
237		49.270s		
238	0+	49.384	2.4 h *1*	ε≥90%, α≤10%
239	(7/2−)	51.086s	≈2.9 h	ε, α<0.1%
240	0+	51.715	27 d *1*	α>99.5%, ε<0.5%, SF 3.9×10^{-6}%
241	1/2+	53.697	32.8 d *2*	ε 99%, α 1%
242	0+	54.798	162.79 d *9*	α, SF 6.2×10^{-6}%
243	5/2+	57.176	29.1 y *1*	α 99.71%, ε 0.29%, SF 5.3×10^{-9}%
244	0+	58.447	18.10 y *2*	α, SF 1×10^{-4}%
245	7/2+	60.999	8500 y *100*	α, SF 6.1×10^{-7}%
246	0+	62.612	4730 y *100*	α 99.97%, SF 0.03%
247	9/2−	65.527	1.56×10^{7} y *5*	α
248	0+	67.385	3.40×10^{5} y *4*	α 91.74%, SF 8.26%
249	1/2(+)	70.743	64.15 m *3*	β−
250	0+	72.982	9700 y	SF≈80%, α≈11%, β−≈9%
251	(1/2+)	76.640	16.8 m *2*	β−
252	0+		<2 d	β−
97 Bk 237		53.213s		
238		54.336s		
239	(7/2+)	54.364s		

Isotope Z El A	Jπ	Δ (MeV)	T1/2 or Abundance	Decay Mode
97 Bk 240		55.655s	4.8 m *8*	ε≈100%, εSF w
241	(7/2+)	56.097s		
242		57.798s	7.0 m *13*	ε
243	(3/2−)	58.685	4.5 h *2*	ε≈99.85%, α≈0.15%
244	(1−)	60.703	4.35 h *15*	ε 99.99%, α 0.006%
245	3/2−	61.809	4.94 d *3*	ε 99.88%, α 0.12%
246	2(−)	63.962	1.80 d *2*	ε, α<0.2%
247	(3/2−)	65.482	1380 y *250*	α≤100%
248	1(−)	68.103	23.7 h *2*	β− 70%, ε 30%, α<0.001%
248	(6+)	68.103	>9 y	α>70%
249	7/2+	69.843	320 d *6*	β−, α 1.4×10^{-3}%, SF 4.7×10^{-8}%
250	2−	72.945	3.217 h *5*	β−
251	(3/2−)	75.220	55.6 m *11*	β−, α≈1.0×10^{-5}%
252		78.527s		
98 Cf 239		58.284s	39 s *+37−12*	α>50%, ε?
240	0+	58.027s	1.06 m *15*	α≈100%
241		59.354s	3.78 m *70*	ε≈90%, α≈10%
242	0+	59.326	3.49 m *12*	α, ε?
243	(1/2+)	60.901s	10.7 m *5*	ε≈86%, α≈14%
244	0+	61.469	19.4 m *6*	α
245	(5/2+)	63.377	45.0 m *15*	ε 64%, α 36%
246	0+	64.085	35.7 h *5*	α, ε<5.0×10^{-4}%, SF 2.0×10^{-4}%
247	(7/2+)	66.128	3.11 h *3*	ε 99.97%, α 0.04%
248	0+	67.233	333.5 d *28*	α, SF 0.0029%
249	9/2−	69.718	351 y *2*	α, SF 5.2×10^{-7}%
250	0+	71.165	13.08 y *9*	α 99.92%, SF 0.08%
251	1/2+	74.127	898 y *44*	α
252	0+	76.027	2.645 y *8*	α 96.91%, SF 3.09%
253	(7/2+)	79.293	17.81 d *8*	β− 99.69%, α 0.31%
254	0+	81.334	60.5 d *2*	SF 99.69%, α 0.31%
255	(9/2+)	84.781s	85 m *18*	β−
256	0+		12.3 m *12*	SF, β−<1%, α≈1.0×10^{-6}%
99 Es 241		63.902s		
242		64.943s		
243		64.861s	21 s *2*	ε≤70%, α≥30%
244		66.027s	37 s *4*	ε 96%, α 4%
245	(3/2−)	66.431s	1.1 m *1*	ε 60%, α 40%
246	(4−,6+)	67.965s	7.7 m *5*	ε 90.1%, α 9.9%
247	(7/2+)	68.603s	4.55 m *26*	ε≈93%, α≈7%
248	(2−,0+)	70.293	27 m *4*	ε>99%, α≈0.25%
249	7/2(+)	71.170s	102.2 m *6*	ε 99.43%, α 0.57%
250	1(−)	73.265s	2.22 h *5*	ε≥99%, α≤1%
250	(6+)	73.265s	8.6 h *1*	ε>97%, α<3%
251	(3/2−)	74.504	33 h *1*	ε 99.51%, α 0.49%
252	(5−)	77.287	471.7 d *19*	α 76%, ε 24%, β−≈0.01%
253	7/2+	79.007	20.47 d *3*	α, SF 8.7×10^{-6}%
254	(7+)	81.988	275.7 d *5*	α, ε<1.0×10^{-6}%, SF<3.0×10^{-6}%, β− 1.7×10^{-6}%
254m	2+	82.066	39.3 h *2*	β− 98%, IT<3%,

Isotope Z El A	Jπ	Δ (MeV)	T1/2 or Abundance	Decay Mode
				α 0.33%, ε 0.08%, SF<0.05%
99 Es 255	(7/2+)	84.081	39.8 d *12*	β− 92%, α 8%, SF 4.1×10^{-3}%
256	(1+)	87.149s	25.4 m *24*	β−
256	(8+)	87.149s	≈7.6 h	β−
257		89.395s		
100 Fm 242	0+		0.8 ms *2*	SF
243		69.398s	0.18 s *+8−4*	α≤100%
244	0+	69.052s	3.7 ms *4*	SF
245		70.214s	4.2 s *13*	α≤100%
246	0+	70.124	1.1 s *2*	α 92%, SF 8%, ε≤1%
247m		71.517s	9.2 s *23*	α≤100%
247m		71.517s	35 s *4*	α≥50%, ε≤50%
248	0+	71.896	36 s *3*	α 99%, ε≈1%, SF≈0.05%
249	(7/2+)	73.610s	2.6 m *7*	ε≈85%, α≈15%
250	0+	74.067	30 m *3*	α>90%, ε<10%, SF≈6.0×10^{-4}%
250m		75.067	1.8 s *1*	IT>80%
251	(9/2−)	75.978	5.30 h *8*	ε 98.2%, α 1.8%
252	0+	76.810	25.39 h *5*	α, SF 0.0023%
253	1/2+	79.340	3.00 d *12*	ε 88%, α 12%
254	0+	80.897	3.240 h *2*	α 99.94%, SF 0.06%
255	7/2+	83.793	20.07 h *7*	α, SF 2.4×10^{-5}%
256	0+	85.479	157.6 m *13*	SF 91.9%, α 8.1%
257	(9/2+)	88.581	100.5 d *2*	α 99.79%, SF 0.21%
258	0+	90.459s	370 μs *43*	SF
259		93.699s	1.5 s *3*	SF
101 Md 247		76.103s	2.9 s *17*	α≤100%
248		77.149s	7 s *3*	ε 80%, α 20%
249		77.315s	24 s *4*	α≈70%, ε≈30%
250		78.699s	52 s *6*	ε 93%, α 7%
251		79.051s	4.0 m *5*	ε≥90%, α≤10%
252		80.695s	2.3 m *8*	α<50%, ε>50%
253		81.301s		α, ε
254		83.580s	10 m *3*	ε
254		83.580s	28 m *8*	ε
255	(7/2−)	84.835	27 m *2*	ε 92%, α 8%
256	(0−,1−)	87.609	76 m *4*	ε 90.7%, α 9.3%, SF<3%
257	(7/2−)	88.990	5.3 h *3*	ε 90%, α 10%, SF<4%
258	(1−)	91.684	60 m *2*	ε
258	(8−)	91.684	55 d *4*	α
259	(7/2−)	93.616s	103 m *12*	SF>97%, α<3%
260		96.596s	31.8 d *5*	SF≈70%, α≤25%, ε<15%, β−<10%
102 No 250	0+		0.25 ms *5*	SF, α≈0.05%
251		82.827s	0.8 s *3*	α≈100%, ε≈1%
252	0+	82.871	2.30 s *22*	α 73.1%, SF 26.9%
253	(9/2−)	84.478s	1.7 m *3*	α≈80%, ε≈20%
254	0+	84.717	55 s *3*	α 90%, ε 10%, SF 0.25%
254m		85.217	0.28 s *4*	IT>80%
255	(1/2+)	86.847	3.1 m *2*	α 61.4%, ε 38.6%
256	0+	87.816	3.3 s *2*	α 99.8%, SF≤0.25%

Isotope Z El A	Jπ	Δ (MeV)	T1/2 or Abundance	Decay Mode
102 No 257	(7/2+)	90.217	25 s *2*	α≈100%
258	0+	91.522s	≈1.2 ms	SF, α 0.001%
259	(9/2+)	94.123s	58 m *5*	α 75%, ε 25%, SF < 10%
260	0+	95.604s	106 ms *8*	SF
103 Lr 252			≈1 s	α≈90%, ε≈10%, SF < 1%
253		88.732s	1.3 s *+6-3*	α 90%, SF < 20%, ε≈1%
254		89.870s	13 s *2*	α 78%, ε 22%, SF < 0.1%
255		90.090s	22 s *4*	α 85%, ε < 30%
256		92.005s	28 s *3*	α > 80%, ε < 20%, SF < 0.03%
257	(9/2+)	92.732s	0.646 s *25*	α
258		94.905s	4.3 s *5*	α > 95%, ε < 5%
259		95.934s	5.4 s *8*	α > 50%, SF < 50%, ε < 0.5%
260		98.339s	180 s *30*	α 75%, ε≈15%, SF < 10%
261		99.615s	39 m *12*	SF
262		102.297s	3.6 h *3*	ε
104 Rf 253			≈1.8 s	α≈50%, SF≈50%
254	0+		0.5 ms *2*	SF, α≈0.3%
255	(9/2−)	94.550s	1.5 s *2*	SF 52%, α 48%
256	0+	94.248	6.7 ms *2*	SF 98%, α 2.2%
257	(7/2+)	96.150s	4.7 s *3*	α 79.6%, ε 18%, SF 2.4%
258	0+	96.392s	12 ms *2*	SF≈87%, α≈13%
259		98.380s	3.1 s *7*	α 93%, SF 7%, ε≈0.3%
260	0+	99.241s	20.1 ms *7*	SF≈98%, α≈2%
261		101.451s	65 s *10*	α > 80%, ε ≤ 10%, SF < 10%
262	0+	102.548s	47 ms *5*	SF
105 Ha 255			1.6 s *+6-4*	α≈80%, SF≈20%
256			2.6 s *+14-8*	α ≤ 90%, SF ≤ 40%, ε≈10%
257		100.462s	1.3 s *+5-3*	α 82%, SF 17%, ε 1%
258		101.840s	4.4 s *+9-6*	α 67%, ε 33%, SF < 1%
258		101.840s	20 s *10*	ε
259		102.204s	? ?	
260		103.801s	1.52 s *13*	α ≥ 90%, SF ≤ 9.6%, ε < 2.5%
261		104.426s	1.8 s *4*	α > 50%, SF < 50%
262		106.535s	34 s *4*	SF 71%, α 26%, ε≈3%
263		107.393s	26 m *2*	α ≤ 100%
106 259	(1/2+)	106.846s	0.48 s *+28-13*	α 90%, SF < 20%
260	0+	106.596	3.6 ms *+9-6*	α 50%, SF 50%
261		108.384s	0.23 s *3*	α 95%, SF < 10%
262	0+	108.603s		
263		110.498s	0.8 s *2*	SF≈70%, α≈30%
107 260				α
261		113.449s	11.8 ms *+53-28*	α 95%, SF < 10%
262		114.681s	102 ms *26*	α ≥ 80%, SF ≤ 20%
262m		114.996s	8.0 ms *21*	α > 70%, SF < 30%
263		114.863s		

Isotope Z	El	A	Jπ	Δ (MeV)	T1/2 or Abundance	Decay Mode
107		264		116.353s		
108		263			?	α
		264	0+	119.821	0.08 ms *+40-4*	α
		265		121.633s	1.8 ms *+22-7*	α≈100%
109		266		128.389s	3.4 ms *+16-13*	α

References

1. *Evaluated Nuclear Structure Data File* – a computer file of evaluated experimental nuclear structure data maintained by the National Nuclear Data Center, Brookhaven National Laboratory. (File as of March 1, 1994.)

2. *Nuclear Data Sheets* – Academic Press, New York. Evaluations published by mass number for *A*=45 to 266. See page ii of any issue for index to A–chains.

3. *Nuclear Physics* – North–Holland Publishing Co., Amsterdam – Evaluations for *A*=5–20 by F. Ajzenberg–Selove and, more recently, by D. R. Tilley, H. R. Weller, and C. M. Cheves.

4. Energy Levels of *A*=21–44 (VII) by P. M. Endt – *Nuclear Physics* A521, 1 (1990).

5. *Chart of the Nuclides* (1988), 14th edition, prepared by F. W. Walker, J. R. Parrington, F. Feiner, Knolls Atomic Power Laboratory, operated by General Electric Co.

6. The 1993 Atomic Masses Evaluation, G. Audi and A. H. Wapstra, *Nuclear Physics* A565, 1 (1993). The computer files giving their recommended as well as experimental values are maintained by the National Nuclear Data Center, Brookhaven National Laboratory.

7. Table of the Isotopes, N. E. Holden, *CRC Handbook of Chemistry and Physics*, 71st edition (1990), 11–33.

8. ENDF/B Summary Documentation, edited by R. Kinsey, Rept. BNL–NCS–17541 (ENDF–201), National Nuclear Data Center, Brookhaven National Laboratory (July 1979).